Classroom Companion: Economics

The Classroom Companion series in Economics includes undergraduate and graduate textbooks alike. It welcomes fundamental textbooks aimed at introducing students to the core concepts, empirical methods, theories and tools of the field, as well as advanced textbooks written for students at the Master and PhD level seeking a deeper understanding of economic theory, mathematical tools and quantitative methods.

Yasuhiro Nakamoto

A Comprehensive Guide to Creating Graphics with MATLAB Programming

Springer

Yasuhiro Nakamoto
Faculty of Informatics
Kansai University
Osaka, Japan

ISSN 2662-2882 ISSN 2662-2890 (electronic)
Classroom Companion: Economics
ISBN 978-981-96-6212-8 ISBN 978-981-96-6213-5 (eBook)
https://doi.org/10.1007/978-981-96-6213-5

This Springer imprint is published by the registered company Springer Nature Singapore Pte Ltd.
The registered company address is: 152 Beach Road, #21-01/04 Gateway East, Singapore 189721, Singapore

Preface

The aim of this book is to help MATLAB users learn how to create graphics in MATLAB. Specifically, it adopts a visual approach to make the learning process more effective. To explain the reason for this approach, I would like to introduce the motivation for writing this MATLAB programming book, titled *A Comprehensive Guide to Creating Graphics with MATLAB Programming*.

I am an economist who uses several software tools, including the numerical computing software MATLAB, in my research and teaching. The primary inspiration for writing this book was my encounter with Michael N. Mitchell's *A Visual Guide to Stata Graphics* in which Stata is a widely used software for statistical analysis and data management. This work highlights the importance of effective data visualization, and its structure allows users to easily locate the figures they need. In addition, the inclusion of programming examples and concise explanations helps the reader to understand the relevant programming techniques without unnecessary complexity. Therefore, this work serves as an excellent guide for efficiently mastering Stata's graphics programming through the extensive use of illustrations. By following this approach, users can improve the quality of the graphs and charts they create, making this book an extremely practical resource in my work.

MATLAB is known for its strength in matrix operations. As a result, most books targeting students, researchers, and engineers with a background in mathematics tend to focus on programming for numerical computation, data analysis, simulations, and other scientific and engineering calculations. Books that focus specifically on MATLAB's graphics capabilities are relatively scarce, even though MATLAB offers many features for creating graphics.

As a longtime MATLAB user, I have often wished for a book with a visual approach similar to Michael N. Mitchell's *A Visual Guide to Stata Graphics*, tailored specifically for MATLAB users. This book fulfills that hope. Furthermore, owing to MATLAB's broad applicability, graph creation can be highly customizable, allowing users to create a wide range of visualizations with ingenuity.

Here are some of the key features of this book.

First, because MATLAB is a programming language, this book is designed to be accessible to students who are new to programming. While most books are structured to be read sequentially from the first chapter, this book adopts a more flexible format. Beginners may find it helpful to browse through the examples of created graphics and identify a figure that appeals to them. Similarly, readers who have a specific idea of the figure they want to create can start by looking for examples that match their vision. This flexible structure is one of the key features of the book.

At the same time, beginners may find it challenging to understand the programs even when they identify a figure they like. To address this, the book provides explanations of the basic syntax for the introduced functions. It also shows common programming errors that beginners may encounter, providing guidance to help them understand and solve such problems effectively.

Next, I explain the structure of this book. ► Chapter 1 provides an overview of the rules that will be followed throughout the book. If you are new to programming, I recommend that you start with ► Chap. 1.

▶ Chapter 2 covers topics such as importing Excel files and organizing rows and columns. Many users who create graphics may use Microsoft Excel files as their data source, however, beginners may find even this step challenging. If so, you may also want to read ▶ Chap. 2.

Next, ▶ Chaps. 3–17 introduce 16 different graphing functions. Although MATLAB provides many toolboxes, many of the functions in the toolboxes require specialized knowledge. Therefore, this book focuses on commonly used functions that are accessible to a broader audience.

Each chapter is divided into six sections. Section 1 of each chapter presents the data used in the programs in a direct input format to help the readers understand the content of the data used. Reviewing the data first may stimulate interest in the programs discussed in this book. The data sets are based on real-world examples familiar to even high school students, making many of the figures in this book relevant to the social sciences. For instance, ▶ Chap. 3 uses Japan's food self-sufficiency rate as its data set, while other chapters use data on GDP, employment and unemployment rates, the percentage of individuals using the Internet, the gross enrollment ratio, crude oil prices, land area, and population density. Section 2 imports the same data using an Excel file.

Sections 3 through 5 focus on graphing programs. Specifically, Section 3 introduces the MATLAB documentation and explains basic syntax. It provides detailed explanations of graphics creation programs, emphasizing the use of matrices and vectors with basic syntax. Section 4 builds on the program introduced in Section 3 by using additional syntax to create more advanced graphics. Section 5 combines various optional functions to create more sophisticated figures. Finally, Section 6 provides exercises based on the programs covered in the chapter to encourage readers to create their own graphics.

The last chapter, ▶ Chap. 18, explains how to use optional functions such as annotation, xline, yline, and text.

Finally, I would like to express my sincere gratitude to everyone who helped me write this book. I am deeply indebted to Professor Kazuo Mino for encouraging me to undertake this book project. My sincere thanks also go to Sneha Arunagiri, Martina Himberger, Juno Kawakami, and Kala Palanisamy of Springer, who patiently listened to my thoughts on the color choices for the figures, and to my longtime friend Katsuya Kadogaki, who provided many suggestions for the programs. I would also like to thank my students, Natsuki Wada and Otoha Inagawa, for their invaluable feedback on the graphics. I would also like to thank Michael N. Mitchell for kindly allowing me to refer to *A Visual Guide to Stata Graphics* in the preface and for his heartwarming message about this book.

Finally, I would like to express my deepest gratitude to my academic advisors, Professors Koichi Futagami, Giacomo Corneo, and Takashi Shiomura, for their invaluable guidance and support. I am also deeply grateful to my parents, Norihiro and Hisako, and my family for their enduring support.

Yasuhiro Nakamoto
Osaka, Japan
March 2025

Contents

1 Introduction ... 1

2 Importing Data ... 15

3 Scatter Function ... 23

4 Scatter3 Function ... 97

5 Plot Function ... 131

6 Plot3 Function ... 175

7 Bar Function ... 201

8 Bar3 Function ... 251

9 Barh Function ... 283

10 Bar3h Function ... 315

11 Histogram Function ... 339

12 Histogram2 Function ... 375

13 Boxchart Function ... 407

14 Piechart and Donutchart Functions ... 439

15 Plotmatrix Function ... 465

16 Bubblechart Function ... 497

17 Bubblechart3 Function ... 533

18 Useful Optional Functions ... 557

About the Author

Yasuhiro Nakamoto is a professor in the Faculty of Informatics at Kansai University. He teaches MATLAB to undergraduate students at Kansai University. He holds a Ph.D. (Top Student) in Economics, awarded by Osaka University in 2009. He has worked at Kyushu Sangyo University and Freie Universität Berlin as an Alexander von Humboldt Fellow. He participated in the 5th Lindau Nobel Laureate Meetings in Economics. His research focuses on two main topics. First, in the area of macroeconomic theory, he analyzes the effects of macroeconomic policies on welfare, income distribution, and natural resources, using MATLAB to effectively present the simulation results. Second, in the area of behavioral economics, he studies decision-making, focusing in particular on catastrophe insurance purchase decisions, the role of future simulations in individual decision-making, and goal-setting behavior. In this case, he uses MATLAB for both statistical graphics and statistical analysis. His manuscripts have been published in Journal of Economic Dynamics and Control, Environmental and Resource Economics, Economic Theory, Journal of Economics, Economic Letters, and International Tax and Public Economics, etc.

List of Figures

Figure 1.1	Program view in MATLAB (.mlx)	3
Figure 1.2	The syntax of the scatter function	4
Figure 1.3	Command Window view in MATLAB	5
Figure 1.4	Error output of double-quoted variable misuse	5
Figure 1.5	Default layout	6
Figure 1.6	Customized layout using the tiledlayout function	7
Figure 1.7	Use of the figure command: A reset of the current tiledlayout	7
Figure 1.8	Execution of an error program by removing the % symbol	8
Figure 1.9	MATLAB warning output	8
Figure 1.10	Empty Workspace	11
Figure 1.11	Error output due to empty Workspace	12
Figure 1.12	Data types of a table variable	13
Figure 1.13	Contents of a table variable	13
Figure 1.14	Execution of "Var1" from a table variable by dot notation	13
Figure 2.1	Contents of Workspace after data import	16
Figure 2.2	Error output due to invalid file path	17
Figure 2.3	Preview of raw dataset in Excel format	18
Figure 2.4	Preview of dataset with missing values	19
Figure 2.5	Preview of dataset after omitting missing values	20
Figure 2.6	Execution of partial data import by range specification	21
Figure 2.7	Preview of a table variable	21
Figure 2.8	Contents of Workspace after variable extraction	22
Figure 3.1	Contents of Workspace	25
Figure 3.2	Contents of the variable "Japan_Food_matrix1"	26
Figure 3.3	Contents of the variable "Japan_Food_matrix2"	26
Figure 3.4	Contents of the variable "Japan_Food_matrix2" imported from Excel	27
Figure 3.5	The syntax of the scatter function	28
Figure 3.6	Scatter plot of yearly data (X-axis: Year)	28
Figure 3.7	Scatter plot of yearly data (Y-axis: Year)	29
Figure 3.8	Scatter plot created from manually input data	29
Figure 3.9	Error output of the scatter function: Missing variable	30
Figure 3.10	Use of a table variable: Scatter plot	30
Figure 3.11	Scatter plot with point sizes determined by the variable "Meat"	31
Figure 3.12	Scatter plot with two groups	31
Figure 3.13	Scatter plot with multiple groups	32
Figure 3.14	Use of a table variable: Scatter plot with multiple groups	32
Figure 3.15	Scatter plot with multiple groups and the creation of the variable "abc"	33
Figure 3.16	Contents of the variable "abc"	33
Figure 3.17	Contents of the element "abc(2)"	33
Figure 3.18	Scatter plot with a small marker size	34
Figure 3.19	Scatter plot with a large marker size	34
Figure 3.20	Scatter plot with different marker sizes	35
Figure 3.21	Scatter plot with extremely large markers	35
Figure 3.22	Error output of the scatter function: Inconsistent vector lengths	36

Figure 3.23 Scatter plot with red markers and a small marker size 36
Figure 3.24 Scatter plot with red markers and a large marker size................. 37
Figure 3.25 Scatter plot with green markers and default size 37
Figure 3.26 Scatter plot with green markers and default size (cf. ◘ Fig. 3.25) 38
Figure 3.27 Error output of the scatter function: Incorrect order 38
Figure 3.28 Use of single quotes: Scatter plot with red markers 39
Figure 3.29 Error output of the scatter function: Missing quotation............... 39
Figure 3.30 Use of the RGB vector: Scatter plot with red color.................... 40
Figure 3.31 Use of the RGB vector (precise numbers): Scatter plot with purple color... 40
Figure 3.32 Scatter plot with filled red markers 41
Figure 3.33 Scatter plot with blue filled markers of various sizes................. 41
Figure 3.34 Name-value pairs: the syntax of the scatter function 42
Figure 3.35 Scatter plot with thick marker edges 43
Figure 3.36 Error output of the name-value pairs: Missing value 43
Figure 3.37 Scatter plot with large markers and thick marker edges 44
Figure 3.38 Error output of the scatter function: Misplaced property value......... 44
Figure 3.39 Scatter plot with blue marker edges using the "MarkerEdgeColor" property 45
Figure 3.40 Scatter plot with blue marker edges without using the "MarkerEdgeColor" property (cf. ◘ Fig. 3.39).................. 45
Figure 3.41 Scatter plot with red filled markers and blue thick marker edges 46
Figure 3.42 Scatter plot with plus markers 46
Figure 3.43 Scatter plot with large cross markers 47
Figure 3.44 Scatter plot with large red cross markers 47
Figure 3.45 Scatter plot with plus markers and the creation of the variable "sc1" 48
Figure 3.46 Contents of the variable "sc1" 48
Figure 3.47 Scatter plot with large plus markers................................. 49
Figure 3.48 Scatter plot with cyan plus markers 49
Figure 3.49 Change of marker shape from plus to square in a scatter plot 50
Figure 3.50 Scatter plot with multiple groups and a large marker size.............. 50
Figure 3.51 Scatter plot with multiple groups and various marker sizes 51
Figure 3.52 Scatter plot with multiple groups and various marker shapes........... 52
Figure 3.53 Scatter plot with the X-axis labeled "Year" 52
Figure 3.54 Scatter plot with numeric years on the X-axis 53
Figure 3.55 Scatter plot with large red X-axis label 54
Figure 3.56 Scatter plot with red left-pointing markers and large green X-axis label.... 54
Figure 3.57 Scatter plot with various X- and Y-axis labels 55
Figure 3.58 Scatter plot with adjusted X-axis limits 56
Figure 3.59 Scatter plot with adjusted Y-axis limits 56
Figure 3.60 Error output of the xlim function: Missing brackets................... 57
Figure 3.61 Scatter plot with default axis limits 57
Figure 3.62 Scatter plot with data labels 58
Figure 3.63 Scatter plot using a for loop to create data labels and adjust label positions 59
Figure 3.64 Scatter plot using a for loop to create bold red data labels and adjust label positions 60
Figure 3.65 Scatter plot using a for loop to create data labels with a legend 61
Figure 3.66 Scatter plot with a default legend.................................... 61

Figure 3.67 Scatter plot with a specified legend . . . 62
Figure 3.68 Scatter plot with less legend entries . . . 63
Figure 3.69 Scatter plot with one missing legend entry. . . . 63
Figure 3.70 Scatter plot with a blank legend entry . . . 64
Figure 3.71 Scatter plot with a legend (bottom-left position). . . . 64
Figure 3.72 Scatter plot with a legend (outside-top-center position) . . . 65
Figure 3.73 Scatter plot with extra legend entries (best position). . . . 66
Figure 3.74 Scatter plot with a legend containing large blue text (outside-right position). . . . 66
Figure 3.75 Scatter plot with a legend containing large cyan text (bottom-left position) . . . 67
Figure 3.76 Scatter plot with a horizontal legend (outside-top-center position) . . . 68
Figure 3.77 Scatter plot with a figure title . . . 68
Figure 3.78 Scatter plot with figure and legend titles . . . 69
Figure 3.79 Scatter plot (2-row layout) . . . 70
Figure 3.80 Scatter plot (1-row layout) . . . 70
Figure 3.81 Scatter plot (2-row layout with an empty bottom plot). . . . 71
Figure 3.82 Scatter plot (2-row layout with an empty frameless bottom space). . . . 72
Figure 3.83 Scatter plot (3-column layout with an empty center plot) . . . 72
Figure 3.84 Scatter plot (2-row layout with a framed bottom plot) . . . 73
Figure 3.85 Scatter plot (2-row layout with gridlines on a bottom plot) . . . 74
Figure 3.86 Scatter plot (2-row layout with thick gridlines on a bottom plot). . . . 75
Figure 3.87 Scatter plot (3-row, 3-column layout with various annotations and text styles) . . . 76
Figure 3.88 Scatter plot (3-row, 3-column layout with a large framed bottom plot) . . . 77
Figure 3.89 Scatter plot (3-row, 3-column layout with the bottom plot replaced due to not using the figure command) . . . 78
Figure 3.90 Scatter plot with cross markers after using the figure command. . . . 79
Figure 3.91 Use of gca: Scatter plot with a large axis font size. . . . 79
Figure 3.92 Use of gca: Scatter plot with outward tick marks . . . 80
Figure 3.93 Use of gca: Scatter plot with large outward tick marks . . . 81
Figure 3.94 Use of gca: Scatter plot with large inward tick marks and a legend . . . 81
Figure 3.95 Use of gca: Scatter plot with thick black frame. . . . 82
Figure 3.96 Use of gca: Scatter plot with red frame and ticks . . . 83
Figure 3.97 Use of gca: Scatter plot with a red X-axis and a black Y-axis. . . . 83
Figure 3.98 Scatter plot with a vertical gray shaded area and a legend . . . 84
Figure 3.99 Scatter plot with two vertical gray shaded areas and a legend . . . 85
Figure 3.100 Scatter plot with various horizontal shaded areas and a legend . . . 86
Figure 3.101 Scatter plot with horizontal and vertical shaded areas . . . 86
Figure 3.102 Scatter plot with a vertical line. . . . 87
Figure 3.103 Scatter plot with labeled horizontal black dashed and horizontal red dash-dot lines . . . 88
Figure 3.104 Scatter plot with a shaded area and two horizontal dashed lines labeled in a large font . . . 89
Figure 3.105 Scatter plot with a shaded area, two horizontal dotted lines, and two arrows . . . 90
Figure 3.106 Scatter plot with a shaded area, two horizontal dashed lines, and a textbox. . . . 90

Figure 3.107 Scatter plot with a shaded area, two horizontal dashed lines, and a fitted textbox . . . 91
Figure 3.108 Scatter plot with a polynomial approximation curve. . . . 92
Figure 3.109 Scatter plot with polynomial approximation curves (2-row, 2-column layout). . . . 94
Figure 3.110 Scatter plot (Question 1). . . . 95
Figure 3.111 Scatter plot (Question 2). . . . 95
Figure 3.112 Scatter plot (Question 3). . . . 96
Figure 4.1 The syntax of the scatter3 function . . . 100
Figure 4.2 Scatter 3D plot with a single point. . . . 101
Figure 4.3 Scatter 3D plot with three points (different colors) . . . 101
Figure 4.4 Scatter 3D plot with three points (the same color) . . . 102
Figure 4.5 Error output of the scatter3 function: Different length vectors . . . 102
Figure 4.6 Scatter 3D plot from three variables. . . . 103
Figure 4.7 Scatter 3D plot from reordered variables (cf. ◘ Fig. 4.6) . . . 103
Figure 4.8 Scatter 3D plot using a subset of variables (cf. ◘ Fig. 4.6). . . . 104
Figure 4.9 Scatter 3D plot using two subsets of variables (cf. ◘ Fig. 4.8) . . . 104
Figure 4.10 Scatter 3D plot using multiple subsets of variables (cf. ◘ Fig. 4.8) . . . 105
Figure 4.11 Scatter 3D plot with a tiny marker size . . . 106
Figure 4.12 Scatter 3D plot with a large marker size . . . 106
Figure 4.13 Scatter 3D plot with red markers and a large marker size. . . . 107
Figure 4.14 Scatter 3D plot with filled red markers . . . 107
Figure 4.15 Scatter 3D plot with large asterisk markers. . . . 108
Figure 4.16 Error output of the scatter3 function: Incorrect order . . . 108
Figure 4.17 Scatter 3D plot with filled magenta markers and black marker edges . . . 109
Figure 4.18 Scatter 3D plot with left-pointing triangle and plus markers . . . 109
Figure 4.19 Scatter 3D plot with various markers. . . . 110
Figure 4.20 Scatter 3D plot with X-, Y-, and Z-axis labels . . . 111
Figure 4.21 Scatter 3D plot from a different angle (cf. ◘ Fig. 4.20). . . . 112
Figure 4.22 Scatter 3D plot with a red X-axis and blue Y- and Z-axes, including colored gridlines . . . 112
Figure 4.23 Scatter 3D plot overlaid with dashed and dotted lines using the plot3 function . . . 113
Figure 4.24 Scatter 3D plot with a colorbar . . . 114
Figure 4.25 Scatter 3D plot with a colorbar, no grid, and various axis colors . . . 115
Figure 4.26 Scatter 3D plot using a for loop to create data labels, with a colorbar . . . 116
Figure 4.27 Flat scatter 3D plot using a for loop to create data labels, with a colorbar (cf. ◘ Fig. 4.27). . . . 117
Figure 4.28 Scatter 3D plot with a text arrow and a colorbar . . . 118
Figure 4.29 Scatter 3D plot with a pink-colored bottom plane and a colorbar . . . 119
Figure 4.30 Scatter 3D plot with a purple-colored plane and a colorbar . . . 119
Figure 4.31 Scatter 3D plot using a for loop to create vertical dashed droplines, with a colorbar . . . 120
Figure 4.32 Scatter 3D plot using a for loop to create red dashed and blue dotted droplines, with a colorbar. . . . 121
Figure 4.33 Scatter 3D plot using a for loop to create dashed droplines from a different angle, with a colorbar. . . . 122

Figure 4.34 Scatter 3D plot using a for loop to create three-directional droplines from a different angle 124
Figure 4.35 Scatter 3D plot grouped by BRICS, ASEAN, and G7 125
Figure 4.36 Scatter 3D plot (3-row layout grouped by BRICS, ASEAN, and G7) 126
Figure 4.37 Scatter 3D plot (Question 1) 127
Figure 4.38 Scatter 3D plot (Question 2) 128
Figure 4.39 Scatter 3D plot (Question 3) 128
Figure 5.1 The syntax of the plot function 133
Figure 5.2 Line plot of yearly data (X-axis: Year) 134
Figure 5.3 Line plot of yearly data (Y-axis: Year) 134
Figure 5.4 Line plot connecting four manually entered data points 135
Figure 5.5 Error output of the plot function: Inconsistent vector lengths 135
Figure 5.6 Scatter plot with four points 136
Figure 5.7 Line plot with default numeric X-axis ticks 136
Figure 5.8 Error output of the plot function: Third argument interpreted as a line style 137
Figure 5.9 Line plot using repeated X-axis data 137
Figure 5.10 Line plot using data series of different lengths 138
Figure 5.11 Line plot with common X-axis data (cf. ◘ Fig. 5.9) 139
Figure 5.12 Use of a table variable: Line plot 139
Figure 5.13 Use of a subset of a table variable: Line plot 140
Figure 5.14 Use of a table variable: Line plot using repeated X-axis data 141
Figure 5.15 Line plot with a dotted line style 141
Figure 5.16 Line plot with a dotted line style specified by single quotes 142
Figure 5.17 Error output of the plot function: Missing quotation mark 142
Figure 5.18 Line plot with a dashed line style 143
Figure 5.19 Line plot with dotted and dashed line styles 143
Figure 5.20 Line plot with a blue dotted line and circle markers in the same color 144
Figure 5.21 Line plot with a green dotted line and circle markers in the same color . . . 144
Figure 5.22 Line plot with a red dashed line and cross markers in the same color 145
Figure 5.23 Use of a table variable: Line plot with black dashed lines and circle markers in the same color 146
Figure 5.24 Line plot with multiple lines, various marker styles, and colors 147
Figure 5.25 Line plot with a thick red line and square markers in the same color 147
Figure 5.26 Line plot with a red line and cyan square markers 148
Figure 5.27 Line plot with a dotted magenta line and large black circle markers 148
Figure 5.28 Line plot with various line styles and markers 149
Figure 5.29 Use of a table variable: Line plot with various line styles and markers . . . 150
Figure 5.30 Use of a table variable: Line plot using a for loop to adjust the sizes of circle markers 150
Figure 5.31 Line plot with overlaid scatter plot 151
Figure 5.32 Line plot with overlaid bar chart and data labels 152
Figure 5.33 Line plot with overlaid grouped bar chart and data labels 153
Figure 5.34 Line plot with wide Y-axis limits and fine gridlines (cf. ◘ Fig. 5.33) 154
Figure 5.35 Scatter and line plots with various line styles and marker types 155
Figure 5.36 Line plot with a text arrow 156
Figure 5.37 Line plot with a text arrow, two horizontal lines, and a legend 157

Figure 5.38 Line plot with two shaded areas, a circle marker, two text arrows, and a legend 158
Figure 5.39 Line plot with two ellipses, a rectangle, a textbox, and fine gridlines 159
Figure 5.40 Line plot with two text arrows, two vertical double arrows, and a circle marker 160
Figure 5.41 Line plot with adjusted X- and Y-axis limits 160
Figure 5.42 Line plot with axis limits matching data range 161
Figure 5.43 Line plot with dual Y-axes (default left Y-axis) 162
Figure 5.44 Line plot with dual Y-axes (default right Y-axis ticks in green) 163
Figure 5.45 Line plot with dual Y-axes (no left ticks) 164
Figure 5.46 Line plot with a top-aligned X-axis via the set function 164
Figure 5.47 Line plot with various line styles, text arrows, and dual Y-axes 165
Figure 5.48 Line plot with the X-axis labeled "Year" in bold blue 166
Figure 5.49 Line plot with a two-line X-axis label containing a text label and a LaTeX expression 167
Figure 5.50 Line plot with a red two-line X-axis label and a Y-axis label 167
Figure 5.51 Use of a table variable: Line plot with multiple lines and a legend (outside-bottom-right position) 168
Figure 5.52 Use of a table variable: Line plot with multiple lines and a frameless legend (bottom-right position) 169
Figure 5.53 Use of a table variable: Line plot with multiple lines and a 3-column legend (bottom-right position) 170
Figure 5.54 Use of a table variable: Line plot (2-row, 3-column layout) 171
Figure 5.55 Use of a table variable Line plot (2-row layout) with two legends 172
Figure 5.56 Line plot (Question 1) 172
Figure 5.57 Line plot (Question 2) 173
Figure 5.58 Line plot (Question 3) 174
Figure 6.1 The syntax of the plot3 function 178
Figure 6.2 3D plot not generated due to a single data point 179
Figure 6.3 3D plot generated due to a marker for a single data point 179
Figure 6.4 3D plot with one line 180
Figure 6.5 3D plot with two lines 180
Figure 6.6 3D plot with two groups of circle markers 181
Figure 6.7 3D plot with dashed and dash-dot lines and different markers 181
Figure 6.8 3D plot with a line from three variables 182
Figure 6.9 3D plot with a line from different three variables (cf. ◘ Fig. 6.8) 182
Figure 6.10 3D plot with two lines created by two subsets of variables 183
Figure 6.11 3D plot using repeated X-, Y-, and Z-axis data for two subsets (cf. ◘ Fig. 6.10) 183
Figure 6.12 3D plot with circle markers 184
Figure 6.13 3D plot with two dotted lines and cross markers 184
Figure 6.14 3D plot with dash-dot triangle and dashed hexagram lines 185
Figure 6.15 3D plot with a red dashed line and green-filled triangle markers 186
Figure 6.16 3D plot with a thick line and large markers (cf. ◘ Fig. 6.15) 186
Figure 6.17 3D plot with one line and cross black markers 187
Figure 6.18 3D plot with a thick line and large markers (cf. ◘ Fig. 6.17) 187
Figure 6.19 3D plot with black frame, gridlines, and axis labels 188

Figure 6.20 3D plot from a different angle, with black frame, gridlines, and axis labels . . . 189
Figure 6.21 3D plot using a for loop to create vertical dotted droplines . . . 190
Figure 6.22 3D plot using a for loop to create vertical dotted droplines and data labels . . . 191
Figure 6.23 3D plot using a for loop to create vertical dotted droplines and data labels for selected points . . . 192
Figure 6.24 Flat 3D plot using a for loop to create data labels for selected points . . . 193
Figure 6.25 Flat 3D plot with data labels, a text arrow, and a textbox . . . 194
Figure 6.26 Flat 3D plot with two horizontal dotted lines . . . 195
Figure 6.27 3D plot with three-directional dotted droplines . . . 196
Figure 6.28 3D plot (2-column layout) . . . 197
Figure 6.29 3D plot (2-row, 2-column layout) . . . 198
Figure 6.30 3D plot (Question 1) . . . 198
Figure 6.31 3D plot (Question 2) . . . 199
Figure 6.32 3D plot (Question 3) . . . 200
Figure 7.1 The syntax of the bar function . . . 203
Figure 7.2 Bar chart (X-axis: Default index) . . . 204
Figure 7.3 Grouped bar chart . . . 205
Figure 7.4 Bar chart of yearly data (X-axis: Year) . . . 205
Figure 7.5 Error output of the bar function: Inconsistent vector lengths . . . 206
Figure 7.6 Bar chart created from manually input data . . . 206
Figure 7.7 Error output of the bar function: Inconsistent vector lengths from manually input data . . . 207
Figure 7.8 Bar chart with a negative value . . . 207
Figure 7.9 Grouped bar chart of yearly data (X-axis: Year) . . . 208
Figure 7.10 Use of a table variable: Grouped bar chart . . . 208
Figure 7.11 Use of a subset of a table variable: Grouped bar chart . . . 209
Figure 7.12 Grouped bar chart at a single X-point . . . 209
Figure 7.13 Bar chart from the elements of variables . . . 210
Figure 7.14 Bar chart with text label at a single X-point . . . 210
Figure 7.15 Contents of the matrix "y1" . . . 211
Figure 7.16 Grouped bar chart at two X-points . . . 211
Figure 7.17 Contents of the vector "y2" . . . 212
Figure 7.18 Error output of the bar function: Mismatch between first and second arguments . . . 212
Figure 7.19 Bar chart using manually entered Y-data, with text labels at two X-points . . . 213
Figure 7.20 Bar chart using variables for Y-data, with text labels at two X-points . . . 213
Figure 7.21 Bar chart in red . . . 214
Figure 7.22 Bar chart in red with blue dashed edges . . . 214
Figure 7.23 Error output of the bar function: Index beyond single bar object . . . 215
Figure 7.24 Grouped bar chart with one thick-edged group . . . 215
Figure 7.25 Grouped bar chart with various edge styles by group . . . 216
Figure 7.26 Grouped bar chart with a reference line . . . 217
Figure 7.27 Grouped bar chart with a thick red reference line . . . 217
Figure 7.28 Error output of the bar function: Invalid use of "BaseLine" property . . . 218
Figure 7.29 Grouped bar chart with a thick dashed red reference line . . . 218

Figure 7.30 Grouped bar chart with a thick dash-dot red reference line 219
Figure 7.31 Grouped bar chart with various edge styles and a magenta reference line 220
Figure 7.32 Stacked bar chart at a single X-point 220
Figure 7.33 Stacked bar chart with a styled bottom segment 221
Figure 7.34 Stacked bar chart with styled middle segments 222
Figure 7.35 Stacked bar chart with negative values 222
Figure 7.36 Stacked bar chart with various segment styles and a reference line 223
Figure 7.37 Stacked bar chart from two variables (X-axis: Year) 224
Figure 7.38 Stacked bar chart from two variables (X-axis: Countries) 224
Figure 7.39 Bar chart with one pink-colored bar (X-axis: Year) 225
Figure 7.40 Grouped bar chart with bars in various colors (X-axis: Year) 226
Figure 7.41 Grouped bar chart with predefined color order 226
Figure 7.42 Grouped bar chart with adjusted Y-axis limits and predefined color order 227
Figure 7.43 Grouped bar chart with specified Y-axis ticks and predefined color order 228
Figure 7.44 Use of gca: Grouped bar chart with large outward tick marks 228
Figure 7.45 Grouped bar chart with a legend 229
Figure 7.46 Grouped bar chart with a horizontal legend (top-center position) 229
Figure 7.47 Grouped bar chart with a blank legend entry (outside-right position) 230
Figure 7.48 Grouped bar chart with a title, axis labels, and a horizontal legend 231
Figure 7.49 Grouped bar chart with various text arrows 232
Figure 7.50 Bar chart with two text arrows and a textbox 233
Figure 7.51 Bar chart with a textbox and a rectangle 234
Figure 7.52 Grouped bar chart with horizontal lines of different styles and colors 234
Figure 7.53 Bar chart with two horizontal shaded areas 235
Figure 7.54 Bar chart with three vertical shaded areas 236
Figure 7.55 Grouped bar chart with data labels 237
Figure 7.56 Grouped bar chart with position-adjusted data labels (cf. ◘ Fig. 7.55) 238
Figure 7.57 Bar chart with Greek and numeric data labels 238
Figure 7.58 Bar chart with two-line and colored text labels 239
Figure 7.59 Bar chart with two-line text labels 240
Figure 7.60 Bar chart with overlaid bars (non-stacked) 241
Figure 7.61 Bar chart with overlaid transparent bars (non-stacked) 241
Figure 7.62 Bar chart with an overlaid line connecting data points 242
Figure 7.63 Bar chart with an overlaid line connecting adjusted data points (cf. ◘ Fig. 7.62) 243
Figure 7.64 Bar chart with overlaid lines connecting data points (1-row, 3-column layout) 245
Figure 7.65 Stacked bar chart with overlaid lines connecting data points 246
Figure 7.66 Stacked bar chart with overlaid lines connecting data points (X-axis: Year) 247
Figure 7.67 Bar chart (Question 1) 248
Figure 7.68 Bar chart (Question 2) 248
Figure 7.69 Bar chart (Question 3) 249
Figure 8.1 The syntax of the bar3 function 254

Figure 8.2 3D bar chart (Y-axis: Default index) ... 255
Figure 8.3 3D bar chart (Y-axis: Year) ... 255
Figure 8.4 3D bar chart with multiple groups (X-axis: Default index; Y-axis: Year) ... 256
Figure 8.5 3D bar chart with multiple groups using a subset of variables (X-axis: Default index; Y-axis: Year) ... 257
Figure 8.6 3D bar chart with multiple groups using a different subset of variables (cf. ◘ Fig. 8.5) ... 257
Figure 8.7 Use of a table variable: 3D bar chart ... 258
Figure 8.8 Use of a table variable: 3D bar chart with two groups ... 259
Figure 8.9 3D bar chart with narrow bars ... 259
Figure 8.10 3D bar chart with narrow red bars ... 260
Figure 8.11 3D bar chart with one group of transparent bars with dashed edges and another group of yellow bars with green edges ... 261
Figure 8.12 3D bar chart with semi-transparent red bars ... 261
Figure 8.13 3D bar chart with "stacked" option (not stacked) ... 262
Figure 8.14 Stacked 3D bar chart at two Y-points ... 263
Figure 8.15 Stacked 3D bar chart (Y-axis: Year) ... 263
Figure 8.16 Stacked 3D bar chart with a semi-transparent red bottom segment ... 264
Figure 8.17 Grouped 3D bar chart ... 265
Figure 8.18 Grouped 3D bar chart (one bar with no fill and dash-dot outline) ... 266
Figure 8.19 3D bar chart with one group having interpolated face colors ... 267
Figure 8.20 3D bar chart with one group having interpolated face colors (cf. ◘ Fig. 8.19) ... 268
Figure 8.21 3D bar chart with one group being invisible ... 268
Figure 8.22 3D bar chart with specified X-tick labels ... 269
Figure 8.23 3D bar chart using a for loop to vary transparency ... 270
Figure 8.24 3D bar chart using a for loop to vary red transparency ... 270
Figure 8.25 3D bar chart with no X-tick labels and a red ellipse ... 271
Figure 8.26 3D bar chart with no X-tick labels and a legend ... 272
Figure 8.27 3D bar chart with no X-tick labels and a legend (top-center position) ... 272
Figure 8.28 3D bar chart from a different angle ... 273
Figure 8.29 3D bar chart from a different angle (cf. ◘ Fig. 8.28) ... 274
Figure 8.30 Flat 3D bar chart from 2D top-down perspective ... 275
Figure 8.31 3D bar chart using a for loop to apply interpolated colors, with labeled Y- and Z-axes and a color bar ... 276
Figure 8.32 3D bar chart with overlaid lines connecting data points ... 277
Figure 8.33 3D bar chart with overlaid bars (hard to distinguish due to identical color) ... 278
Figure 8.34 3D bar chart with overlaid bars (cf. ◘ Fig. 8.33) ... 279
Figure 8.35 3D bar chart (Question 1) ... 279
Figure 8.36 3D bar chart (Question 2) ... 280
Figure 8.37 3D bar chart (Question 3) ... 281
Figure 9.1 The syntax of the barh function ... 286
Figure 9.2 Horizontal bar chart (Y-axis: Default index) ... 287
Figure 9.3 Horizontal bar chart with two groups (Y-axis: Default index) ... 287
Figure 9.4 Horizontal bar chart with multiple groups (cf. ◘ Fig. 9.3) ... 288
Figure 9.5 Horizontal bar chart with two groups (Y-axis: Year) ... 288

Figure 9.6 Use of a table variable: Horizontal bar chart ... 289
Figure 9.7 Horizontal bar chart with multiple groups (Y-axis: Year) ... 289
Figure 9.8 Horizontal bar chart with Y-axis text labels ... 290
Figure 9.9 Error output of the barh function: Inconsistent vector lengths ... 290
Figure 9.10 Grouped horizontal bar chart with Y-axis text labels ... 291
Figure 9.11 Grouped horizontal bar chart with narrow bars ... 291
Figure 9.12 Grouped horizontal bar chart with one group of dashed-outlined green bars ... 292
Figure 9.13 Grouped horizontal bar chart with one group of dashed-outlined green bars and another group of magenta bars ... 293
Figure 9.14 Grouped horizontal bar chart with a reference line ... 293
Figure 9.15 Grouped horizontal bar chart with a dash-dot red reference line ... 294
Figure 9.16 Horizontal stacked bar chart ... 295
Figure 9.17 Horizontal stacked bar chart with Y-axis text labels ... 295
Figure 9.18 Horizontal stacked bar chart with one yellow segment outlined in thick dotted black ... 296
Figure 9.19 Horizontal stacked bar chart with one pale segment outlined in thick magenta ... 297
Figure 9.20 Horizontal stacked bar chart with a reference line ... 297
Figure 9.21 Use of the RGB vector: Horizontal bar chart ... 298
Figure 9.22 Grouped horizontal bar chart with manually spaced Y-axis labels ... 299
Figure 9.23 Grouped horizontal bar chart with manually spaced Y-axis labels using text rotation (cf. ◘ Fig. 9.22) ... 299
Figure 9.24 Grouped horizontal bar chart with partially labeled Y-axis ... 300
Figure 9.25 Grouped horizontal bar chart with X- and Y-axis labels and X-axis limits ... 301
Figure 9.26 Grouped horizontal bar chart with a yellow shaded area and a legend ... 301
Figure 9.27 Grouped horizontal bar chart using a for loop to vary red transparency ... 302
Figure 9.28 Grouped horizontal bar chart with arrows and text arrows ... 303
Figure 9.29 Grouped horizontal bar chart using a for loop to vary blue transparency, with an arrow, a text arrow, and a vertical dotted line ... 304
Figure 9.30 Grouped horizontal bar chart using a for loop to vary blue transparency, with vertical reference lines and shaded areas ... 305
Figure 9.31 Grouped horizontal bar chart using a for loop to vary blue transparency, with a horizontal legend and specified X-ticks ... 306
Figure 9.32 Grouped horizontal bar chart with lines connecting data points and a legend (outside-right position) ... 307
Figure 9.33 Grouped horizontal bar chart with lines connecting data points and a horizontal legend (cf. ◘ Fig. 9.32) ... 308
Figure 9.34 Horizontal bar chart with overlaid bars and customized edges (no stacked) ... 309
Figure 9.35 Horizontal bar chart with blue thick dashed edges and a two-line legend (no stacked) ... 310
Figure 9.36 Horizontal bar charts (2-row layout with outward ticks and right Y-axis in top chart) ... 311
Figure 9.37 Horizontal bar chart (Question 1) ... 312
Figure 9.38 Horizontal bar chart (Question 2) ... 312
Figure 9.39 Horizontal bar chart (Question 3) ... 313

Figure 10.1 The syntax of the bar3h function. 317
Figure 10.2 Horizontal 3D bar chart (Z-axis: Default index). 318
Figure 10.3 Horizontal 3D bar chart (Z-axis: Year) 319
Figure 10.4 Horizontal 3D bar chart with multiple groups (Z-axis: Year). 319
Figure 10.5 Horizontal 3D bar chart with multiple groups using a subset of variables 320
Figure 10.6 Use of a table variable: Horizontal 3D bar chart 321
Figure 10.7 Horizontal 3D bar chart with narrow bars 321
Figure 10.8 Horizontal 3D bar chart with narrow bars having thick green edges 322
Figure 10.9 Horizontal 3D bar chart with one group of semi-transparent red bars having dashed black edges 323
Figure 10.10 Horizontal 3D bar chart using a for loop to vary yellow transparency, with dashed black edges 324
Figure 10.11 Horizontal 3D bar chart with “stacked” option (not stacked) 325
Figure 10.12 Stacked horizontal 3D bar chart 325
Figure 10.13 Stacked 3D bar chart with red semi-transparent and yellow segments. 326
Figure 10.14 Grouped horizontal 3D bar chart with one group of dash-dot outlined transparent bars 327
Figure 10.15 Horizontal 3D bar chart with one group of bars having interpolated face colors. 328
Figure 10.16 Horizontal 3D bar chart with two groups having interpolated face colors by a for loop 329
Figure 10.17 Horizontal 3D bar chart with semi-transparent purple bars, circle markers, and dotted edges. 329
Figure 10.18 Horizontal 3D bar chart with black frame, labeled X-ticks, and adjusted axes ticks 330
Figure 10.19 Horizontal 3D bar chart from a different angle (cf. ◘ Fig. 10.18) 331
Figure 10.20 3D bar chart from a different angle, with a text arrow, an arrow, and an ellipse 332
Figure 10.21 Flat horizontal 3D bar chart from 2D top-down perspective 333
Figure 10.22 3D bar chart with overlaid bars (not stacked) 334
Figure 10.23 3D bar chart with data labels 335
Figure 10.24 3D bar chart (Question 1) 335
Figure 10.25 3D bar chart (Question 2) 336
Figure 10.26 3D bar chart (Question 3) 337
Figure 11.1 Contents of Workspace 343
Figure 11.2 Bar chart comparing age distribution by gender (2-column layout) 344
Figure 11.3 Error output of the histogram function: Use of string input 344
Figure 11.4 Histogram using aggregate data (inappropriate input) 345
Figure 11.5 Histogram using aggregate data (inappropriate input), with adjusted X-axis limits 346
Figure 11.6 The syntax of the histogram function 346
Figure 11.7 Histogram using one variable. 347
Figure 11.8 Histogram using merged data of two variables 348
Figure 11.9 Error output of the histogram function: Inconsistent vector lengths 348
Figure 11.10 Histogram using vertically concatenated column vectors (cf. ◘ Fig. 11.9). 349

Figure 11.11 Histogram using merged column vectors of two variables (cf. ◘ Fig. 11.8). . . . 349
Figure 11.12 Use of a table variable: Histogram. . . . 350
Figure 11.13 Histogram using fine bin division. . . . 350
Figure 11.14 Histogram using fine data division and merged column vectors . . . 351
Figure 11.15 Histogram with wide–narrow–wide bin intervals . . . 351
Figure 11.16 Histogram with adjusted bin edges including a negative interval . . . 352
Figure 11.17 Error output of the histogram function: Simultaneous use of both number of bins and bin edges. . . . 352
Figure 11.18 Histogram with manually specified bin ranges and frequencies . . . 353
Figure 11.19 Error output of the histogram function: Mismatch between number of bin edges and bin counts . . . 353
Figure 11.20 Histogram with negative bin edges. . . . 354
Figure 11.21 Histogram with negative bin edges and semi-transparent red bins . . . 354
Figure 11.22 Histogram using "BinLimits" property to restrict range. . . . 355
Figure 11.23 Histogram using "BinLimits" property to expand range. . . . 355
Figure 11.24 Histogram with white bins and red edges . . . 356
Figure 11.25 Histogram with white bins and thick blue edges . . . 357
Figure 11.26 Histogram with white large bins and thick blue edges. . . . 357
Figure 11.27 Histogram with white bins and no internal edges . . . 358
Figure 11.28 Histogram with white bins, no internal edges, and dotted border. . . . 358
Figure 11.29 Histogram with integer bins . . . 359
Figure 11.30 Histogram using "BinMethod" property set to "scott" for automatic bin width . . . 360
Figure 11.31 Horizontal histogram . . . 360
Figure 11.32 Horizontal histogram with red bins and blue edges . . . 361
Figure 11.33 Overlaid histograms with a legend . . . 362
Figure 11.34 Overlaid histograms with dashed edges and a legend . . . 362
Figure 11.35 Overlaid histograms with fine gridlines and a legend (bottom-left position) . . . 363
Figure 11.36 Histogram (3-row layout) with different settings of "Normalization" properties. . . . 364
Figure 11.37 Histogram overlaid with a red line connecting the data points . . . 365
Figure 11.38 Histogram with data labels. . . . 366
Figure 11.39 Histogram with two vertical dashed lines and a double arrow . . . 367
Figure 11.40 Overlaid histograms with vertical dashed and dot-dashed lines . . . 368
Figure 11.41 Overlaid horizontal histograms with two horizontal lines and a double arrow . . . 369
Figure 11.42 Histogram with two categorical age groups . . . 369
Figure 11.43 Histogram with three categorical age groups and data labels . . . 370
Figure 11.44 Overlaid histograms with two categorical age groups and data labels . . . 371
Figure 11.45 Histogram (Question 1) . . . 372
Figure 11.46 Histogram (Question 2) . . . 373
Figure 11.47 Histogram (Question 3) . . . 374
Figure 12.1 The syntax of the histogram2 function . . . 378
Figure 12.2 3D histogram from two variables . . . 379
Figure 12.3 Error output of the histogram2 function: Incorrect use of four variables . . . 379

Figure 12.4	3D histogram using merged column vectors	380
Figure 12.5	3D histogram using merged 2-column matrix	380
Figure 12.6	3D histogram using fine bin division	381
Figure 12.7	3D histogram using fine bin division and merged column vectors	381
Figure 12.8	3D histogram using fine bin division and merged column vectors (X- and Y-data reversed, cf. ◘ Fig. 12.7)	382
Figure 12.9	3D histogram with different bin numbers in X- and Y-axes	382
Figure 12.10	3D histogram with different bin edges in X- and Y-axes	383
Figure 12.11	Error output of the histogram2 function: Incomplete bin edge specification	384
Figure 12.12	3D histogram with manually specified bin ranges and frequencies	384
Figure 12.13	3D histogram with narrow bins	385
Figure 12.14	3D histogram with different bin widths in X- and Y-axes	385
Figure 12.15	3D histogram with extended bin limits in X-axis	386
Figure 12.16	3D histogram with visible empty bins	386
Figure 12.17	3D histogram with no face lighting and visible empty bins	387
Figure 12.18	3D histogram with semi-transparent red bins and dashed black edges	388
Figure 12.19	3D histogram with colors based on frequency	388
Figure 12.20	3D histogram and the creation of the variable "h"	389
Figure 12.21	Contents of the bin counts using the variable "h"	389
Figure 12.22	Tiled 3D histogram with a colorbar	390
Figure 12.23	Tiled 3D histogram with small tiles and a colorbar	390
Figure 12.24	Tiled 3D histogram with elongated tiles and a colorbar	391
Figure 12.25	3D histogram with various axes labels	392
Figure 12.26	3D histogram from a different angle, with colors based on frequency	392
Figure 12.27	3D histogram from a different angle, with adjusted number of bins	393
Figure 12.28	3D histogram from a different angle, with adjusted number of X-axis bins	394
Figure 12.29	3D histogram with more Y-axis bins and fewer X-axis bins	394
Figure 12.30	Tiled 3D histogram with a labeled colorbar (outside-bottom-center position)	395
Figure 12.31	Tiled 3D histogram with a red-labeled, reversed color scale (outside-bottom-center position)	396
Figure 12.32	Tiled 3D histogram with a tick-labeled colorbar (outside-left position)	396
Figure 12.33	Tiled 3D histogram with a tick-labeled colorbar and a text arrow	397
Figure 12.34	Overlaid tiled 3D histograms for 2010 and 2020	398
Figure 12.35	Overlaid tiled 3D histograms with adjusted tile sizes	399
Figure 12.36	Tiled 3D histogram with bin counts as text labels in red	400
Figure 12.37	Tiled 3D histogram with percentages as text labels in red	401
Figure 12.38	Overlaid tiled 3D histograms with bin counts as text labels	402
Figure 12.39	3D histogram (Question 1)	403
Figure 12.40	3D histogram (Question 2)	404
Figure 12.41	3D histogram (Question 3)	404
Figure 13.1	The syntax of the boxchart function	410
Figure 13.2	Box plot of a single variable	410
Figure 13.3	Box plots of two variables	411
Figure 13.4	Error output of the boxchart function: Incorrect use of data	411

Figure 13.5 Error output of the boxchart function: Incorrect use of data (two arguments) . . . 412
Figure 13.6 Box plots with two categorical groups . . . 412
Figure 13.7 Box plots with only median lines . . . 413
Figure 13.8 Box plots with multiple categorical groups . . . 414
Figure 13.9 Error output of the boxchart function: Invalid matrix input . . . 414
Figure 13.10 Error output of the boxchart function: Inconsistent input lengths. . . . 415
Figure 13.11 Box plots with multiple categorical groups from combined variables. . . . 415
Figure 13.12 Contents of a categorical variable . . . 416
Figure 13.13 Box plots due to incorrect data extraction. . . . 416
Figure 13.14 Contents of a logical index. . . . 417
Figure 13.15 Box plots with extra categories on the X-axis . . . 418
Figure 13.16 Use of a table variable: Box plots. . . . 418
Figure 13.17 Box plots with red boxes. . . . 419
Figure 13.18 Box plots with cyan boxes, black median lines, and dashed blue whiskers . . . 420
Figure 13.19 Box plots with red boxes, black median lines, and dotted whiskers . . . 420
Figure 13.20 Error output of the boxchart function: Unsupported use of different colors for each box . . . 421
Figure 13.21 Two-tick single box plots using "GroupByColor" property . . . 421
Figure 13.22 Multiple-tick single box plots using "GroupByColor" property. . . . 422
Figure 13.23 Single-tick grouped box plots using "GroupByColor" property. . . . 423
Figure 13.24 Notched box plots . . . 423
Figure 13.25 Notched box plots with red median lines and black whiskers. . . . 424
Figure 13.26 Box plots with large blue outliers. . . . 425
Figure 13.27 Box plots with jittered-cross outliers . . . 425
Figure 13.28 Horizontal box plots. . . . 426
Figure 13.29 Box plots with manually created X-axis labels. . . . 427
Figure 13.30 Single-tick grouped box plots with a two-line X-axis label and a legend. . . . 427
Figure 13.31 Box plots with a large space between boxes. . . . 428
Figure 13.32 Two-tick grouped box plots with a large space between groups . . . 429
Figure 13.33 Box plots with text labels . . . 430
Figure 13.34 Box plots with text labels (maximum, mean, and minimum values) . . . 431
Figure 13.35 Box plots with two boxes connected by a dotted line and mean value labels. . . . 432
Figure 13.36 Box plots with lines connecting maximum, mean, and minimum values. . . . 433
Figure 13.37 Box plots (2-column layout) . . . 434
Figure 13.38 Box plot (Question 1) . . . 435
Figure 13.39 Box plot (Question 2) . . . 436
Figure 13.40 Box plot (Question 3) . . . 437
Figure 14.1 The syntax of the piechart function. . . . 441
Figure 14.2 The syntax of the donutchart function . . . 441
Figure 14.3 Pie chart showing proportions within a single variable . . . 442
Figure 14.4 The sum of the "Production" variable in Command Window . . . 442
Figure 14.5 Pie chart showing proportions for three categories . . . 443
Figure 14.6 Pie chart showing proportions for three categories (cf. ◘ Fig. 14.5) . . . 443
Figure 14.7 Pie chart showing proportions for multiple categories . . . 444

Figure 14.8 Partial pie chart showing proportions for two categories 444
Figure 14.9 Donut chart showing proportions within a single variable 445
Figure 14.10 Partial donut chart showing proportions for two categories 445
Figure 14.11 Pie chart showing proportions for multiple categories with labels 446
Figure 14.12 Donut chart showing proportions for multiple categories with labels 446
Figure 14.13 Pie chart with red edges 447
Figure 14.14 Pie chart with thick black edges 447
Figure 14.15 Pie chart with thick green edges 448
Figure 14.16 Pie chart with large magenta labels 448
Figure 14.17 Small semi-transparent pie chart at specified position. 449
Figure 14.18 Pie chart in counterclockwise slice order 449
Figure 14.19 Pie chart with one separated slice. 450
Figure 14.20 Pie chart with multiple separated slices 450
Figure 14.21 Pie chart with rotated and multiple separated slices 451
Figure 14.22 Donut chart with separated slices in counterclockwise order 451
Figure 14.23 Partial pie chart with no labels. 452
Figure 14.24 Partial pie chart with manually specified labels 452
Figure 14.25 Pie chart with space between specified categories and percentage labels 453
Figure 14.26 Error output of the donutchart function: Unsupported use of the "hold on" command 454
Figure 14.27 Donut chart with thick black edges and rotated slices. 454
Figure 14.28 Pie chart with separated slices and predefined color order ("meadow") 455
Figure 14.29 Pie chart with separated slices and user-specified color order. 455
Figure 14.30 Pie chart with separated slices and predefined color order ("reef") 456
Figure 14.31 Donut chart with a textbox placed in the center hole and text arrows 457
Figure 14.32 Error output of the donutchart function: Unsupported use of the text function. 457
Figure 14.33 Donut chart with a textbox placed in the center hole and three textboxes on the donut 458
Figure 14.34 Donut chart with a yellow textbox placed in the center hole. 459
Figure 14.35 Pie chart with a blue textbox placed at the center 460
Figure 14.36 Pie charts (2-column layout). 461
Figure 14.37 Donut charts with large and small holes (2-column layout) 462
Figure 14.38 Pie chart (Question 1). 462
Figure 14.39 Donut chart (Question 2) 463
Figure 14.40 Pie chart (Question 3). 464
Figure 15.1 The syntax of the plotmatrix function. 469
Figure 15.2 Scatter plot of two variables using the plotmatrix function 469
Figure 15.3 Use of table variables: Scatter plot using the plotmatrix function 470
Figure 15.4 Histogram using the plotmatrix function. 470
Figure 15.5 Scatter plots using the plotmatrix function (2-column layout) 471
Figure 15.6 Error output of the plotmatrix function: Use of three variables. 471
Figure 15.7 Scatter plots using the plotmatrix function (2-row layout) 472
Figure 15.8 Scatter plots using the plotmatrix function (2-row, 2-column layout). 472
Figure 15.9 Scatter plots using the plotmatrix function (3-row, 3-column layout). 473
Figure 15.10 Scatter plot using two merged column vectors. 474

Figure 15.11 Histograms and scatter plots using the plotmatrix function (3-row, 3-column layout). 475
Figure 15.12 Scatter plots using the plotmatrix function (4-row, 4-column layout). 476
Figure 15.13 Histograms and scatter plots using the plotmatrix function (4-row, 4-column layout). 477
Figure 15.14 Scatter plots with default markers on the left and red crosses on the right (2-column layout) 478
Figure 15.15 Scatter plots with upward triangles on the top and green stars on the bottom (2-row layout) 478
Figure 15.16 Histograms and scatter plots with black asterisks (3-row, 3-column layout). 479
Figure 15.17 Histograms and black line plots (3-row, 3-column layout) 480
Figure 15.18 Histograms and multi-colored scatter plots (3-row, 3-column layout) 481
Figure 15.19 Various histograms and scatter plots (3-row, 3-column layout) 482
Figure 15.20 Pink histograms with thick edges and scatter plots (3-row, 3-column layout). 483
Figure 15.21 Histograms and scatter plots with a large red title (3-row, 3-column layout). 484
Figure 15.22 Histograms and scatter plots with subplot labels (4-row, 4-column layout). 485
Figure 15.23 Two-column display of histograms and scatter plots (3-row, 3-column layout). 486
Figure 15.24 Histograms and scatter plots with a bold black numeric label (3-row, 3-column layout). 487
Figure 15.25 Histograms and scatter plots with red-boxed labels (3-row, 3-column layout). 488
Figure 15.26 Histograms and scatter plots with an empty upper triangle and legends (3-row, 3-column layout). 490
Figure 15.27 Histograms and scatter plots with two subplots outlined in red and blue (4-row, 4-column layout) 491
Figure 15.28 Histograms and scatter plots with black-boxed numeric labels placed in an empty upper triangle (4-row, 4-column layout) 492
Figure 15.29 Histograms and scatter plots (Question 1). 493
Figure 15.30 Histograms and scatter plots (Question 2). 494
Figure 15.31 Histograms and scatter plots (Question 3). 495
Figure 16.1 The syntax of the bubblechart function. 500
Figure 16.2 A single bubble created by the bubblechart function. 500
Figure 16.3 Two grouped bubbles 501
Figure 16.4 Two grouped bubbles (cf. ◘ Fig. 16.3). 501
Figure 16.5 Bubbles from three variables 502
Figure 16.6 Bubbles from three variables (cf. ◘ Fig. 16.5) 502
Figure 16.7 Error output of the bubblechart function: Not enough input variables 503
Figure 16.8 Two grouped bubbles in red and cyan 503
Figure 16.9 Three grouped bubbles in red, cyan, and black 504
Figure 16.10 Single grouped bubbles with various colors. 504
Figure 16.11 Green bubbles with thick black edges 505
Figure 16.12 Light purple bubbles with thin black edges 506

Figure 16.13 Bubbles with a large bold title 506
Figure 16.14 Use of the bubblechart function before applying the bubblesize function 507
Figure 16.15 Results of the bubblesize function 507
Figure 16.16 Bubbles with large bubble sizes 508
Figure 16.17 Bubbles with small bubble sizes 508
Figure 16.18 Use of the bubblechart function before applying the bubblelim function 509
Figure 16.19 Results of the bubblelim function 509
Figure 16.20 Bubbles with adjusted upper bubble size limit 510
Figure 16.21 Bubbles with adjusted lower and upper bubble size limits 510
Figure 16.22 Bubbles with black and red text arrows 511
Figure 16.23 Bubbles with a size-fitted textbox in the lower right 512
Figure 16.24 Bubbles with a size-fitted textbox in the upper right and a semi-transparent red rectangle 512
Figure 16.25 Bubbles with a vertical gray shaded area 513
Figure 16.26 Bubbles with adjusted sizes, a legend, and a vertical purple shaded area 514
Figure 16.27 Bubbles with horizontal yellow and vertical gray shaded areas, and a legend 514
Figure 16.28 Bubbles with a vertical red dash-dot line and a horizontal black dashed line 515
Figure 16.29 Bubbles with a horizontal thick blue dashed line and a vertical black dash-dot line 516
Figure 16.30 Bubbles with horizontal and vertical gray shaded areas below zero, a legend, and two lines 517
Figure 16.31 Bubbles with several vertical shaded areas and a horizontal legend 518
Figure 16.32 Bubbles with a bubble legend 518
Figure 16.33 Bubbles with a two-size bubble legend 519
Figure 16.34 Bubbles with a bubble legend (outside-right position) and two text arrows 520
Figure 16.35 Bubbles with small bubble sizes, a bubble legend (outside-right position), and two text arrows (cf. ◘ Fig. 16.34) 521
Figure 16.36 Red bubbles with a bubble legend (outside-right position), a title, and axis labels (cf. ◘ Fig. 16.35) 522
Figure 16.37 Pink bubbles with thick blue edges, a bubble legend (outside-right position), a title, and axis labels 523
Figure 16.38 Bubble chart with large bubble sizes, a title, a subtitle, a bubble legend (outside-right position), and modified x-axis ticks 524
Figure 16.39 Bubbles with a bubble legend in telescopic style 524
Figure 16.40 Bubbles with a horizontal bubble legend 525
Figure 16.41 Two grouped bubbles with a legend 526
Figure 16.42 Two grouped bubbles with a legend (bottom-right position), and gray shaded areas under zero on both axes 526
Figure 16.43 Two grouped bubbles with horizontally aligned group and bubble legends (outside-right position) 527
Figure 16.44 Two grouped bubbles with vertically aligned group and bubble legends (outside-right position) 528

Figure 16.45 Bubbles (Question 1). 529
Figure 16.46 Bubbles (Question 2). 530
Figure 16.47 Bubbles (Question 3). 531
Figure 17.1 The syntax of the bubblechart3 function 536
Figure 17.2 A single 3D bubble created by the bubblechart3 function 536
Figure 17.3 Two grouped 3D bubbles 537
Figure 17.4 3D bubbles from four variables 537
Figure 17.5 Two grouped 3D bubbles from variable subsets 538
Figure 17.6 Two grouped 3D bubbles from matrix data 538
Figure 17.7 Red 3D bubbles 539
Figure 17.8 Green 3D bubbles with thick black edges 539
Figure 17.9 Green 3D bubbles with semi-transparent black edges 540
Figure 17.10 Two grouped 3D bubbles with one black-edged group in green 541
Figure 17.11 Two grouped 3D bubbles with one blue-edged group and one transparent red-edged group 541
Figure 17.12 3D bubbles with X-, Y-, and Z-axis labels 542
Figure 17.13 3D bubbles with red X-, blue Y-, green Z-axes, and red gridlines 543
Figure 17.14 3D bubbles with red dashed gridlines and thick black axes 543
Figure 17.15 3D bubbles with no gridlines and adjusted X- and Y-axis ticks 544
Figure 17.16 3D bubbles from a different angle, with no gridlines and adjusted X- and Y-axis ticks 544
Figure 17.17 3D bubbles from a different angle, with a bubble legend 545
Figure 17.18 3D bubbles in various colors, with a bubble legend 546
Figure 17.19 3D bubbles with some bubbles outside of the frame and a bubble legend 547
Figure 17.20 3D bubbles with fine gridlines, a large red title, and a legend 548
Figure 17.21 3D bubbles with a blue dashed line and a red dotted line 549
Figure 17.22 3D bubbles with vertical surfaces in red and purple 550
Figure 17.23 3D bubbles with a yellow text label and a rectangle 550
Figure 17.24 3D bubbles from a different angle, with text labels added by a for loop 551
Figure 17.25 3D bubbles (2-row layout) 552
Figure 17.26 3D bubbles (2-column layout) 553
Figure 17.27 3D bubbles (Question 1) 553
Figure 17.28 3D bubbles (Question 2) 554
Figure 17.29 3D bubbles (Question 3) 555
Figure 18.1 The syntax of the annotation function 558
Figure 18.2 Single-point scatter with multiple arrows and a line 559
Figure 18.3 Single-point 3D scatter with multiple arrows and a line 560
Figure 18.4 Donut chart with multiple arrows and a line 560
Figure 18.5 Single-point 3D scatter with a blue dashed text arrow 561
Figure 18.6 Donut chart with a red dotted double-headed arrow in the central hole 561
Figure 18.7 Single-point scatter with multiple rectangles and ellipses 562
Figure 18.8 Single-point 3D scatter with multiple rectangles and ellipses 563
Figure 18.9 Single-point scatter with a blue-filled rectangle and red dotted edge 563
Figure 18.10 Single-point scatter with a black dash-dot textbox containing red text 564
Figure 18.11 Single-point scatter with a fitted textbox (cf. ◘ Fig. 18.10) 565

Figure 18.12 Single-point scatter with red text in a dash-dot textbox over a yellow rectangle . . . 565
Figure 18.13 The syntax of the text function . . . 566
Figure 18.14 Two text labels placed at center and top-right corner . . . 566
Figure 18.15 Three text labels aligned vertically at center. . . . 567
Figure 18.16 Multiple red text lines at center . . . 567
Figure 18.17 Text labels placed on multiple diagonal coordinates . . . 568
Figure 18.18 Single-point 3D scatter with two text labels. . . . 568
Figure 18.19 Single-point 3D scatter with two text labels (cf. ◘ Fig. 18.18) . . . 569
Figure 18.20 Single-point 3D scatter with red, blue, and black text labels at three corners . . . 570
Figure 18.21 LaTeX formula and rotated text. . . . 570
Figure 18.22 3D scatter with numeric labels using a for loop. . . . 571
Figure 18.23 Scatter points with labels of increasing font size using a for loop. . . . 572
Figure 18.24 The syntax of the xline function . . . 572
Figure 18.25 The syntax of the yline function . . . 572
Figure 18.26 Three vertical lines and one horizontal line . . . 573
Figure 18.27 Three vertical dashed lines in blue and one horizontal line in red. . . . 573
Figure 18.28 Four lines with text labels. . . . 574
Figure 18.29 One vertical line with a text label placed at bottom left (cf. ◘ Fig. 18.28). . . . 575
Figure 18.30 One horizontal dotted line with a text label placed below its left end. . . . 575
Figure 18.31 Three vertical lines with text labels placed at bottom, middle, and top. . . . 576
Figure 18.32 Three vertical lines with text labels placed to the left, on, and to the right of the lines . . . 576
Figure 18.33 Error output of the xline function: Invalid specification of multiple line styles as an array. . . . 577
Figure 18.34 Vertical and horizontal lines with two-line text labels . . . 577
Figure 18.35 Annotation function (Question 1) . . . 578
Figure 18.36 Text function (Question 2) . . . 579
Figure 18.37 Xline and yline functions (Question 3). . . . 580

Introduction

Contents

1.1 The mlx File – 3

1.2 Structure from ► Chaps. 3–17 – 4

1.3 Single and Double Quotation Marks – 5

1.4 The Figure Command and the Tiledlayout Function – 6

1.5 Errors and Warnings – 8

1.6 Row and Column Vectors – 9

1.7 Semicolon – 10

1.8 Multiplication – 11

1.9 The Data Used in This Book and an Error to Address Before Running the Program – 11

1.10 Table Variables – 12

Y. Nakamoto, *A Comprehensive Guide to Creating Graphics with MATLAB Programming*, Classroom Companion: Economics, https://doi.org/10.1007/978-981-96-6213-5_1

1

Learning Objectives

This chapter provides tips for using this book effectively. By reading this chapter, you will understand the rules used to help beginners avoid difficulties.

Overview and Guidelines

In this book, you can find various types of figures created in MATLAB R2024b. This chapter provides tips for using this book effectively; you should read this chapter first before proceeding to the other chapters. Here are some tips to help you read this book.

- While conventional books are often read sequentially from cover to cover, this book is designed with each chapter being independent. For example, ▶ Chap. 3 provides detailed instructions for using the **scatter** function which is useful for creating scatter charts. However, if you want to create pie charts ▶ Chap. 3 might not be helpful, while ▶ Chap. 14, which explains the **piechart** function, would be more relevant. Therefore, if you have specific goals for creating charts it is better to refer to the appropriate chapter. Alternatively, some readers may have data but not know what type of chart to create, in which case it may be efficient to browse through the figures in this book and find one that catches your interest while you skip the programs and explanations and just focus on the figures. If a particular figure captures your interest, you can refer to the corresponding program and explanation. Additionally, to fully understand the program that creates the figure, it is recommended that you review the corresponding program along with the preceding and following programs which will allow you to efficiently find the figure you want to create. Of course, this will also deepen your understanding of the MATLAB programs.
- For beginners, we recommend starting with this chapter and then proceeding to either ▶ Chaps. 2 or 3. ▶ Chapter 2 explains how to load Excel files, as many readers may receive data in Excel format rather than entering it directly into MATLAB programs. If you need to handle Excel data, refer to ▶ Chap. 2. ▶ Chapter 3 is the first chapter that introduces graphical programming; it explains the state of the Workspace after executing a program as well as the use of more detailed functions and optional functions. For beginners, it is recommended to study ▶ Chap. 3 before moving on to the other chapters.
- Throughout this book, when we explain programming, we use quotation marks 「 」 for variables, file names, and the programming itself. This is because single quotes, double quotes, and parentheses are used in MATLAB programs, which might confuse beginners. Since quotation marks 「」 are not used in MATLAB programs, they are suitable for distinguishing the programming sections from their explanations.
- Function names are in bold. For example, when referring to the **scatter** function used in ▶ Chap. 3, we write it as **scatter** in bold. We do not use italics or underlined characters.

1.1 The mlx File

MATLAB beginners may start by entering programs in the Command Window but it is recommended to use the mlx file format. Note that the Command Window allows you to enter individual statements at the command line, indicated by the prompt (>>). When you input a command, the result is displayed immediately. For example, typing 「abc=1」 and pressing Enter creates the variables 「abc」 in the Workspace. Additionally, the mlx files enable you to create interactive documents that include MATLAB code, explanatory text, formulas, graphs, and figures. For instance, in ◘ Fig. 1.1, which is part of an actual mlx file, the gray background section contains MATLAB code; a screenshot image, ► Fig. 3.4, is attached below. The mlx files not only display MATLAB code and its execution results on the same page but also allow you to create comprehensive documents.

```
file = "alldata.xlsx";
sheet = "Chap3_Food_self_sufficiency";
Japan_Food_matrix2 = readtable(file,"Sheet",sheet);
```

Figure 3.4

11x5 table

	1 Year	2 Rice	3 Vegetables	4 Meat	5 Milk	6
1	1935	73	101	93	107	
2	1955	109	100	100	90	
3	1960	102	100	91	89	
4	1970	106	99	89	89	
5	1980	100	97	81	82	
6	1990	100	91	70	78	
7	2000	95	81	52	68	
8	2005	95	79	54	68	
9	2010	97	81	56	67	
10	2015	98	80	54	62	
11	2020	97	80	53	61	
12						

◘ **Fig. 1.1** Program view in MATLAB (.mlx)

1.2 Structure from ▶ Chaps. 3–17

From ▶ Chaps. 3–17 in this book, the chapter structure remains the same. ▶ Section 1 in each chapter introduces and enters raw data. ▶ Section 2 involves importing the same data from an Excel file named 「alldata.xlsx」. Therefore, the data created in ▶ Sections 1 and 2 will ultimately be the same. If readers are not interested in importing data from the Excel file 「alldata.xlsx」 and prefer to enter it directly, they can skip ▶ Section 2 and move on to ▶ Section 3. Using this data, ▶ Sections 3–5 introduce the functions in each chapter and present various figures. Therefore, if you want to recreate a figure introduced in ▶ Sections 3–5 of a particular chapter, you need to import the data into MATLAB using the method from either ▶ Sections 1 or 2 of the same chapter. You can check if the data has been successfully imported by looking at the Workspace.

▶ Section 3 introduces the basic syntax of the functions introduced in each chapter. This section is geared toward beginners. Specifically, ◘ Fig. 1.2 shows the syntax of the **scatter** function introduced in ▶ Chap. 3. This information can be found in the MATLAB documentation. As shown in ◘ Fig. 1.3, if you type 「doc scatter」 in the Command Window and press Enter, you will see the syntax of the **scatter** function as shown in ◘ Fig. 1.2. The basic syntax is enclosed in a black square. Next, in ▶ Section 4, figures are created using other syntaxes of the **scatter** function. In ▶ Section 5, figures are created using the function introduced in the chapter along with optional functions such as **text**, **xline**, **yline**, and **annotation** functions. While ▶ Sections 3 and 4 are relatively easy to understand, ▶ Section 5 contains some more complex programs. Finally, ▶ Section 6 provides exercises for creating graphics using the functions introduced in each chapter.

```
scatter(x,y)
scatter(x,y,sz)
scatter(x,y,sz,c)
scatter(___,"filled")
scatter(___,mkr)

scatter(tbl,xvar,yvar)
scatter(tbl,xvar,yvar,"filled")

scatter(ax,___)
scatter(___,Name,Value)
s = scatter(___)
```

◘ **Fig. 1.2** The syntax of the scatter function

Fig. 1.3 Command Window view in MATLAB

1.3 Single and Double Quotation Marks

In MATLAB, a string enclosed in double quotation marks (" ") is treated as a string array. A string array is a container that holds multiple pieces of text, allowing you to store and manage multiple text values in an organized manner. For example, if you type 「t="abc"」, MATLAB recognizes it as a string of 「abc」 in a 1-by-1 string array in the Workspace, which is stored in a variable named 「t」. Using 「t(1)」 will output the first element of the variable 「t」. The following is a description of errors. For details on how errors are written, refer to ▶ Sect. 1.5.

Attention

Error Program: When we execute 「t(2)」, an error occurs. See Fig. 1.4 for the error message: 「Index exceeds the number of array elements. Index must not exceed 1」. In other words, this error occurs because 「t」 is a variable with only one element, and specifying 「t(2)」 tries to access a non-existent element. Typing 「t(1)」 generates 「ans="abc"」, implying that the first element of 「t」 contains the string 「abc」.

```
t="abc";
t(1)
```

```
ans =
"abc"
```

```
%%% Error program
%t(2)
```

```
Index exceeds the number of array elements. Index must not exceed 1.
```

Fig. 1.4 Error output of double-quoted variable misuse

On the other hand, if you type ⌜'def'⌟ in single quotation marks, MATLAB recognizes it as a 1-by-3 char array (see Workspace), with each character as a single element. A char array is a data type used to store text as a sequence of individual characters. In the following program, the string ⌜tt='def'⌟ in single quotation marks is stored in a variable named ⌜tt⌟. Then, typing ⌜tt(1)⌟, ⌜tt(2)⌟, and ⌜tt(3)⌟ returns the first, second, and third elements of the variable ⌜tt⌟. When executed, no errors occur and the results are displayed as expected. You may now understand the difference between single and double quotation marks. Here is an important point: In MATLAB functions used to create figures, many options can be executed using either single or double quotation marks. In this book, we use double quotation marks throughout.

```
tt= 'def';
tt(1)
```

```
ans =
'd'
```

```
tt(2)
```

```
ans =
'e'
```

```
tt(3)
```

```
ans =
'f'
```

1.4 The Figure Command and the Tiledlayout Function

When creating figures in MATLAB, it is common to use the **figure** command at the beginning. In this book, however, the **figure** command is omitted to save space unless it would confuse readers. In some cases, the **figure** command is used to reset the spacing and layout settings to their defaults. We will explain why the **figure** command is used in specific cases. One of these cases is the creation of spaces. For instance, when you create a scatter plot or other type of figure, a figure is automatically created in a designated space. See ◘ Fig. 1.5.

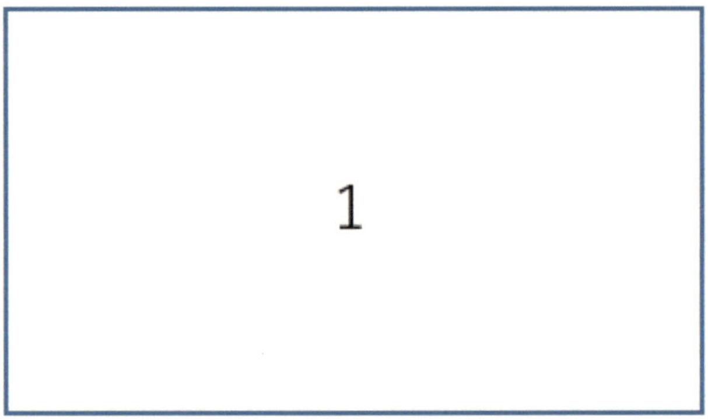

◘ **Fig. 1.5** Default layout

In this book, we sometimes use the **tiledlayout** function to divide the space for figures. For example, if you use 「tiledlayout(1,2)」, the space for figures will be divided into 1 row and 2 columns. See ◘ Fig. 1.6.

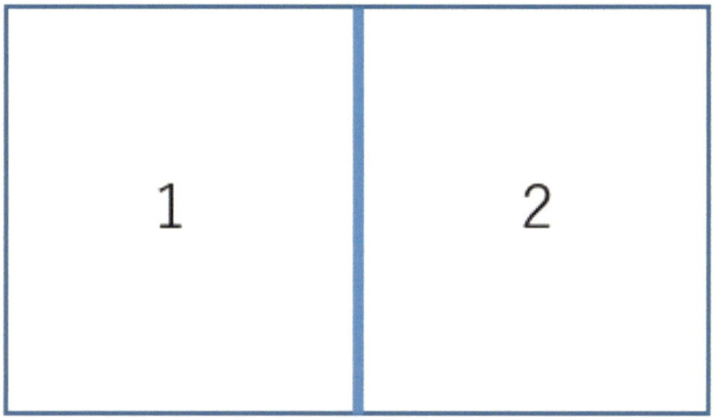

◘ **Fig. 1.6** Customized layout using the **tiledlayout** function

In the following programs, the division of space created by the **tiledlayout** function remains intact. If readers run the programs in order from the top, they will continue to create figures within the divided space, which may cause confusion. To prevent this, we use the **figure** command before and after the **tiledlayout** function to reset the figure space settings to their defaults. However, some readers may create figures on a different page without using the **figure** command after the **tiledlayout** function to divide the space. In this case, you should use the **figure** command because the space remains divided. ◘ Figure 1.7 illustrates how the space is reset when the **figure** command is used after executing 「tiledlayout(1,2)」.

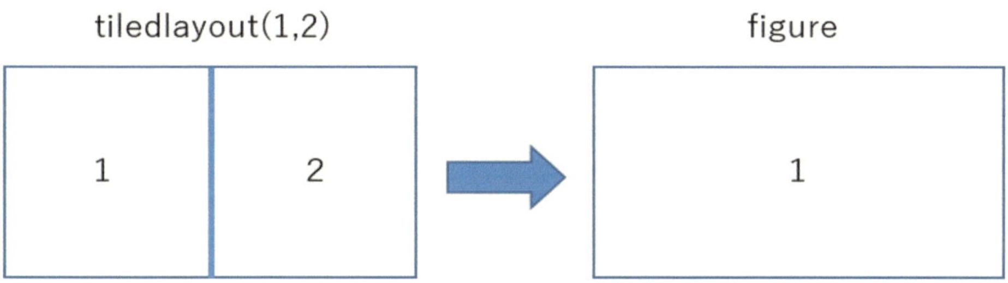

◘ **Fig. 1.7** Use of the figure command: A reset of the current **tiledlayout**

1.5 Errors and Warnings

Throughout this book, we describe common programming errors that beginners may encounter. These errors are commented out with 「%」 throughout the book, as shown in the following program. The reason for using 「%」 is that running them as-is could interfere with the execution of subsequent programs.

```
%%% Error program
% plotmatrix(BEL_Fertality,BEL_Mortality,BEL_pop);
```

If you want to execute the error program, you need to remove the 「%」 in front of the program. Compare the previous program with the one shown in ◘ Fig. 1.8. Running the error program will result in an error and display an error message along with a exclamation mark, as shown in ◘ Fig. 1.8.

```
%%% Error program

plotmatrix(BEL_Fertality,BEL_Mortality,BEL_Pop);
```

◘ **Fig. 1.8** Execution of an error program by removing the % symbol

Next, when you run MATLAB programs, you may sometimes get warning messages. These warnings alert you to deprecated operations, potential issues, or performance degradation. Unlike errors, warnings are not critical and do not stop program execution. Since these warnings are not errors, you can usually ignore them and proceed with the program. See ◘ Fig. 1.9 for an example. This is part of the program from ► Chap. 3. Although this warning appears, you can still run the program and generate the figure. In MATLAB, you can turn warnings on or off and suppress specific warnings. Refer to 「doc warning」 in the MATLAB documentation for details.

```
scatter(Year,Vegetables);
p = polyfit(Year,Vegetables,3);
```

```
Warning: Polynomial is badly conditioned. Add points with distinct
X values, reduce the degree of the polynomial, or try centering and
scaling as described in HELP POLYFIT.
```

◘ **Fig. 1.9** MATLAB warning output

1.6 Row and Column Vectors

Understanding row and column vectors is extremely important. A row vector is a 1-by-n array, meaning that it has one row and multiple columns. A column vector is an n-by-1 array, meaning that it has multiple rows and one column. For example, even when you create a vector with elements 1, 2, 3, 4, and 5, entering 「[1,2,3,4,5]」 and 「[1;2;3;4;5]」 results in different interpretations. The input 「[1,2,3,4,5]」 (or 「[1 2 3 4 5]」) represent a row vector, while 「[1;2;3;4;5]」 represents a column vector. Refer to the following program and results.

```
[1,2,3,4,5]
```

```
ans = 1×5
     1     2     3     4     5
```

```
[1;2;3;4;5]
```

```
ans = 5×1
     1
     2
     3
     4
     5
```

For instance, in ▶ Sect. 3.1, we explain row and column vectors using data on Japan's food self-sufficiency rate. The variables are defined as follows: 「Year」 represents the year, 「Vegetables」 represents the self-sufficiency rate of vegetables, 「Meat」 represents the self-sufficiency rate of meat, 「Milk」 represents the self-sufficiency rate of milk and dairy products, and 「Rice」 represents the self-sufficiency rate of rice. See the following program. The data for each variable is represented as a column vector. The apostrophe 「'」 symbol after the bracket] indicates transposition, converting a row vector into a column vector.

```
Year = [1935,1955,1960,1970,1980,1990,2000,2005,2010,2015,2020]';

Vegetables = [101,100,100,99,97,91,81,79,81,80,80]';

Meat = [93,100,91,89,81,70,52,54,56,54,53]';

Milk = [107,90,89,89,82,78,68,68,67,62,61]';

Rice = [73,109,102,106,100,100,95,95,97,98,97]';
```

Next, in ▶ Chap. 5, we use data on the urbanization rate, an indicator showing the proportion of the population living in urban areas within a specific region (such as a country, a region, or the world). In this chapter, we combine multiple column vectors into a single long column vector or matrix. The variables are defined as follows: 「Year1」 represents the year, 「Global」 represents the global urbanization rate, and 「Advanced」 represents the urbanization rate in advanced regions.

```
Year1 = [1950,1970,1990,2015]';
Global = [29.6,36.6,43,53.9]';
Advanced = [54.8,66.8,72.4,78.1]';
```

1

The variables, 「Advanced」 and 「Global」, are column vectors with 4 rows and 1 column each. Therefore, the variable 「col_combined」 creates an 8-by-1 vector, while 「row_combined」 creates a 4-by-2 matrix. This concept may be slightly advanced, but try to check and confirm the difference between them. Next, to further explore the differences between 「col_combined」 and 「row_combined」, we use the **max** function. Since 「col_combined」 is an 8-by-1 vector, the **max** function returns a single number, which is the maximum value in that column vector. Therefore, 「col_max」 returns a single value of 78.1. On the other hand, 「row_combined」 is a 4-by-2 matrix, so the **max** function returns the maximum values for each column. As a result, the variable 「row_max」 has two values: 78.1 and 53.9.

```
col_combined = [Advanced;Global]
```

```
col_combined = 8×1
   54.8000
   66.8000
   72.4000
   78.1000
   29.6000
   36.6000
   43.0000
   53.9000
```

```
row_combined = [Advanced,Global]
```

```
row_combined = 4×2
   54.8000   29.6000
   66.8000   36.6000
   72.4000   43.0000
   78.1000   53.9000
```

```
col_max = max(col_combined)
```

```
col_max = 78.1000
```

```
row_max = max(row_combined)
```

```
row_max = 1×2
   78.1000   53.9000
```

1.7 Semicolon

A semicolon has uses beyond creating column vectors. When placed at the end of a program line, it suppresses the display of the result, allowing only the created figures to appear. To display the result of a particular line of code, you should remove the semicolon at the end of the line. In this book, semicolons are generally placed at the end of almost every line. Sometimes semicolons are placed even when the result would not be displayed anyway. This is because we emphasize the places where we specifically want to show the results. Therefore, readers do not need to be overly concerned about the presence or absence of a semicolon at the end of each line.

1.8 Multiplication

In this book, although multiplication is not used very often, it is important to know about multiplication in the context of vectors. In MATLAB, 「.*」 denotes element-wise multiplication, while 「*」 denotes matrix multiplication. Here are some specific examples.

```
[1,2,3,4,5].*[1;2;3;4;5]
```

```
ans = 5×5
     1     2     3     4     5
     2     4     6     8    10
     3     6     9    12    15
     4     8    12    16    20
     5    10    15    20    25
```

```
[1,2,3,4,5]*[1;2;3;4;5]
```

```
ans = 55
```

1.9 The Data Used in This Book and an Error to Address Before Running the Program

In this book, we use data such as world population, GDP, gold and silver prices, Internet penetration rates, wages, food self-sufficiency rates, and oil prices-data that even high school students might find interesting. This approach is intended to help young beginners, such as high school students, become familiar with MATLAB. By working with real-world data, they can better understand how MATLAB is used to analyze trends in various fields, making the learning process more engaging and practical. While we do not provide detailed explanations of the data used in each chapter, if you want to advance in your programming studies, it may be important to create various figures and analyze the meanings derived from the figures you create.

Next, we explain a common error often made by beginners: attempting to read a variable when the Workspace is empty. Suppose that the Workspace contains no variables, as shown in Fig. 1.10.

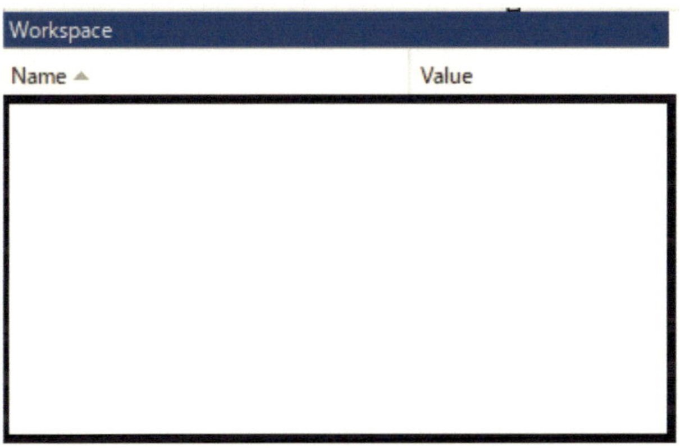

Fig. 1.10 Empty Workspace

In this state, if you enter the program as shown in Fig. 1.11 using the **scatter** function, an error will occur because the variables 「Year」 and 「Rice」 do not exist. If you encounter an error such as 「Unrecognized function or variable 'Year'」, check whether the variable 「Year」 exists in the Workspace.

```
scatter(Year,Rice)
```

```
Unrecognized function or variable 'Year'.
Did you mean:
scatter(year,rica)
Fix
```

Fig. 1.11 Error output due to empty Workspace

1.10 Table Variables

In this book, we generally use variables that are directly created, called as Workspace variables in MATLAB, because the programs are shorter and simpler, making them easier to understand for beginners. On the other hand, throughout the book, we also introduce some programs that utilize table variables. Table variables must have the same number of rows for all variables. However, having different data types does not cause any issues. Specifically, we use the **readtable** function to read data from the Excel file 「alldata.xlsx」 and store it in a table variable. A commonly used method is dot notation. It follows the format 「T.varname」, where 「T」 is the name of the table variable, and 「varname」 is a variable within the table 「T」.

▶ Example of Table Variables

In the following program, we create a table variable directly without reading an Excel file. In this program, the table name is 「T_Chap1」. Figure 1.12 demonstrates that when using the **varfun** function to check the data type of each column, the first and second columns are of type double, the third column is logical, and the fourth column is string. In other words, the table variable 「T_Chap1」 contains different data types. Figure 1.13 shows the contents of the table variable 「T_Chap1」, which was created in the Workspace. The column names are 「Var1」 to 「Var4」. By entering 「T_Chap1.Var1」, you can view its data: 1935, 1955, 1960. Figure 1.14 displays this result when 「T_Chap1.Var1」 is entered in the Command Window.

```
T_Chap1 = table([1935;1955;1960],[73;109;102],...
    [true;false;true],["X";"Y";"Z"]);
varfun(@class, T_Chap1)
```

ans = 1×4 table

	class_Var1	class_Var2	class_Var3	class_Var4
1	double	double	logical	string

Fig. 1.12 Data types of a table variable

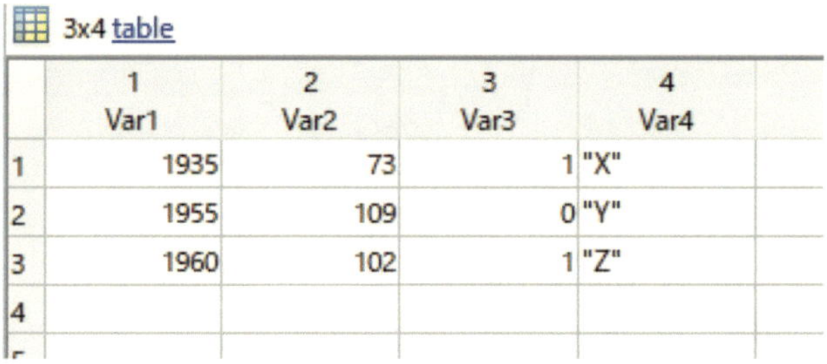
3x4 table

	1 Var1	2 Var2	3 Var3	4 Var4
1	1935	73	1	"X"
2	1955	109	0	"Y"
3	1960	102	1	"Z"
4				

Fig. 1.13 Contents of a table variable

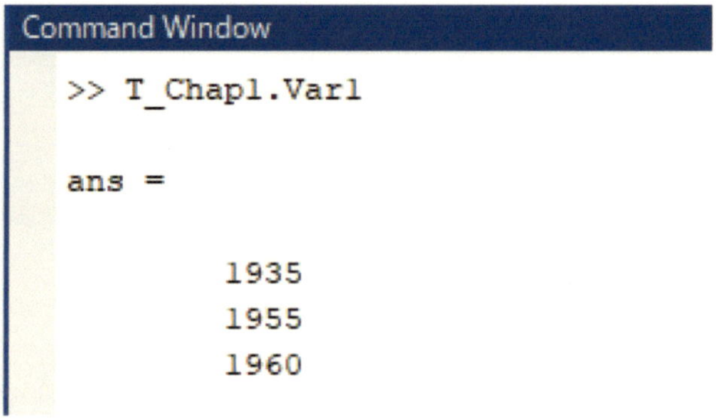

Fig. 1.14 Execution of "Var1" from a table variable by dot notation

◀

Importing Data

Contents

2.1 Importing Excel Data – 16

2.2 Handling Table Data – 21

Y. Nakamoto, *A Comprehensive Guide to Creating Graphics with MATLAB Programming*, Classroom Companion: Economics, https://doi.org/10.1007/978-981-96-6213-5_2

Learning Objectives

This chapter provides tips on importing Excel data into MATLAB. By reading this chapter, you will learn how to use Excel data in MATLAB.

Overview

After importing files such as Excel, many readers analyze data using MATLAB. However, Excel files often contain not only data but also text and other elements. Therefore, it is necessary to arrange the imported data. This chapter explains how to import and manipulate Excel data.

2.1 Importing Excel Data

The **readtable** function is used to read data in table format. Using the **readtable** function, we can import data from Excel files, text files, XML files, HTML files, Word files, and so on. The data is then represented in a table format. We can use the syntax 「T =readtable(filename,Name,Value)」 of the **readtable** function. By using the second and third arguments, corresponding to 「Name」 and 「Value」, you can specify whether the first row of the file should be used as variable names or as data. In the following program, we will read the 「alldata.xlsx」 file. Since the 「alldata.xlsx」 file contains multiple sheets, we need to specify which sheet. In the **readtable** function, we specify 「filename」 as 「alldata.xlsx (=file111)」, 「Name」 as 「Sheet」, and 「Value」 as 「Chap2_Food_self_sufficiency_NaN (=sheet111)」.

```
file111 = "alldata.xlsx";
sheet111 = "Chap2_Food_self_sufficiency_NaN";
data1 = readtable(file111,"Sheet",sheet111);
```

You can confirm that the imported data 「data1」 is given by an 86-by-5 table. See ◘ Fig. 2.1.

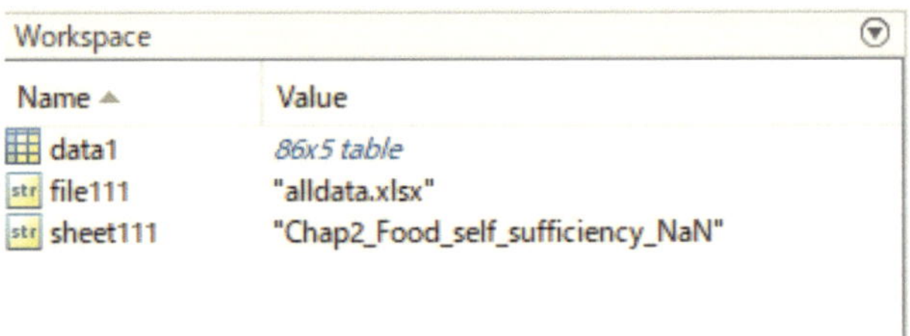

◘ **Fig. 2.1** Contents of Workspace after data import

If the Excel file is in the current folder, you only need to write the file name in the **readtable** function. However, if the Excel file is not in the current folder, you cannot read the Excel file using only the file name, and an error will occur. If the Excel file is not in the current folder, we need to specify the directory path where the Excel file is located. In the following program, the 「alldata.xlsx」 file is located in 「"C:\class"」.

```
file222 = "C:¥class¥alldata.xlsx";
sheet222 = "Chap2_Food_self_sufficiency_NaN";
data222 = readtable(file222,"Sheet",sheet222);
```

Attention

Error Program: In the following program, the readtable function tries to read a data file from the location specified in 「file 333」 (C:\class\classsssss\alldata.xlsx), but this location is incorrect. That is, the folder specified as C:\class\classsssss does not contain the Excel file 「alldata.xlsx」. As a result, an error occurs when using the readtable function. The error message is displayed in Fig. 2.2.

```
%%% Error Program
% file333 = "C:¥class¥classsssss¥alldata.xlsx";
% sheet333 = "Chap2_Food_self_sufficiency_NaN";
% data333 = readtable(file333,"Sheet",sheet333);
```

```
Error using readtable (line 517)
Unable to find or open 'C:\class\classsssss\alldata.xlsx'. Check the path and filename or file permissions.
```

Fig. 2.2 Error output due to invalid file path

The 「alldata.xlsx」 file has a sheet named 「Chap2_Food_self_sufficiency_NaN」 as shown in Fig. 2.3.

	A	B	C	D	E
1	Year	Rice	Vegetable	Meat	Milk
2	1935	73	101	93	107
3	1936				
4	1937				
5	1938				
6	1939				
7	1940				
8	1941				
9	1942				
10	1943				
11	1944				
12	1945				
13	1946				
14	1947				
15	1948				
16	1949				
17	1950				
18	1951				
19	1952				
20	1953				
21	1954				
22	1955	109	100	100	90

Fig. 2.3 Preview of raw dataset in Excel format

In Fig. 2.3, some cells do not contain any values. These are called missing values (NaN), and missing values can cause significant problems in data analysis and processing. By using 「data_nan=readtable(file_nan,"Sheet",sheet_nan)」 without the semicolon, the results stored in the variable 「data_nan」 are displayed directly. See Fig. 2.4.

```
file_nan="alldata.xlsx";
sheet_nan="Chap2_Food_self_sufficiency_NaN";
data_nan=readtable(file_nan,"Sheet",sheet_nan)
```

data_nan = 86×5 table

	Year	Rice	Vegetables	Meat	Milk
1	1935	73	101	93	107
2	1936	NaN	NaN	NaN	NaN
3	1937	NaN	NaN	NaN	NaN
4	1938	NaN	NaN	NaN	NaN
5	1939	NaN	NaN	NaN	NaN
6	1940	NaN	NaN	NaN	NaN
7	1941	NaN	NaN	NaN	NaN
8	1942	NaN	NaN	NaN	NaN
9	1943	NaN	NaN	NaN	NaN
10	1944	NaN	NaN	NaN	NaN
11	1945	NaN	NaN	NaN	NaN
12	1946	NaN	NaN	NaN	NaN
13	1947	NaN	NaN	NaN	NaN
14	1948	NaN	NaN	NaN	NaN

⋮

◘ **Fig. 2.4** Preview of dataset with missing values

Here, we specify 「omitrow」 with 「MissingRule」. As a result, entire rows containing cells with missing values are deleted. The result is stored in the variable 「data_nan_delete_row」. Since the semicolons have been removed, as in 「data_nan_delete_row=readtable(file_nan,"Sheet",sheet_nan,MissingRule="omitrow")」, the results are displayed directly. See ◘ Fig. 2.5.

```
file_nan="alldata.xlsx";
sheet_nan="Chap2_Food_self_sufficiency_NaN";
data_nan_delete_row=readtable(file_nan,"Sheet",sheet_nan,MissingRule="omitrow")
```

	Year	Rice	Vegetables	Meat	Milk
1	1935	73	101	93	107
2	1955	109	100	100	90
3	1960	102	100	91	89
4	1970	106	99	89	89
5	1980	100	97	81	82
6	1990	100	91	70	78
7	2000	95	81	52	68
8	2005	95	79	54	68
9	2010	97	81	56	67
10	2015	98	80	54	62
11	2020	97	80	53	61

Fig. 2.5 Preview of dataset after omitting missing values

Here, we use 「Range="A1:B5"」 to store only the specified range of data in the variable 「data_range」. Specifically, we extract the data from column A, row 1 to column B, row 5. The variable 「data_range」 contains a 4-by-2 data set, since the first row is interpreted as variable names by default in the **readtable** function. See Fig. 2.6.

```
file_nan="alldata.xlsx";
sheet_nan="Chap2_Food_self_sufficiency_NaN";
data_range=readtable(file_nan,"Sheet",sheet_nan,Range="A1:B5")
```

data_range = 4×2 table

	Year	Rice
1	1935	73
2	1936	NaN
3	1937	NaN
4	1938	NaN

Fig. 2.6 Execution of partial data import by range specification

2.2 Handling Table Data

The data extracted by the **readtable** function is of the table type. A table variable is effective for extracting tabular data, with each column labeled similarly to Excel. Note that tabular is generally used to represent data where each column corresponds to a specific variable or attribute, and each row represents an individual observation or measurement. The upper left of Fig. 2.7 shows 「11-by-5 table」, indicating that the table type has 11 rows and 5 columns.

data ×

11x5 table

	1 Year	2 Rice	3 Vegetables	4 Meat	5 Milk
1	1935	73	101	93	107
2	1955	109	100	100	90
3	1960	102	100	91	89
4	1970	106	99	89	89
5	1980	100	97	81	82
6	1990	100	91	70	78
7	2000	95	81	52	68
8	2005	95	79	54	68
9	2010	97	81	56	67
10	2015	98	80	54	62
11	2020	97	80	53	61

Fig. 2.7 Preview of a table variable

2

To extract specific data from a table, we use the period 「.」. In the program preceding ◘ Fig. 2.8, the variable will be represented in the form 「data_table.Year」, where the period separates the table name from the column name. Let us look at the following program. First, we use the **readtable** function to read the data from the sheet named 「Chap2_Food_self_sufficiency_NaN」. The result is stored in the variable 「data_table」.

```
file_table = "alldata.xlsx";
sheet_table = "Chap2_Food_self_sufficiency_NaN";
data_table = readtable(file_table,"Sheet",sheet_table);
```

► Example of Table Variables

We then extract the data of the columns 「Year」, 「Rice」, 「Vegetables」, 「Meat」, and 「Milk」 from this table format, 「data_table」. The data in each column is stored in 「Year1」, 「Rice1」, 「Vegetables1」, 「Meat1」, and 「Milk1」. Finally, we combine the variables 「Year1」 and 「Milk1」 to create the array 「YM」. ◘ Figure 2.8 shows the Workspace after running the following program.

```
Year1 = data_table.Year;
Rice1 = data_table.Rice;
Vege1 = data_table.Vegetables;
Milk1 = data_table.Milk;
YM = [Year1,Milk1];
```

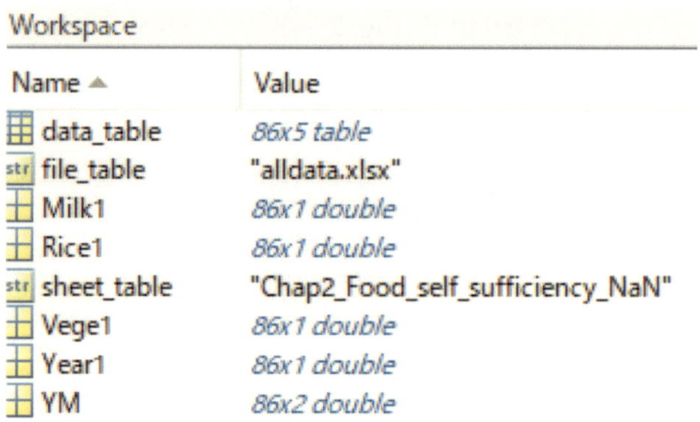

◘ **Fig. 2.8** Contents of Workspace after variable extraction

◄

Scatter Function

Contents

3.1 Details of the Data Used in this Chapter – 24

3.2 Excel Data of ► Section 3.1 – 26

3.3 Basic Syntax of the Scatter Function – 27

3.4 Setting and Controlling the Scatter Function – 34

3.5 Combining Optional Functions and the Scatter Function – 52

3.6 Questions – 94

Supplementary Information The online version contains supplementary material available at https://doi.org/10.1007/978-981-96-6213-5_3.

Y. Nakamoto, *A Comprehensive Guide to Creating Graphics with MATLAB Programming*, Classroom Companion: Economics, https://doi.org/10.1007/978-981-96-6213-5_3

Learning Objectives

This chapter creates scatter plots using the **scatter** function. This chapter is the first chapter that introduces functions for creating figures. Therefore, it provides a more detailed explanation compared to ▶ Chaps. 4–17. By reading this chapter, you will learn how to use the **scatter** function as well as other optional functions.

Overview and Guidelines

In this chapter, we will create scatter plots using the **scatter** function. Scatter plots are used to visually represent the relationship between two variables, where each variable is typically placed on the horizontal axis (x-axis) and vertical axis (y-axis), and each data point is plotted at coordinates (x,y). Additionally, in this chapter, we will explain in detail optional functions such as **legend**, **title**, **tiledlayout**, **annotation**, **text**, **xlabel**, **ylabel**, and more. We explained the rules of this book in ▶ Chap. 1. However, for readers starting from this chapter, we explain the minimum rules below. Please read these rules as you proceed through this chapter.

- When creating figures, there are two main methods for importing data. The first is to enter the data directly into MATLAB in each ▶ Section 1, and the second is to import data from sources like Excel in each ▶ Section 2. In this book, ▶ Chaps. 3–17 employ both methods to load data. For example, in this chapter, we create scatter plots using the **scatter** function. In ▶ Sect. 3.1, data is entered directly. In ▶ Sect. 3.2, we import data from Excel. Finally, ▶ Sects. 3.3–3.5 create various scatter plots using the **scatter** function. ▶ Section 3.6 provides exercises. The same structure is followed in ▶ Chap. 4 and beyond.
- From ▶ Chaps. 3–17 of this book, you will notice that many common functions are used as options for the figures created in MATLAB. For example, the **tiledlayout** and **title** functions are used repeatedly in many chapters. These optional functions can be used with various functions introduced in other chapters. Specifically, the **annotation**, **text**, **xline**, and **yline** functions are explained in detail in ▶ Chap. 18.
- Many readers may receive data in Excel format rather than entering it directly into MATLAB programs. For information on using Excel in such cases, see ▶ Chap. 2.
- Throughout this book, we list common programming mistakes that beginners often make. We also provide advice on how to correct these errors.
- In this book, when we explain programming, we use quotation marks 「」 for variables and the programming itself. On the other hand, function names are written in bold. For example, we use bold text as shown in the **scatter** function.

3.1 Details of the Data Used in this Chapter

In this chapter, we use data on Japan's food self-sufficiency rate. 「Year」 represents the fiscal year. 「Vegetables」, 「Meat」, 「Milk」, and 「Rice」 represent the self-sufficiency rates of vegetables, meat, milk and dairy products, and rice, respectively. An apostrophe ('), placed after a vector, indicates a transposition, which means that a row vector is converted into a column vector. See ◘ Fig. 3.1.

```
Year1 = [1935,1955,1960,1970,1980,1990,2000,2005,2010,2015,2020];

Year = [1935,1955,1960,1970,1980,1990,2000,2005,2010,2015,2020]';

Vegetables1 = [101,100,100,99,97,91,81,79,81,80,80];

Vegetables = [101,100,100,99,97,91,81,79,81,80,80]';

Meat1 = [93,100,91,89,81,70,52,54,56,54,53];

Meat = [93,100,91,89,81,70,52,54,56,54,53]';

Milk1 = [107,90,89,89,82,78,68,68,67,62,61];

Milk = [107,90,89,89,82,78,68,68,67,62,61]';

Rice1 = [73,109,102,106,100,100,95,95,97,98,97];

Rice = [73,109,102,106,100,100,95,95,97,98,97]';
```

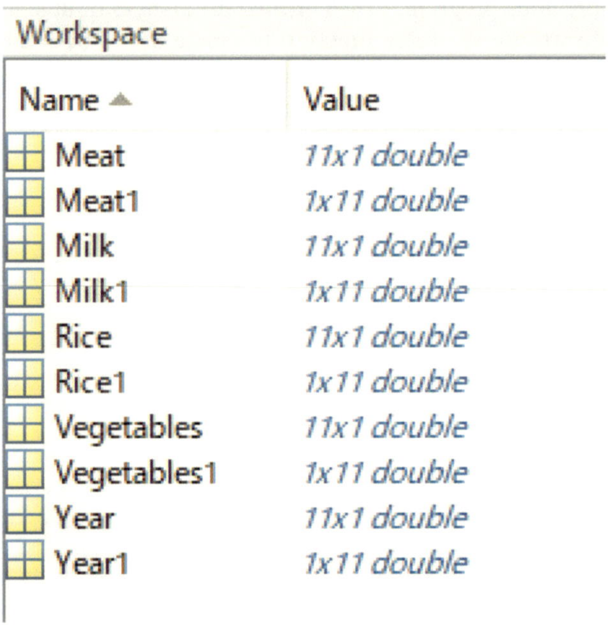

Fig. 3.1 Contents of Workspace

When including column vectors in brackets 「[ ]」 as in the following program, we create an 11-by-5 double matrix. See Fig. 3.2.

```
Japan_Food_matrix1 = [Year,Vegetables,Meat,Milk,Rice];
```

3

11x5 double

	1	2	3	4	5
1	1935	73	101	93	107
2	1955	109	100	100	90
3	1960	102	100	91	89
4	1970	106	99	89	89
5	1980	100	97	81	82
6	1990	100	91	70	78
7	2000	95	81	52	68
8	2005	95	79	54	68
9	2010	97	81	56	67
10	2015	98	80	54	62
11	2020	97	80	53	61
12					

Fig. 3.2 Contents of the variable "Japan_Food_matrix1"

Unlike the previous program, which created an 11-by-5 double matrix shown in Fig. 3.2, here we use the **table** function to create an 11-by-5 table format. See Fig. 3.3.

```
Japan_Food_matrix2 = table(Year,Vegetables,Meat,Milk,Rice);
```

11x5 table

	1 Year	2 Rice	3 Vegetables	4 Meat	5 Milk
1	1935	73	101	93	107
2	1955	109	100	100	90
3	1960	102	100	91	89
4	1970	106	99	89	89
5	1980	100	97	81	82
6	1990	100	91	70	78
7	2000	95	81	52	68
8	2005	95	79	54	68
9	2010	97	81	56	67
10	2015	98	80	54	62
11	2020	97	80	53	61
12					

Fig. 3.3 Contents of the variable "Japan_Food_matrix2"

3.2 Excel Data of ▶ Section 3.1

We explain how to import data recorded in Excel into the Workspace. First, we need to specify the file and sheet names. The created variable, 「Japan_Food_matrix2」, is in table format (see Fig. 3.4). See also ▶ Chap. 2 for details on the **readtable** function.

```
file = "alldata.xlsx";
sheet = "Chap3_Food_self_sufficiency";
Japan_Food_matrix2 = readtable(file,"Sheet",sheet);
```

11x5 table

	1 Year	2 Rice	3 Vegetables	4 Meat	5 Milk	6
1	1935	73	101	93	107	
2	1955	109	100	100	90	
3	1960	102	100	91	89	
4	1970	106	99	89	89	
5	1980	100	97	81	82	
6	1990	100	91	70	78	
7	2000	95	81	52	68	
8	2005	95	79	54	68	
9	2010	97	81	56	67	
10	2015	98	80	54	62	
11	2020	97	80	53	61	
12						
13						

Fig. 3.4 Contents of the variable "Japan_Food_matrix2" imported from Excel

Then, each column can be extracted from the variables of the table array.

```
Year = Japan_Food_matrix2.Year;
Vegetables = Japan_Food_matrix2.Vegetables;
Meat = Japan_Food_matrix2.Meat;
Milk = Japan_Food_matrix2.Milk;
Rice = Japan_Food_matrix2.Rice;
```

3.3 Basic Syntax of the Scatter Function

Basic Syntax:

The **scatter** function is used to create a two-way scatter plot. Various syntax variations can be viewed by typing 「doc scatter」 in the Command Window. Figure 3.5 shows the syntaxes for the **scatter** function. The first syntax inside the black rectangle is the basic syntax of the **scatter** function, 「scatter(x,y)」. Based on the basic syntax, the **scatter** function requires two types of data, 「x」 and 「y」

3

Syntax

```
scatter(x,y)
scatter(x,y,sz)
scatter(x,y,sz,c)
scatter( __ ,"filled")
scatter( __ ,mkr)

scatter(tbl,xvar,yvar)
scatter(tbl,xvar,yvar,"filled")

scatter(ax, __ )
scatter( __ ,Name,Value)
s = scatter( __ )
```

Fig. 3.5 The syntax of the scatter function

Based on the basic syntax of the **scatter** function, we create the scatter plot shown in Fig. 3.6, where 「x」 and 「y」 in the basic syntax 「scatter(x,y)」 represent 「Year」 and 「Rice」 data, respectively.

```
scatter(Year,Rice);
```

Fig. 3.6 Scatter plot of yearly data (X-axis: Year)

A different scatter plot can be obtained by swapping 「Rice」 and 「Year」 on the x-axis and y-axis, respectively. See ◘ Fig. 3.7 for the created figure.

```
scatter(Rice,Year);
```

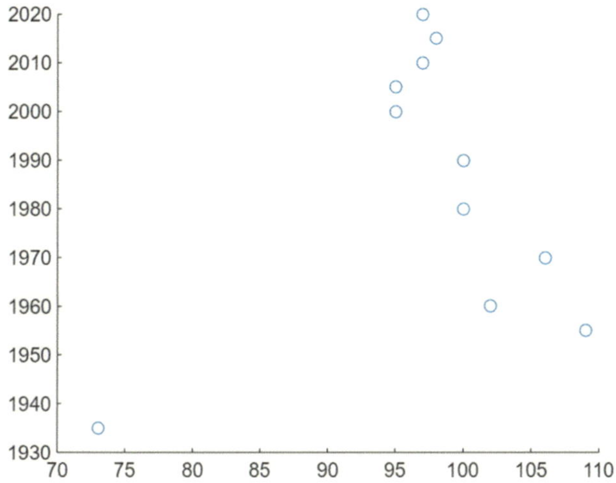

◘ **Fig. 3.7** Scatter plot of yearly data (Y-axis: Year)

Here, we enter vectors directly into the **scatter** function without using variables such as 「Year」 or 「Rice」. Here, ellipses 「...」 are used to break a long program into multiple lines, allowing the statement to continue on the next line. The resulting figure is displayed in ◘ Fig. 3.8.

```
scatter([1935,1955,1960,1970,1980,1990,2000,2005,2010,2015,2020],...
[73,109,102,106,100,100,95,95,97,98,97]);
```

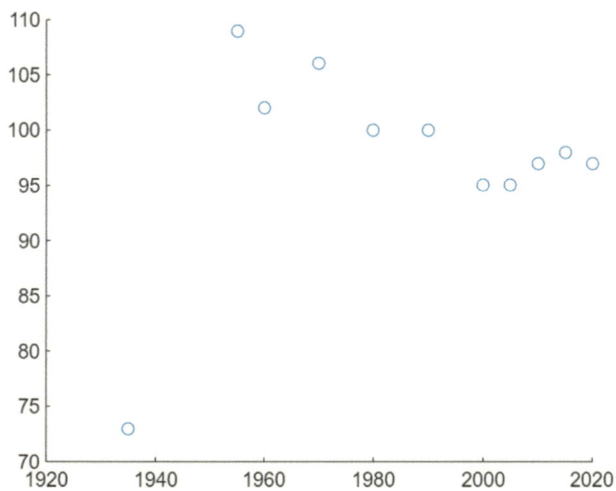

◘ **Fig. 3.8** Scatter plot created from manually input data

Attention

Error Program: Here, we do not follow the basic syntax of 「scatter(x,y)」, which generates an error. The resulting error message is shown in Fig. 3.9.

```
%%% Error program
% scatter(Rice)
```

```
Error using scatter
Not enough input arguments.
```

Fig. 3.9 Error output of the scatter function: Missing variable

▶ Example of Table Variables

Here, by using variables from the table array in the **scatter** function, we create a scatter plot. See Fig. 3.10.

```
scatter(Japan_Food_matrix2.Year,Japan_Food_matrix2.Rice);
```

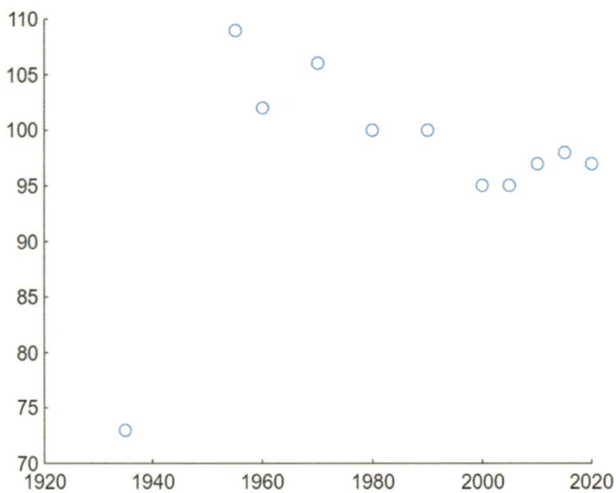

Fig. 3.10 Use of a table variable: Scatter plot

◀

In the basic syntax, 「scatter(x,y)」, the second argument should contain the y-axis data. In contrast, using 「scatter(Year,Rice,Meat)」 will not create a scatter plot that includes the 「Meat」 data, because 「Meat」 is in the third argument. Here, we cannot create the scatter plot for the variable 「Meat」. Note that the data in 「Meat」 is used to determine the circle sizes in the scatter plot of 「Year」 and 「Rice」 because this follows a different syntax of 「scatter(x,y,sz)」. See Fig. 3.11 for the resulting plot. To create a scatter plot that includes both 「Rice」 and 「Meat」 data, the matrix 「[Rice,Meat]」 should be used as the second argument, as shown in Fig. 3.12.

```
scatter(Year,Rice,Meat);
```

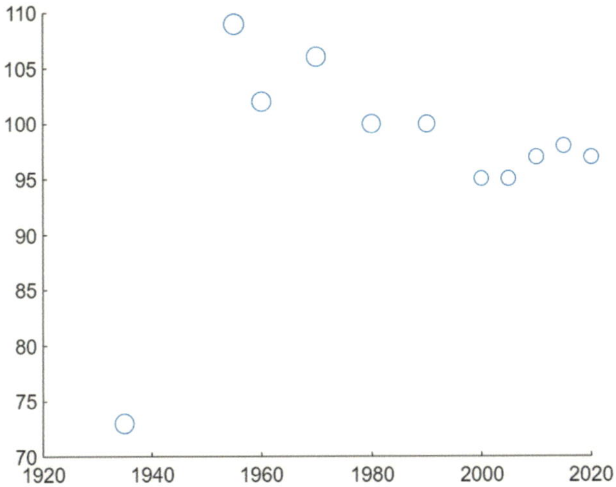

Fig. 3.11 Scatter plot with point sizes determined by the variable "Meat"

Here, to create two scatter plots, we include two vectors of 「Rice」 and 「Meat」 in the second argument as a matrix 「[Rice,Meat]」. In this case, two scatter plots, 「(x,y)=(「Year」, 「Rice」)」 and 「(x,y)=(「Year」, 「Meat」)」, are displayed, as in Fig. 3.12.

```
scatter(Year,[Rice,Meat]);
```

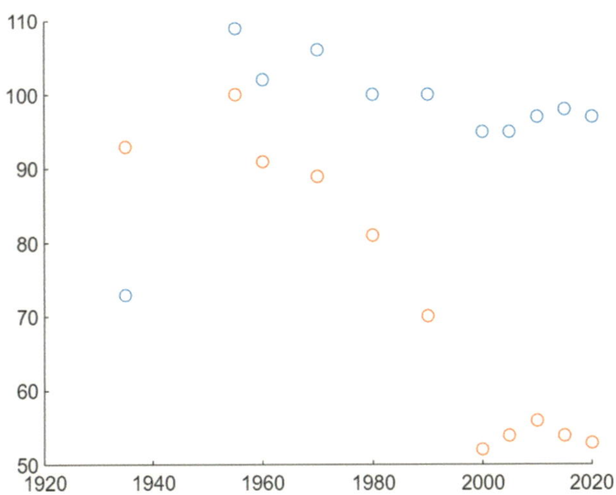

Fig. 3.12 Scatter plot with two groups

Here, the four vectors of 「Rice」, 「Meat」, 「Milk」, and 「Vegetables」 are included into the second argument. As a result, four scatter plots, 「(x,y)=(「Year」,

「Rice」)」, 「(x,y)=(「Year」, 「Meat」)」, 「(x,y)=(「Year」, 「Milk」)」, and 「(x,y)=(「Year」, 「Vegetables」)」, are displayed, as shown in Fig. 3.13.

```
scatter(Year,[Rice,Meat,Milk,Vegetables]);
```

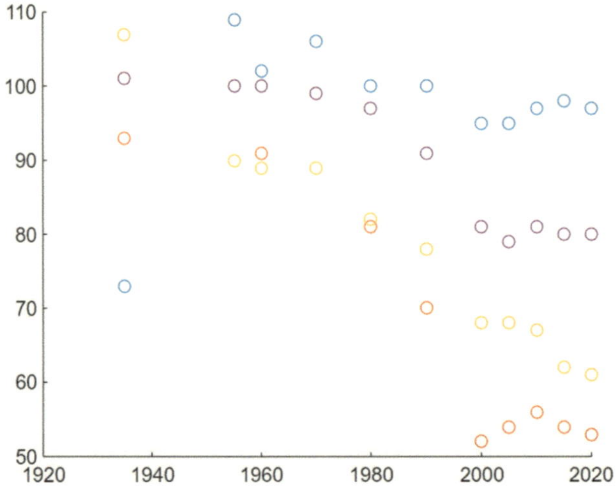

Fig. 3.13 Scatter plot with multiple groups

▶ Example of Table Variables

As shown in Fig. 3.13, we create four scatter plots here. However, unlike the previous program that created Fig. 3.13, we use the table array 「Japan_Food_matrix2」. Compare the program that created Fig. 3.13 with the following program that creates Fig. 3.14.

```
scatter(Japan_Food_matrix2.Year,...
    [Japan_Food_matrix2.Rice,Japan_Food_matrix2.Meat,Japan_Food_matrix2.Milk]);
```

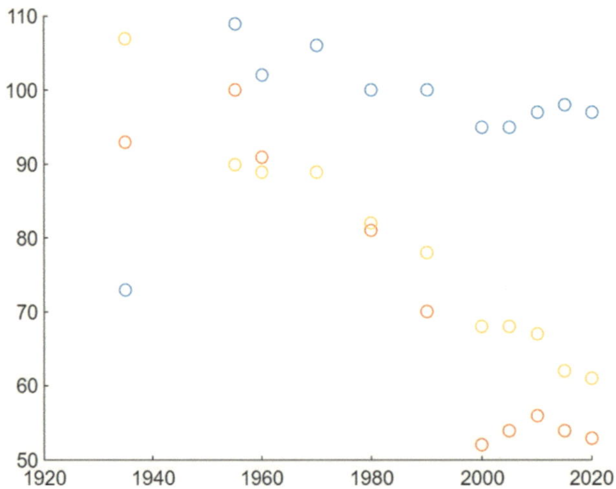

Fig. 3.14 Use of a table variable: Scatter plot with multiple groups

◀

Here, we use a variable 「abc」 for a 1-by-4 cell array. The resulting plot is shown in ◘ Fig. 3.15. Here, 「abc(1)」, 「abc(2)」, 「abc(3)」, and 「abc(4)」 contain the properties of 「Rice」, 「Meat」, 「Milk」, and 「Vegetables」, respectively. The following code demonstrates the properties of both 「abc」 and 「abc(2)」 after using the **scatter** function. See ◘ Figs. 3.16 and 3.17.

```
abc = scatter(Year,[Rice,Meat,Milk,Vegetables]);
```

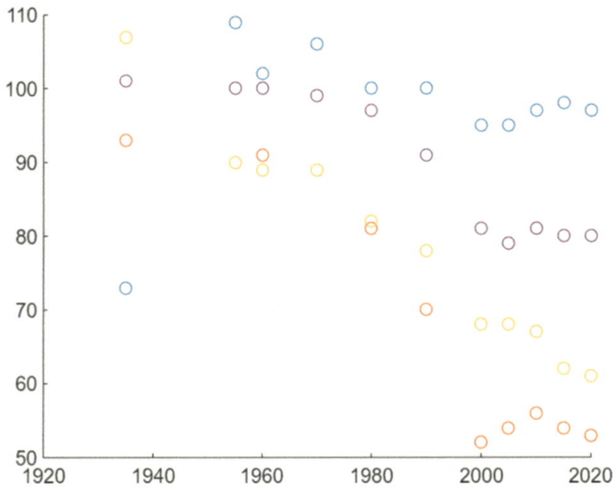

◘ **Fig. 3.15** Scatter plot with multiple groups and the creation of the variable "abc"

```
abc
```

```
abc =
  1x4 Scatter array:

    Scatter    Scatter    Scatter    Scatter
```

◘ **Fig. 3.16** Contents of the variable "abc"

```
abc(2)
```

```
ans =
  Scatter with properties:

              Marker: 'o'
     MarkerEdgeColor: 'flat'
     MarkerFaceColor: 'none'
           LineWidth: 0.5000
               XData: [1935 1955 1960 1970 1980 1990 2000 2005 2010 2015 2020]
               YData: [93 100 91 89 81 70 52 54 56 54 53]
            SizeData: 36
               CData: [0.8500 0.3250 0.0980]

  Show all properties
```

◘ **Fig. 3.17** Contents of the element "abc(2)"

3

3.4 Setting and Controlling the Scatter Function

Here, the sizes of the circle markers are modified by introducing a scalar in the third argument. To represent circles of the same size, a scalar value can be used as shown below. ◻ Figures 3.18 and 3.19 display circles with scalar values of 10 and 50, respectively.

```
scatter(Year,Rice,10);
```

◻ **Fig. 3.18** Scatter plot with a small marker size

```
scatter(Year,Rice,50);
```

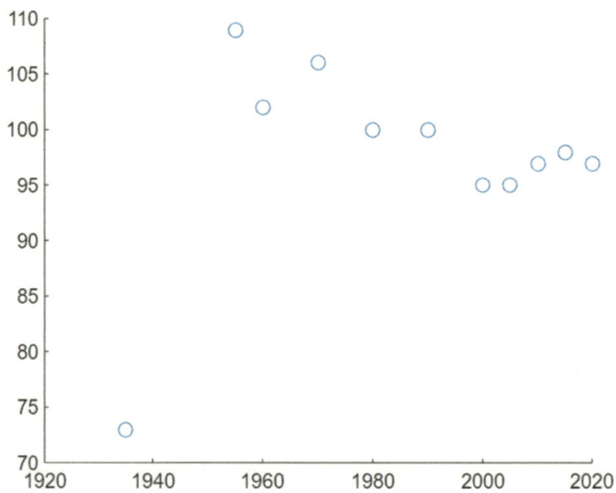

◻ **Fig. 3.19** Scatter plot with a large marker size

To plot each circle with a different size, a vector ⌜[10,10,10,10,10,10,10,10,10,10, 50]⌟ should be used in the third argument. In the following program, the last circle has a size of 50, while all other circles have a size of 10. The resulting plot is presented in ◘ Fig. 3.20.

```
scatter(Year,Rice,[10,10,10,10,10,10,10,10,10,10,50]);
```

◘ **Fig. 3.20** Scatter plot with different marker sizes

Here, we use various values in the third argument. The resulting figure contains circles of different sizes. See ◘ Fig. 3.21.

```
scatter(Year,Rice,[50,30,10,0.1,100,200,1000,2000,3000,4000,5000]);
```

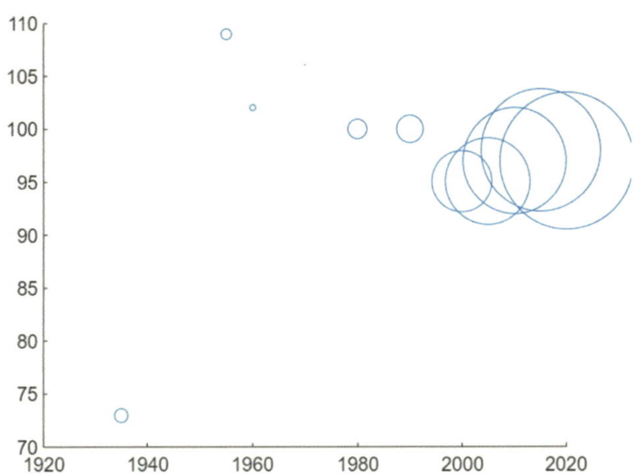

◘ **Fig. 3.21** Scatter plot with extremely large markers

3

Attention

Error Program: Since each vector in 「Year」 and 「Rice」 contains 11 elements, a vector with 11 elements should be used in the third argument. If the number of elements in the vector does not match the number of elements in the variable, an error occurs. For instance, the following example demonstrates an error because the number of elements in the third argument is three, which is less than 11. The error message is presented in Fig. 3.22.

```
%%% Error program
% scatter(Year,Rice,[10,10,10])
```

```
Error using scatter
X, Y, and size arguments must be vectors of the same length, matrices of the same size,
or a mix of vectors and matrices where the length of the vectors matches either the
number of rows or columns of the matrices.
```

Fig. 3.22 Error output of the scatter function: Inconsistent vector lengths

The color of the marker can be controlled by specifying the color in the fourth argument. For instance, to create red circles, either 「red」 (Fig. 3.23) or the shorthand 「r」 (Fig. 3.24) can be used in the fourth argument.

```
scatter(Year,Rice,30,"red");
```

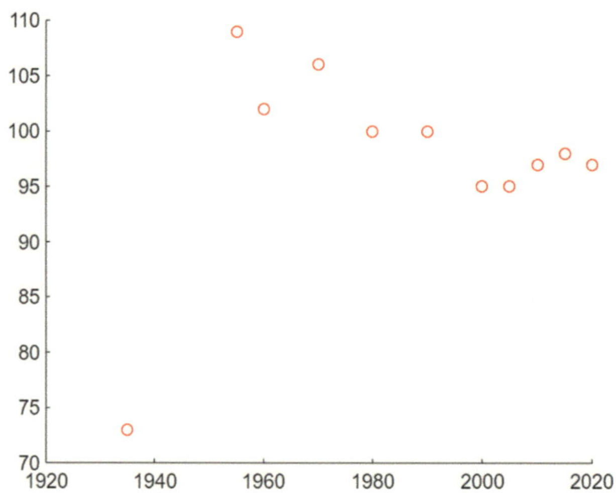

Fig. 3.23 Scatter plot with red markers and a small marker size

```
scatter(Year,Rice,50,"r");
```

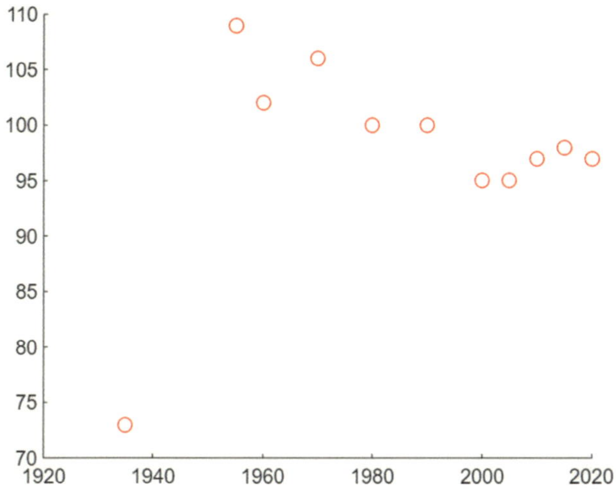

Fig. 3.24 Scatter plot with red markers and a large marker size

To keep the size of the circle as the default and only change the color to 「green」, an empty cell can be used in the third argument, as in the following program that creates Fig. 3.25. Alternatively, the third argument (circle size) can be removed as in the program, which creates Fig. 3.26. The scatter plots generated by these two methods (Fig. 3.25 and Fig. 3.26) are identical.

```
scatter(Year,Rice,[],"green");
```

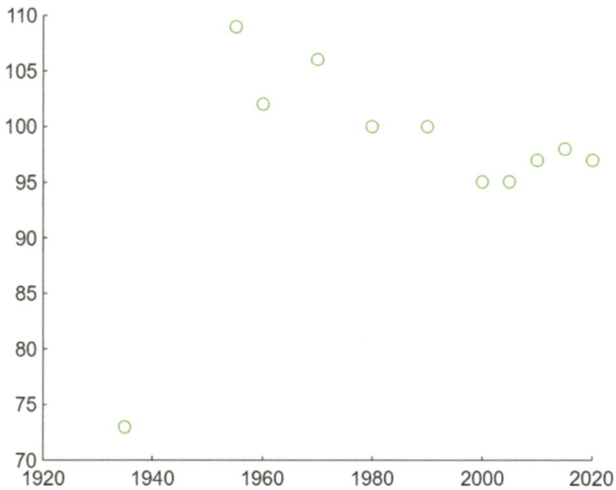

Fig. 3.25 Scatter plot with green markers and default size

```
scatter(Year,Rice,"g");
```

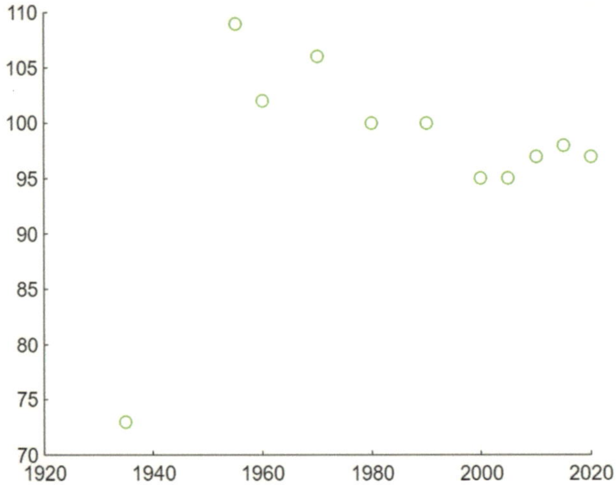

Fig. 3.26 Scatter plot with green markers and default size (cf. Fig. 3.25)

Attention

Error Program: Although the basic syntax is 「scatter(x,y)」, using the first or second argument to specify a color instead of the 「x」 and 「y」 positions will result in an error. The error message is shown in Fig. 3.27.

```
%%% Error program
% scatter(Year,"magenta",Rice);
```

```
Error using scatter
Incorrect number of inputs for property-value pairs.
```

Fig. 3.27 Error output of the scatter function: Incorrect order

The color can be changed using single quotation marks instead of double quotation marks. The resulting plot is displayed in Fig. 3.28.

```
scatter(Year,Rice,30,'red');
```

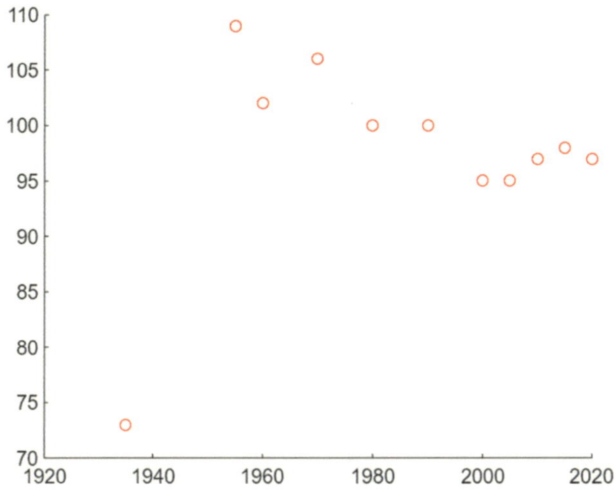

Fig. 3.28 Use of single quotes: Scatter plot with red markers

Attention

Error Program: Note that the absence of single or double quotation marks results in an error. The error message is presented in Fig. 3.29.

```
%%% Error program
% scatter(Year,Rice,30,red)
```

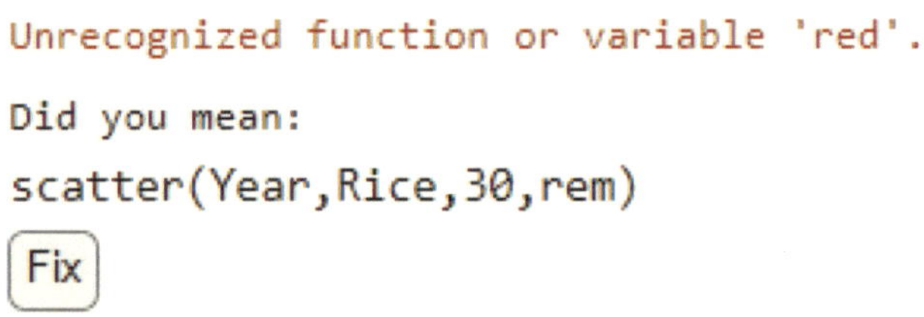

Fig. 3.29 Error output of the scatter function: Missing quotation

The color can also be changed using numeric values instead of words. For example, red is 「[1 0 0]」 and green is 「[0 1 0]」. See Fig. 3.30.

```
scatter(Year,Rice,30,[1 0 0]);
```

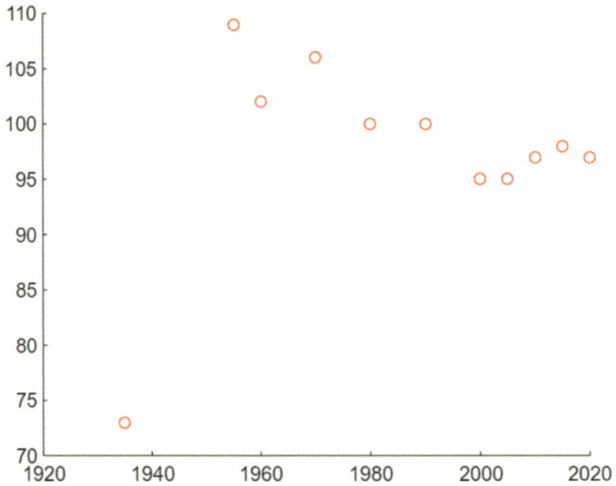

Fig. 3.30 Use of the RGB vector: Scatter plot with red color

The numbers, corresponding to the colors, range from 0 to 1, allowing precise color specification, as shown in Fig. 3.31.

```
scatter(Year,Rice,100,[0.4976 0.00001 0.50002]);
```

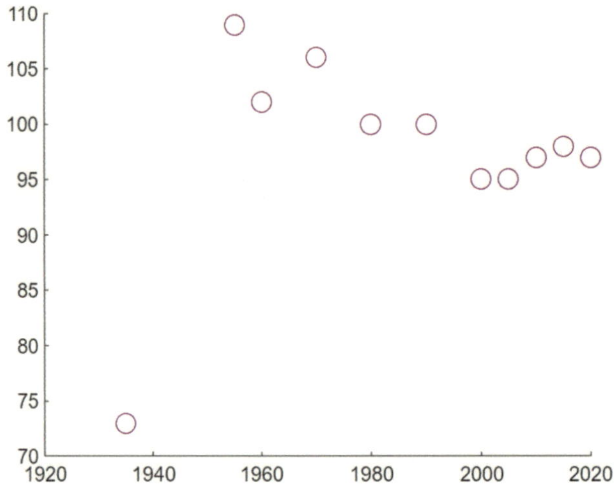

Fig. 3.31 Use of the RGB vector (precise numbers): Scatter plot with purple color

Here, after setting the color to 「red」, we use 「filled」 to create filled red scatter plots. See Fig. 3.32.

```
scatter(Year,Rice,30,"red","filled");
```

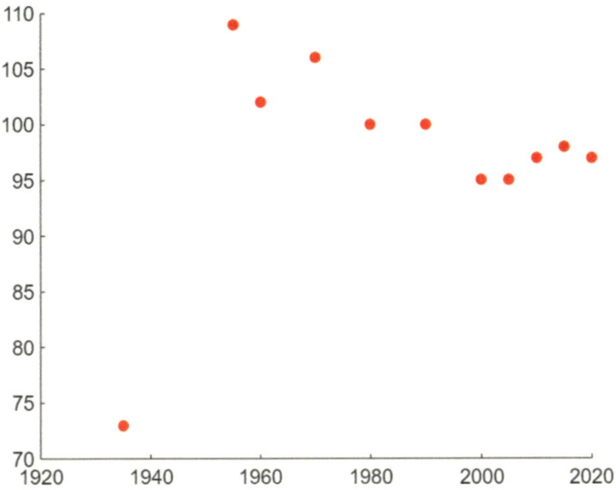

Fig. 3.32 Scatter plot with filled red markers

Here, we use the vector 「[50,30,10,0.1,100,200,1000,2000,3000,4000,5000]」 in the third argument to change the sizes of circles. Even if the circle sizes are not specified consistently, the 「filled」 option still creates circles with a filled color. See Fig. 3.33.

```
scatter(Year,Rice,[50,30,10,0.1,100,200,1000,2000,3000,4000,5000],"filled");
```

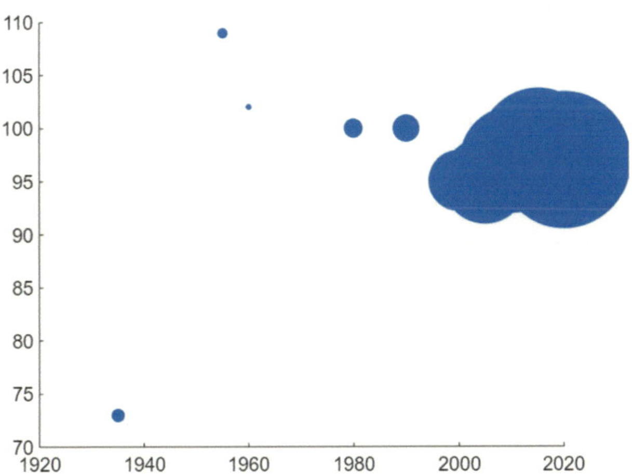

Fig. 3.33 Scatter plot with blue filled markers of various sizes

The syntax 「scatter(___,Name,Value)」 in 「doc scatter」 modifies the scatter plot using one or more Name-Value arguments to set properties, where both 「Name」 and 「Value」 must be used together. See the rectangle in Fig. 3.34. For instance, in the syntax 「scatter(___,Name,Value)」, 「Name」 shows a property of the scatter plot that you want to customize. 「Value」 refers to the value assigned to that property. There are some possible 「Name」 options, such as 「MarkerFaceColor」 for setting the fill color of markers, 「MarkerEdgeColor」 for defining the color of the marker outline, and 「LineWidth」 for specifying the thickness of plotted lines. These Name-Value pairs are used not only in the **scatter** function but also in other plotting functions in MATLAB.

Syntax

```
scatter(x,y)
scatter(x,y,sz)
scatter(x,y,sz,c)
scatter(___,"filled")
scatter(___,mkr)

scatter(tbl,xvar,yvar)
scatter(tbl,xvar,yvar,"filled")

scatter(ax,___)
scatter(___,Name,Value)
s = scatter(___)
```

Fig. 3.34 Name-value pairs: the syntax of the scatter function

Here, we set 「LineWidth (Name)」 to 2.5 (Value) to make the marker outlines wider. That is, 「scatter(x,y,"LineWidth",2.5)」 creates a scatter plot with 2.5-point marker outlines, whereas the default is 0.5. See Fig. 3.35 for the resulting figure.

```
scatter(Year,Rice,"LineWidth",2.5);
```

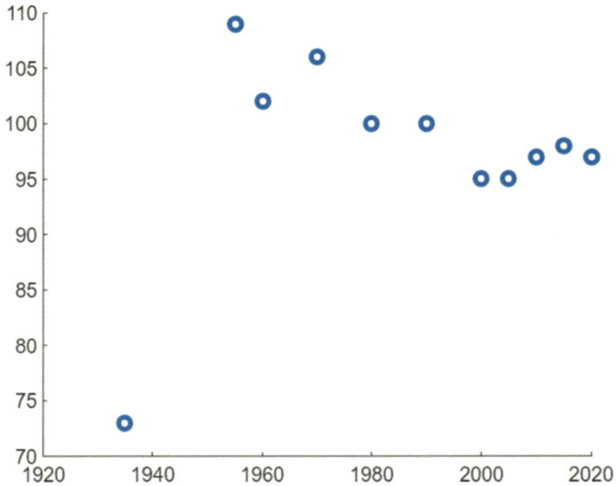

Fig. 3.35 Scatter plot with thick marker edges

Attention

Error Program: In general, Name-Value arguments (such as setting 「LineWidth」 to 2.5) can be used after the basic syntax. It is particularly important to use both 「Name」 and 「Value」 together. Note that using 「LineWidth」 as 「Name」 and leaving 「Value」 empty will result in an error. The error message is presented in Fig. 3.36.

```
%%% Error
% scatter(Year,Rice,"LineWidth")
```

```
Error using scatter
Error in color/linetype argument.
```

Fig. 3.36 Error output of the name-value pairs: Missing value

The circle size and the outline width can be changed simultaneously. Specifically, in the following program, the circle size is set to 100. Moreover, the outline width is changed to 2.5 (Value) using 「LineWidth (Name)」. The resulting plot is displayed in Fig. 3.37.

```
scatter(Year,Rice,100,"LineWidth",2.5);
```

3

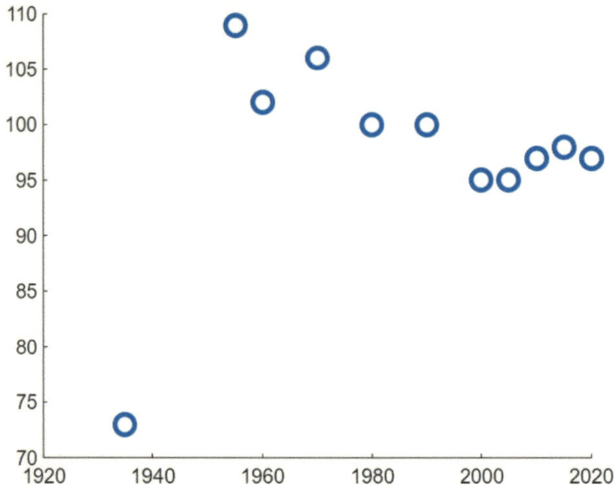

Fig. 3.37 Scatter plot with large markers and thick marker edges

Attention

Error Program: If the number 100 in the program that created Fig. 3.37 is moved to a later position, it results in a syntax error. The error message is shown in Fig. 3.38

```
%%% Error
% scatter(Year,Rice,"LineWidth",2.5,100)
```

Fig. 3.38 Error output of the scatter function: Misplaced property value

Here, we set the circle size to 200. Additionally, we set「MarkerEdgeColor」to「blue」and「LineWidth」to 2, creating large circles with a blue, thicker outline. The resulting plot is shown in Fig. 3.39.

```
scatter(Year,Rice,200,"MarkerEdgeColor","blue","LineWidth",2);
```

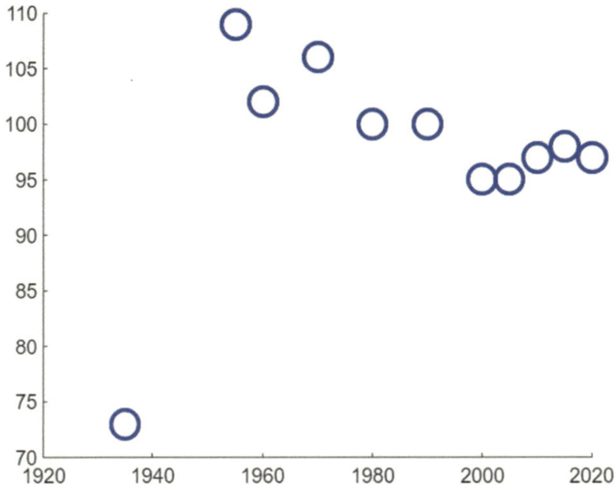

Fig. 3.39 Scatter plot with blue marker edges using the "MarkerEdgeColor" property

The following program removes only 「MarkerEdgeColor」 from the previous program that created Fig. 3.39. The resulting figure is essentially the same as in the previous program. Compare Fig. 3.39 with Fig. 3.40.

```
scatter(Year,Rice,200,"blue","LineWidth",2);
```

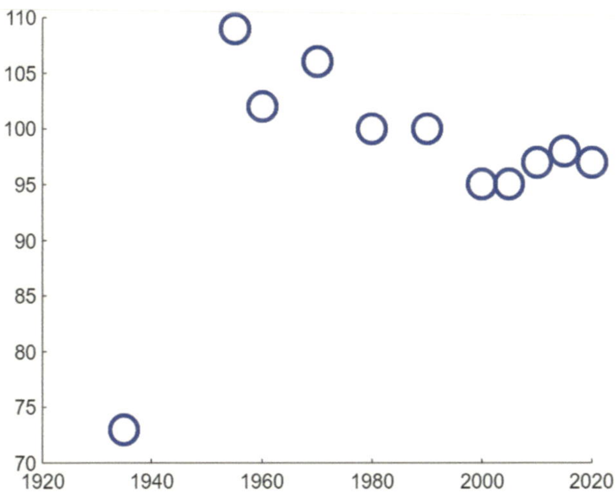

Fig. 3.40 Scatter plot with blue marker edges without using the "MarkerEdgeColor" property (cf. Fig. 3.39)

The 「filled」 option can be used to fill the interior of markers. In the following program, the outline color 「MarkerEdgeColor」 is set to 「blue」, and the circles are filled with 「red」. Additionally, we set 「LineWidth」 to 2, which changes the width of the marker edges. The resulting figure is presented in Fig. 3.41.

```
scatter(Year,Rice,200,"red","filled","MarkerEdgeColor","blue","LineWidth",2);
```


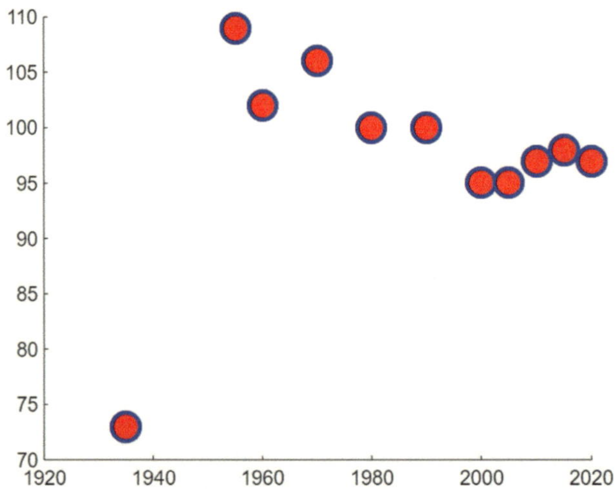

Fig. 3.41 Scatter plot with red filled markers and blue thick marker edges

The marker symbol can also be customized. The following examples (Figs. 3.42, 3.43, and 3.44) use the plus 「+」 and cross 「x」 markers, respectively, to create plus and cross markers. Additionally, note that other markers, such as the asterisk 「*」, dot 「.」, horizontal line 「_」, and square 「square」, are also available in MATLAB.

```
scatter(Year,Rice,"+");
```

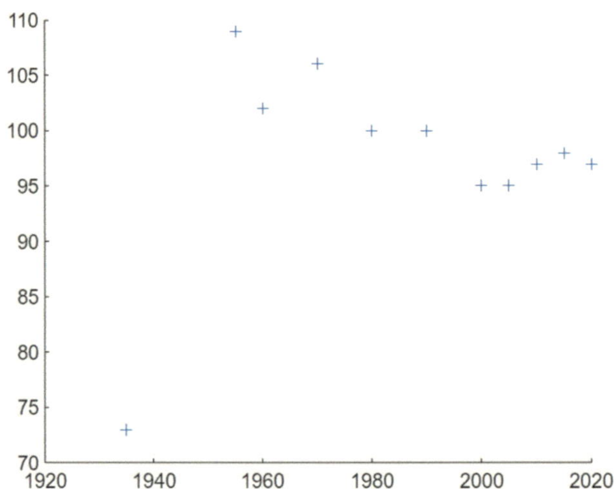

Fig. 3.42 Scatter plot with plus markers

```
scatter(Year,Rice,"x","LineWidth",50);
```

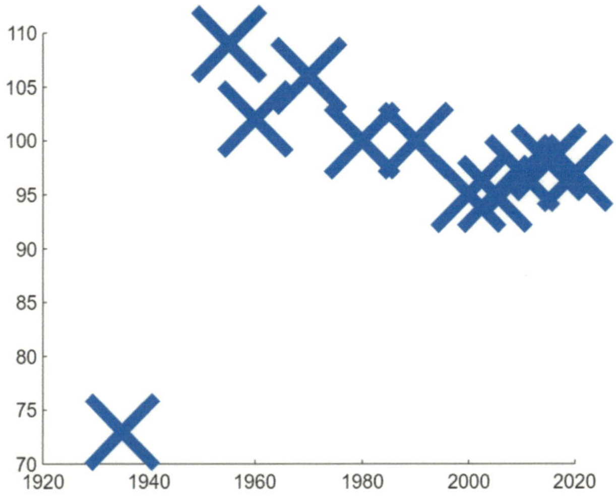

Fig. 3.43 Scatter plot with large cross markers

```
scatter(Year,Rice,"red","x","LineWidth",20);
```

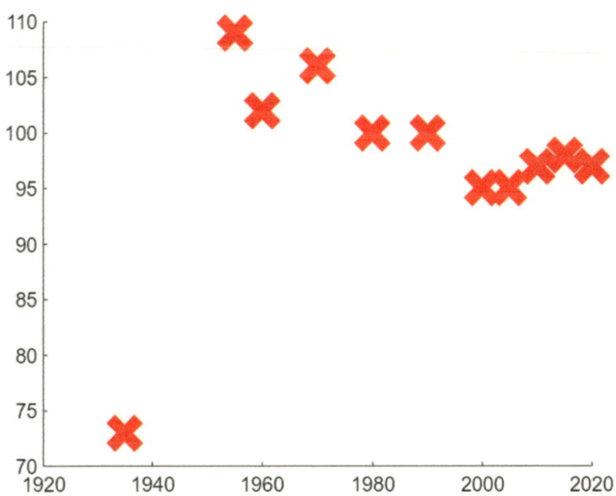

Fig. 3.44 Scatter plot with large red cross markers

In some cases, the property of the **scatter** function is stored into an arbitrary name like 「sc1」 for convenience. This makes it easy to change the size and color of the markers. It also helps prevent the program from becoming too long. In this case, the properties of the scatter plot are stored in the variable, 「sc1」. The resulting plot is presented in Fig. 3.45. The information stored in the variable 「sc1」, that is, the properties of the scatter plot, can be reviewed using the following code, as shown in ◘ Fig. 3.46. The 「Marker」 property can be changed to 「+」.

```
sc1 = scatter(Year,Rice,"+");
```


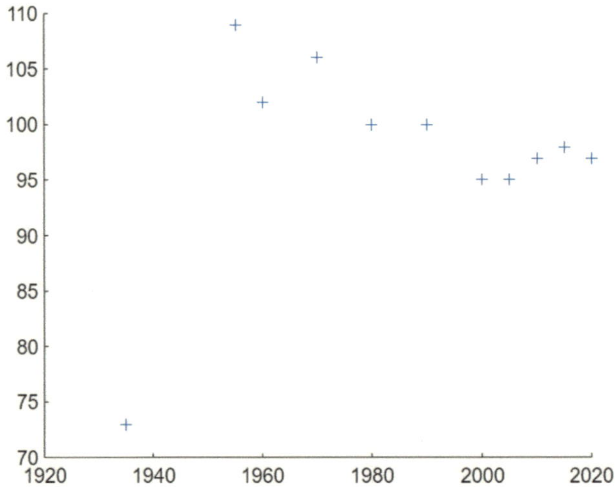

◘ **Fig. 3.45** Scatter plot with plus markers and the creation of the variable "sc1"

```
sc1
```

```
sc1 =
  Scatter with properties:

              Marker: '+'
     MarkerEdgeColor: 'flat'
     MarkerFaceColor: 'none'
           LineWidth: 0.5000
               XData: [1935 1955 1960 1970 1980 1990 2000 2005 2010 2015 2020]
               YData: [73 109 102 106 100 100 95 95 97 98 97]
            SizeData: 36
               CData: [0 0.4470 0.7410]

  Show all properties
```

◘ **Fig. 3.46** Contents of the variable "sc1"

After the properties of the **scatter** function are stored in the variable 「sc1」, we change the width of the marker edges by setting 「LineWidth」 to 50. The resulting figure is presented in ◘ Fig. 3.47.

```
sc1 = scatter(Year,Rice,"+");
    sc1.LineWidth = 50;
```

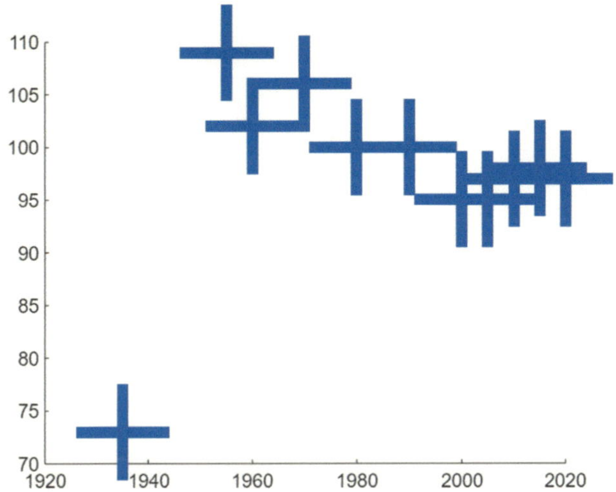

Fig. 3.47 Scatter plot with large plus markers

Here, using 「MarkerEdgeColor」, we specify the color of the marker edges as 「cyan」. The resulting plot is shown in Fig. 3.48.

```
sc1 = scatter(Year,Rice,"+");
    sc1.MarkerEdgeColor = "cyan";
```

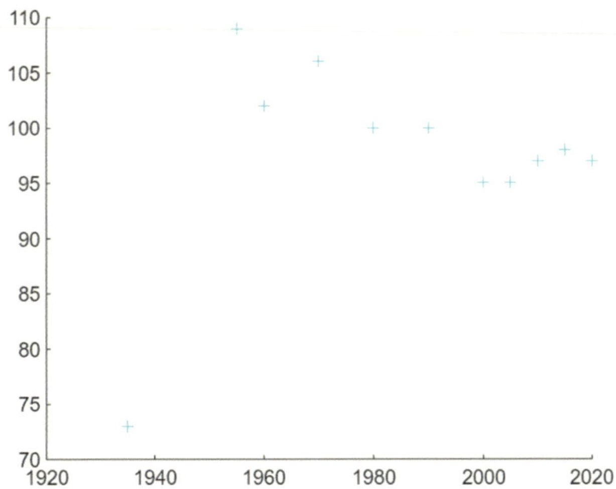

Fig. 3.48 Scatter plot with cyan plus markers

Here, the initial marker was a 「+」 symbol, but you can change the marker again if necessary. When we set 「Marker」 to 「square」, the marker changes to a square with a transparent interior. See Fig. 3.49.

```
sc1 = scatter(Year,Rice,"+");
    sc1.Marker = "square";
```

Fig. 3.49 Change of marker shape from plus to square in a scatter plot

The size of the markers in a scatter plot can be changed by specifying a scalar value. This scalar value is applicable to all markers and changes their size uniformly, as shown in Fig. 3.50.

```
scatter(Year,[Rice,Meat,Milk,Vegetables],100);
```

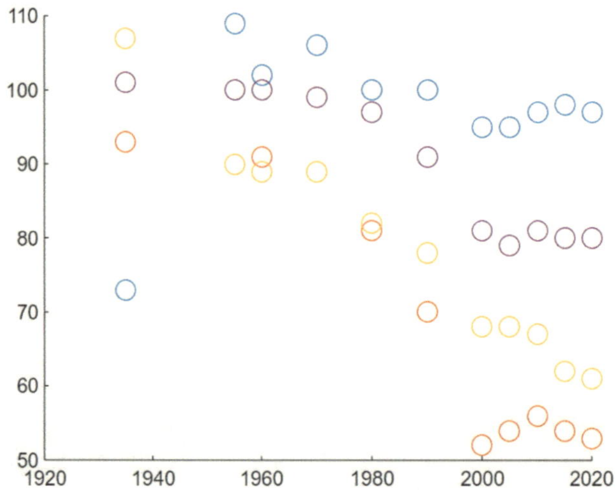

Fig. 3.50 Scatter plot with multiple groups and a large marker size

The sizes of specific data points, such as the fifth and eleventh points, can be adjusted by specifying a vector as the third argument, where each element of the vector corresponds to the size of a data point. For instance, the value 10 in the first element of the vector in the third argument represents a size of 10 for the first data point. The resulting plot is presented in ◘ Fig. 3.51.

```
scatter(Year,[Rice,Meat,Milk,Vegetables],[10,10,10,10,100,10,10,10,10,10,150]);
```

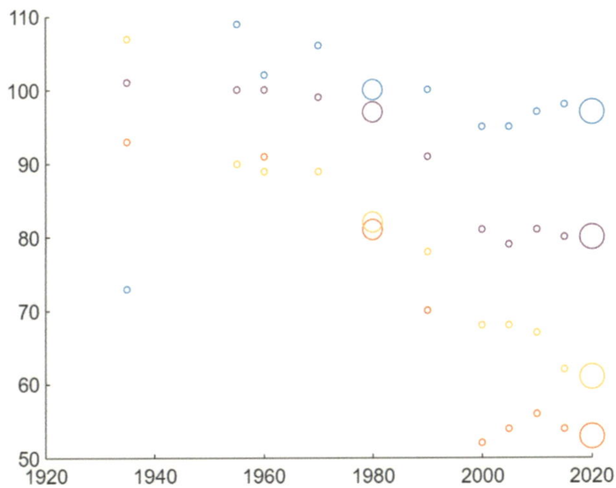

◘ **Fig. 3.51** Scatter plot with multiple groups and various marker sizes

The specific properties of the four types of data (⌜Rice⌟, ⌜Meat⌟, ⌜Milk⌟, and ⌜Vegetables⌟) stored in ⌜abc⌟ can be changed. Note that ⌜abc(1)⌟, ⌜abc(2)⌟, ⌜abc(3)⌟, and ⌜abc(4)⌟ correspond to ⌜Rice⌟, ⌜Meat⌟, ⌜Milk⌟, and ⌜Vegetables⌟, respectively.

In the following example, for the data in ⌜abc(2)⌟, we set ⌜Marker⌟ to ⌜square⌟ and ⌜LineWidth⌟ to 5, creating square markers with a thick outline. Next, for the data in ⌜abc(4)⌟, we fill the markers with black by setting ⌜MarkerFaceColor⌟ to ⌜black⌟. For the data in ⌜abc(1)⌟, we set ⌜MarkerEdgeColor⌟ to ⌜[0 1 0]⌟ to change the edge color to green. The resulting plot is presented in ◘ Fig. 3.52.

```
abc = scatter(Year,[Rice,Meat,Milk,Vegetables]);
    abc(2).Marker = "square";
    abc(2).LineWidth = 5;
    abc(4).MarkerFaceColor = "black";
    abc(1).MarkerEdgeColor = [0 1 0];
```

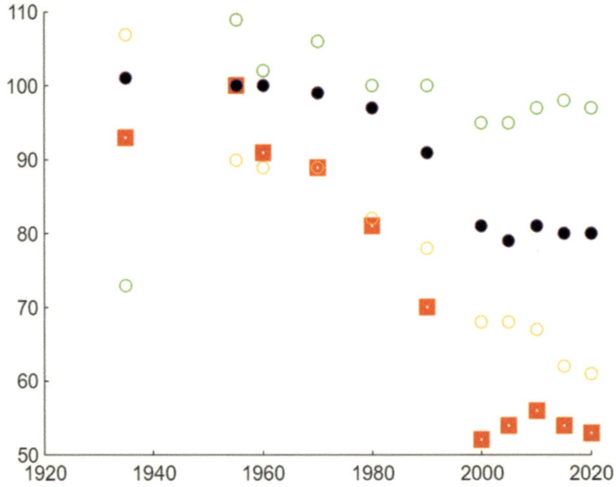

Fig. 3.52 Scatter plot with multiple groups and various marker shapes

3.5 Combining Optional Functions and the Scatter Function

A plot can be well explained by labeling the x-axis and y-axis. Therefore, labeling the axes is critical when creating a plot. Here, we explain how to label the x-axis and y-axis in MATLAB. The x-axis and y-axis labels can be specified using the **xlabel** and **ylabel** functions, respectively. Since their use is the same, we will explain the **xlabel** function specifically. The created figure is shown in Fig. 3.53.

```
scatter(Year,Rice,50,"red","filled");
xlabel("Year");
```

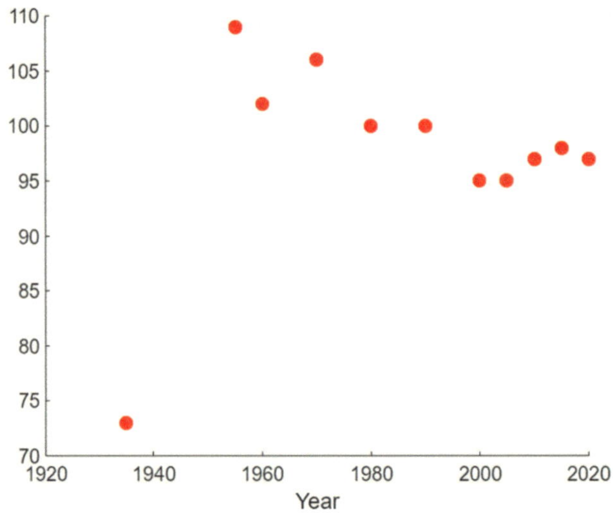

Fig. 3.53 Scatter plot with the X-axis labeled "Year"

Removing the double quotes around the text in the **xlabel** function displays the data values instead of the text, as shown in ◘ Fig. 3.54.

```
scatter(Year,Rice,50,"red","filled");
xlabel(Year);
```

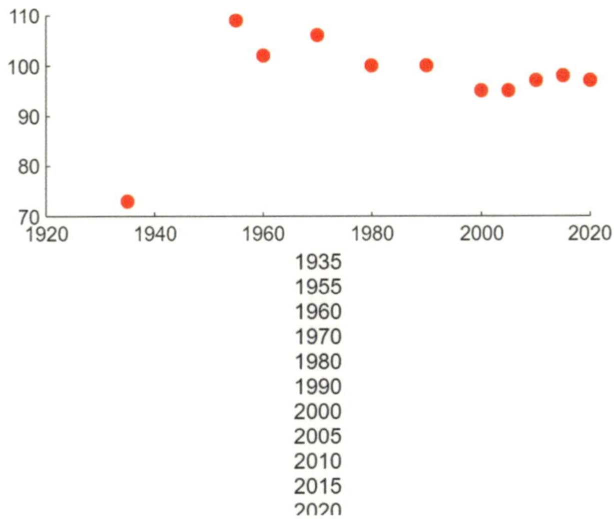

◘ **Fig. 3.54** Scatter plot with numeric years on the X-axis

The font size and color of the text characters in the **xlabel** function can be changed. 「r」 is an abbreviation for red. The color red can also be specified using 「red」, 「#FF0000」, or 「[1 0 0]」. Additionally, 「FontSize」 is set to 20. This is a Name-Value pair argument. See ◘ Fig. 3.55 for the created figure.

```
scatter(Year,Rice,50,"red","filled");
xlabel("Year","Color","r","FontSize",20);
```

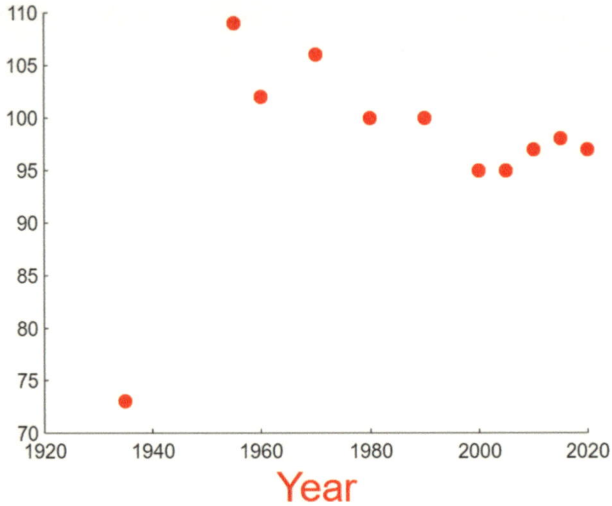

Fig. 3.55 Scatter plot with large red X-axis label

In the following example, the properties of the **scatter** and **xlabel** functions are assigned to the variables 「sc2」 and 「XL」, respectively. Specifically, we set 「XL. Color」 to 「g (= green)」, creating an x-axis label in green. Additionally, we set 「sc2. Marker」 to 「<」, creating left-pointing triangle markers. The resulting plot is presented in Fig. 3.56.

```
sc2 = scatter(Year,Rice,50,"red","filled");
XL = xlabel("Year","Color","r","FontSize",20);
    XL.Color = "g";
    sc2.Marker = "<";
```

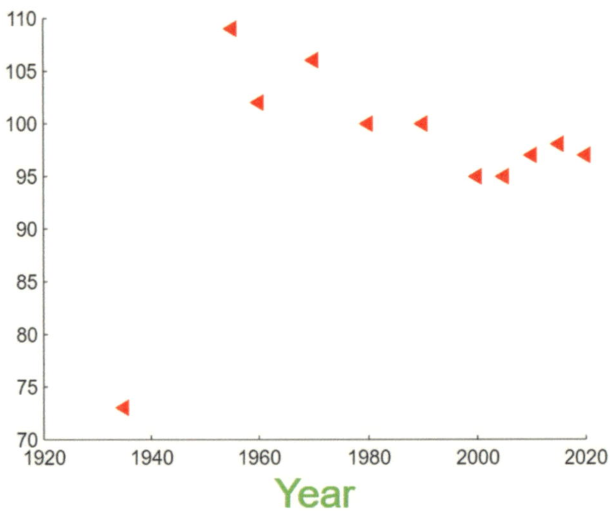

Fig. 3.56 Scatter plot with red left-pointing markers and large green X-axis label

The following program demonstrates a more complex use of the **xlabel** function. The variable 「XL」 stores the result of the **xlabel** function, which creates an x-axis label with multiple lines of text. The input to the **xlabel** function is a cell array containing three lines. The first line displays the word 「YEAR」 and the second line displays the word 「Year」. The third line uses subscripts and superscripts: 「Y_{ear}」 displays 「ear」 as a subscript to 「Y」, and 「Y^{ear}」 displays 「ear」 as a superscript to 「Y」. We also change the color, font size, and font weight. The resulting plot is shown in ◘ Fig. 3.57.

```
scatter(Year,Rice,50,"red","filled","MarkerEdgeColor","blue");
XL = xlabel({"YEAR","Year","Y_{ear}          Y^{ear}"});
YL = ylabel("Rice");
    XL.Color = [0 0 0];
    XL.FontSize = 20;
    YL.Color = "#000000";
    YL.FontSize = 30;
    YL.FontWeight = "bold";
```

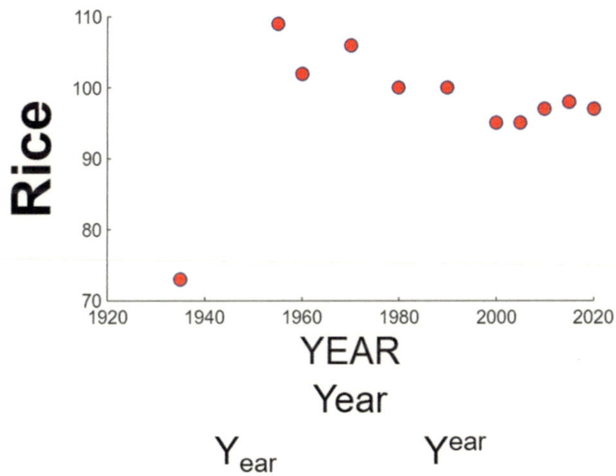

◘ **Fig. 3.57** Scatter plot with various X- and Y-axis labels

The range on the x-axis can be manually set using the **xlim** function, which is given by a two-element vector in the form of 「[xmin, xmax]」, where xmax is larger than xmin. The resulting plot is presented in ◘ Fig. 3.58.

```
scatter(Year,Rice,50);
xlim([0 2020]);
```

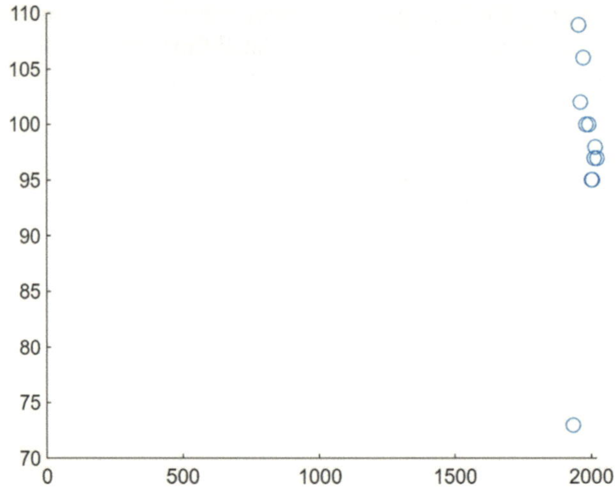

Fig. 3.58 Scatter plot with adjusted X-axis limits

The **ylim** function is used to specify the range on the y-axis. The resulting plot is presented in Fig. 3.59.

```
scatter(Year,Rice,50);
ylim([0 200]);
```

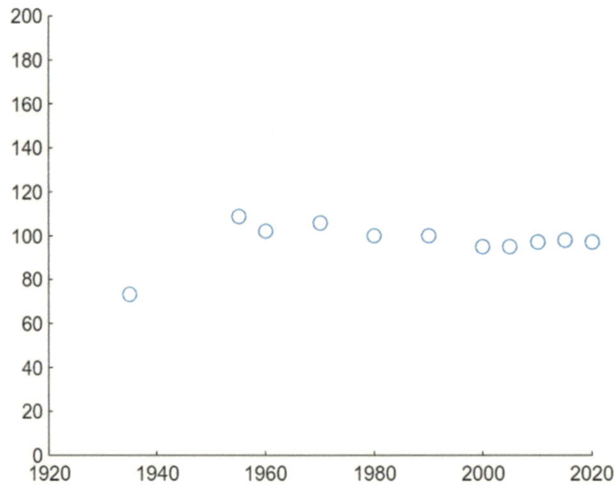

Fig. 3.59 Scatter plot with adjusted Y-axis limits

Attention

Error Program: Here, an error occurs if the vector form in the **xlim** function is not used. In the MATLAB documentation, 「xlim(limits)」 sets the x-axis limits for the current axes or chart. The limits should be specified as a two-element vector using square brackets like 「[xmin xmax]」, where xmax is larger than xmin. The error message is presented in ◘ Fig. 3.60.

```
%%% Error program
%scatter(Year,Rice,50);
%xlim(0 2020);
```

```
Invalid expression. Check for missing multiplication operator, missing or
unbalanced delimiters, or other syntax error. To construct matrices, use
brackets instead of parentheses.
```

◘ **Fig. 3.60** Error output of the xlim function: Missing brackets

Before creating this scatter plot to display two sets of data (「Rice」 and 「Meat」), we first clear the range settings for both the **xlim** and **ylim** functions in the previous program using the **clear** command. The resulting plot is displayed in ◘ Fig. 3.61.

```
clear xlim;
clear ylim;
scatter(Year,[Rice,Meat]);
```

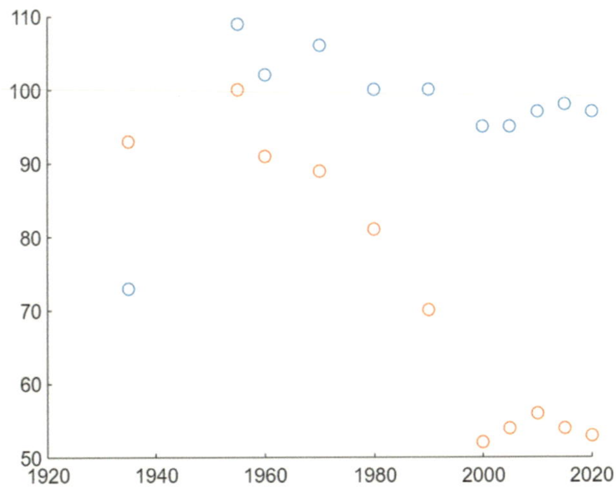

◘ **Fig. 3.61** Scatter plot with default axis limits

Here, we use the **text** function to display the values of 「Rice」 next to the 11 data points. Additionally, the following four aspects are noteworthy. The first is the repetition of the **for** loop. When using the **for** function, we need to specify the number of repetitions, which is given by 「for i=1:length(Rice)」. As it repeats, 「i」 increases from 1 to 2, 3, 4, and so on, until it reaches 「i=length(Rice)」. The second is the **length** function. This function is often used in **for** loops to specify the number of repetitions. Specifically, since 「Rice」 is an 11-by-1 vector, 「length(Rice)」 equals

3

11. The third is the **text** function. The **text** function can be used to place data on the figure. Specifically, the first argument of the **text** function specifies the x-axis, the second argument specifies the y-axis, and the third argument places the value of 「Rice」 on the figure. Fourth, it is necessary to use the **num2str** function. Since the variable 「Rice」 is of type double, it needs to be converted to a string. Then the value is placed on the figure. See ◘ Fig. 3.62. For more details on the **text** function, see ► Sect. 18.2.

```
scatter(Year,Rice,50,"filled");
for i = 1:length(Rice)
    text(Year(i),Rice(i),num2str(Rice(i)));
end
```

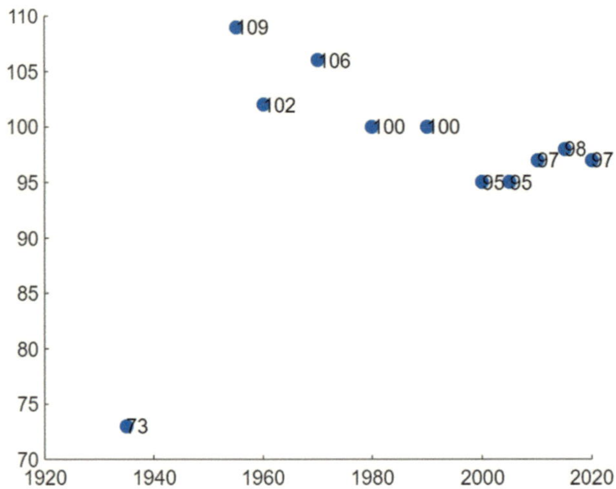

◘ **Fig. 3.62** Scatter plot with data labels

Here, we use the previous program to create ◘ Fig. 3.62. Then, we use the Name-Value pair of the **text** function to specify the position where the numbers are placed. First, we use 「VerticalAlignment」 to determine where to align the text relative to the data point. 「VerticalAlignment」 adjusts the vertical positioning. In this case, it is set to 「bottom」, so that the text is aligned below the data point. Other options include 「top」 and 「middle」.

Subsequently, 「HorizontalAlignment」 determines where the text is horizontally aligned relative to the data point. In this case, it is set to 「right」, so the text is aligned to the right of the data point. Other options include 「left」 and 「center」. For more details on the **text** function, see ► Sect. 18.2. Compare ◘ Fig. 3.62 with ◘ Fig. 3.63.

```
scatter(Year,Rice,50,"filled");
for i = 1:length(Rice)
    text(Year(i),Rice(i),num2str(Rice(i)),...
        "VerticalAlignment","bottom", "HorizontalAlignment","right");
end
```

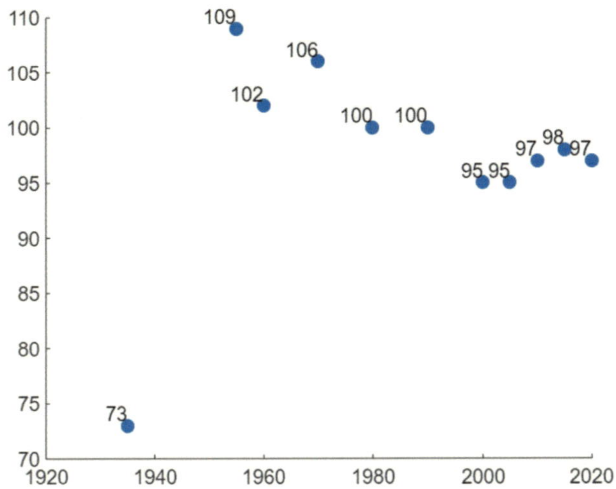

Fig. 3.63 Scatter plot using a for loop to create data labels and adjust label positions

Here, the properties of the **text** function are stored in the variable 「txt」. Pay attention to the second argument of the **text** function. In the second argument, 「Rice(i)-5」 is used. This means that the y-axis coordinate is set to the value of 「Rice(i)」 minus 5. Then, the text characters are modified. Specifically, we set 「Color」 to 「red」, 「FontSize」 to 15, and 「FontWeight」 to 「bold」, creating bold red text with a larger size. For more details on the **text** function, see ▶ Sect. 18.2. Additionally, the **ylim** function is used to set the minimum and maximum values of the y-axis. The resulting figure is displayed in Fig. 3.64.

```
scatter(Year,Rice,50,"filled");
for i = 1:length(Rice)
    txt = text(Year(i),Rice(i)-5,num2str(Rice(i)));
        txt.HorizontalAlignment = "center";
        txt.Color = "red";
        txt.FontSize = 15;
        txt.FontWeight = "bold";
end
ylim([60 120]);
```

3

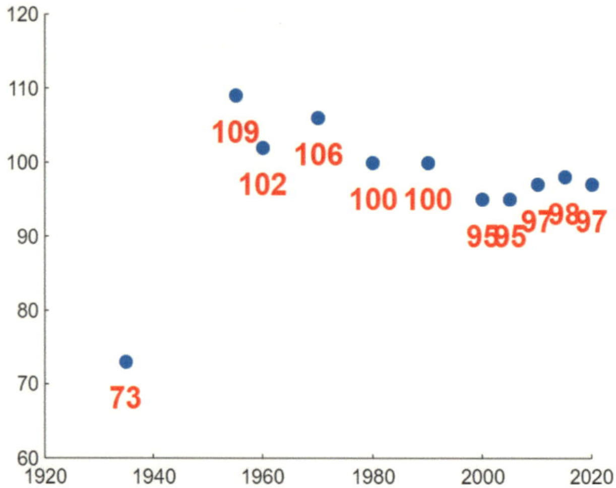

Fig. 3.64 Scatter plot using a for loop to create bold red data labels and adjust label positions

Here, the **hold on** command is used to overlay two scatter plots. Additionally, the **text** function is used to display the values of both 「Rice」 and 「Meat」 on the figure, with the **for** loop repeating the process. For more details on the **text** function, see ► Sect. 18.2. The **legend** function is used to create a legend, which is particularly important when dealing with multiple data series. Including a legend makes it easy to clarify which data points correspond to each series in a scatter plot. In addition, using the **legend** function simplifies visual comparisons between different data series. Finally, be sure to enter the **hold off** command after completing the overlay with the **hold on**. If you forget to enter the **hold off**, all subsequent plots will be overlaid. See Fig. 3.65 for the resulting figure.

```
scatter(Year,Rice,50,"filled","b");
hold on
scatter(Year,Meat,50,"filled","r");
for i = 1:length(Rice)
    text(Year(i),Rice(i)-2,num2str(Rice(i)),...
        "HorizontalAlignment","center","Color","b");
end
for i = 1:length(Meat)
    text(Year(i),Meat(i)-2,num2str(Meat(i)),...
        "HorizontalAlignment","center","Color","r");
end
legend("Rice","Meat");
hold off
```

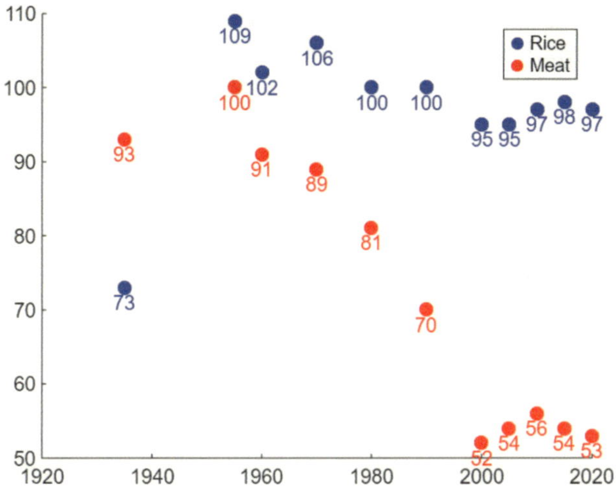

■ **Fig. 3.65** Scatter plot using a for loop to create data labels with a legend

As shown in the following program, because we use the **legend** function after the **scatter** function, a legend is automatically created within the scatter plot figure. Since the names of the four data sets are not specified, they are labeled as 「data1」, 「data2」, 「data3」, and 「data4」 as seen in ■ Fig. 3.66. According to the legend, the entries 「data1」 through 「data4」 correspond to, from top to bottom, 「Rice」, 「Meat」, 「Milk」, and 「Vegetables」 in the second argument of the **scatter** function.

```
scatter(Year,[Rice,Meat,Milk,Vegetables]);
legend;
```

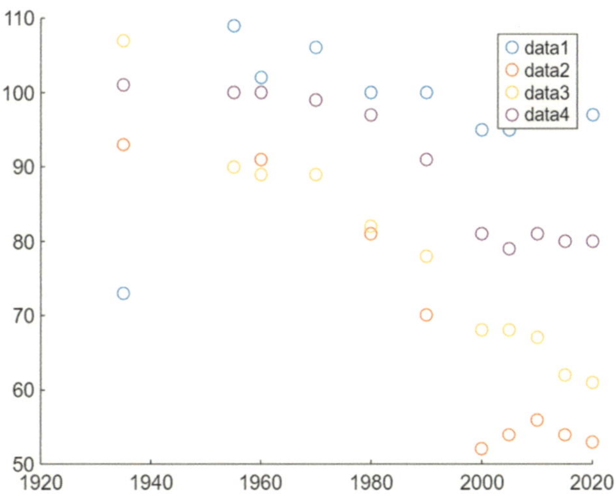

■ **Fig. 3.66** Scatter plot with a default legend

When text strings are entered into the **legend** function, they appear in the legend in the order in which they are entered. See Fig. 3.67 for the created figure.

```
scatter(Year,[Rice,Meat,Milk,Vegetables]);
legend("Rice","Meat","Milk","Vegetables");
```


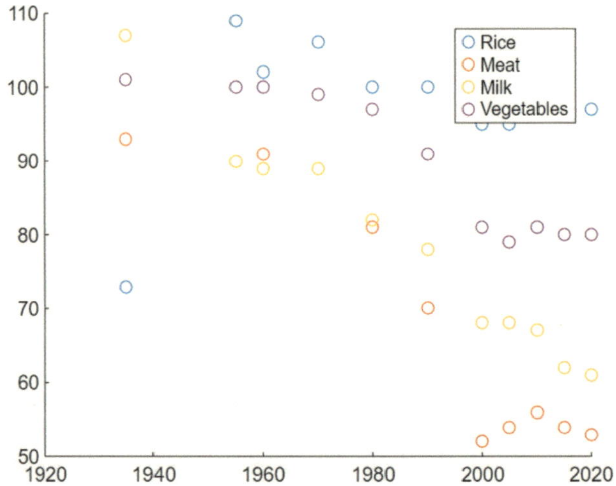

Fig. 3.67 Scatter plot with a specified legend

When only two of the four data types are specified in the **legend** function, the legend describes only the first two data entries. In this example, as only 「Rice」 and 「Meat」 are specified in the **legend** function, the legend contains the entries for 「Rice」 and 「Meat」, with no entries for the third and fourth data types. See Fig. 3.68.

```
scatter(Year,[Rice,Meat,Milk,Vegetables]);
legend("Rice","Meat");
```

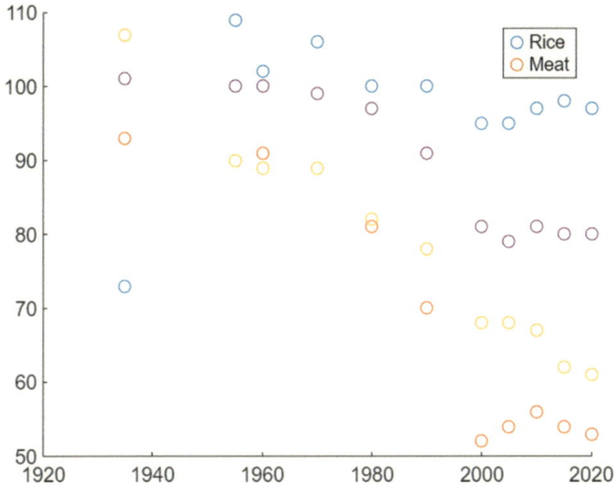

Fig. 3.68 Scatter plot with less legend entries

If there is data that should not be shown in the legend, it can be excluded by using empty double quotation marks, making it appear blank in the legend. The resulting plot is presented in Fig. 3.69.

```
scatter(Year,[Rice,Meat,Milk,Vegetables]);
legend("Rice","","Milk","Vegetables");
```

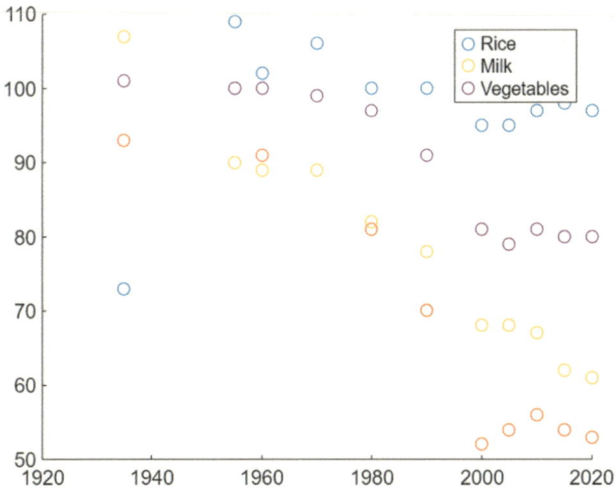

Fig. 3.69 Scatter plot with one missing legend entry

Unlike the program that created Fig. 3.69, if there is a space inside the double quotation marks, the corresponding data type is displayed without a name. Compare Fig. 3.69 with Fig. 3.70.

```
scatter(Year,[Rice,Meat,Milk,Vegetables]);
legend("Rice","      ","Milk","Vegetables");
```

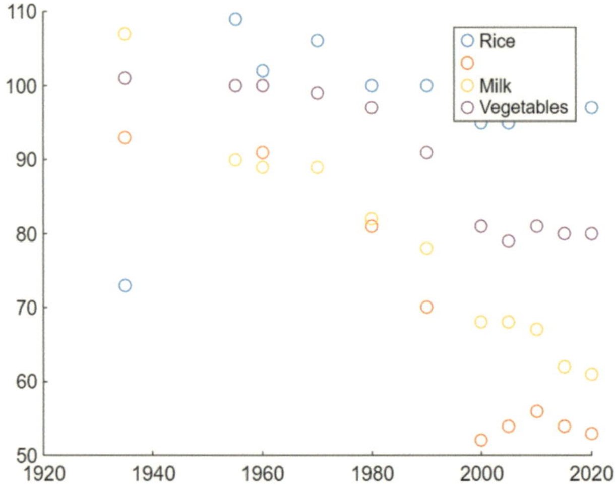

Fig. 3.70 Scatter plot with a blank legend entry

To specify the location of the legend, we use the「Location」 property, and pass the desired location name as the next argument. The following program moves the legend to the「southwest」 position. The resulting plot is shown in Fig. 3.71. Other available locations include「north」,「south」,「east」,「west」, and combinations like「northeast」. For more information, see the MATLAB documentation by typing「doc legend」.

```
scatter(Year,[Rice,Meat,Milk,Vegetables]);
legend("Rice","Meat","Location","southwest");
```

Fig. 3.71 Scatter plot with a legend (bottom-left position)

The legend can be placed outside the figure at a specific location by adding 「outside」 after the location. As shown in Fig. 3.72, 「northoutside」 places the legend outside the figure at the north location.

```
scatter(Year,[Rice,Meat,Milk,Vegetables]);
legend("Rice","Meat","Location","northoutside");
```

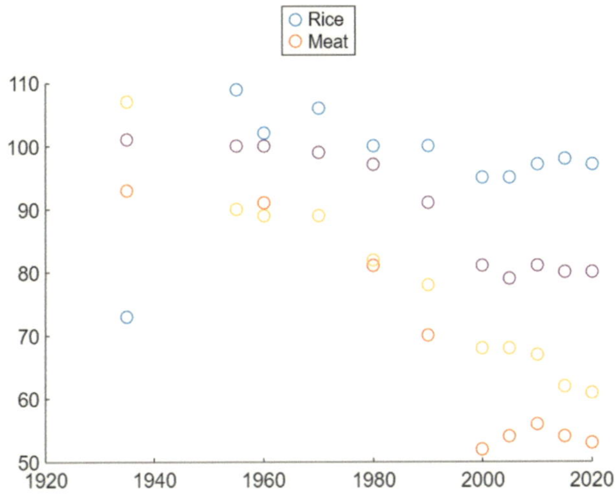

Fig. 3.72 Scatter plot with a legend (outside-top-center position)

Here, the data in 「y」 of the basic syntax 「scatter(x,y)」 consists of two types: 「Rice」 and 「Meat」. In contrast, the **legend** function uses four data names (「1111」, 「2222」, 「3333」, and 「4444」). In such cases, a warning message appears, while the legend is created with only the first two (「1111」 and 「2222」). In addition, when we set 「Location」 to 「best」, the legend is placed inside the axis range where it has the least conflict with the plot data. In Fig. 3.73, the legend is positioned between 「southwest」 and 「south」 because it has the fewest conflicts with the plot data.

```
scatter(Year,[Rice,Meat],50);
legend("1111","2222","3333","4444","Location","best");
```

```
Warning: Ignoring extra legend entries.
```

3

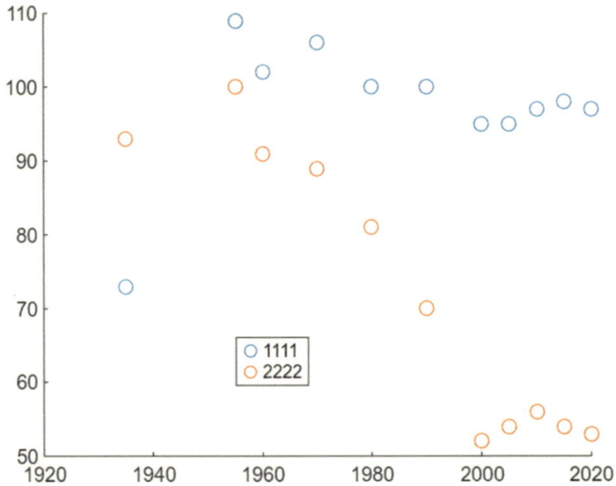

Fig. 3.73 Scatter plot with extra legend entries (best position)

The properties of the **legend** function can be stored in a variable. Here, the properties are stored in a variable named 「lgd1」. The resulting legend, with 「FontSize」 of 20 and 「TextColor」 of blue, is displayed in Fig. 3.74.

```
scatter(Year,[Rice,Meat],50);
lgd1 = legend("Rice","Meat","Location","eastoutside");
    lgd1.FontSize = 20;
    lgd1.TextColor = "blue";
```

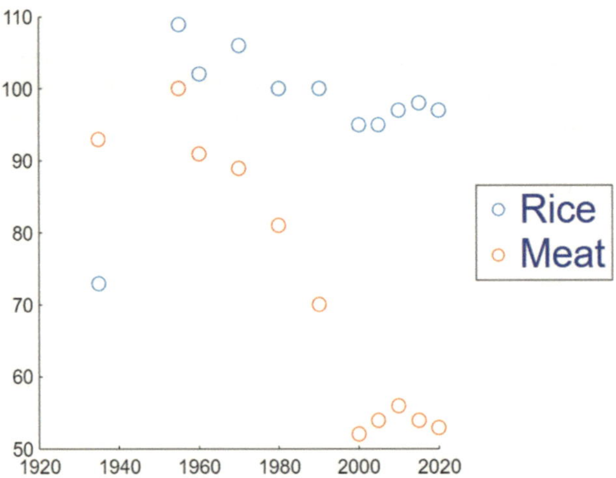

Fig. 3.74 Scatter plot with a legend containing large blue text (outside-right position)

「NumColumns」 (a property of the **legend** function) can be used to split the legend into multiple columns. This property specifies the number of columns in the legend and must be a positive integer. By default, 「NumColumns」 is set to 1, resulting in a single column. Here, we set 「NumColumns」 to 2, which contains two columns. With four data types, this arrangement creates a layout with two rows and two columns. See ◘ Fig. 3.75.

```
scatter(Year,[Rice,Meat,Milk,Vegetables],50);
lgd2 = legend("Rice","      ","Milk","Vegetables","Location","southwest");
    lgd2.FontSize = 50;
    lgd2.TextColor = "cyan";
    lgd2.NumColumns = 2;
```

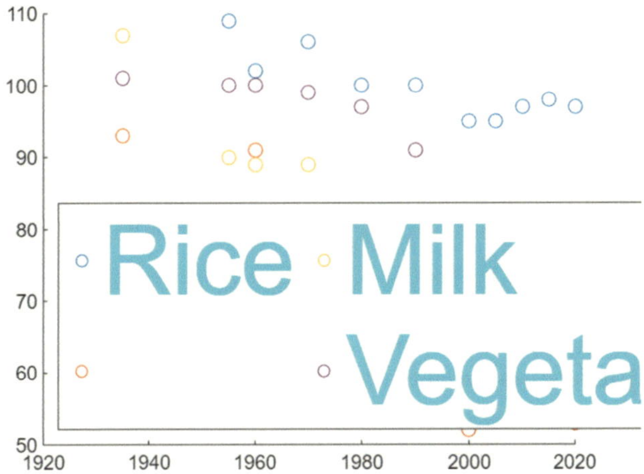

◘ **Fig. 3.75** Scatter plot with a legend containing large cyan text (bottom-left position)

Here, we set 「NumColumns」 to 4, which contains four columns. As a result, the four items, 「Rice」, 「Meat」, 「Milk」, and 「Vegetables」 are arranged horizontally. Additionally, we set 「Location」 to 「northoutside」, placing it outside the top of the figure. The resulting figure is presented in ◘ Fig. 3.76.

```
scatter(Year,[Rice,Meat,Milk,Vegetables],50);
lgd3 = legend("Rice","Meat","Milk","Vegetables","Location","northoutside");
    lgd3.TextColor = "black";
    lgd3.NumColumns = 4;
```

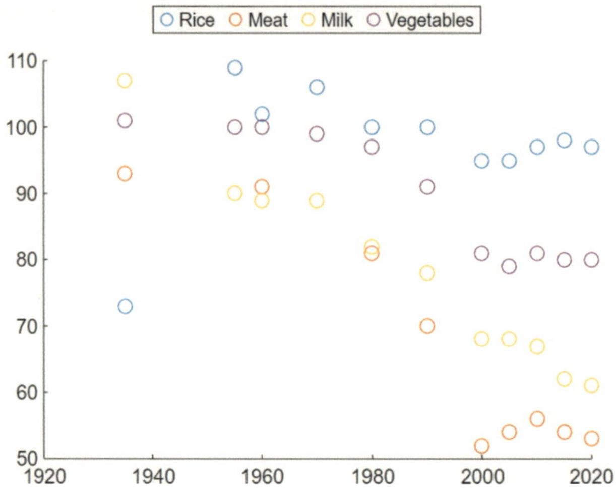

Fig. 3.76 Scatter plot with a horizontal legend (outside-top-center position)

Here, we use the **title** function to add a specific title to the figure. The basic syntax of the **title** function from the MATLAB documentation is 「title(titletext)」. Type 「doc title」 in the Command Window. The resulting plot titled 「Title: Food self-sufficiency in Japan」 is shown in Fig. 3.77.

```
scatter(Year,[Rice,Meat],50);
title("Title: Food self-sufficiency in Japan");
```

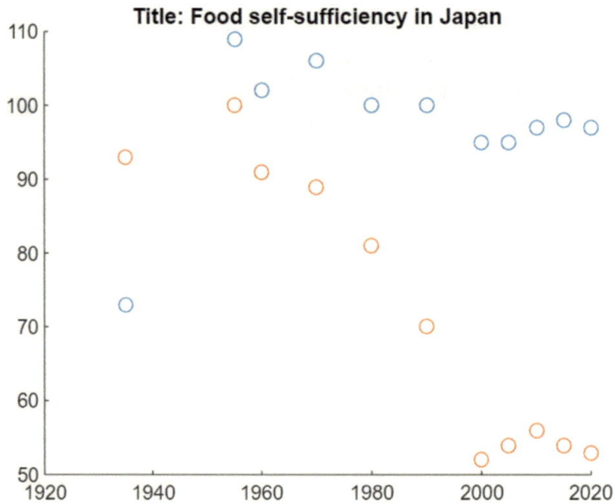

Fig. 3.77 Scatter plot with a figure title

Here, the first use of the **title** function is used to add a title to the figure, as in the previous program that created Fig. 3.77. Next, the second use of the **title** function is used to add a title to the legend. In this case, the properties of the **legend** function are stored in a variable called 「lgdlgd」. This variable is then used as the first argument in the **title** function, such as 「title(lgdlgd,.....) 」. The resulting plot is presented in ◘ Fig. 3.78.

```
scatter(Year,[Rice,Meat],50);
title("Title: Food self-sufficiency in Japan");
lgdlgd = legend("Rice","Meat","Location","southwest");
title(lgdlgd,"Legend: Food self-sufficiency in Japan");
```

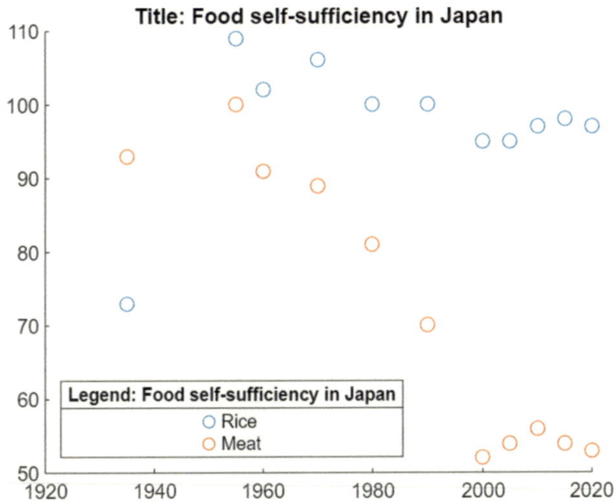

◘ **Fig. 3.78** Scatter plot with figure and legend titles

Here, using the **tiledlayout** function, we create a tiled chart layout for displaying multiple plots. For example, 「tiledlayout(m,n)」 creates an m-by-n tiled layout. In the following program, we use 「tiledlayout(2,1)」 to create a two-row, one-column tiled layout. After using 「nexttile」, the plot for 「scatter(Year,Vegetables)」 is placed in the first position (first row, first column). The next use of 「nexttile」 ensures that 「scatter(Year,Milk)」 is placed on the next empty tile. The resulting plot is presented in ◘ Fig. 3.79.

```
figure
tiledlayout(2,1);
nexttile;
scatter(Year,Vegetables);
nexttile;
scatter(Year,Milk);
```

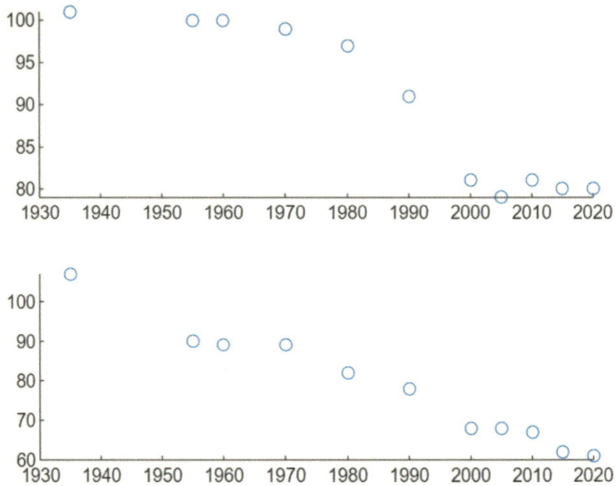

Fig. 3.79 Scatter plot (2-row layout)

Here, we use 「nexttile」 after the **scatter** function. Therefore, the following program uses 「nexttile」 differently from the previous program that created Fig. 3.79. After the first scatter plot 「scatter(Year,Vegetables)」 is displayed as a single figure, as in Fig. 3.80, we use 「nexttile」. The first use of 「nexttile」 ensures that the scatter plot is placed in the first empty tile of the 2-by-1 tiled layout. Then, the plot for 「scatter(Year,Milk)」 is placed in the first tile of the 2-by-1 layout. Next, we use the second 「nexttile」; however, we do not use the **scatter** function after that. Therefore, only the axes are displayed in the second row of the 2-by-1 layout. See Fig. 3.81.

```
figure
tiledlayout(2,1);
scatter(Year,Vegetables);
nexttile;
```

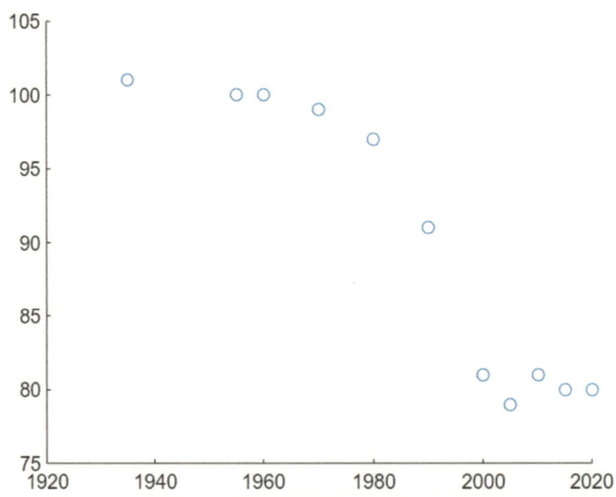

Fig. 3.80 Scatter plot (1-row layout)

```
scatter(Year,Milk);
nexttile;
```

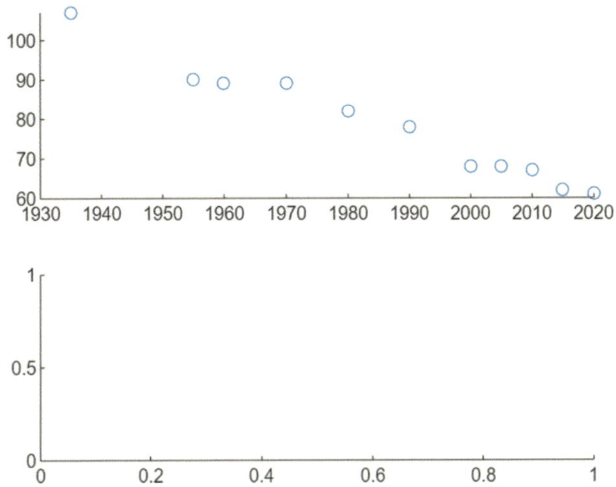

Fig. 3.81 Scatter plot (2-row layout with an empty bottom plot)

Here, we use the **annotation** function to visually indicate the blank layout. First, using the **tiledlayout** function, we create a 2-by-1 tiled layout. The first ⌜nexttile⌟ places the figure for ⌜scatter(Year,Vegetables)⌟ in the position of the 1-by-1 tile. However, because the second ⌜nexttile⌟ is not used, the figure is not created at the position of the 2-by-1 tile, and the axes are not created. Nevertheless, because ⌜tiledlayout(2,1)⌟ is executed, the tile spaces for the 2-by-1 layout are reserved. To indicate the reserved tile space clearly, we use the **annotation** function to create a text arrow by ⌜TextArrow⌟. For more details on the **annotation** function, see ▶ Sect. 18.1. The resulting figure is shown in Fig. 3.82.

```
figure
tiledlayout(2,1);
nexttile;
scatter(Year,Vegetables);
x = [0.6 0.5];
y = [0.45 0.3];
ta = annotation("textarrow",x,y,"String","a space for the 2-by-1 layout");
    ta.Color = "red";
```

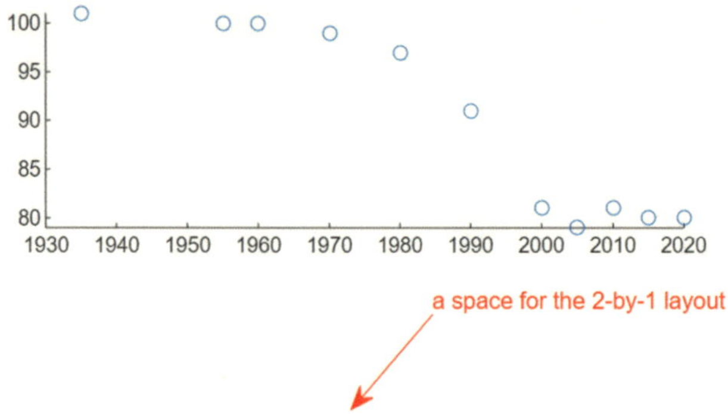

Fig. 3.82 Scatter plot (2-row layout with an empty frameless bottom space)

Here, we create a 1-by-3 tiled layout. As the **scatter** function is not used after the second 「nexttile」, only the axes are created at the position of the 1-by-2 tile, as shown in Fig. 3.83.

```
figure
tiledlayout(1,3);
nexttile;
scatter(Year,Vegetables);
nexttile;
nexttile;
scatter(Year,Milk);
```

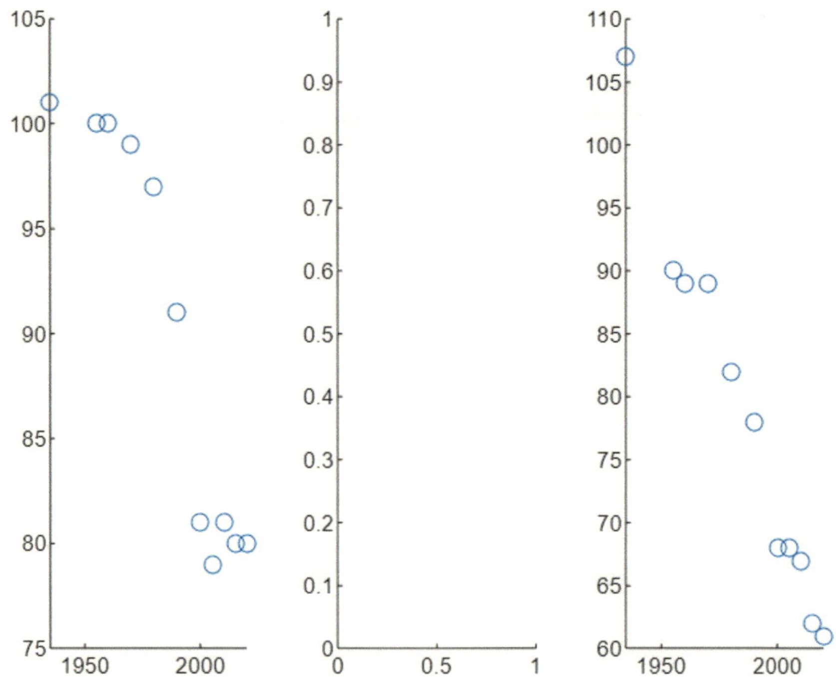

Fig. 3.83 Scatter plot (3-column layout with an empty center plot)

We use the **title** function based on the syntax ⌜title(target, ___)⌟. For more details, see the MATLAB documentation for the **title** function. First, we create a 2-by-1 tiled layout using ⌜tiledlayout(2,1)⌟. Then, by using ⌜ax1=nexttile⌟ and ⌜ax2=nexttile⌟, we store the properties of the 1-by-1 and 2-by-1 tiles in ⌜ax1⌟ and ⌜ax2⌟, respectively.

Using the **title** function, we create titles for each figure: ⌜title(ax1, ...)⌟ in the 1-by-1 tile and ⌜title(ax2, ...)⌟ in the 2-by-1 tile. Additionally, we use ⌜box(ax2, "on")⌟ to draw a box outline around the axes of the lower plot. The resulting figure is shown in ◘ Fig. 3.84.

```
figure
tiledlayout(2,1);
ax1 = nexttile;
scatter(Year,Vegetables);
ax2 = nexttile;
scatter(Year,Milk);
title(ax1,"Top Scatter");
title(ax2,"Bottom Scatter");
box(ax2,"on");
```

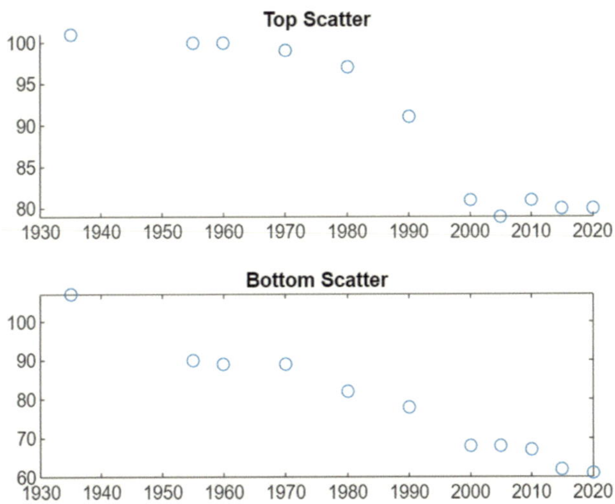

◘ **Fig. 3.84** Scatter plot (2-row layout with a framed bottom plot)

We use the **grid on** command to add axis grid lines to the second plot in the tiled layout. The **grid on** command is executed after the second ⌜nexttile⌟, adding axis grid lines only to the plot in the 2-by-1 tile, as shown in ◘ Fig. 3.85.

```
figure
tiledlayout(2,1);
nexttile;
scatter(Year,Vegetables);
nexttile;
scatter(Year,Milk);
grid on
```

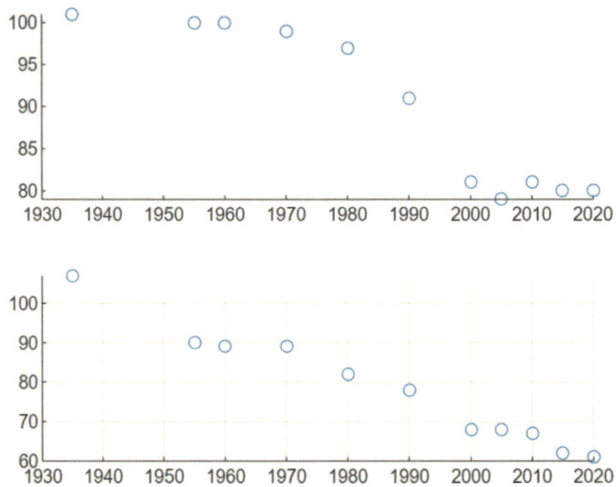

Fig. 3.85 Scatter plot (2-row layout with gridlines on a bottom plot)

We create a 2-by-1 tiled layout. For the plot in the 1-by-1 tile, we add finer axis grid lines. To do this, after the first 「nexttile」, instead of using the **grid on** command, we use the **grid minor** command. For the second 「nexttile」, we use the **grid on** command. After that, we execute 「ga=gca」 to store the axis properties of that plot in a variable called 「ga」. 「gca」 stands for 「Get Current Axes」. Finally, we set 「LineWidth」 to 5 to widen the axis lines. The resulting plot is displayed in Fig. 3.86.

```
figure
tiledlayout(2,1);
nexttile;
scatter(Year,Vegetables);
grid minor
nexttile;
scatter(Year,Milk);
grid on
ga = gca;
    ga.LineWidth = 5;
```

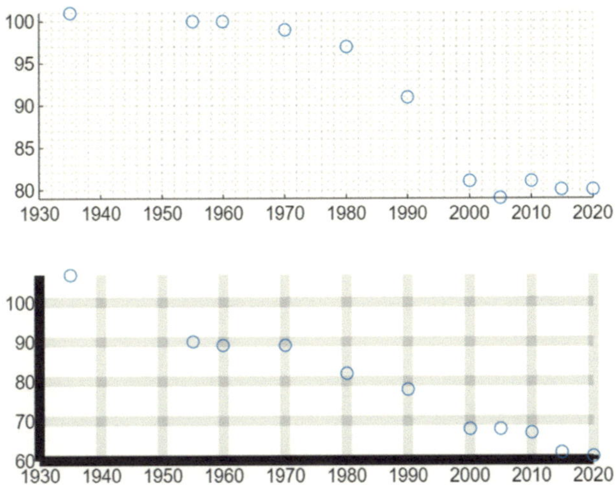

Fig. 3.86 Scatter plot (2-row layout with thick gridlines on a bottom plot)

Here, using the **tiledlayout** function, we create a 3-by-3 tiled layout and store its properties in a variable named 「t1」. Then, we specify the first 「nexttile」 as 「nexttile(5)」. Thus, the plot is placed in the fifth position. Note that the positions are assigned in the order of 1-by-1, 1-by-2, 1-by-3, 2-by-1, 2-by-2, and so on. Since the fifth position corresponds to the 2-by-2 tile, a scatter plot 「scatter(ax1,Year,Rice)」 is created in the 2-by-2 tile. Next, 「nexttile(1)」 creates a scatter plot 「scatter(ax2, Year,Meat)」 in the 1-by-1 tile. 「nexttile(8)」 creates a scatter plot 「scatter(ax3,Year ,Vegetables)」 in the eighth position (3-by-2 tile).

The **title** function is used to provide titles for each of the figures. Additionally, we specify 「TextArrow」 and 「TextBox」 in the **annotation** functions to create two text arrows, one in blue and the other in red, and a text box in magenta. For more details on the **annotation** function, see ▶ Sect. 18.1. Fig. 3.87 shows the created figure.

```
figure
t1 = tiledlayout(3,3);
ax1 = nexttile(5);
scatter(ax1,Year,Rice);
ax2 = nexttile(1);
scatter(ax2,Year,Meat);
ax3 = nexttile(8);
scatter(ax3,Year,Vegetables);
title(t1,"Food self-sufficiency in Japan","FontSize",15,"Color","red");
title(ax1,"Rice");
title(ax2,"Meat");
title(ax3,"Vegetables","Color","blue");
x1 = [0.1 0.2];
y1 = [0.5 0.5];
annotation("TextArrow",x1,y1,"String","A layout","Color","blue");
x2 = [0.1 0.2];
y2 = [0.2 0.2];
annotation("TextArrow",x2,y2,"String","A layout","Color","red");
x3 = [0.45 0.55];
y3 = [0.85 0.8];
annotation("TextArrow",x3,y3,"String","A layout","Color","black");
dim = [0.65 0.1 0.2 0.8];
annotation("TextBox",dim,"String","Layouts","Color","magenta","FontSize",20);
```

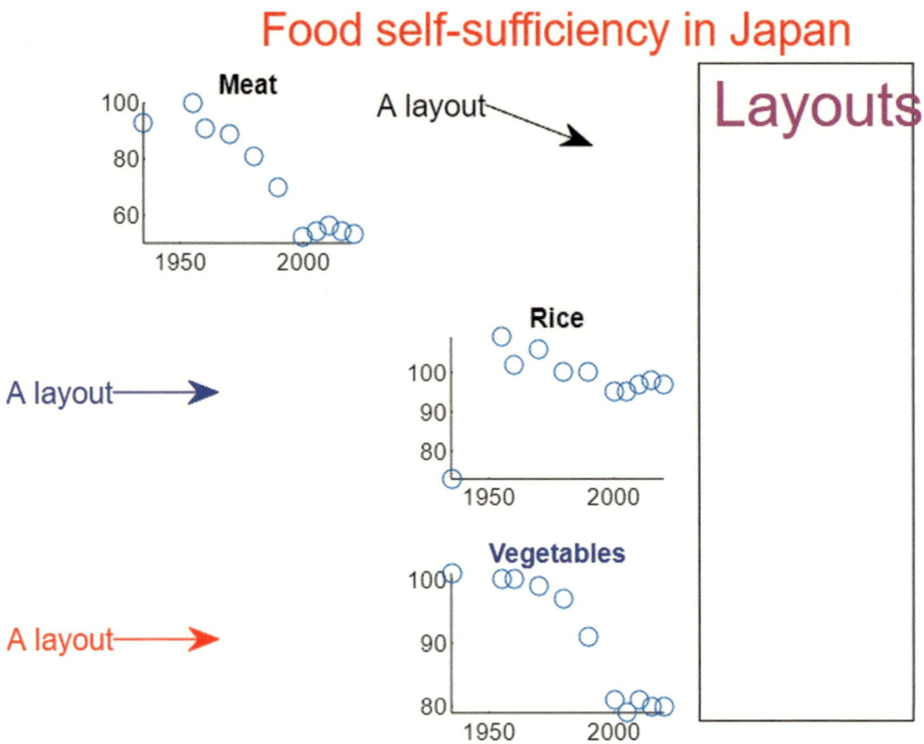

Fig. 3.87 Scatter plot (3-row, 3-column layout with various annotations and text styles)

Here, we create a 3-by-3 tiled layout using the **tiledlayout** function. We then use 「ax4=nexttile([2,3])」 to create a large plot that spans two rows and three columns.

Next, we create an outline around the larger plot using 「box(ax4,"on")」. The **subtitle** function is used to add a subtitle to 「ax4」, as shown in ◘ Fig. 3.88.

```
figure
t1 = tiledlayout(3,3);
ax1 = nexttile;
scatter(ax1,Year,Rice);
ax2 = nexttile;
scatter(ax2,Year,Meat);
ax3 = nexttile;
scatter(ax3,Year,Vegetables);
ax4 = nexttile([2,3]);
scatter(ax4,Year,Milk);
box(ax4,"on");
ylabel(ax4,"Milk");
title(ax4,"Milk");
subtitle(ax4,"Milk & Dairy Products");
```

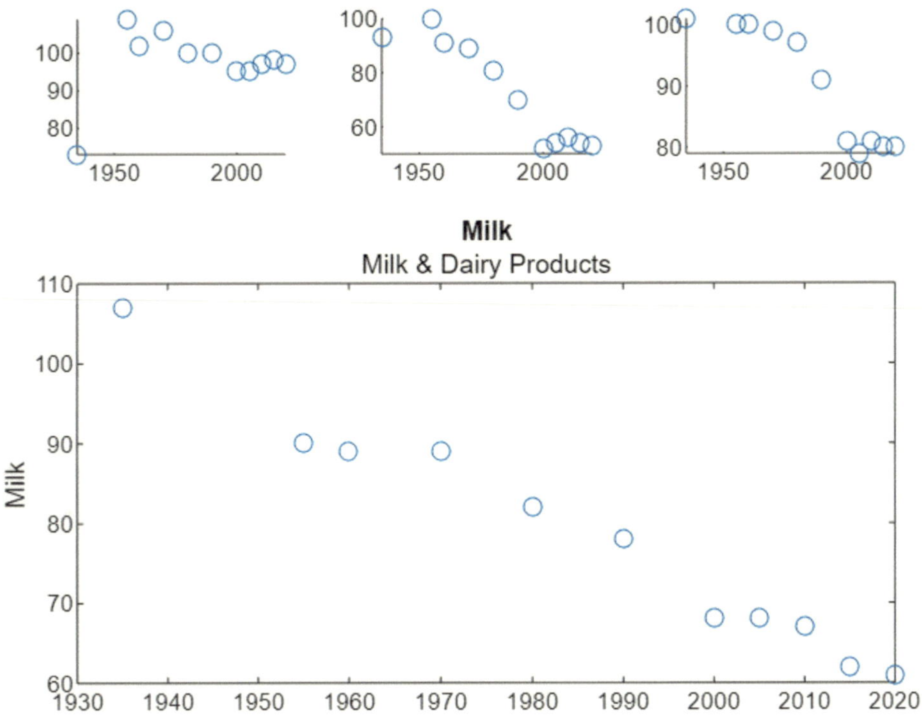

◘ **Fig. 3.88** Scatter plot (3-row, 3-column layout with a large framed bottom plot)

The plot shown in ◘ Fig. 3.89 consists of a 3-by-3 tiled layout, created by 「tiledlayout(3,3)」. In the figure, the final plot for 「ax4」 in the previous program (◘ Fig. 3.88) is overwritten by the following program in 「scatter(Year,[Rice,Meat, Vegetables],100, "x")」. See ◘ Fig. 3.89. To create a new space, we need to use the **figure** command. Compare the following program with the program that creates ◘ Fig. 3.90.

```
scatter(Year,[Rice,Meat,Vegetables],100,"x");
```

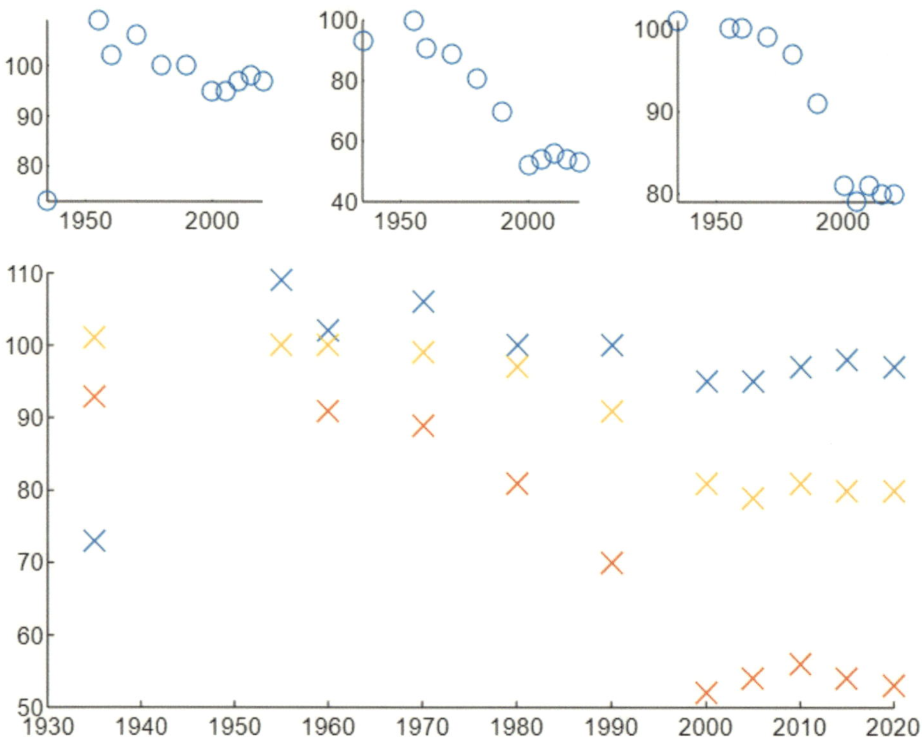

Fig. 3.89 Scatter plot (3-row, 3-column layout with the bottom plot replaced due to not using the figure command)

To reset the layout created by the **tiledlayout** function in Fig. 3.89, we need to use the **figure** command, which creates a space for a new figure. Compare Fig. 3.89 with Fig. 3.90.

```
figure
scatter(Year,[Rice,Meat,Vegetables],100,"x");
```

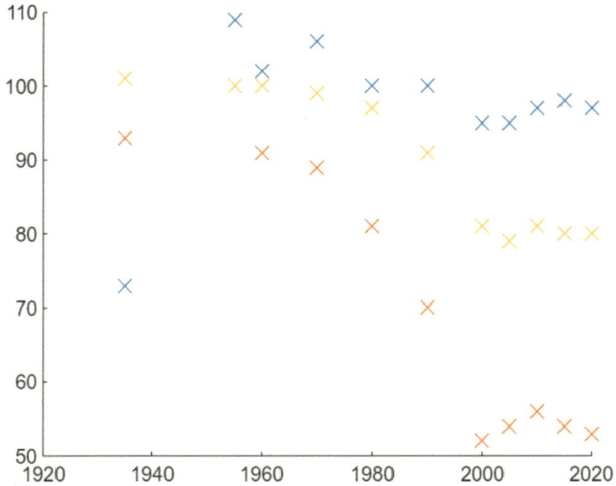

Fig. 3.90 Scatter plot with cross markers after using the figure command

Here, we use 「gca」 to obtain the properties of the axes, which are stored in the variable 「ax」. By using 「ax」, we can change the properties of the current axes. Specifically, to increase the font size of the axes, we set 「FontSize」 to 20. The resulting figure is presented in Fig. 3.91.

```
figure
scatter(Year,[Rice,Meat],50);
ax = gca;
    ax.FontSize = 20;
```

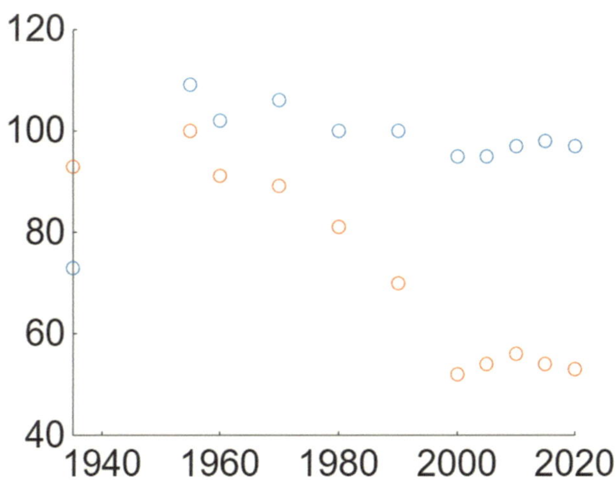

Fig. 3.91 Use of gca: Scatter plot with a large axis font size

Here, we use 「gca」 to change the properties of the ticks. Specifically, we set 「TickDir」 to 「out」 to orient the ticks outward. By default, 「TickDir」 is set to 「in」, which orients the ticks inward. We also set 「FontSize」 to 20, making the ticks larger. See ◘ Fig. 3.92.

```
figure
scatter(Year,[Rice,Meat],50);
ax = gca;
    ax.TickDir = "out";
    ax.FontSize = 20;
```


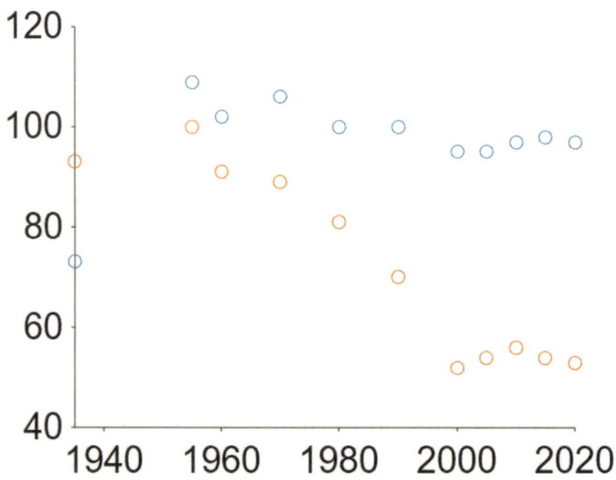

◘ **Fig. 3.92** Use of gca: Scatter plot with outward tick marks

Here, we set 「TickDir」 to 「out」 to orient the ticks outward, and set 「TickLength」 to 「[0.1 0.1]」 to change the length of the ticks. Specifically, note that 「TickLength」 must be specified as a two-element vector 「[2Dlength,3Dlength]」. Since the scatter plot generated by the **scatter** function is 2D, we only change the first argument from the default 0.01 to 0.1, resulting in longer tick marks. The value of the second argument is not related to the length of the 2D figure; however, based on the syntax of 「TickLength」, a two-element vector must be provided, so we set 0.1 as the second argument. The resulting figure is presented in ◘ Fig. 3.93.

```
figure
scatter(Year,[Rice,Meat],50);
ga1 = gca;
    ga1.TickDir = "out";
    ga1.TickLength = [0.1 0.1];
```

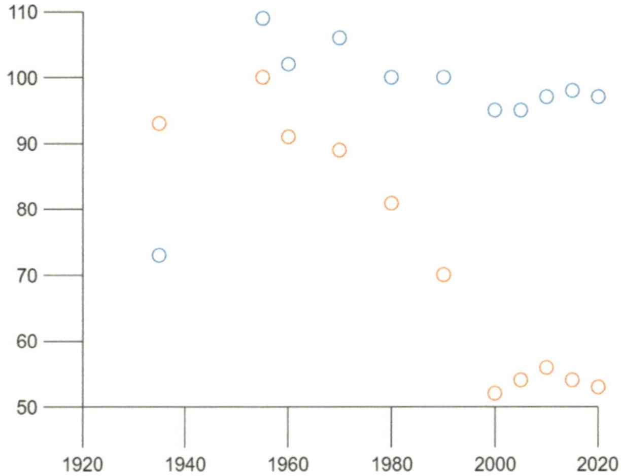

Fig. 3.93 Use of gca: Scatter plot with large outward tick marks

We leave 「TickDir」 at its default setting of 「in」 and adjust 「TickLength」 to change the length of the tick marks. Additionally, we use the **legend** function to create a legend. See Fig. 3.94.

```
figure
scatter(Year,[Rice,Meat],50);
legend("Rice","Meat","Location","southwest");
ga1 = gca;
    ga1.TickLength = [0.1 0.1];
```

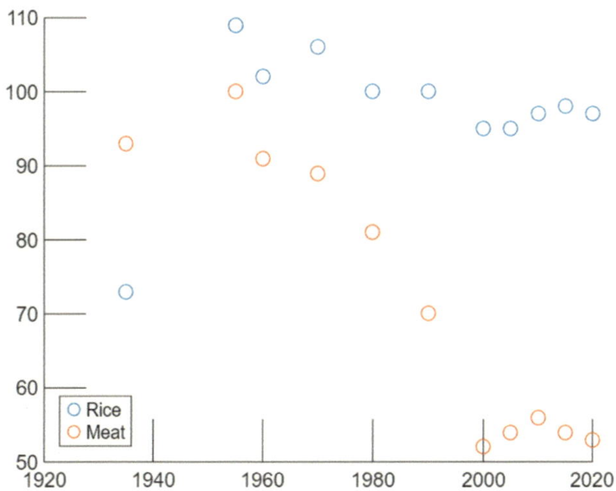

Fig. 3.94 Use of gca: Scatter plot with large inward tick marks and a legend

Here, we use the **box on** command to display the outer borders of the axes. Then, we store the properties of the current axes in the variable 「ax」. Next, we set 「LineWidth」 to 2.5, which makes the outer borders of the axes thicker. See Fig. 3.95.

3

```
figure
scatter(Year,[Rice,Meat],50);
box on
ax = gca;
    ax.LineWidth = 2.5;
```

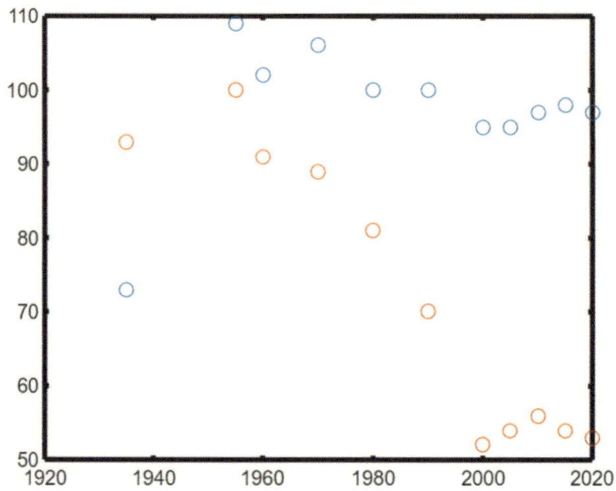

Fig. 3.95 Use of gca: Scatter plot with thick black frame

Here, we use the **box on** command to display the outer borders of the axes. Then, we set 「XColor」 and 「YColor」 to 「red」. These color the x-axis and y-axis, including their tick marks, in red. See Fig. 3.96.

```
figure
scatter(Year,[Rice,Meat],50);
box on
ax = gca;
    ax.XColor = "red";
    ax.YColor = "red";
```

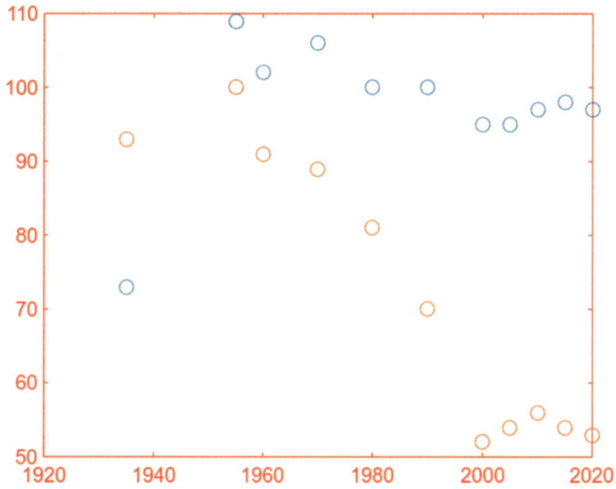

■ **Fig. 3.96** Use of gca: Scatter plot with red frame and ticks

Here, after obtaining the properties of the current axes, we set 「XColor」 to 「red」 and 「YColor」 to 「black」, changing the color of the x-axis to red and the y-axis to black. See ■ Fig. 3.97.

```
figure
scatter(Year,[Rice,Meat],50);
box on
ax = gca;
    ax.XColor = "red";
    ax.YColor = "black";
```

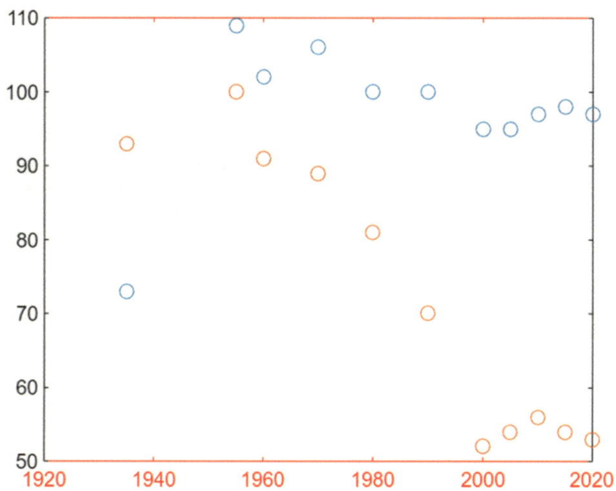

■ **Fig. 3.97** Use of gca: Scatter plot with a red X-axis and a black Y-axis

Here, we use the **xregion** function to create a filled region. Specifically, the range from x=1930 to x=1940 is filled with the default color. Additionally, we use the **legend** function, which displays not only the scatter plot points for 「Rice」, 「Meat」, 「Milk」, and 「Vegetables」 in the created legend but also the filled region labeled as 「data1」. The resulting figure is shown in Fig. 3.98.


```
scatter(Year,[Rice,Meat,Milk,Vegetables]);
legend("Rice","Meat","Milk","Vegetables");
xregion([1930 1940]);
```

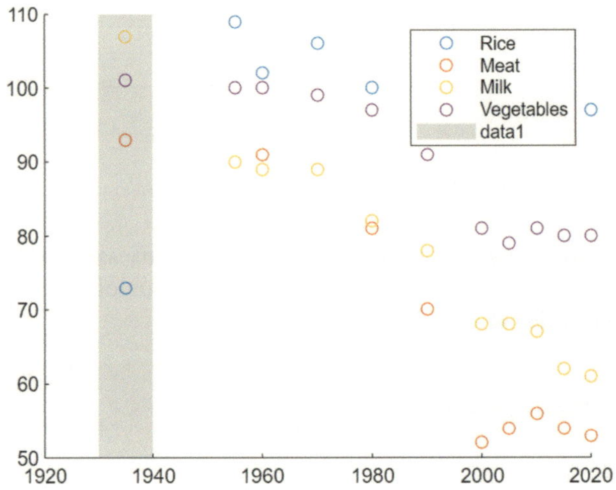

Fig. 3.98 Scatter plot with a vertical gray shaded area and a legend

Here, using the **xregion** function, we fill two regions: the first filled region is from x=1930 to x=1940; the second filled region is from x=1980 to x=2020. The properties of the **xregion** function are stored in the vector 「xr」, represented as a 2-by-1 ConstantRegion. See the Workspace. In particular, 「xr(1)」 and 「xr(2)」 correspond to the first and second filled regions. 「DisplayName」 allows us to specify the text labels of the filled regions in the legend. See Fig. 3.99 for the resulting figure.

```
scatter(Year,[Rice,Meat,Milk,Vegetables]);
legend("Rice","Meat","Milk","Vegetables");
xr = xregion([1930 1980],[1940 2020]);
    xr(1).DisplayName ="1930-1940";
    xr(2).DisplayName ="1980-2020";
```

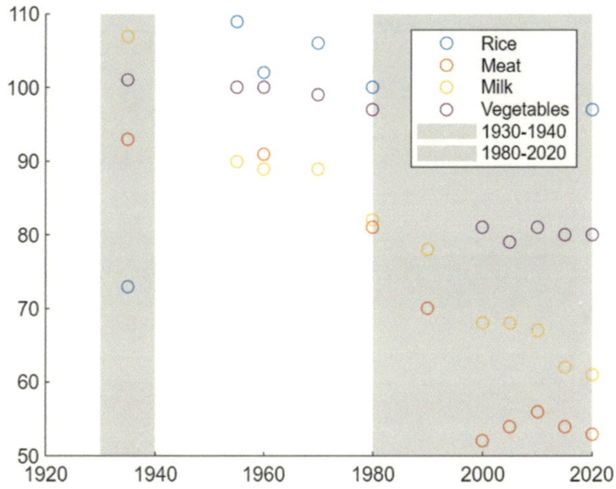

Fig. 3.99 Scatter plot with two vertical gray shaded areas and a legend

Here, we use the **yregion** function to create three filled regions. Specifically, 「yr(1)」, 「yr(2)」, and 「yr(3)」 correspond to filled regions from y=50 to y=60, from y=70 to y=80, and from y=90 to y=100, respectively.

For 「yr(1)」, we set 「FaceColor」 to 「red」 and 「EdgeColor」 to 「black」, ma king the red filled region with a black border. For 「yr(2)」, we set 「FaceColor」 to 「green」, making the filled region green. For 「yr(3)」, we set 「LineStyle」 to 「--」, 「LineWidth」 to 2, 「FaceColor」 to 「yellow」, and 「EdgeColor」 to 「blue」, res ulting in the yellow filled region with a thicker blue dashed border. See Fig. 3.100.

```
scatter(Year,[Rice,Meat,Milk,Vegetables]);
legend("Rice","Meat","Milk","Vegetables","NumColumns",4);
yr = yregion([50 70 90],[60 80 100]);
    yr(1).FaceColor = "red";
    yr(1).EdgeColor = "black";
    yr(1).DisplayName = "Low";
    yr(2).FaceColor = "green";
    yr(2).DisplayName = "Middle";
    yr(3).FaceColor = "yellow";
    yr(3).EdgeColor = "blue";
    yr(3).LineStyle = "--";
    yr(3).LineWidth = 2;
    yr(3).DisplayName = "High";
```

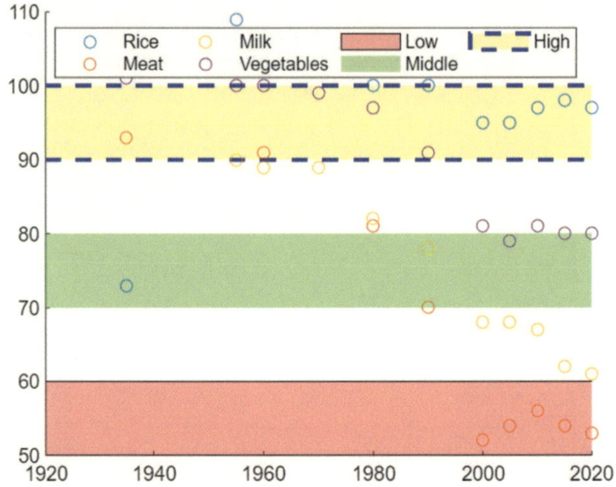

Fig. 3.100 Scatter plot with various horizontal shaded areas and a legend

Here, we use the **xregion** and **yregion** functions simultaneously. Although the colors are the default, the overlapping area of the two filled regions appears darker due to the layering of the fills. See Fig. 3.101 for the created figure.

```
scatter(Year,[Rice,Meat,Milk,Vegetables]);
xregion([1930  1960]);
yregion([50 80]);
```

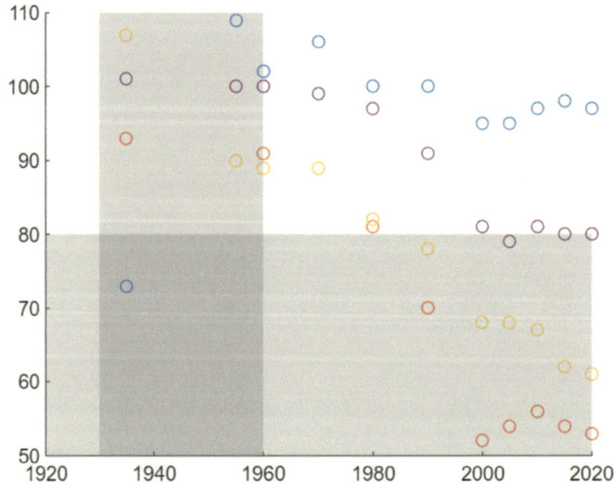

Fig. 3.101 Scatter plot with horizontal and vertical shaded areas

Here, we use the **xline** function to draw a line at x=1950. See Fig. 3.102. For more details on the **xline** function, see ► Sect. 18.3.

```
scatter(Year,[Rice,Meat,Milk,Vegetables]);
xline(1950);
```

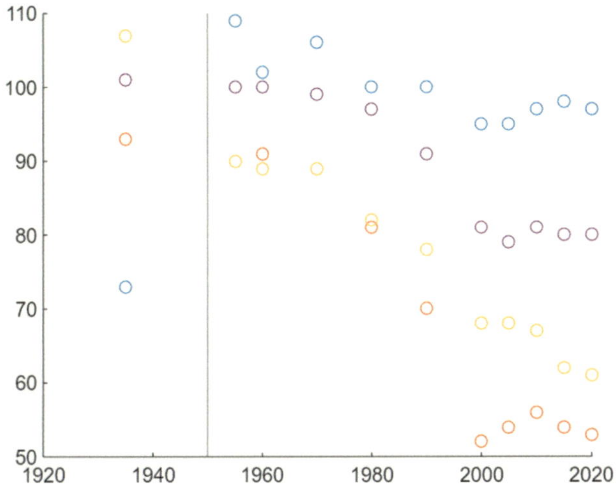

Fig. 3.102 Scatter plot with a vertical line

Here, we use the **yline** function to draw a dashed line with a width of 3 points at y=70 and a red dash-dotted line at y=100. We use 「LabelHorizontalAlignment」 and 「LabelVerticalAlignment」 to adjust the horizontal and vertical alignment of the label. For y=70, the text 「Low level」 is centered below the dashed line at y=70. Meanwhile, for y=100, the text 「High level」 is placed above the dash-dotted line at y=100 and aligned to the left. For more details on the **yline** function, see ▶ Sect. 18.3. The resulting figure is shown in Fig. 3.103.

```
scatter(Year,[Rice,Meat,Milk,Vegetables]);
yl1 = yline(70,"--",{"Low level"});
    yl1.LabelHorizontalAlignment = "center";
    yl1.LabelVerticalAlignment = "bottom";
    yl1.LineWidth = 3;
yl2 = yline(100,"-.",{"High level"});
    yl2.LabelHorizontalAlignment = "left";
    yl2.LabelVerticalAlignment = "top";
    yl2.Color = "red";
```

3

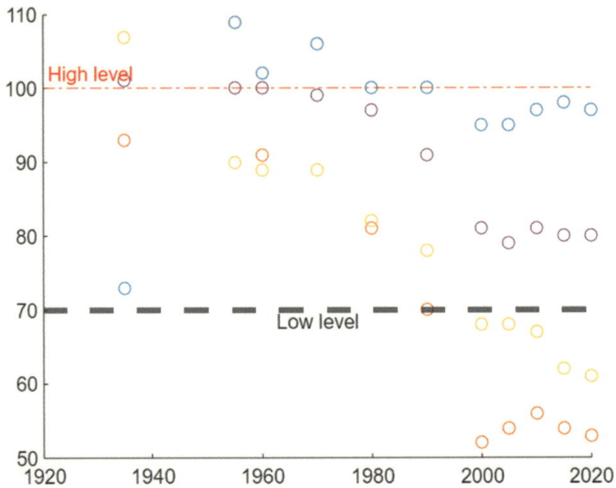

Fig. 3.103 Scatter plot with labeled horizontal black dashed and horizontal red dash-dotted lines

Here, we use the **yline** and **yregion** functions simultaneously. Specifically, the **yregion** function is used to create a filled region from y=70 to y=100. Additionally, the **yline** function is used to draw a dashed line with a width of 3 points at y=70. The text 「Low level」, with a larger font size, is centered below the dashed line. Next, another dashed line is drawn at y=100 with a width of 3. The text 「High level」, also with a larger font size, is placed above the dashed line and aligned to the right. For more details on the **yline** function, see ▶ Sect. 18.3. The resulting figure is presented in Fig. 3.104.

```
scatter(Year,[Rice,Meat,Milk,Vegetables]);
yl1 = yline(70,"--",{"Low level"});
    yl1.LabelHorizontalAlignment = "center";
    yl1.LabelVerticalAlignment = "bottom";
    yl1.LineWidth = 3;
    yl1.FontSize = 20;
yl2 = yline(100,"--",{"High level"});
    yl2.LabelHorizontalAlignment = "right";
    yl2.LabelVerticalAlignment = "top";
    yl2.LineWidth = 3;
    yl2.FontSize = 20;
yregion([70 100]);
```

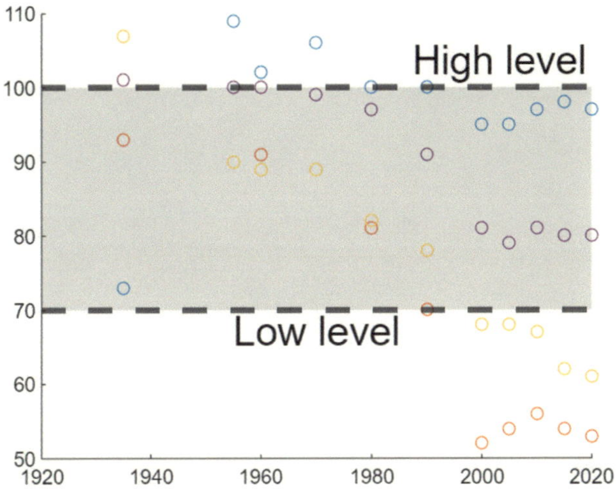

Fig. 3.104 Scatter plot with a shaded area and two horizontal dashed lines labeled in a large font

Here, we use the **yregion** and **annotation** functions simultaneously. Specifically, the **yregion** function is used to create a filled region from y=70 to y=100 with a thicker dotted border line. Next, using the **annotation** function, we create two arrows with text labels (「High level」 and 「Low level」) using 「TextArrow」. For more details on the **annotation** function, see ▶ Sect. 18.1. See ◘ Fig. 3.105.

```
scatter(Year,[Rice,Meat,Milk,Vegetables]);
yr = yregion([70 100]);
    yr.EdgeColor = "black";
    yr.LineWidth = 2;
    yr.LineStyle = ":";
x1 = [0.75 0.5];
y1 = [0.9 0.8];
annotation("TextArrow",x1,y1,"String","High level");
x2 = [0.25 0.5];
y2 = [0.27 0.37];
ta = annotation("TextArrow",x2,y2,"String","Low level");
    ta.Color = "red";
```

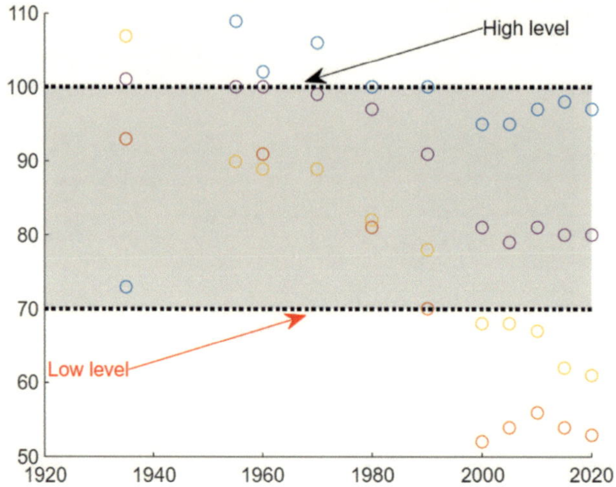

Fig. 3.105 Scatter plot with a shaded area, two horizontal dotted lines, and two arrows

Here, we use the **yregion** function to create a filled region. Within this region, we use the **annotation** function, specifying the first argument as 「TextBox」 to create a rectangular box containing text. As confirmed in Fig. 3.106, note that the box does not fit the text perfectly. For more details on the **annotation** function, see ▶ Sect. 18.1.

```
scatter(Year,[Rice,Meat,Milk,Vegetables]);
yr = yregion([70 100]);
    yr.EdgeColor = "blue";
    yr.LineWidth = 2;
    yr.LineStyle = "--";
    yr.FaceColor = "yellow";
dim = [0.5 0.6 0.15 0.15];
annotation("TextBox",dim,"String","Middle range");
```

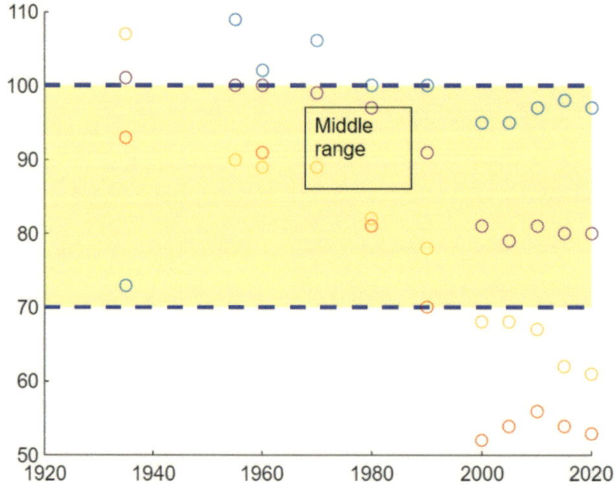

Fig. 3.106 Scatter plot with a shaded area, two horizontal dashed lines, and a textbox

Here, the program is the same as the previous program that created ◘ Fig. 3.106, except for one point. The difference is that in the final **annotation** function, 「FitBoxToText」 is set to 「on」. As a result, the box containing the text is automatically adjusted to fit the text. Compare ◘ Fig. 3.106 with ◘ Fig. 3.107. For more details on the **annotation** function, see ► Sect. 18.1.

```
scatter(Year,[Rice,Meat,Milk,Vegetables]);
yr = yregion([70 100]);
    yr.EdgeColor = "blue";
    yr.LineWidth = 2;
    yr.LineStyle = "--";
    yr.FaceColor = "yellow";
dim = [0.5 0.6 0.15 0.15];
annotation("TextBox",dim,"String","Middle range","FitBoxToText","on");
```

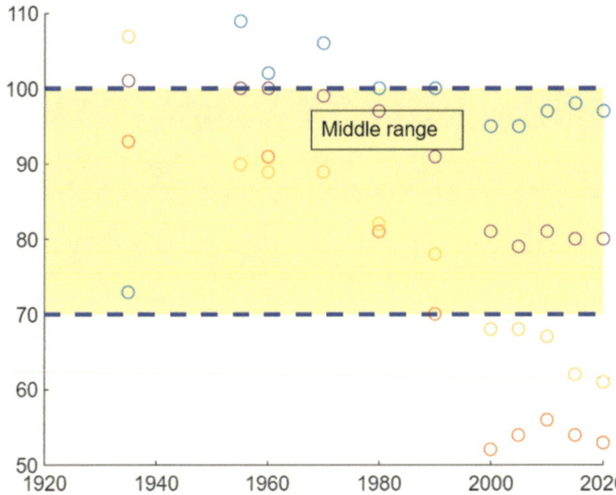

◘ **Fig. 3.107** Scatter plot with a shaded area, two horizontal dashed lines, and a fitted textbox

Here, we use the **scatter** function to create points. Then, we use the **polyfit** function to create a polynomial approximation curve. According to the MAT LAB documentation, the basic syntax is 「p=polyfit(x,y,n)」, which fits an n-th degree polynomial to the data (x,y) and returns the coefficients 「p」.

Here, we use 「p=polyfit(Year,Vegetables,3)」 to obtain the coefficients 「p」 for a cubic polynomial for the data (「Year」, 「Vegetables」). Using the **polyval** function, we calculate 「y_p」 for each corresponding point in 「x_p」 based on the obtained coefficients 「p」. Finally, using 「plot(x_p,y_p)」, we plot the cubic approximation curve. For more details of the **plot** function, see ► Chap. 5. The resulting figure is shown in ◘ Fig. 3.108.

3

```
scatter(Year,Vegetables);
p = polyfit(Year,Vegetables,3);
```

```
Warning: Polynomial is badly conditioned. Add points with distinct X values, reduce the degree of the
polynomial, or try centering and scaling as described in HELP POLYFIT.
```

```
x_p = linspace(min(Year),max(Year));
y_p = polyval(p,x_p);
hold on
plot(x_p,y_p,"LineWidth",2);
hold off
```

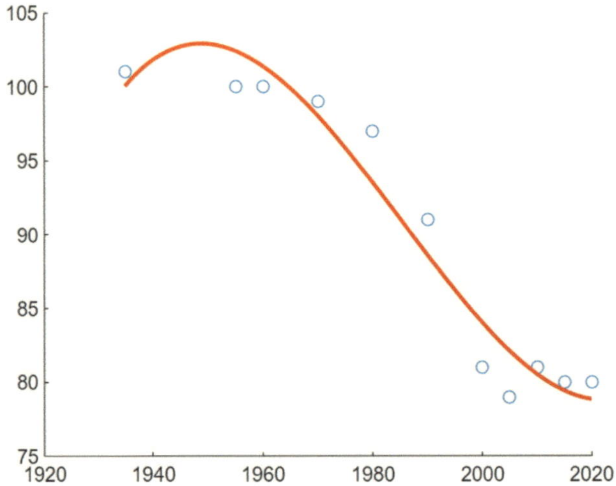

Fig. 3.108 Scatter plot with a polynomial approximation curve

Here, we use the **tiledlayout** function where we display polynomial approximation curves. The first-order approximation is shown in the 1-by-1 tile, the second-order approximation is shown in the 1-by-2 tile, the third-order approximation is shown in the 2-by-1 tile, and the fourth-order approximation is shown in the 2-by-2 tile. See ▶ Chap. 5 for more details on the **plot** function. See Fig. 3.109 for the created figure.

```
figure
tiledlayout(2,2);
x_p = linspace(min(Year),max(Year));
nexttile(1);
scatter(Year,Vegetables);
p_v = polyfit(Year,Vegetables,1);
y_v = polyval(p_v,x_p);
hold on
plot(x_p,y_v,"LineWidth",2);
hold off
title("Vegetables");
grid on
nexttile(2);
scatter(Year,Meat);
p_meat = polyfit(Year,Meat,2);
y_meat = polyval(p_meat,x_p);
hold on
plot(x_p,y_meat,"LineWidth",2,"Color","green");
hold off
title("Meat");
grid on
nexttile(3);
scatter(Year,Milk);
p_milk = polyfit(Year,Milk,3);
```

```
Warning: Polynomial is badly conditioned. Add points with distinct X values, reduce the degree of the
polynomial, or try centering and scaling as described in HELP POLYFIT.
```

```
y_milk = polyval(p_milk,x_p);
hold on
plot(x_p,y_milk,"LineWidth",2,"Color","black");
hold off
title("Milk");
grid on
nexttile(4);
scatter(Year,Rice);
p_r = polyfit(Year,Rice,4);
```

```
Warning: Polynomial is badly conditioned. Add points with distinct X values, reduce the degree of the
polynomial, or try centering and scaling as described in HELP POLYFIT.
```

```
y_r = polyval(p_r,x_p);
hold on
plot(x_p,y_r,"--c","LineWidth",1);
hold off
title("Rice");
grid on
```

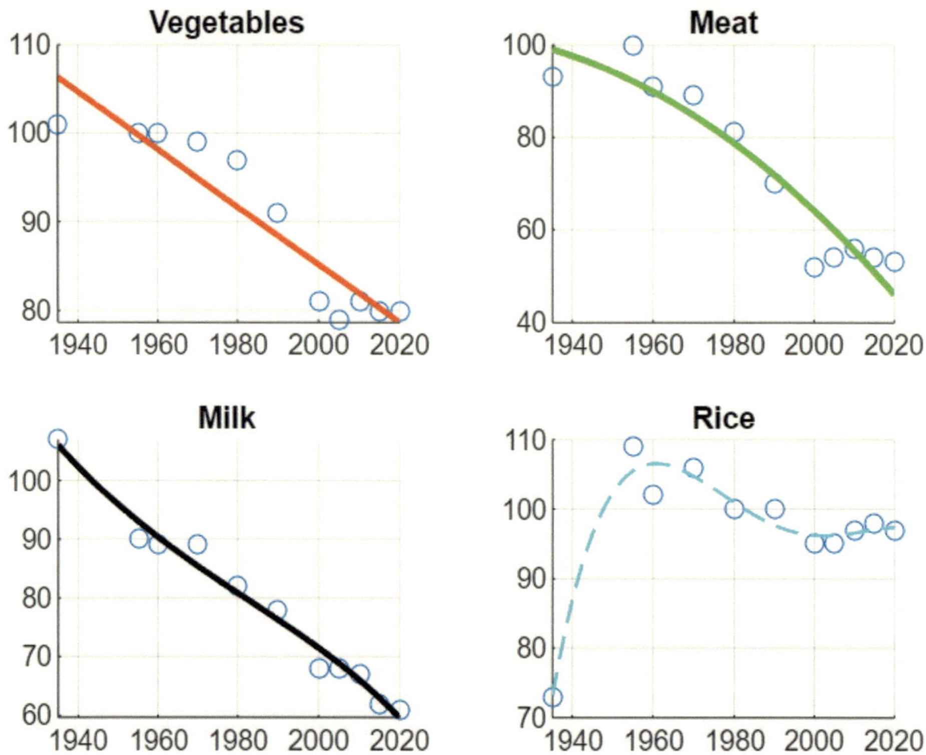

Fig. 3.109 Scatter plot with polynomial approximation curves (2-row, 2-column layout)

3.6 Questions

Question 1

Create Fig. 3.110 using the **scatter** function, to display a scatter plot of rice (「Rice」) and vegetables (「Vegetables」), with the point size set to 20, where the x-axis uses the year (「Year」).

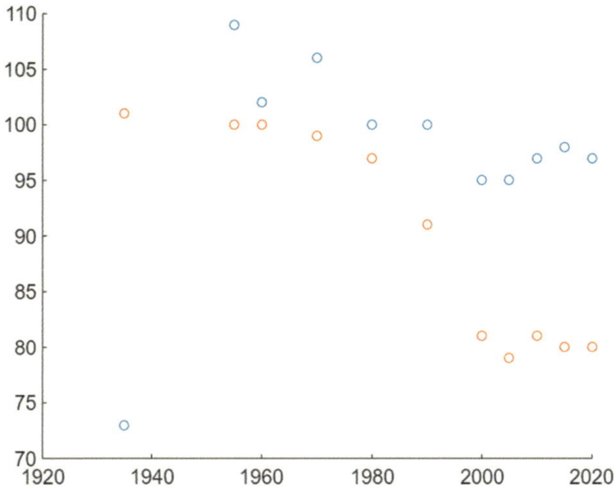

Fig. 3.110 Scatter plot (Question 1)

Answer 1

```
figure
scatter(Year,[Rice,Vegetables],20);
```

Question 2

Create Fig. 3.111 by filling in the inside of the points in the figure created in Question 1.

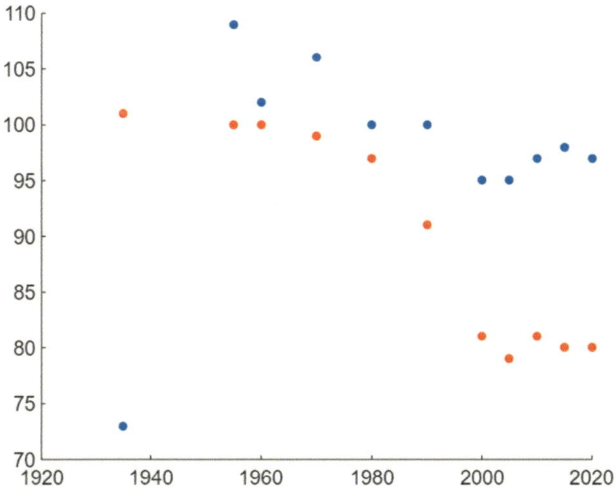

Fig. 3.111 Scatter plot (Question 2)

Answer 2

```
figure
scatter(Year,[Rice,Vegetables],20,"filled");
```

Question 3

Create Fig. 3.112 using 「tiledlayout(3,1)」. Use the variable 「Vegetables」 for the y-axis in the 1-by-1 tile, the variable 「Rice」 for the y-axis in the 2-by-1 tile, and the variable 「Milk」 for the y-axis in the 3-by-1 tile.

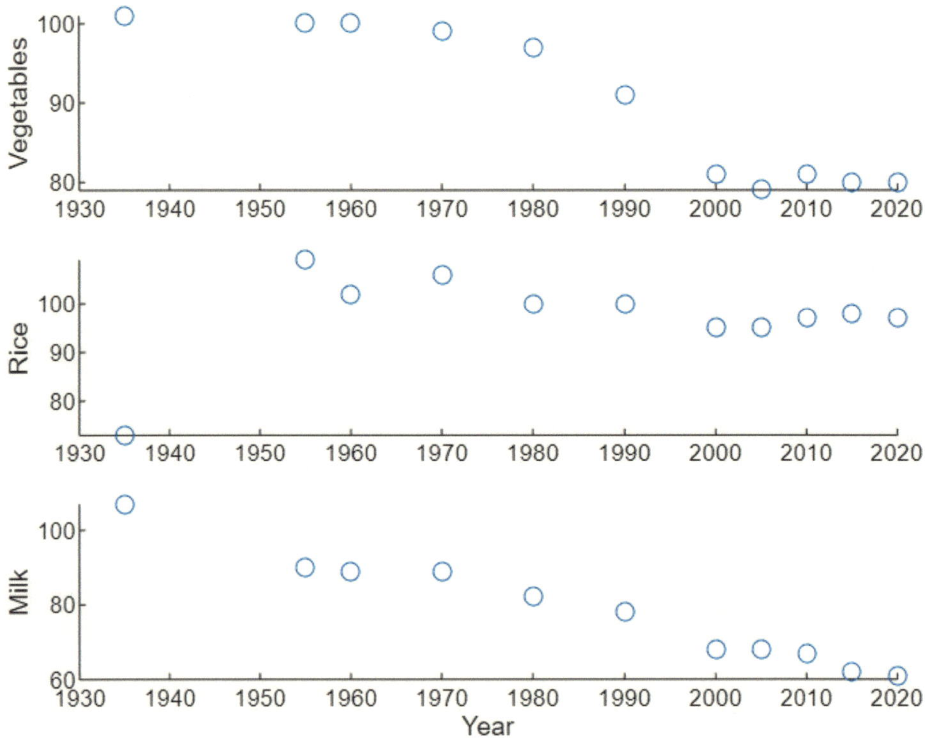

Fig. 3.112 Scatter plot (Question 3)

Answer 3

```
figure
tiledlayout(3,1);
nexttile;
scatter(Year,Vegetables);
ylabel("Vegetables");
nexttile;
scatter(Year,Rice);
ylabel("Rice");
nexttile;
scatter(Year,Milk);
ylabel("Milk");
xlabel("Year");
```

Scatter3 Function

Contents

4.1 Details of the Data Used in this Chapter – 98

4.2 Excel Data of ► Section 4.1 – 99

4.3 Basic Syntax of the Scatter3 Function – 100

4.4 Setting and Controlling the Scatter3 Function – 105

4.5 Combining Optional Functions and the Scatter3 Function – 111

4.6 Questions – 127

Supplementary Information The online version contains supplementary material available at https://doi.org/10.1007/978-981-96-6213-5_4.

Y. Nakamoto, *A Comprehensive Guide to Creating Graphics with MATLAB Programming*, Classroom Companion: Economics, https://doi.org/10.1007/978-981-96-6213-5_4

Learning Objectives

This chapter creates 3D scatter plots using the **scatter3** function. By reading this chapter, you will learn how to use the **scatter3** function and understand the similarities and differences in the created figures, syntax, and the type of data between the **scatter3** function and the **scatter** function (covered in ► Chap. 3).

Overview

The **scatter3** function is used to create a scatterplot in 3D space. Unlike the **scatter** function, which visualizes data in 2D, the **scatter3** function positions data points along three axes (x-axis, y-axis, and z-axis) to create a 3D figure. An advantage of 3D plots is to represent complex relationships between three variables simultaneously, potentially providing deeper insight into data patterns that might be missed in a 2D plot. However, a disadvantage of 3D plots is that the addition of a third axis can make it more difficult to fully understand the relationships between variables. In this case, the optional function introduced in ► Sect. 4.5 may be helpful.

4.1 Details of the Data Used in this Chapter

Here, we use labor-related data from the G7, ASEAN, and BRICS countries. The variable 「wage_2022」 represents the percentage of wage and salaried workers as a share of total employment in 2022. That is, it shows the proportion of employed individuals who work under formal employment contracts and receive regular remuneration. Next, 「young_employment_2022」 represents the share of employed individuals aged 15 to 24 in 2022. Lastly, 「unemployment_2022」 represents the unemployment rate, expressed as a percentage of the unemployed in the total labor force in 2022. The variable 「Country」 represents the name of the country, and the variable 「Group」 indicates whether each country belongs to G7, ASEAN, or BRICS. Finally, a table variable named 「wage_etal」 is created.

```
wage_2022 = [67.87,92.14,47.67,86.43,54.24,91.25,86.91, ...
    46.56,23.13,78.46,90.31,21.14,35.53,75.26,63.94,93.39, ...
    86.38,47.96,84.33,93.72,46.21,71.01]';

young_employment_2022 = [44.73,32.09,57.25,57.94,41.23,...
    50.40,33.29,38.93,23.75,19.83,46.56,45.30,...
    40.16,39.82,31.09,24.96,34.59,37.03,48.64,...
    46.44,41.13,14.96]';

unemployment_2022 = [9.23,5.19,0.25,5.28,4.98,...
    3.14,7.31,3.46,4.82,8.07,2.60,1.21,...
    2.83,3.93,2.60,3.87,3.59,0.94,3.73,...
    3.65,1.54,28.84]';

Country = {"Brazil","Brunei Darussalam","Cambodia","Canada",...
    "China","Germany","France","Indonesia","India","Italy",...
    "Japan","Lao PDR","Myanmar","Malaysia","Philippines",...
    "Russian Federation","Singapore","Thailand","United Kingdom",...
    "United States","Viet Nam","South Africa"}';

Group = {"BRICS","ASEAN","ASEAN","G7","BRICS","G7","G7",...
    "ASEAN","BRICS","G7","G7","ASEAN","ASEAN","ASEAN","ASEAN",...
    "BRICS","ASEAN","ASEAN","G7","G7","ASEAN","BRICS"}';

wage_etal = table(Country,Group,wage_2022,young_employment_2022,...
    unemployment_2022);
```

4.2 Excel Data of ▶ Section 4.1

In ▶ Sect. 4.2, we import the data from an Excel file. First, we use the file named 「alldata.xlsx」 to import the data specifically from the sheet named 「Chap4_wage_etal」. After importing the data into the table variable 「wage_etal」, we convert the columns 「Country」 and 「Group」 to string format, using the **string** function. Then, we rename three columns as follows: 「WageAndSalariedWorkers」, 「EmploymentToPopulationRatio_ages15_24」, and 「Unemployment」 to 「wage_2022」, 「young_employment_2022」, and 「unemployment_2022」, respectively. Finally, we extract individual columns from the table variable 「wage_etal」 and assign them to the corresponding variables: 「Country」, 「Group」, 「wage_2022」, 「young_employment_2022」, and 「unemployment_2022」.

```
wage_etal = readtable("alldata.xlsx","Sheet","Chap4_wage_etal");
```

```
Warning: Column headers from the file were modified to make them valid MATLAB identifiers before creating
variable names for the table. The original column headers are saved in the VariableDescriptions property.
Set 'VariableNamingRule' to 'preserve' to use the original column headers as table variable names.
```

```
wage_etal.Country = string(wage_etal.Country);
wage_etal.Group = string(wage_etal.Group);
wage_etal = renamevars(wage_etal, ["WageAndSalariedWorkers",...
    "EmploymentToPopulationRatio_ages15_24","Unemployment"],...
    ["wage_2022","young_employment_2022","unemployment_2022"]);
Country = wage_etal.Country;
Group = wage_etal.Group;
wage_2022 = wage_etal.wage_2022;
young_employment_2022 = wage_etal.young_employment_2022;
unemployment_2022 = wage_etal.unemployment_2022;
```


4.3 Basic Syntax of the Scatter3 Function

Basic Syntax:
The **scatter3** function in MATLAB is used to create a 3D scatter plot. By typing 「doc scatter3」in the Command Window, we can see the syntaxes of the **scatter3** function, as shown in Figure 4.1. Specifically, the first syntax inside the black box is the basic syntax of the **scatter3** function,「scatter3(X,Y, Z)」. Based on this syntax, three types of data are required:「X」,「Y」, and 「Z」. It is important to note that the variables「X」,「Y」, and 「Z」 must be specified as vectors or matrices of the same size.

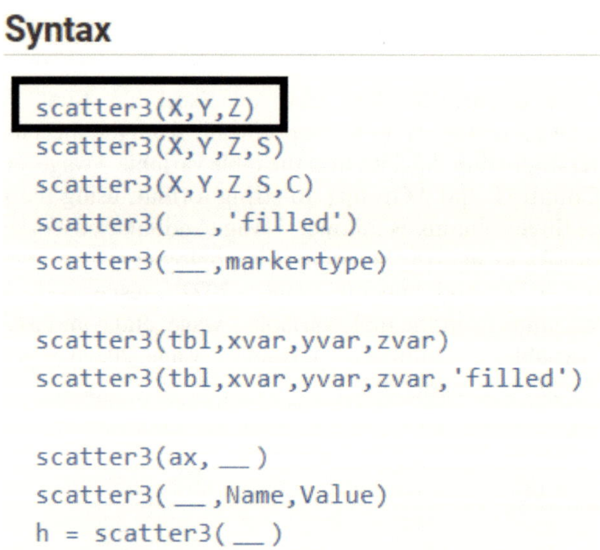

Fig. 4.1 The syntax of the scatter3 function

Here, based on the basic syntax 「scatter3(X,Y,Z)」, we enter a single number for each variable. As a result, the figure generated displays a single point. See Fig. 4.2.

```
scatter3(3,4,5);
```

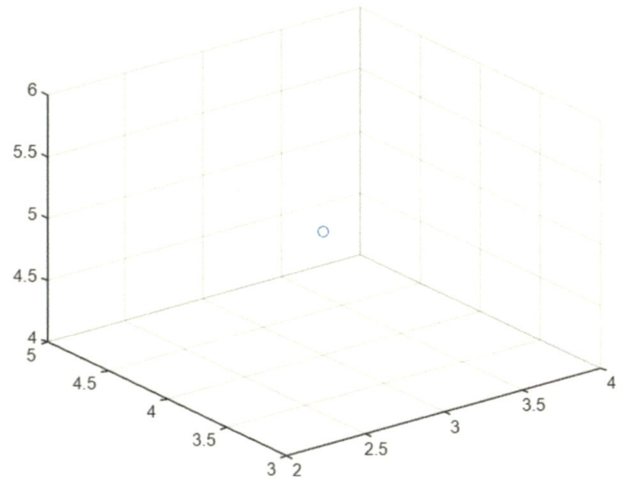

■ **Fig. 4.2** Scatter 3D plot with a single point

Here, we use the same values for the second and third arguments as in the previous program that created ■ Fig. 4.2. Alternatively, in the first argument of the **scatter3** function, we use a 1-by-3 row vector. As a result, three points are created. Specifically, the first point is 「(X,Y,Z)=(5,4,5)」, the second point is 「(X,Y,Z)=(10,4,5)」, and the third point is 「(X,Y,Z)=(15,4,5)」. See ■ Fig. 4.3.

```
scatter3([5,10,15],4,5);
```

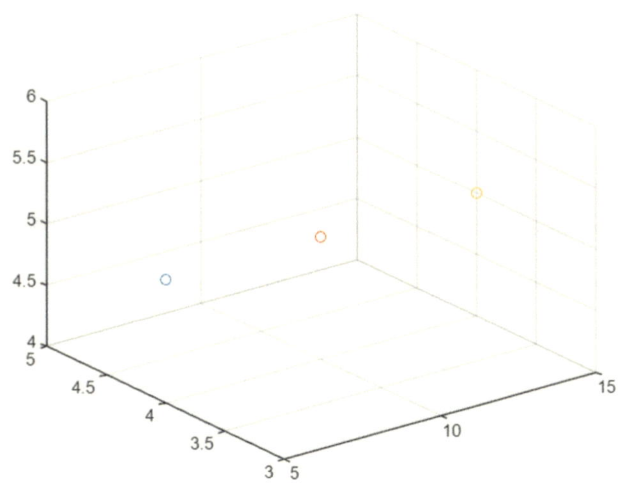

■ **Fig. 4.3** Scatter 3D plot with three points (different colors)

Here, we use column vectors for the first and second arguments, while we use a row vector for the third argument. Since each argument contains only three elements, we can generate a 3D scatter plot using the **scatter3** function. See ■ Fig. 4.4.

```
scatter3([2;3;4],[10;20;30],[50,30,10]);
```

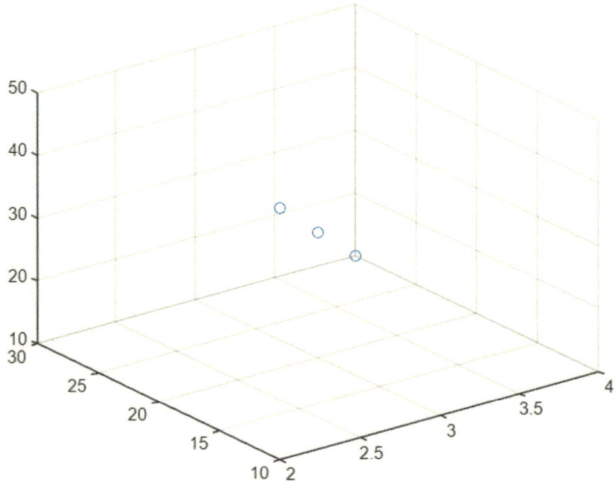

Fig. 4.4 Scatter 3D plot with three points (the same color)

Attention

Error Program: Here, we use row vectors for the first, second, and third arguments. However, since the number of elements in the first and third arguments is two, and the number of elements in the second argument is three, the number of elements is not the same. As a result, an error occurs. See Fig. 4.5 for the concrete error message.

```
%%% Error program
% scatter3([1,2],[3,4,5],[1,2]);
```

```
Error using scatter3 (line 58)
X, Y, Z, and size arguments must be vectors of the same length, matrices of the same size, or a mix of vectors and
matrices where the length of the vectors matches either the number of rows or columns of the matrices.
```

Fig. 4.5 Error output of the scatter3 function: Different length vectors

Here, we use the variable 「unemployment_2022」 in the first argument of the **scatter3** function, 「wage_2022」 in the second argument, and 「young_employment_2022」 in the third argument. All of these variables are 22-by-1 column vectors. As a result, a 3D scatter plot is created. See Fig. 4.6.

```
scatter3(unemployment_2022,wage_2022,young_employment_2022);
```

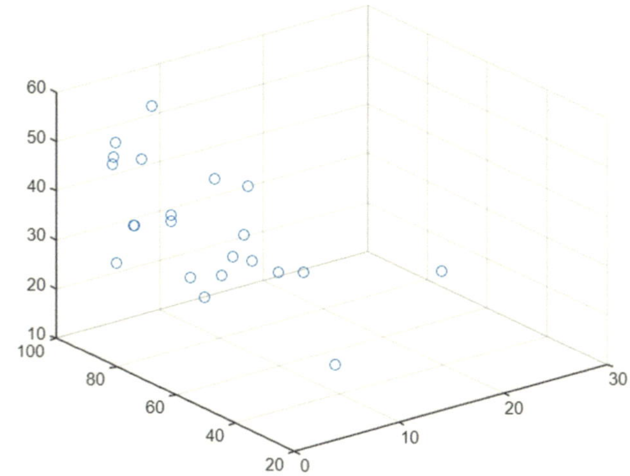

Fig. 4.6 Scatter 3D plot from three variables

Here, we use the same variables as those used to create Fig. 4.6. However, in Fig. 4.6, the first argument was 「unemployment_2022」 and the second argument was 「wage_2022」, but in the following program, we swap the first and second arguments. As a result, changing the order of the arguments creates a figure with a different point of view. The resulting figure is presented in Fig. 4.7.

```
scatter3(wage_2022,unemployment_2022,young_employment_2022);
```

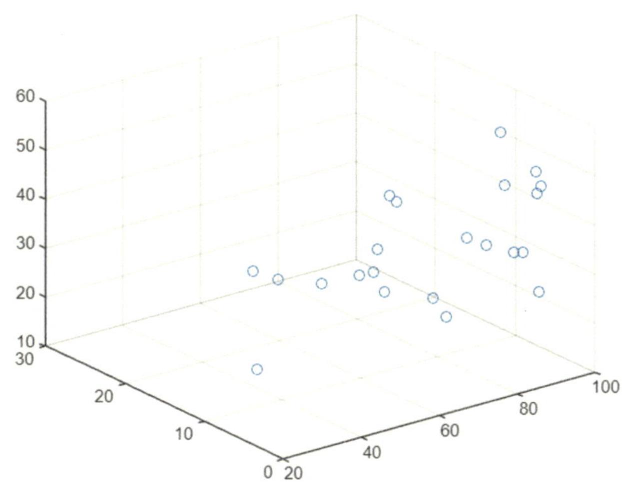

Fig. 4.7 Scatter 3D plot from reordered variables (cf. Fig. 4.6)

As long as the number of elements is the same, you can create a 3D scatter plot using a portion of the vectors. The following program uses the first eleven data points from each vector. See Fig. 4.8.

```
scatter3(unemployment_2022(1:11),wage_2022(1:11),young_employment_2022(1:11));
```

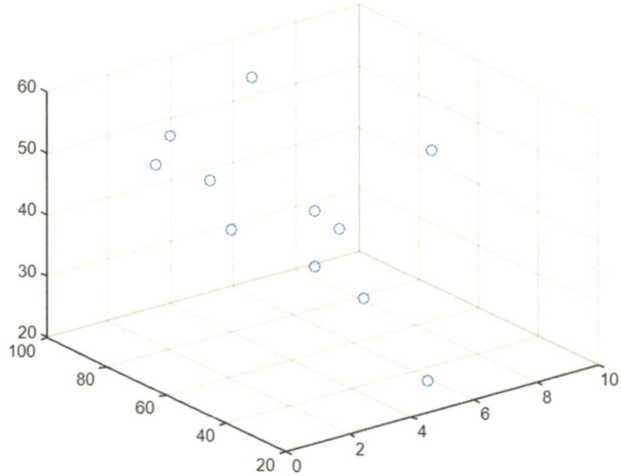

Fig. 4.8 Scatter 3D plot using a subset of variables (cf. Fig. 4.6)

Here, each of the variables 「unemployment_mat」, 「wage_mat」, and 「young_mat」 is an 11-by-2 matrix. In this case, the data in the first column is used as the first group, and the data in the second column is used as the second group, to create a 3D scatter plot. The result consists of two types of scatter plots: blue and red. See Fig. 4.9.

```
unemployment_mat = [unemployment_2022(1:11),unemployment_2022(12:22)];
wage_mat = [wage_2022(1:11),wage_2022(12:22)];
young_mat = [young_employment_2022(1:11),young_employment_2022(12:22)];
scatter3(unemployment_mat,wage_mat,young_mat);
```

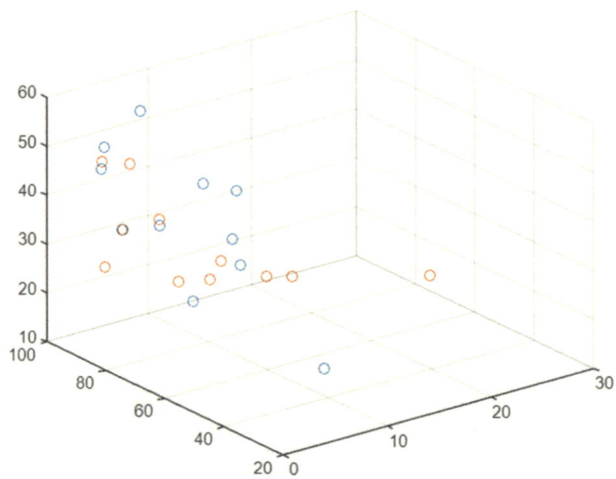

Fig. 4.9 Scatter 3D plot using two subsets of variables (cf. Fig. 4.8)

Here, 「unemployment_mat」, 「wage_mat」, and 「young_mat」 are 5-by-4 matrices. In this case, we can observe scatter plots of four groups, each represented by a different color. See Fig. 4.10.

```
unemployment_mat = [unemployment_2022(1:5),unemployment_2022(6:10),...
    unemployment_2022(11:15),unemployment_2022(16:20)];
wage_mat = [wage_2022(1:5),wage_2022(6:10),...
    wage_2022(11:15),wage_2022(16:20)];
young_mat = [young_employment_2022(1:5),young_employment_2022(6:10),...
    young_employment_2022(11:15),young_employment_2022(16:20)];
scatter3(unemployment_mat,wage_mat,young_mat);
```

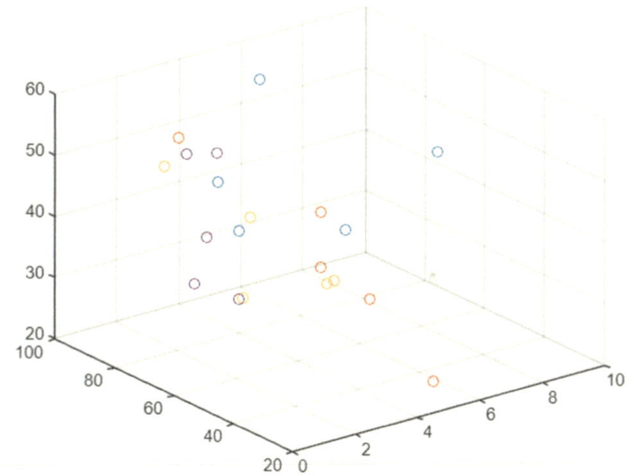

Fig. 4.10 Scatter 3D plot using multiple subsets of variables (cf. Fig. 4.8)

4.4 Setting and Controlling the Scatter3 Function

Here, we set the fourth argument to 1, specifying the size of the scatter circles as 1. See Fig. 4.11.

```
scatter3(wage_2022,unemployment_2022,young_employment_2022,1);
```

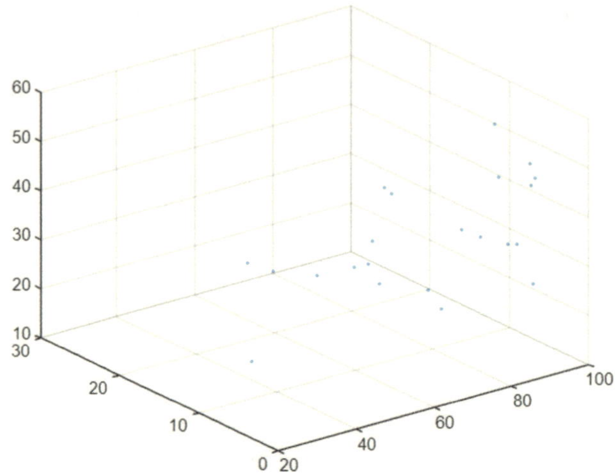

Fig. 4.11 Scatter 3D plot with a tiny marker size

Here, we set the fourth argument to 100, increasing the size of the scatter circles. See Fig. 4.12.

```
scatter3(wage_2022,unemployment_2022,young_employment_2022,100);
```

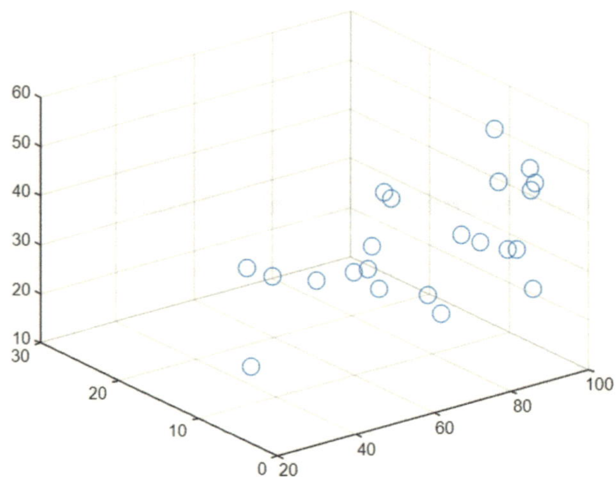

Fig. 4.12 Scatter 3D plot with a large marker size

Here, we set the fourth argument to 100 and the fifth argument to 「red」, creating large red circles. See Fig. 4.13.

```
scatter3(wage_2022,unemployment_2022,young_employment_2022,100,"red");
```

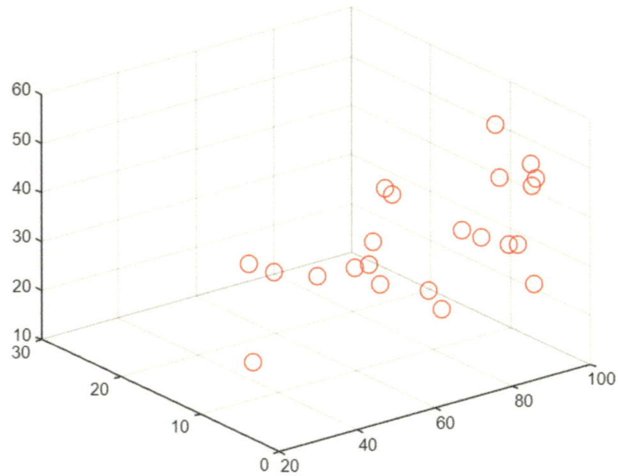

Fig. 4.13 Scatter 3D plot with red markers and a large marker size

Here, by setting the fourth argument to 100, the fifth argument to 「red」, and the sixth argument to 「filled」, we create large scatter circles filled with red. The resulting figure is shown in Fig. 4.14.

```
scatter3(wage_2022,unemployment_2022,young_employment_2022,100,"red","filled");
```

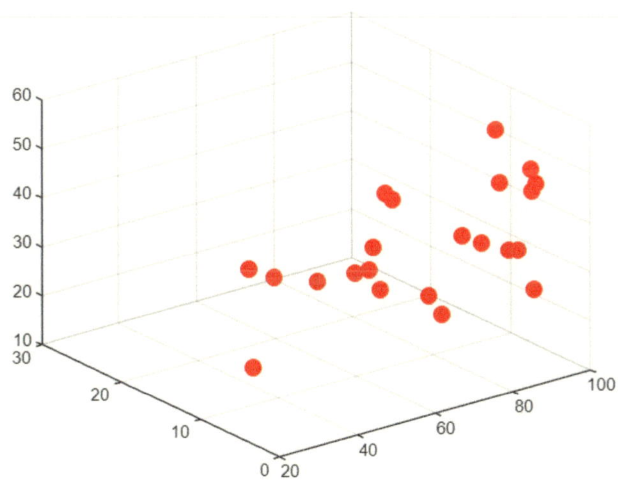

Fig. 4.14 Scatter 3D plot with filled red markers

Here, we set the fourth argument to 100 and the fifth argument to 「*」, creating large asterisks. See Fig. 4.15.

```
scatter3(wage_2022,unemployment_2022,young_employment_2022,100,"*");
```

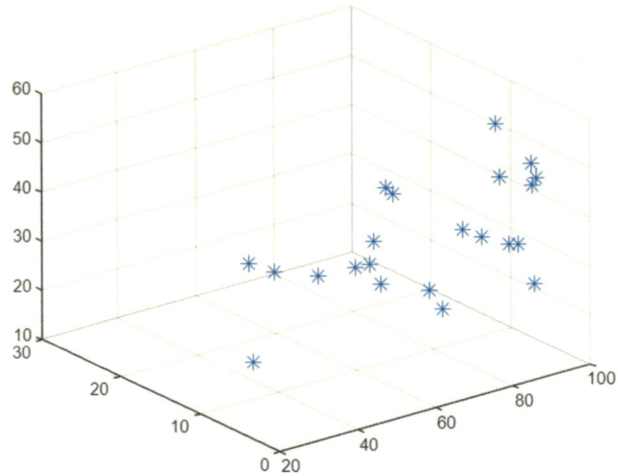

■ **Fig. 4.15** Scatter 3D plot with large asterisk markers

Attention

Error Program: Switching the fourth and fifth arguments in the previous program that created ■ Fig. 4.15 results in an error. See ■ Fig. 4.16 for the error message.

```
%%% Error program
% scatter3(wage_2022,unemployment_2022,young_employment_2022,"*",100);
```

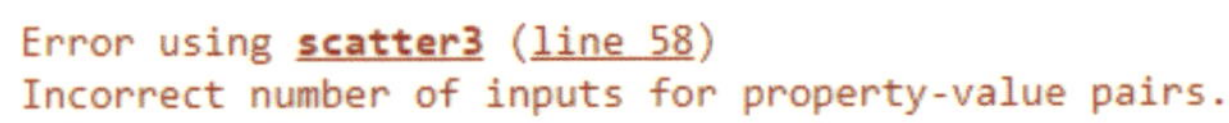

■ **Fig. 4.16** Error output of the scatter3 function: Incorrect order

Here, we store the properties of the **scatter3** function in the variable 「sc3」. Then, we set 「MarkerEdgeColor」 to 「black」, 「LineWidth」 to 2, and 「MarkerFace Color」 to 「magenta」, resulting in circles with a thick black outline and filled with magenta. The resulting figure is presented in ■ Fig. 4.17.

```
sc3 = scatter3(wage_2022,unemployment_2022,young_employment_2022,"filled");
    sc3.MarkerEdgeColor = "black";
    sc3.LineWidth = 2;
    sc3.MarkerFaceColor = "magenta";
```

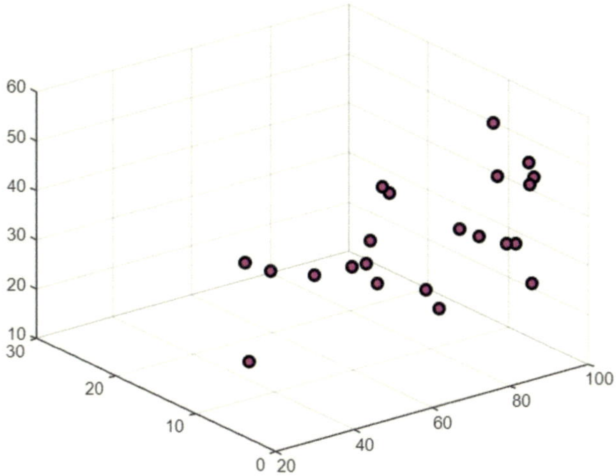

Fig. 4.17 Scatter 3D plot with filled magenta markers and black marker edges

Here, each of 「unemployment_mat」, 「wage_mat」, and 「young_mat」 is an 11-by-2 array. Therefore, using the **scatter3** function creates two types of scatter plots. We store their properties in the variable 「sc3」. We then set 「sc3(1).Marker」 to 「+」 to make the marker for the first group a plus sign, and set 「sc3(2).Marker」 to 「<」 to make the marker for the second group a left-pointing triangle. The resulting figure is presented in Fig. 4.18.

```
unemployment_mat = [unemployment_2022(1:11),unemployment_2022(12:22)];
wage_mat = [wage_2022(1:11),wage_2022(12:22)];
young_mat = [young_employment_2022(1:11),young_employment_2022(12:22)];
sc3 = scatter3(unemployment_mat,wage_mat,young_mat);
    sc3(1).Marker = "+";
    sc3(2).Marker = "<";
```

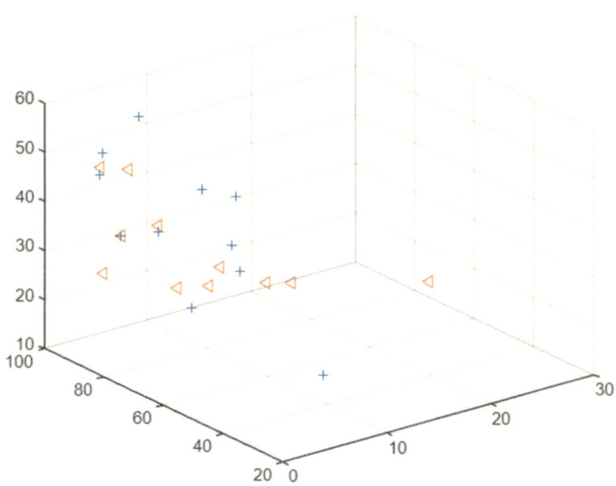

Fig. 4.18 Scatter 3D plot with left-pointing triangle and plus markers

Here, each of「unemployment_mat」,「wage_mat」, and「young_mat」is a 5-by-4 matrix. Therefore, four types of scatter plots are created. We store the properties of the **scatter3** function in「sc3」.

Then, for the first group, we set「Marker」to「+」, making the marker for the first group a plus sign. Next, for the second group, we set「Marker」to「^」,「MarkerFaceColor」to「green」,「MarkerEdgeColor」to「black」, and「LineWidth」to 2, creating an upward-pointing triangle filled with green and outlined with a thick black border. For the third group, we set「Marker」to「pentagram」,「MarkerFaceColor」to「yellow」, and「MarkerEdgeColor」to「black」, creating a pentagram filled with yellow and outlined in black. See ◘ Fig. 4.19.

```
unemployment_mat = [unemployment_2022(1:5),unemployment_2022(6:10),...
    unemployment_2022(11:15),unemployment_2022(16:20)];
wage_mat = [wage_2022(1:5),wage_2022(6:10),...
    wage_2022(11:15),wage_2022(16:20)];
young_mat = [young_employment_2022(1:5),young_employment_2022(6:10),...
    young_employment_2022(11:15),young_employment_2022(16:20)];
sc3 = scatter3(unemployment_mat,wage_mat,young_mat,100);
    sc3(1).Marker = "+";
    sc3(2).Marker = "^";
    sc3(2).MarkerFaceColor = "green";
    sc3(2).MarkerEdgeColor = "black";
    sc3(2).LineWidth = 2;
    sc3(3).Marker = "pentagram";
    sc3(3).MarkerFaceColor = "yellow";
    sc3(3).MarkerEdgeColor = "black";
```

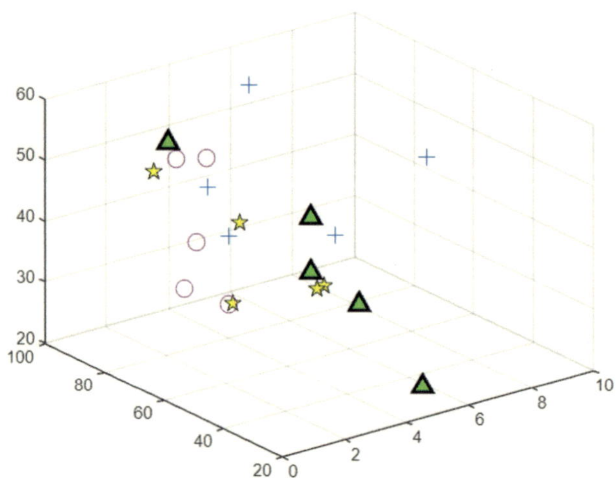

◘ **Fig. 4.19** Scatter 3D plot with various markers

4.5 Combining Optional Functions and the Scatter3 Function

Here, we use the **xlabel**, **ylabel**, and **zlabel** functions. First, we store the properties of the **xlabel** function in 「xl」. Second, we set 「Color」 to 「red」 and 「FontSize」 to 20, creating a large red text (「wage rate」) on the x-axis. See ◘ Fig. 4.20.

```
scatter3(wage_2022,young_employment_2022,unemployment_2022);
xl = xlabel("wage rate");
    xl.Color = "red";
    xl.FontSize = 20;
ylabel("young employment share");
zlabel("unemployment rate");
box on
```

◘ **Fig. 4.20** Scatter 3D plot with X-, Y-, and Z-axis labels

The program differs from the one used to create ◘ Fig. 4.20 in the following point; here, we use 「view([20 60])」 to change the viewing angle of the 3D scatter plot. Compare ◘ Fig. 4.20 with ◘ Fig. 4.21.

```
scatter3(wage_2022,young_employment_2022,unemployment_2022);
view([20 60]);
xl = xlabel("wage rate");
    xl.Color = "red";
    xl.FontSize = 20;
ylabel("young employment share");
zlabel("unemployment rate");
```

Fig. 4.21 Scatter 3D plot from a different angle (cf. Fig. 4.20)

Here, we use 「view([60 20])」 to adjust the viewing angle of the 3D scatter plot. Additionally, we store the axis properties in the variable 「ax」 using 「gca」. We then set 「XColor」 to 「red」, 「YColor」 to 「blue」, 「ZColor」 to 「blue」, and 「LineWidth」 to 2, making each axis thicker and changing the x-axis color to red, while setting the y-axis and z-axis colors to blue. The resulting figure is shown in Fig. 4.22.

```
scatter3(wage_2022,young_employment_2022,unemployment_2022,"black","filled");
view([60 20]);
ax = gca;
    ax.XColor = "red";
    ax.YColor = "blue";
    ax.ZColor = "blue";
    ax.LineWidth = 2;
```

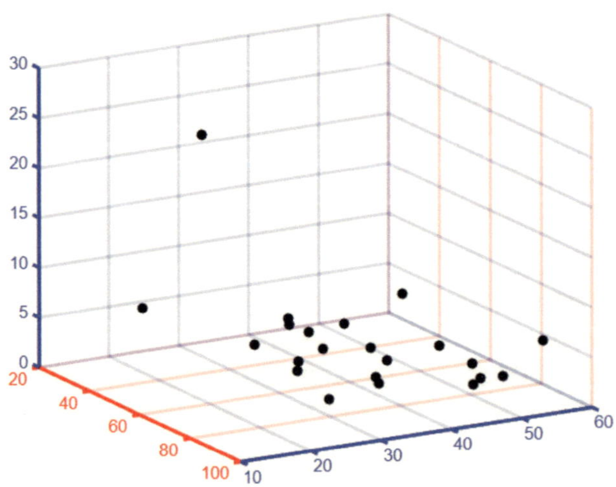

Fig. 4.22 Scatter 3D plot with a red X-axis and blue Y- and Z-axes, including colored gridlines

Here, using the **hold on** command, we overlay a plot created with the **scatter3** function and two plots created with the **plot3** function. Specifically, using the **plot3** function, we connect two of the points.

To do this, we obtain the coordinates of these points, where 「idx1」, 「idx2」, 「idx3」, and 「idx4」 specify the 10th (= 「idx1」), 22nd (= 「idx2」), 1st (= 「idx3 」), and 4th (= 「idx4」) points, and connect them. In particular, we connect the two points at 「idx1=10」 and 「idx2=22」 with a black dotted line and the two points at 「idx3=1」 and 「idx4=4」 with a blue dash-dotted line. See Fig. 4.23.

```
figure
scatter3(wage_2022,young_employment_2022,unemployment_2022,"black");
idx1 = 10;
idx2 = 22;
idx3 = 1;
idx4 = 4;
hold on
plot3([wage_2022(idx1),wage_2022(idx2)],...
      [young_employment_2022(idx1),young_employment_2022(idx2)],...
      [unemployment_2022(idx1),unemployment_2022(idx2)],"k:","LineWidth",1);
plot3([wage_2022(idx3),wage_2022(idx4)],...
      [young_employment_2022(idx3),young_employment_2022(idx4)],...
      [unemployment_2022(idx3),unemployment_2022(idx4)],"b-.","LineWidth",1);
view([80 20]);
hold off
```

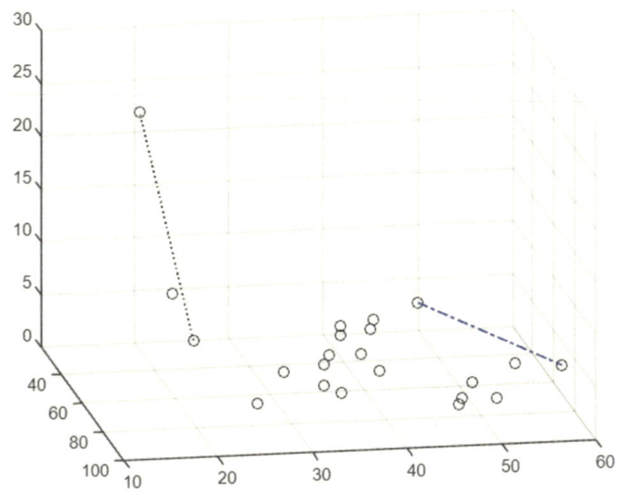

Fig. 4.23 Scatter 3D plot overlaid with dash-dotted and dotted lines using the plot3 function

The fifth argument of the **scatter3** function is set to 「wage_2022」. As a result, 「colorbar」 is generated based on the fifth argument, 「wage_2022」, with corresponding colors and values. For example, if the fifth argument of the **scatter3** function is changed to 「young_employment_2022」, the colorbar will be created based on this variable. See Fig. 4.24.

```
sc3 = scatter3(wage_2022,young_employment_2022,unemployment_2022,100,...
    wage_2022,"filled");
    sc3.MarkerEdgeColor = "black";
colorbar;
```

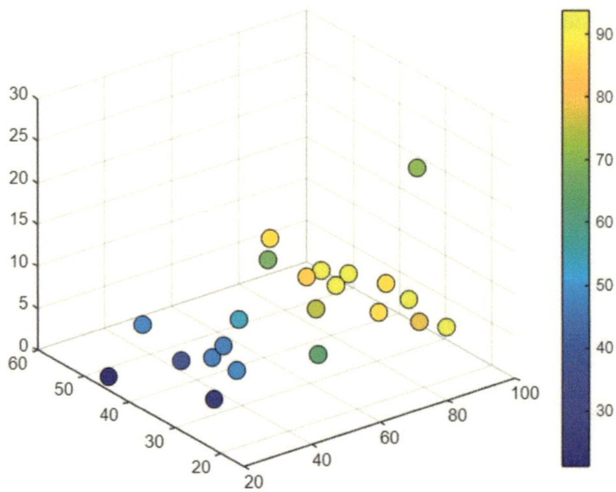

Fig. 4.24 Scatter 3D plot with a colorbar

Here, the fifth argument in the **scatter3** function is set to 「young_employment_2022」. As a result, 「colorbar」 is generated based on the fifth argument, 「young_employment_2022」. Additionally, using the **gca** command, we store the axis properties in the variable 「ax」. We then set 「XColor」 to 「red」, 「YColor」 to 「blue」, and 「ZColor」 to 「green」, changing the colors of the respective axes. Furthermore, the grid lines are removed by setting 「grid off」. The resulting figure is displayed in Fig. 4.25.

```
sc3 = scatter3(wage_2022,young_employment_2022,unemployment_2022,100,...
    young_employment_2022,"filled");
    sc3.MarkerEdgeColor = "black";
colorbar;
ax = gca;
    ax.XColor = "red";
    ax.YColor = "blue";
    ax.ZColor = "green";
    ax.LineWidth = 5;
grid off
```

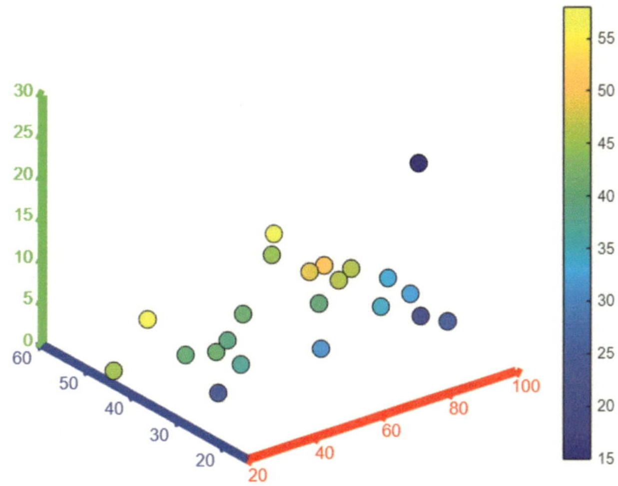

Fig. 4.25 Scatter 3D plot with a colorbar, no grid, and various axis colors

Since the fifth argument in the **scatter3** function is set to 「unemployment_2022」, 「colorbar」 is displayed based on this variable. In addition, we use the **text** function to display the values of 「unemployment_2022」 above each scatter plot point.

Specifically, we use a **for** loop 「for i=1:22」 since there are 22 data points. In the **text** function, the first argument specifies the x-coordinate, the second specifies the y-coordinate, and the third specifies the z-coo-rdinate, which means that 「(X,Y,Z)=(wage_2022(i),young_employment_2022(i),unemployment_2022(i)+1)」. In particular, the third argument is set to 「unem-ployment_2022(i)+1」, to ensure that the text appears slightly above each point by adding 1 to the z-coordinate. This small shift in the z-axis prevents overlap between the text and the scatter points. Since the fourth argument of the **text** function is 「num2str(unemployment_2022(i))」, we convert the numerical value of 「unemployment_2022」 at index 「i」 into a string, so that it can be displayed as text at the specified point 「(X,Y,Z)=(wage_2022(i),young_employment_2022(i),unemployment_2022(i)+1)」.

Finally, by setting 「HorizontalAlignment」 to 「center」, 「FontSize」 to 6, and 「FontWeight」 to 「bold」, the bold text with a font size of 6 points is centered horizontally. For more details on the **text** function, see ▶ Sect. 18.2. The resulting figure is presented in Fig. 4.26.

```
sc3 = scatter3(wage_2022,young_employment_2022,unemployment_2022,50,...
    unemployment_2022,"filled");
    sc3.MarkerEdgeColor = "black";
colorbar;
for i = 1:22
    txt = text(wage_2022(i),young_employment_2022(i),unemployment_2022(i)+1,...
        num2str(unemployment_2022(i)));
        txt.HorizontalAlignment = "center";
        txt.FontSize = 6;
        txt.FontWeight = "bold";
end
```

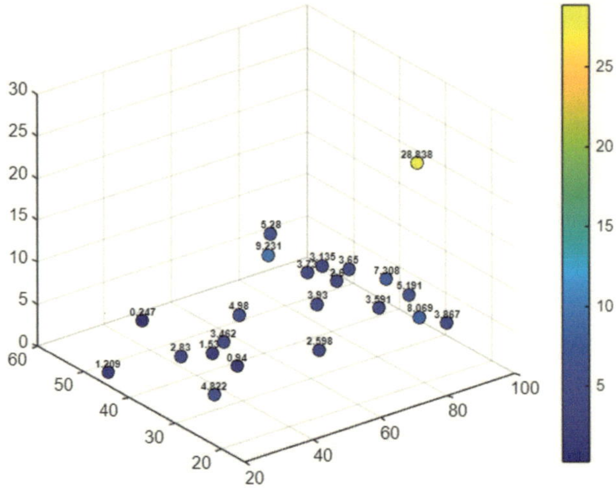

Fig. 4.26 Scatter 3D plot using a for loop to create data labels, with a colorbar

The following program is almost the same as the previous one used to create Fig. 4.26. However, the difference is the addition of 「view([0 90])」 at the end of the following program. This changes the 3D scatter plot to a 2D view. For more details on the **text** function, see ▶ Sect. 18.2. See Fig. 4.27.

```
figure
sc3 = scatter3(wage_2022,young_employment_2022,unemployment_2022,50,...
    unemployment_2022,"filled");
    sc3.MarkerEdgeColor = "black";
colorbar;
for i = 1:22
    txt = text(wage_2022(i),young_employment_2022(i)+2,unemployment_2022(i),...
        num2str(unemployment_2022(i)));
        txt.HorizontalAlignment = "center";
        txt.FontSize = 10;
        txt.FontWeight = "bold";
end
xlabel("wage");
ylabel("young employment");
view([0 90]);
```

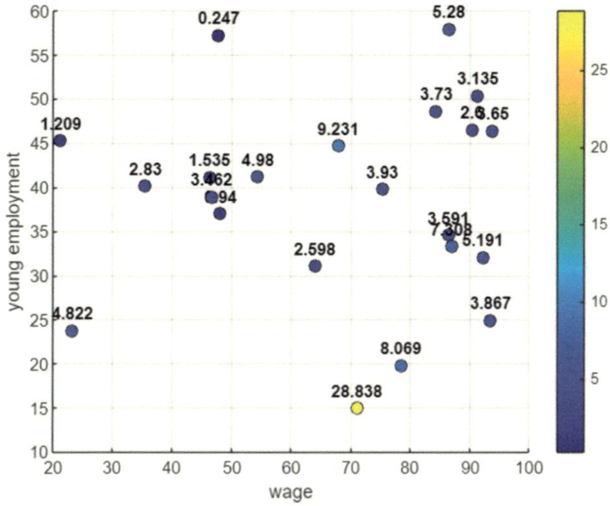

Fig. 4.27 Flat scatter 3D plot using a for loop to create data labels, with a colorbar (cf. **Fig. 4.27**)

Here, using the **max** function, we find the maximum value of the variable 「unemployment_2022」 and store the index of this maximum value in the variable 「pointIndex」. Specifically, we obtain 「pointIndex=22」. Note that we do not use the actual maximum value itself but only the index. Since the value of the maximum itself is not needed, the tilde 「~」 in 「[~, pointIndex]=max(unemployment_2022)」 is used to ignore it.

Finally, an annotation is added to the plot using the **annotation** function. The annotation is of the type 「TextArrow」. The positions for the arrow are specified by the vectors 「x=[0.5 0.62]」 and 「y=[0.8 0.67]」. The text of the annotation represents the maximum value of 「unemployment_2022」, where the **num2str** function is used to convert the numerical value to a string. For more details on the **annotation** function, see ► Sect. 18.1. See Fig. 4.28.

```
sc3 = scatter3(wage_2022,young_employment_2022,unemployment_2022,100,...
    unemployment_2022,"filled");
    sc3.MarkerEdgeColor = "black";
[~, pointIndex] = max(unemployment_2022);
colorbar;
x = [0.5 0.62];
y = [0.8 0.67];
annotation("TextArrow",x,y,"String",num2str(unemployment_2022(pointIndex)));
```

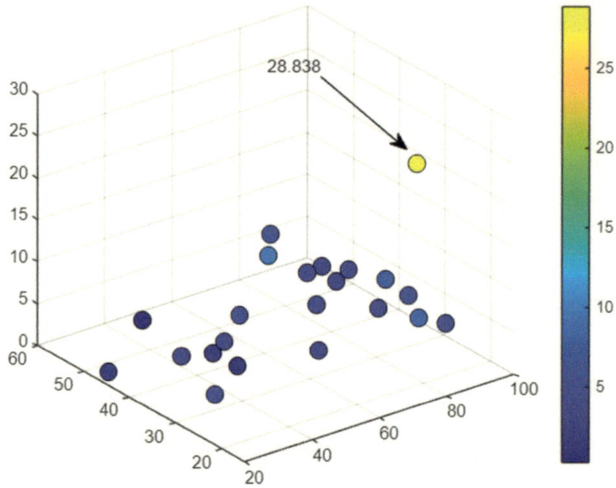

Fig. 4.28 Scatter 3D plot with a text arrow and a colorbar

Here, using the **patch** function, we add a flat patch in the x-y plane at z=0. The patch is a quadrilateral filled with a semi-transparent red color. Specifically, the variable 「x_patch」 defines the x-coordinates of the four corners of the patch, and the variable 「y_patch」 defines the y-coordinates of the four corners. Finally, the variable 「z_patch」 defines the z-coordinates for all four points of the patch; specifically, since all values are 0, the patch lies flat in the x-y plane at z=0. The color of the patch is set to 「red」 with a transparency level of 0.3 using the property 「FaceAlpha」. See Fig. 4.29.

```
sc3 = scatter3(wage_2022,young_employment_2022,unemployment_2022,100,...
    unemployment_2022,"filled");
    sc3.MarkerEdgeColor = "black";
colorbar;
x_patch = [0 100 100 0];
y_patch = [0 0 100 100];
z_patch = [0 0 0 0];
patch(x_patch,y_patch,z_patch,"red","FaceAlpha",0.3);
```

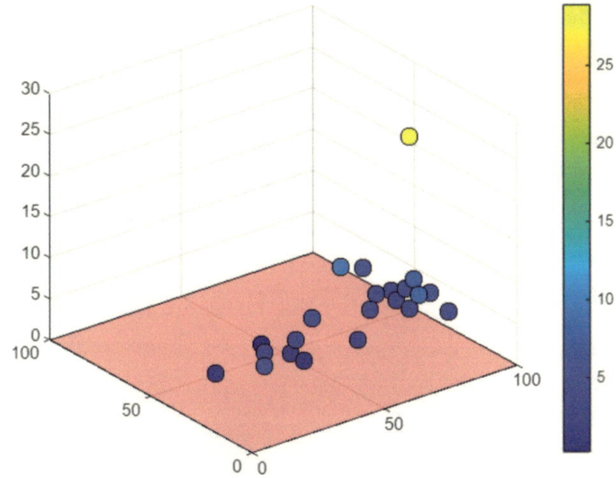

Fig. 4.29 Scatter 3D plot with a pink-colored bottom plane and a colorbar

In the following program, the position of the quadrilateral created by the **patch** function differs from the one in the previous program that created Fig. 4.29. Specifically, we set 「z_patch=[20 20 20 20]」. This means that the patch is not at z=0 as in Fig. 4.29; instead, we shift the patch upwards along the z-axis to z=20. Compare Fig. 4.29 with Fig. 4.30 to see the change in the position of the patch.

```
sc3 = scatter3(wage_2022,young_employment_2022,unemployment_2022,100,...
    unemployment_2022,"filled");
    sc3.MarkerEdgeColor = "black";
colorbar;
x_patch = [0 100 100 0];
y_patch = [0 0 100 100];
z_patch = [20 20 20 20];
patch(x_patch,y_patch,z_patch,"blue","FaceAlpha",0.2);
```

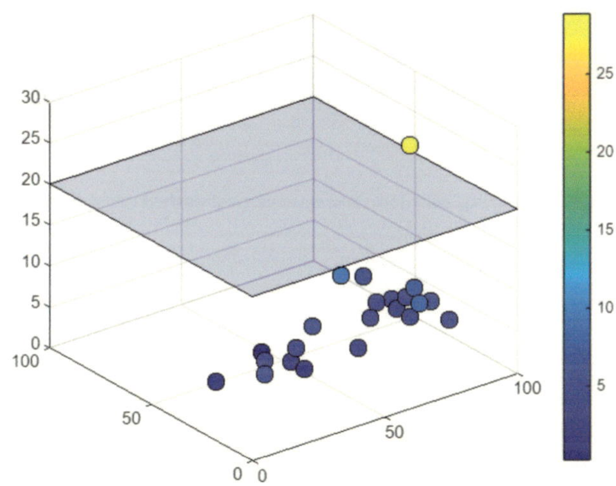

Fig. 4.30 Scatter 3D plot with a purple-colored plane and a colorbar

Here, using the **hold on** command, we overlay descent lines on the points created in the 3D scatter plot. Specifically, using a **for** loop, we iteratively utilize the **plot3** function to draw descent lines. The first and second arguments of the **plot3** function are 「[wage_2022(i),wage_2022(i)]」 and 「[young_employment_2022(i),young_employment_2022(i)]」. However, for the third argument of the **plot3** function, we use 「[unemployment_2022(i),0]」, which means that we draw a descent line for the z-coordinate from each point to 0. Additionally, the line style is set to black dashed lines using 「k--」. See Fig. 4.31.


```
sc3 = scatter3(wage_2022,young_employment_2022,unemployment_2022,100,...
    unemployment_2022,"filled");
    sc3.MarkerEdgeColor = "black";
colorbar;
hold on
for i = 1:length(wage_2022)
    plot3([wage_2022(i),wage_2022(i)],...
        [young_employment_2022(i),young_employment_2022(i)],...
        [unemployment_2022(i), 0],"k--","LineWidth",1);
end
hold off
title("droplines");
```

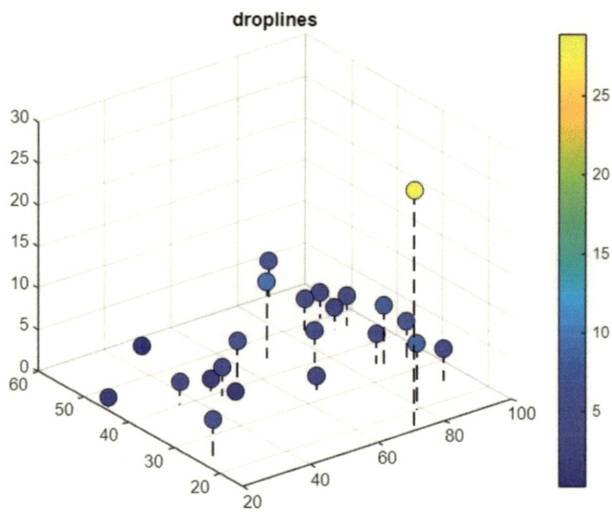

Fig. 4.31 Scatter 3D plot using a for loop to create vertical dashed droplines, with a colorbar

There are three main differences from the previous program that generated Fig. 4.31. First, the number of iterations in the **for** loop is set to 5. Specifically, we create descent lines for only the first five points. Second, the **plot3** function is used twice within the **for** loop. In the first **plot3** function, the third argument is set to 「[unemployment_2022(i),max(unemployment_2022)]」, drawing a line from each point to the maximum value of 「unemployment_2022」, near the top of the figure. In the second **plot3** function, the third argument is 「[unemployment_2022(i),0]」, drawing a line from each point to z=0. This is the same as the previous program that created Fig. 4.31. Finally, we use 「view([10 10])」 to adjust the viewing angle. Compare Fig. 4.31 with Fig. 4.32.

```
figure
sc3 = scatter3(wage_2022,young_employment_2022,unemployment_2022,100,...
    unemployment_2022,"filled");
    sc3.MarkerEdgeColor = "black";
colorbar;
hold on
for i = 1:5
    plot3([wage_2022(i),wage_2022(i)],...
        [young_employment_2022(i),young_employment_2022(i)],...
        [unemployment_2022(i),max(unemployment_2022)],...
        "b:","LineWidth",0.5);
    plot3([wage_2022(i),wage_2022(i)],...
        [young_employment_2022(i),young_employment_2022(i)],...
        [unemployment_2022(i),0],"r--","LineWidth",1);
end
hold off
view([10 10]);
title("droplines");
```

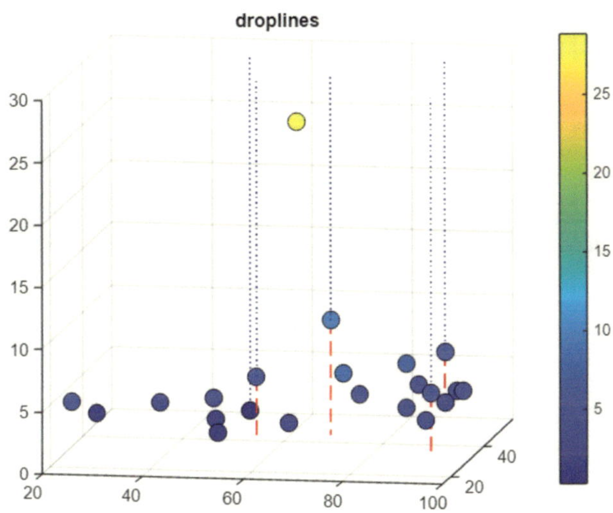

Fig. 4.32 Scatter 3D plot using a for loop to create red dashed and blue dotted droplines, with a colorbar

Here, as in the previous program that generated ◘ Fig. 4.32, we draw descent lines. Additionally, in the **plot3** function within the **for** loop, the third argument is 「[unemployment_2022(i),unemployment_2022(i)]」, meaning that we do not create a descent line for the z-coordinate; alternatively, since the second argument is 「[young_employment_2022(i),0]」, a descent line is created along the y-axis. See ◘ Fig. 4.33.


```
sc3 = scatter3(wage_2022,young_employment_2022,unemployment_2022,100,...
    unemployment_2022,"filled");
    sc3.MarkerEdgeColor = "black";
colorbar;
hold on
for i = 1:length(wage_2022)
    plot3([wage_2022(i),wage_2022(i)],...
        [young_employment_2022(i),0],...
        [unemployment_2022(i),unemployment_2022(i)],...
        "k--","LineWidth",1);
end
view([20, 10]);
hold off
title("droplines");
```

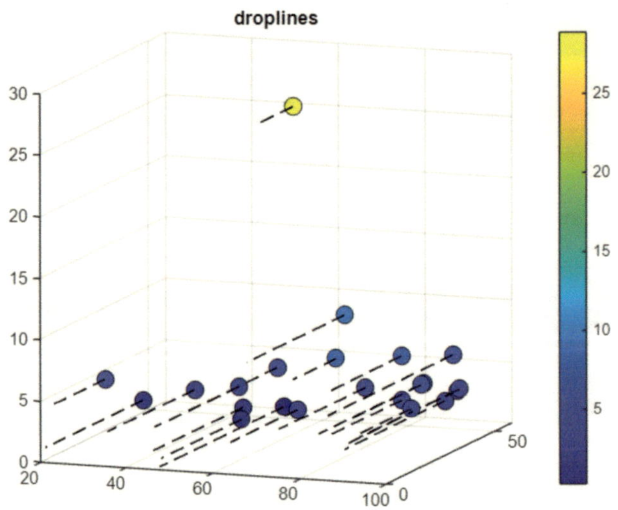

◘ **Fig. 4.33** Scatter 3D plot using a for loop to create dashed droplines from a different angle, with a colorbar

In the previous programs that created Figs. 4.31 and 4.33, we drew descent lines along the z-coordinate (Fig. 4.31) and the y-coordinate (Fig. 4.33). Here, for each point, we simultaneously create descent lines along the x-, y-, and z-coordinates. In the **for** loop, from 「for i=1:length(wage_2022(1:5))」, we create descent lines for each of the first five data points in the 3D scatter plot.

In the first **plot3** function, the first argument is set to 「[wage_2022(i),0]」, creating descent lines along the x-coordinate. Concretely, the x-values are connected between the point and 0, while the y-coordinate and z-coordinate remain fixed at the values for the current point. Next, in the second **plot3** function, the second argument is set to 「[young_employment_2022(i),0]」, creating descent lines along the y-coordinate. Finally, in the third **plot3** function, the third argument is set to 「[unemployment_2022(i),0]」, creating descent lines along the z-coordinate. The resulting figure is shown in Fig. 4.34.

```
figure
sc3 =
scatter3(wage_2022(1:5),young_employment_2022(1:5),unemployment_2022(1:5),100,...
    unemployment_2022(1:5),"filled");
    sc3.MarkerEdgeColor = "black";
hold on
for i = 1:length(wage_2022(1:5))
    plot3([wage_2022(i),0],...
        [young_employment_2022(i),young_employment_2022(i)],...
        [unemployment_2022(i),unemployment_2022(i)],...
        "b--","LineWidth",1);
    plot3([wage_2022(i),wage_2022(i)],...
        [young_employment_2022(i),0],...
        [unemployment_2022(i),unemployment_2022(i)],...
        "k--","LineWidth",1);
    plot3([wage_2022(i),wage_2022(i)],...
        [young_employment_2022(i),young_employment_2022(i)],...
        [unemployment_2022(i),0],...
        "r--","LineWidth",1);
end
view([60, 70]);
box on
hold off
title("droplines");
```

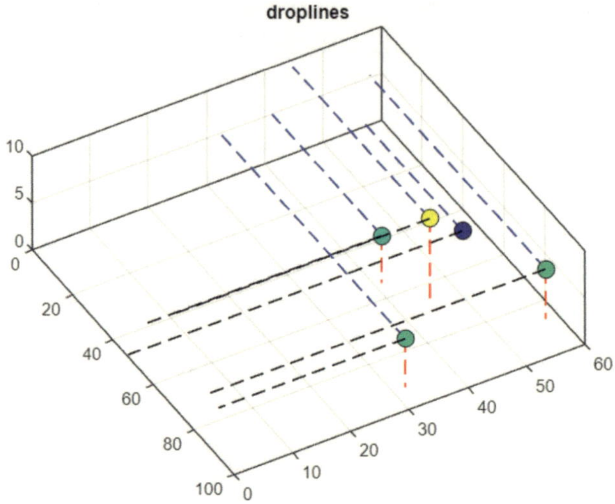

Fig. 4.34 Scatter 3D plot using a for loop to create three-directional droplines from a different angle

Here, we assign different colors to each group: 「BRICS」, 「ASEAN」, and 「G7」. First, we use the **string** function to convert the variable 「Group」 into a string type. Then, we create logical indices for each group. Specifically, the logical variable 「brics_idx」 indicates the 「BRICS」 group as TRUE (=1) and everything else as FALSE (=0). Similarly, we create logical variables 「asean_idx」 and 「g7_idx」 for the 「ASEAN」 and 「G7」 groups, respectively.

Afterward, we use the **scatter3** function three times. In the first **scatter3** function, the first argument is 「wage_2022(brics_idx)」, which extracts only the data of 「brics_idx=1 (= TRUE) 」 from the 「wage_2022」 variable. The second and third arguments work in the same way. By setting the fifth argument to 「r (= red)」, the points corresponding to 「BRICS」 countries are plotted in red. Similarly, in the next **scatter3** function, we extract only the data where 「asean_idx=1 (= TRUE)」, plotting points corresponding to ASEAN countries. Finally, in the third **scatter3** function, we extract only the data where 「g7_idx=1 (= TRUE)」, plotting points corresponding to G7 countries. The resulting figure is shown in Fig. 4.35.

```
Group = string(Group);
brics_idx = Group == "BRICS";
asean_idx = Group == "ASEAN";
g7_idx = Group == "G7";
scatter3(wage_2022(brics_idx),young_employment_2022(brics_idx),...
    unemployment_2022(brics_idx),100,"r","filled");
hold on
scatter3(wage_2022(asean_idx),young_employment_2022(asean_idx),...
    unemployment_2022(asean_idx),100,"g","filled");
scatter3(wage_2022(g7_idx),young_employment_2022(g7_idx),...
    unemployment_2022(g7_idx),100,"b","filled");
xlabel("Wage 2022");
ylabel("Young Employment 2022");
zlabel("Unemployment 2022");
legend({"BRICS", "ASEAN", "G7"},"Location","best");
grid on
hold off
```

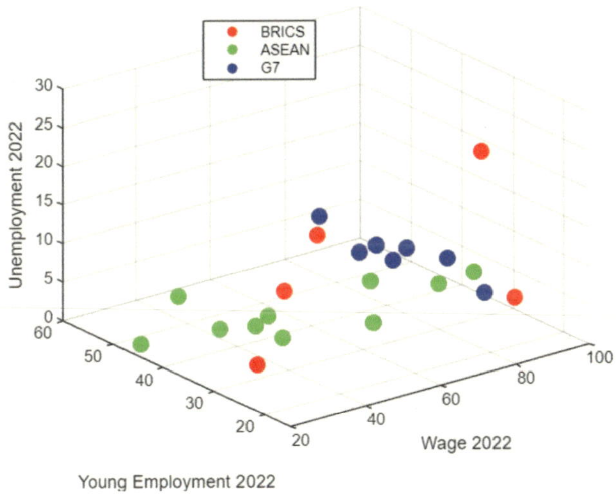

Fig. 4.35 Scatter 3D plot grouped by BRICS, ASEAN, and G7

Here, as in the previous program that created Fig. 4.35, we group the data into the 「BRICS」, 「ASEAN」, and 「G7」 groups. For the explanation of creating the variables 「brics_idx」, 「asean_idx」, and 「g7_idx」, refer to the description of the program for Fig. 4.35. Here, we create a 3-by-1 layout by using 「tiledlayout(3,1)」. Additionally, we create one 3D scatter plot for each tile. Specifically, in the 1-by-1 tile, we create a 3D scatter plot based on the data extracted by the logical variable 「brics_idx」. In the 2-by-1 tile, we create a 3D scatter plot using the data extracted by 「asean_idx」, and in the 3-by-1 tile, we create a 3D scatter plot using the data extracted by 「g7_idx」. The resulting figure is shown in Fig. 4.36.

```
Group = string(Group);
brics_idx = Group == "BRICS";
asean_idx = Group == "ASEAN";
g7_idx = Group == "G7";
zmax = max(unemployment_2022);
tiledlayout(3,1);
nexttile;
scatter3(wage_2022(brics_idx),young_employment_2022(brics_idx),...
    unemployment_2022(brics_idx),100,"r","filled");
view([50 20]);
title("BRICS");
zlabel("unemployment");
zlim([0 zmax]);
nexttile;
scatter3(wage_2022(asean_idx),young_employment_2022(asean_idx),...
    unemployment_2022(asean_idx),100,"g","filled");
view([50 20]);
title("ASEAN");
zlabel("unemployment");
zlim([0 zmax]);
nexttile;
scatter3(wage_2022(g7_idx),young_employment_2022(g7_idx),...
    unemployment_2022(g7_idx),100,"b","filled");
title("G7");
xlabel("Wage");
ylabel("young");
zlabel("unemployment");
zlim([0 zmax]);
view([50 20]);
```

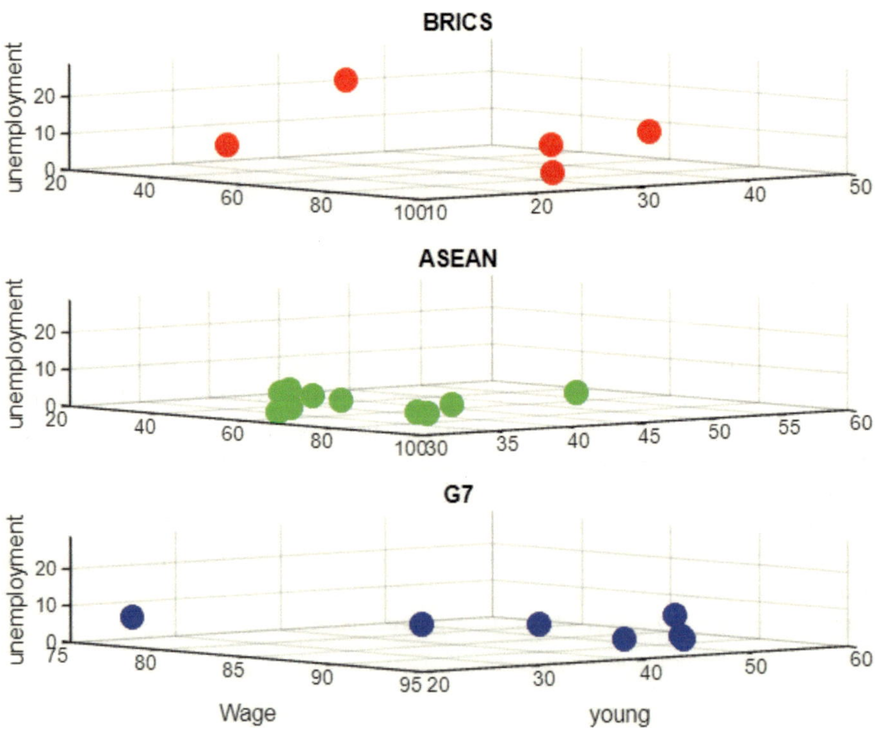

Fig. 4.36 Scatter 3D plot (3-row layout grouped by BRICS, ASEAN, and G7)

4.6 Questions

Question 1

Create Fig. 4.37 using the **scatter3** function where 「unemployment_2022」 is the x-axis data, 「wage_2022」 is the y-axis data, and 「young_employment_2022」 is the z-axis data. Additionally, specify the size of the circles as 50, and make them filled with green color. Finally, use the **xlabel**, **ylabel**, and **zlabel** functions to label each axis as 「unemployment」, 「wage」, and 「employment」, respectively.

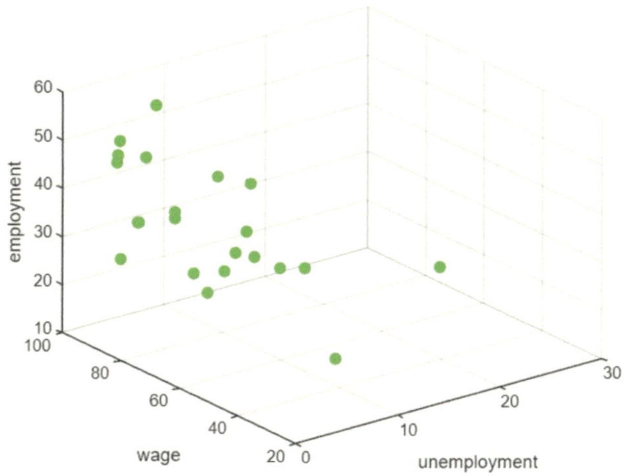

Fig. 4.37 Scatter 3D plot (Question 1)

Answer 1

```
figure
scatter3(unemployment_2022,wage_2022,young_employment_2022,50,"green","filled");
xlabel("unemployment");
ylabel("wage");
zlabel("employment");
```

Question 2

Create Fig. 4.38 using green-filled circles, each with a size of 50, as in Question 1, where 「unemployment_2022」 is the x-axis data, 「wage_2022」 is the y-axis data, and 「young_employment_2022」 is the z-axis data. Then, store the properties of the **scatter3** function into a variable 「Ex2」, and subsequently, change 「Marker」 to be an upward-pointing triangle (「^」). Finally, use the **view** function to change the viewing angle of the plot.

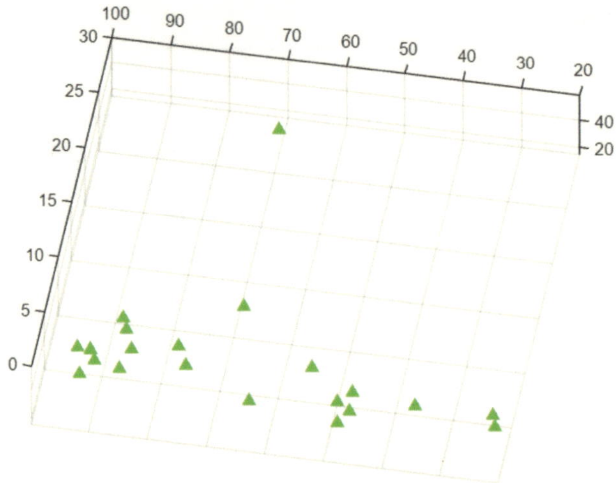

Fig. 4.38 Scatter 3D plot (Question 2)

Answer 2

```
figure
Ex2 =
scatter3(unemployment_2022,wage_2022,young_employment_2022,50,"green","filled");
Ex2.Marker = "^";
view([100,100]);
```

Question 3

Create the 3D figure with green-filled circles with a size of 50, as in Question 2, where 「unemployment_2022」 is the x-axis data, 「wage_2022」 is the y-axis data, and 「young_employment_2022」 is the z-axis data. Then, create Fig. 4.39.

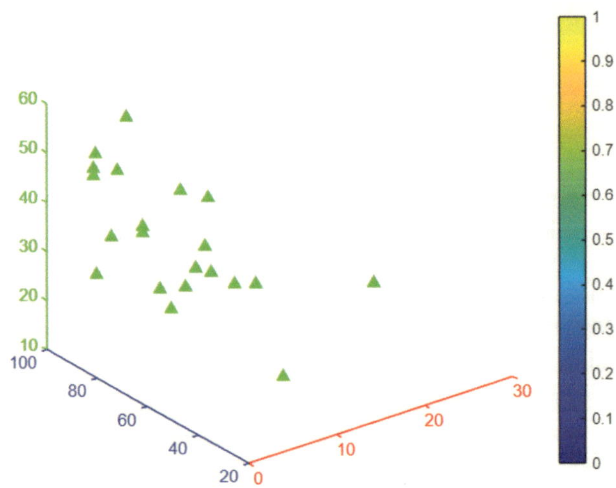

Fig. 4.39 Scatter 3D plot (Question 3)

Answer 3

```
figure
Ex3 = scatter3(unemployment_2022,wage_2022,young_employment_2022,...
    50,"green","filled");
    Ex3.Marker = "^";
colorbar;
ax = gca;
    ax.XColor = "red";
    ax.YColor = "blue";
    ax.ZColor = "green";
grid off
```

Plot Function

Contents

5.1 Details of the Data Used in this Chapter – 132

5.2 Excel Data of ► Section 5.1 – 133

5.3 Basic Syntax of the Plot Function – 133

5.4 Setting and Controlling the Plot Function – 141

5.5 Combining Optional Functions and the Plot Function – 151

5.6 Questions – 172

Supplementary Information The online version contains supplementary material available at https://doi.org/10.1007/978-981-96-6213-5_5.

Y. Nakamoto, *A Comprehensive Guide to Creating Graphics with MATLAB Programming*, Classroom Companion: Economics, https://doi.org/10.1007/978-981-96-6213-5_5

Learning Objectives

This chapter creates 2D line plots using the **plot** function. By reading this chapter, you will learn how to use the **plot** function. In particular, you will understand that although the data type (time-series data) in this chapter is the same as that in the **scatter** function (covered in ▶ Chap. 3), it can be represented as a line plot. This will further enhance your understanding of the differences and characteristics of the **plot** and **scatter** functions when creating figures.

Overview

5

In this chapter, we use the **plot** function to create 2D line plots. The **scatter** function, introduced in ▶ Chap. 3, plots individual data points to visualize their distribution or relationships. In contrast, the **plot** function connects individual data points with lines to show continuous trends. For instance, when visualizing changes over time in time series data, the **plot** function is often more appropriate.

5.1 Details of the Data Used in this Chapter

This chapter uses data on the urbanization rate, which is an indicator that shows the proportion of the population living in urban areas within a specific region (such as a country, a region, or the world). The urbanization rate is usually expressed as a percentage, calculated by dividing the urban population by the total population. 「Year1」 represents the year, 「Global」 represents the global urbanization rate, and 「Advanced」 represents the urbanization rate in advanced regions. Additionally, we also use regional data on urbanization rates for specific areas: Asia, Africa, and Europe. For the urbanization rates for Asia, Africa, and Europe, we use the variables 「Asia」, 「Africa」, and 「Europe」, respectively.

```
Year1 = [1950,1970,1990,2015]';

Global = [29.6,36.6,43,53.9]';

Advanced = [54.8,66.8,72.4,78.1]';

Asia = [17.5,23.7,32.3,48.0]';

Africa = [14.3,22.6,31.5,41.2]';

Europe = [51.7,63.1,69.9,73.9]';
```

Using the **table** function, we create a table variable 「urban_mat」 that consists of the six variables 「Year1」, 「Asia」, 「Africa」, 「Europe」, 「Global」, and 「Advanced」. When using the **renamevars** function, we rename 「Year1」 to 「Year」.

```
urban_mat = table(Year1,Asia,Africa,Europe,Global,Advanced);

urban_mat = renamevars(urban_mat,"Year1","Year");
```

5.2 Excel Data of ► Section 5.1

In ► Sect. 5.2, we import the data from an Excel file. To do this, we specify the file name (「alldata.xlsx」) and the sheet name (「Chap5_Urban_Population_Rate」). See ► Chap. 2 for detailed instructions on importing Excel data. After importing the data, we extract each column from the table variable 「urban_mat」.

```
file = "alldata.xlsx";
sheet = "Chap5_Urban_Population_Rate";
urban_mat = readtable(file,"Sheet",sheet);
Year1 = urban_mat.Year;
Asia = urban_mat.Asia;
Africa = urban_mat.Africa;
Europe = urban_mat.Europe;
Global = urban_mat.Global;
Advanced = urban_mat.Advanced;
```

5.3 Basic Syntax of the Plot Function

Basic Syntax:

The **plot** function in MATLAB is used to create a 2D line plot. Typing 「doc plot」 in the Command Window displays the syntaxes of the **plot** function as shown in Figure 5.1. Specifically, the first syntax inside the black rectangle is the basic syntax of the **plot** function, 「plot(X,Y)」. Based on the basic syntax, two types of data are required: 「X」 and 「Y」 where 「X」 represents the x-axis data and 「Y」 represents the y-axis data. To connect a set of coordinates with line segments, the vectors 「X」 and 「Y」 must have the same length.

Syntax

```
plot(X,Y)
plot(X,Y,LineSpec)
plot(X1,Y1,...,Xn,Yn)
plot(X1,Y1,LineSpec1,...,Xn,Yn,LineSpecn)
plot(Y)
plot(Y,LineSpec)

plot(tbl,xvar,yvar)
plot(tbl,yvar)

plot(ax,___)
plot(___,Name,Value)
p = plot(___)
```

Fig. 5.1 The syntax of the plot function

The basic syntax of the **plot** function requires two types of data. The first argument, 「X」, corresponds to the x-axis data, and the second argument, 「Y」, corresponds to the y-axis data. Here, the x-axis data is the year data of 「Year1」, while the y-axis data is the global urbanization rate of 「Global」. See ◘ Fig. 5.2 for the resulting figure.

```
plot(Year1,Global);
```


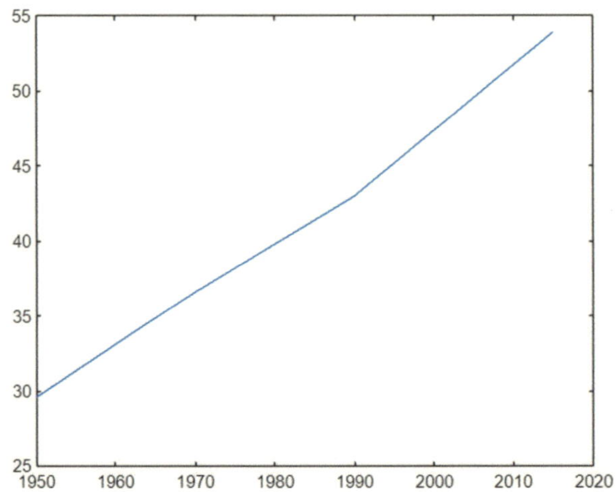

◘ **Fig. 5.2** Line plot of yearly data (X-axis: Year)

If you switch the first and second arguments of the **plot** function used in ◘ Fig. 5.2, the x-axis data becomes 「Global」 and the y-axis data becomes 「Year1」. Compare ◘ Fig. 5.2 with ◘ Fig. 5.3.

```
plot(Global,Year1);
```

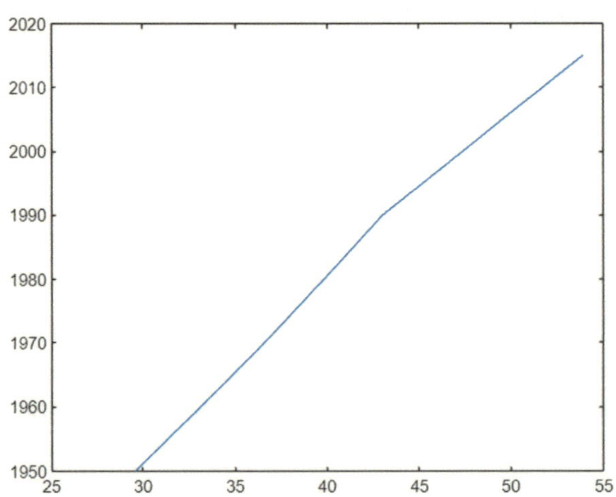

◘ **Fig. 5.3** Line plot of yearly data (Y-axis: Year)

Here, the row vectors, 「[1950,1970,1990,2015]」 and 「[29.6,36.6,43,53.9]」, are entered directly as the first and second arguments of the **plot** function. The resulting figure is shown in Fig. 5.4.

```
plot([1950,1970,1990,2015],[29.6,36.6,43,53.9]);
```

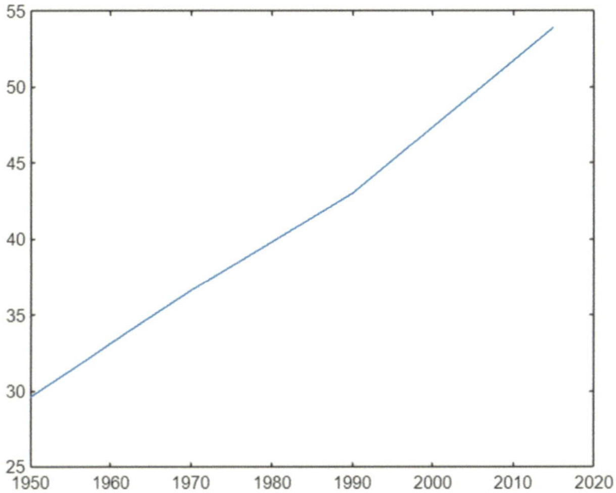

Fig. 5.4 Line plot connecting four manually entered data points

Attention

Error Program: If the lengths of 「X」 and 「Y」 in the basic syntax 「plot(X,Y)」 is different, an error occurs. In the following program, 「X」 has four elements, while 「Y」 has only three, meaning that the two vectors have different lengths. The error message is shown in Fig. 5.5.

```
%%% Error Program
%plot([1,2,3,4],[1,2,3]);
```

```
Error using plot
Vectors must be the same length.
```

Fig. 5.5 Error output of the plot function: Inconsistent vector lengths

Note that both the 「Year1」 and 「Global」 vectors contain four elements. Therefore, using the **scatter** function introduced in ▶ Chap. 3, we can display the four data points. The **plot** function, on the other hand, connects these points with straight lines. Compare Fig. 5.6 (the **scatter** function) with Fig. 5.4 (the **plot** function).

```
scatter(Year1,Global);
```

Fig. 5.6 Scatter plot with four points

The basic syntax of the **plot** function, 「plot(X,Y)」, requires two types of data, 「X」 and 「Y」. However, it is also possible to execute the **plot** function with just one type of data. In this case, the x-axis data is automatically assigned. In the following program, the y-axis data is represented by 「Year1」, and the x-axis data is automatically assigned. See Fig. 5.7 for the created figure.

```
plot(Year1);
```

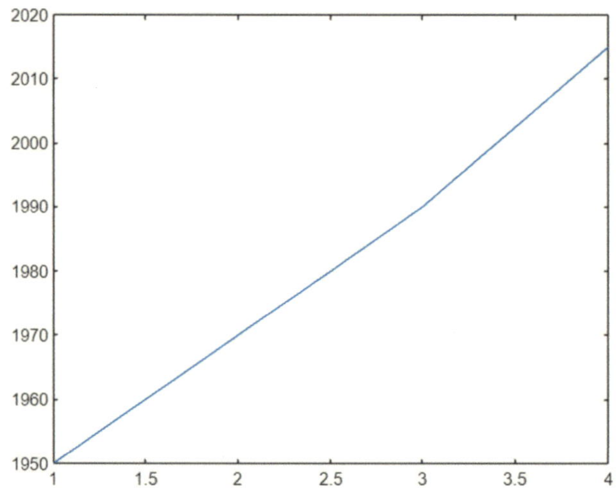

Fig. 5.7 Line plot with default numeric X-axis ticks

Attention

Error Program: Suppose that you want to create two line graphs for 「(X,Y)=(「Year1」, 「Global」)」 and 「(X,Y)=(「Year1」, 「Advanced」)」. As in the following program, if you use 「Advanced」 as the third argument, an error will occur, because the third argument in the **plot** function is used to specify the line style or color. See ▶ Sect. 5.4 for a detailed description of line style and so on. See also ◘ Fig. 5.8 for the error message.

```
%%% Error program
%plot(Year1,Global,Advanced);
```

```
Error using plot
Data must be a single input of y-values or one or more pairs of x- and y-values.
```

◘ **Fig. 5.8** Error output of the plot function: Third argument interpreted as a line style

Here, we plot the data pairs of (「Year1」, 「Global」) and (「Year1」, 「Advanced」) on the same figure. This corresponds to the second basic syntax 「plot(X1,Y1,X2,Y2)」. Refer to ◘ Fig. 5.1. The first and second arguments contain the x-coordinate and y-coordinate data for the first pair, while the third and fourth arguments contain the x-coordinate and y-coordinate data for the second pair. The resulting figure is presented in ◘ Fig. 5.9.

```
plot(Year1,Global,Year1,Advanced);
```

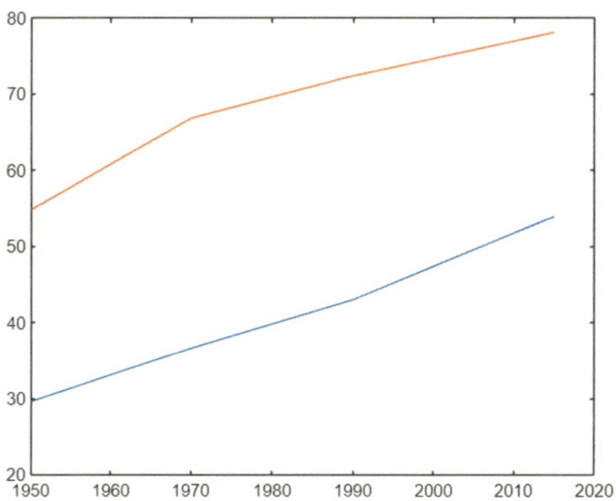

◘ **Fig. 5.9** Line plot using repeated X-axis data

Here, the first data set is (「Year1」, 「Global」), with x-axis data ranging from 1950 to 2015 with four data points. The second data set has x-axis data as 「[1900 2000 2100]」 and y-axis data as 「[50 100 300]」. There are two differences between these pairs. First, the number of data points in the first set is different from that in the second set. Second, the range of the x-axis data in the first set is different from that in the second set. The resulting figure is presented in ◘ Fig. 5.10.

```
plot(Year1,Global,[1900 2000 2100],[50 100 300]);
```


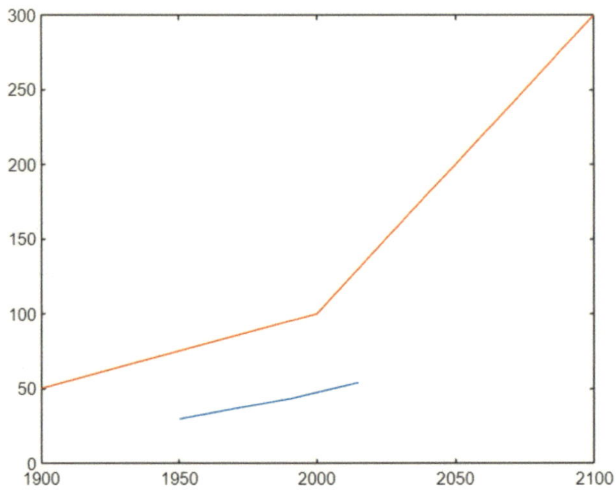

◘ **Fig. 5.10** Line plot using data series of different lengths

Here, unlike the previous program that created ◘ Fig. 5.9, we represent two line graphs in a different way. Specifically, we use the vector 「[Global, Advanced]」 for the second argument. Consequently, the first line graph is the pair (「Year1」, 「Global」), and the second line graph is the pair (「Year1」, 「Advanced」). Compare the following program that creates ◘ Fig. 5.11 with the program that created ◘ Fig. 5.9.

```
plot(Year1,[Global,Advanced]);
```

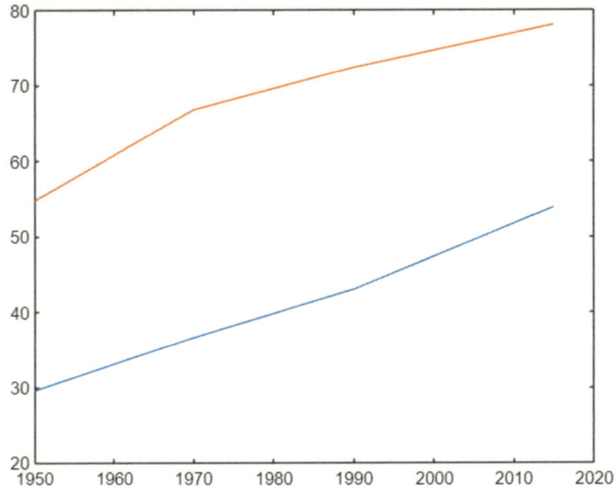

Fig. 5.11 Line plot with common X-axis data (cf. Fig. 5.9)

▶ Example of Table Variables

Here, we use 「urban_mat.Year」 in the first argument and 「urban_mat.Asia」 in the second argument of the **plot** function. The resulting figure is shown in Fig. 5.12.

```
plot(urban_mat.Year,urban_mat.Asia);
```

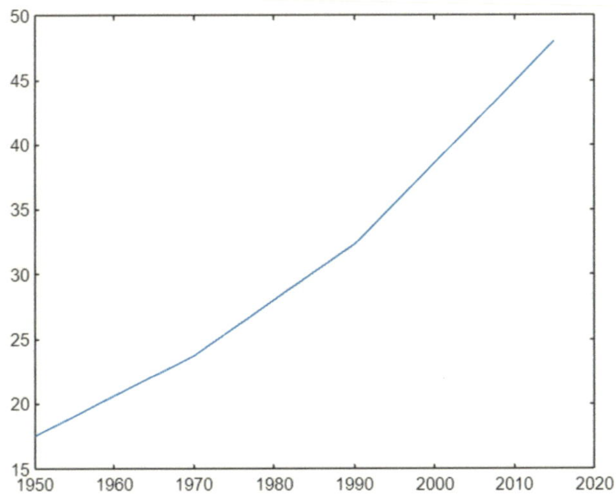

Fig. 5.12 Use of a table variable: Line plot

◀

▶ Example of Table Variables

Here, we use the table variable 「urban_mat」. Specifically, 「urban_mat.Year(1:3)」 in the first argument represents data from the first to the third row of the 「Year」 column. Similarly, 「urban_mat.Asia(1:3)」 in the second argument represents the data from the first to the third row of the 「Asia」 column. The resulting figure is presented in ◘ Fig. 5.13.

```
plot(urban_mat.Year(1:3),urban_mat.Asia(1:3));
```

5

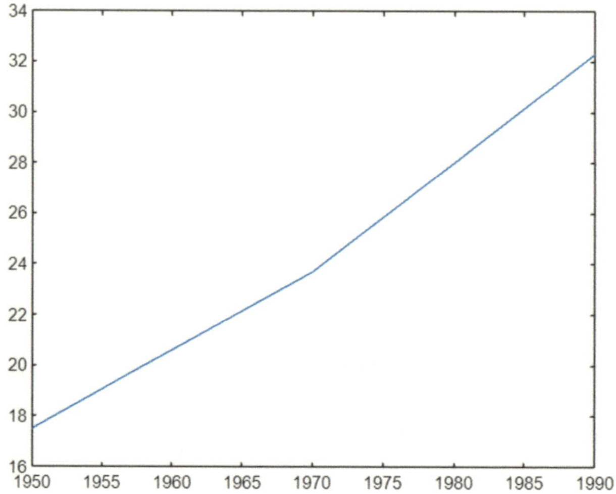

◘ **Fig. 5.13** Use of a subset of a table variable: Line plot

◀

▶ Example of Table Variables

Here, we use the table variable 「urban_mat」 to represent two line graphs. The basic approach is the same as in ◘ Fig. 5.9. The first and second arguments contain the x-coordinate and y-coordinate data for the first pair, while the third and fourth arguments contain the x-coordinate and y-coordinate data for the second pair. See ◘ Fig. 5.14.

```
plot(urban_mat.Year,urban_mat.Asia,urban_mat.Year,urban_mat.Africa);
```

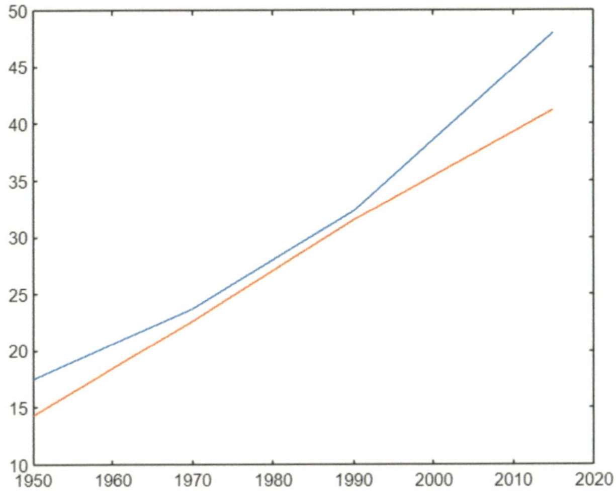

Fig. 5.14 Use of a table variable: Line plot using repeated X-axis data

◄

5.4 Setting and Controlling the Plot Function

Here, using 「:」 in the third argument of the **plot** function, we create a dotted line. See Fig. 5.15 for the created line.

```
plot(Year1,Global,":");
```

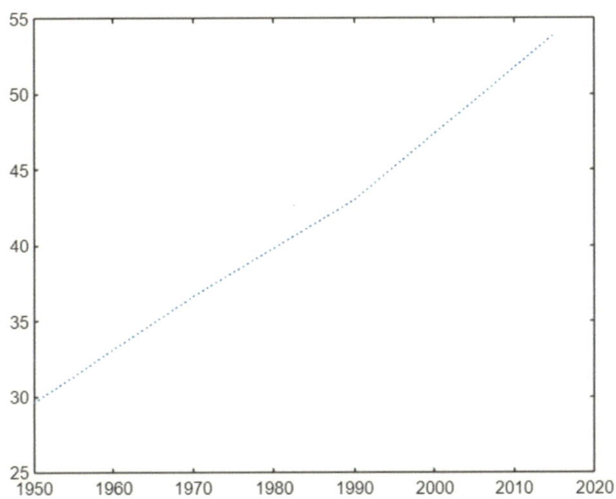

Fig. 5.15 Line plot with a dotted line style

Here, we run the previous program that created ◘ Fig. 5.15, with a single quotation mark. See ◘ Fig. 5.16.

```
plot(Year1,Global,':');
```

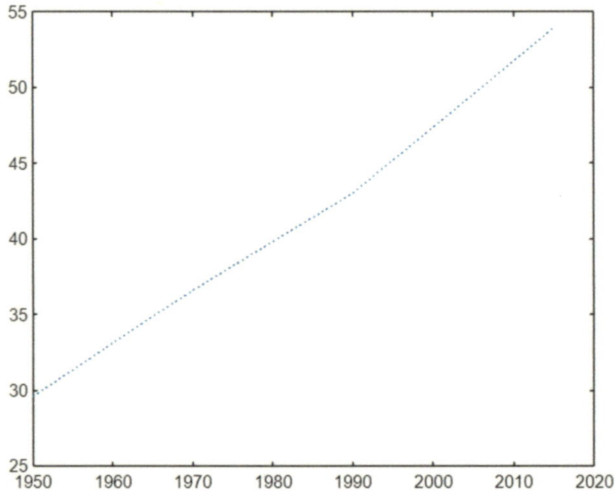

◘ **Fig. 5.16** Line plot with a dotted line style specified by single quotes

Attention

Error Program: If you do not use either single or double quotation marks for a colon in the third argument, an error will occur. The error message is presented in ◘ Fig. 5.17.

```
%%% Error program
%plot(Year1,Global,:);
```

```
Unrecognized function or variable 'plot'.
```

◘ **Fig. 5.17** Error output of the plot function: Missing quotation mark

Here, we set the line style to 「--」, creating a dashed line. Additionally, it is possible to create a dash-dotted line using 「-.」. See ◘ Fig. 5.18 for an example of a dashed line.

```
plot(Year1,Global,"--");
```

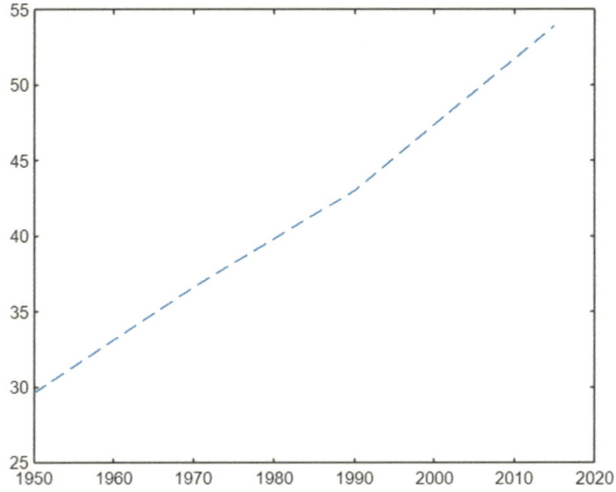

Fig. 5.18 Line plot with a dashed line style

Here, the first data set is represented by a dashed line, and the second data set is represented by a dotted line. In sum, the third argument is 「--」 for creating the dashed line, and the sixth argument is 「:」 for creating the dotted line. See Fig. 5.19 for the created figure.

```
plot(Year1,Global,"--",Year1,Advanced,":");
```

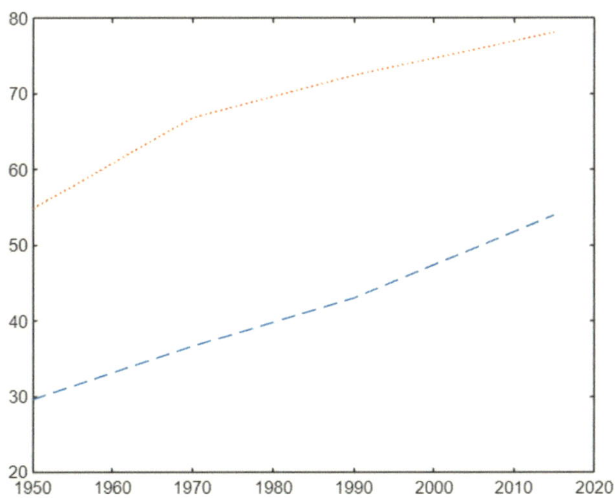

Fig. 5.19 Line plot with dotted and dashed line styles

In the third argument of the following program, we set the line style to 「:」, the color to 「b (= blue)」, and the marker to 「o」, creating a blue dotted line with circle markers. The resulting figure is shown in ◘ Fig. 5.20.

```
plot(Year1,Global,"b:o");
```

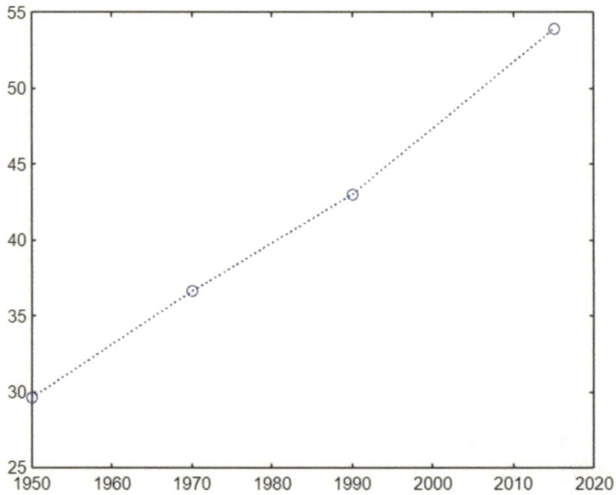

◘ **Fig. 5.20** Line plot with a blue dotted line and circle markers in the same color

In the third argument of the following program, we set the line style to 「:」, the color to 「g (= green)」, and the marker to 「o」, creating a green dotted line with circle markers. See ◘ Fig. 5.21.

```
plot(Year1,Global,"g:o");
```

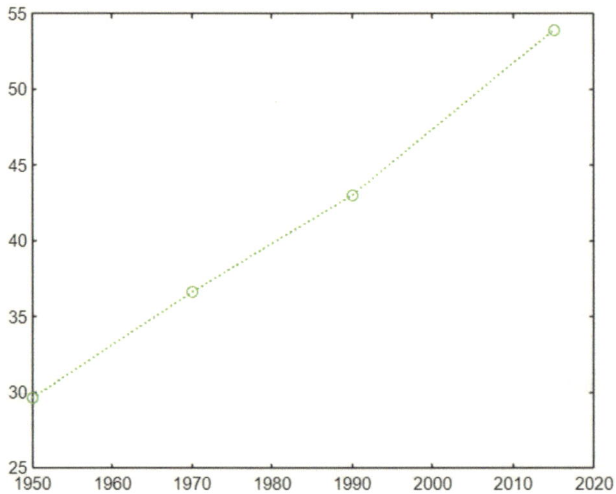

◘ **Fig. 5.21** Line plot with a green dotted line and circle markers in the same color

There is no specific order for specifying the line style, marker, and color in the third argument. In the following program, the marker is specified as 「x」 first, then the color is set to 「r (= red)」, and finally, the line style is changed to dashed by 「--」. See ◘ Fig. 5.22 for the created figure.

```
plot(Year1,Global,"xr--");
```

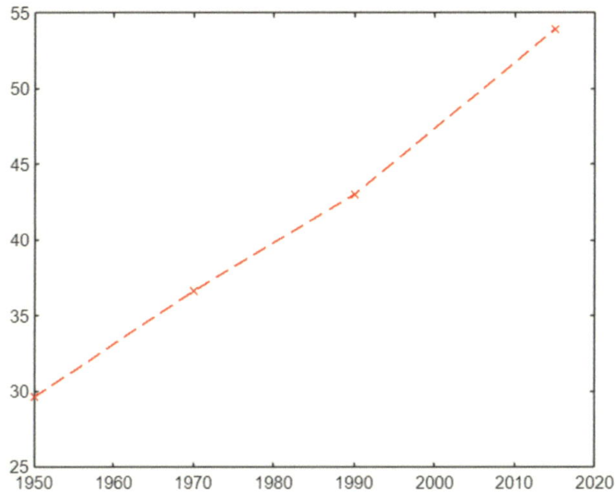

◘ **Fig. 5.22** Line plot with a red dashed line and cross markers in the same color

▶ Example of Table Variables

Here, for all three lines, corresponding to 「(X,Y)=(「urban_mat.Year」, 「urban_mat.Asia」)」, 「(X,Y)=(「urban_mat.Year」, 「urban_mat.Africa」)」, and 「(X,Y)=(「urban_mat.Year」, 「urban_mat.Europe」)」, we specify a marker as circular using 「o」, a line style as dashed using 「--」, and a color as black using 「k (= black)」. The resulting figure is displayed in ◘ Fig. 5.23.

```
plot(urban_mat.Year,[urban_mat.Asia,urban_mat.Africa,urban_mat.Europe],"--ok");
```

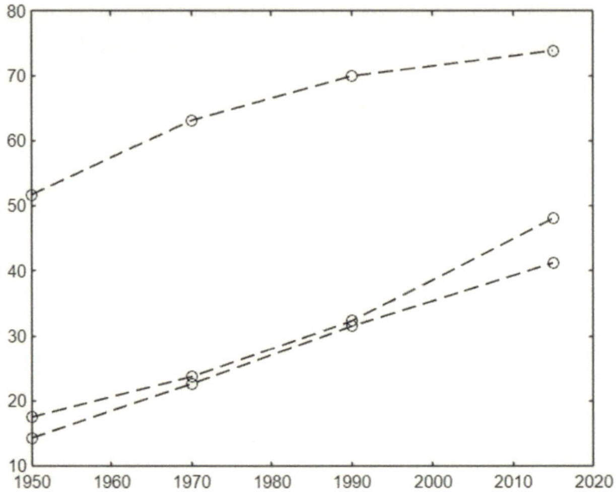

Fig. 5.23 Use of a table variable: Line plot with black dashed lines and circle markers in the same color

◀

▶ Example of Table Variables

We store the properties of the **plot** function in the variable 「pl」. After that, for the first line represented by 「pl(1)」, we set 「LineStyle」 to 「--」, 「Marker」 to 「x」, and 「MarkerSize」 to 15, creating a blue dashed line with large cross markers. Next, for 「pl(2)」, we set 「Marker」 to 「o」 and 「MarkerSize」 to 15, creating an orange solid line with large circle markers. Finally, for 「pl(3)」, we set 「Marker」 to 「s」, 「MarkerEdgeColor」 to 「black」, 「MarkerFaceColor」 to 「blue」, and 「MarkerSize」 to 15, creating a yellow solid line with square markers filled in blue.

Note that 「pl(1)」, 「pl(2)」, and 「pl(3)」 correspond to the pairs 「(X,Y)=(「urban_mat.Year」, 「urban_mat.Asia」)」, 「(X,Y)=(「urban_mat.Year」, 「urban_mat.Africa」)」, and 「(X,Y)=(「urban_mat.Year」, 「urban_mat.Europe」)」, respectively. The resulting figure is presented in Fig. 5.24.

```
pl = plot(urban_mat.Year,[urban_mat.Asia,urban_mat.Africa,urban_mat.Europe]);
    pl(1).LineStyle = "--";
    pl(1).Marker = "x";
    pl(1).MarkerSize = 15;
    pl(2).Marker = "o";
    pl(2).MarkerSize = 15;
    pl(3).Marker = "s";
    pl(3).MarkerEdgeColor = "black";
    pl(3).MarkerFaceColor = "blue";
    pl(3).MarkerSize = 15;
```

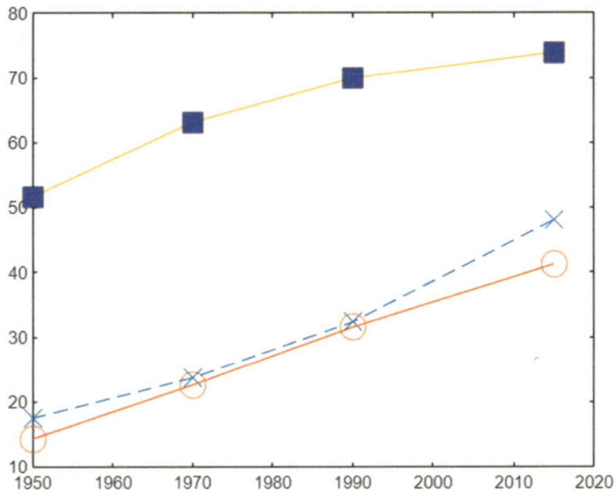

Fig. 5.24 Line plot with multiple lines, various marker styles, and colors

◀

Here, we represent the solid line with red color and square markers using 「r-s」 in the third argument. Additionally, we set 「LineWidth」 to 5 and 「MarkerSize」 to 20, creating a thicker line with large square markers. See Fig. 5.25 for the created figure.

```
plot(Year1,Global,"r-s","LineWidth",5,"MarkerSize",20);
```

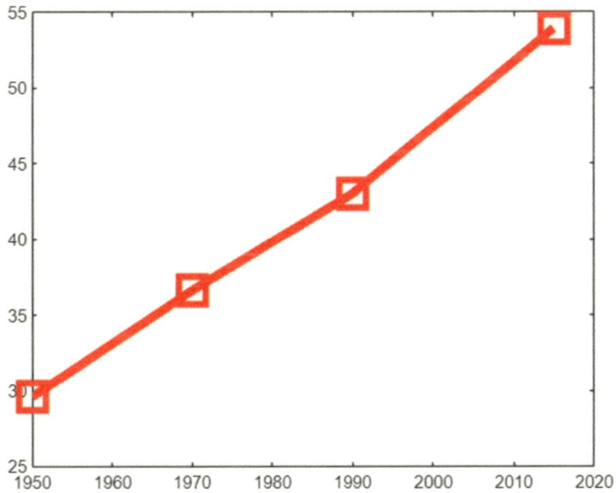

Fig. 5.25 Line plot with a thick red line and square markers in the same color

Here, using 「r-s」 in the third argument, we represent the solid line with a red color and square markers. We also set 「LineWidth」 to 2, 「MarkerSize」 to 20, and 「MarkerEdgeColor」 to 「black」, creating a thicker line with large square markers outlined in black. Finally, we set 「MarkerFaceColor」 to 「cyan」, making the inside of the markers cyan. The resulting figure is presented in Fig. 5.26.

```
plot(Year1,Global,"r-s","LineWidth",2,"MarkerSize",20,...
    "MarkerEdgeColor","black","MarkerFaceColor","cyan");
```

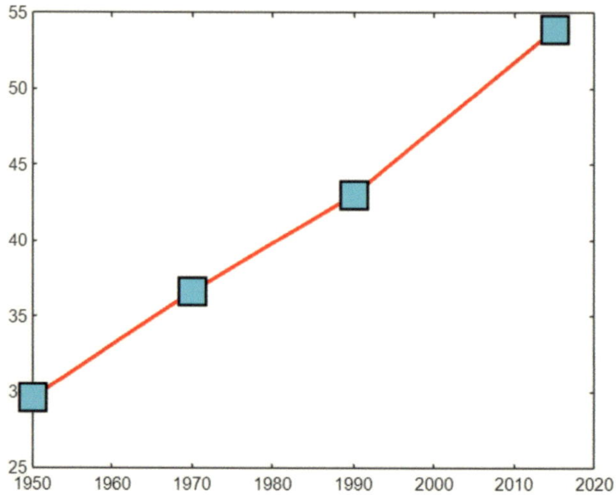

Fig. 5.26 Line plot with a red line and cyan square markers

Here, we store the properties of the **plot** function in the variable 「p_G」. We then set 「LineWidth」 to 0.1, 「MarkerSize」 to 20, and 「MarkerFaceColor」 to 「k (= black)」, creating a thinner dotted line with large circle markers filled with black. See Fig. 5.27.

```
p_G = plot(Year1,Global,"m:o");
    p_G.LineWidth = 0.1;
    p_G.MarkerSize = 20;
    p_G.MarkerFaceColor = "k";
```

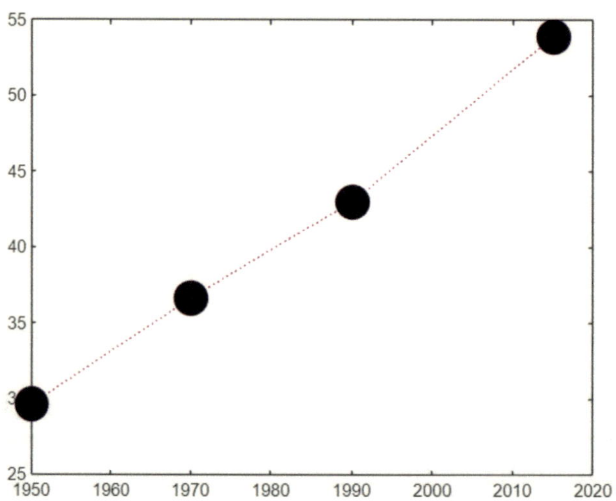

Fig. 5.27 Line plot with a dotted magenta line and large black circle markers

Here, we store the properties of three plotted lines in the variable 「p_GA」. For the third set 「(X,Y)」 of data entered as vectors, where 「X=[1950,2000,2050]」 and 「Y=[100,50,0]」, we set 「Marker」 to 「>」, 「MarkerSize」 to 25, 「Color」 to 「blue」, 「MarkerFaceColor」 to 「c (= cyan)」, and 「MarkerEdgeColor」 to 「k (= black)」, creating a blue line with right-pointing triangle markers filled with cyan and outlined in black. See ◘ Fig. 5.28.

```
p_GA = plot(Year1,Global,"r-s",Year1,Advanced,"k:+",[1950,2000,2050],[100,50,0]);
    p_GA(3).Marker = ">";
    p_GA(3).MarkerSize = 25;
    p_GA(3).Color = "blue";
    p_GA(3).MarkerFaceColor = "c";
    p_GA(3).MarkerEdgeColor = "k";
```

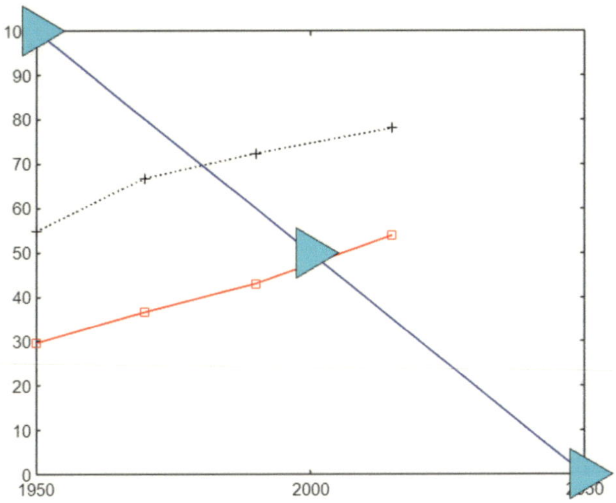

◘ **Fig. 5.28** Line plot with various line styles and markers

▶ **Example of Table Variables**

Here, using 「:sr」 in the third argument of the **plot** function, we set the properties to unify all three lines as dotted red lines with square markers. After that, we store the properties of the **plot** function in the variable 「p_AAE」. For 「p_AAE(1)」, we set 「MarkerSize」 to 20, creating large square markers. For 「p_AAE(3)」, we set 「MarkerFaceColor」 to 「blue」, creating the square markers filled in blue. See ◘ Fig. 5.29.

```
p_AAE = plot(urban_mat.Year,...
    [urban_mat.Asia,urban_mat.Africa,urban_mat.Europe],":sr");
    p_AAE(1).MarkerSize = 20;
    p_AAE(3).MarkerFaceColor = "blue";
```

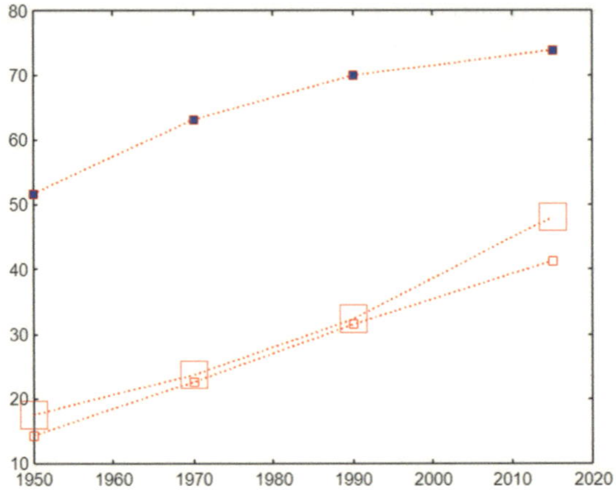

Fig. 5.29 Use of a table variable: Line plot with various line styles and markers

◀

5

▶ Example of Table Variables

Here, we use a **for** loop to change 「MarkerSize」. Specifically, the program inside the **for** loop is repeated three times. In the first iteration, we set 「i=1」, so 「p_AAE(1).MarkerSize=1*15」 is executed. Then, 「MarkerSize」 of the 「urban_mat.Asia」 line is 15. In the second iteration, we set 「i=2」, so 「p_AAE(2).MarkerSize=2*15」 is executed. Then, 「MarkerSize」 of the 「urban_mat.Africa」 line is 30. In the third iteration, 「i=3」, so 「p_AAE(3).MarkerSize=3*15」 is executed. Then, 「MarkerSize」 of the 「urban_mat.Europe」 line is 45. See Fig. 5.30 for the resulting figure.

```
p_AAE = plot(urban_mat.Year,...
    [urban_mat.Asia,urban_mat.Africa,urban_mat.Europe],"o-");
for i = 1:3
    p_AAE(i).MarkerSize = i*15;
end
```

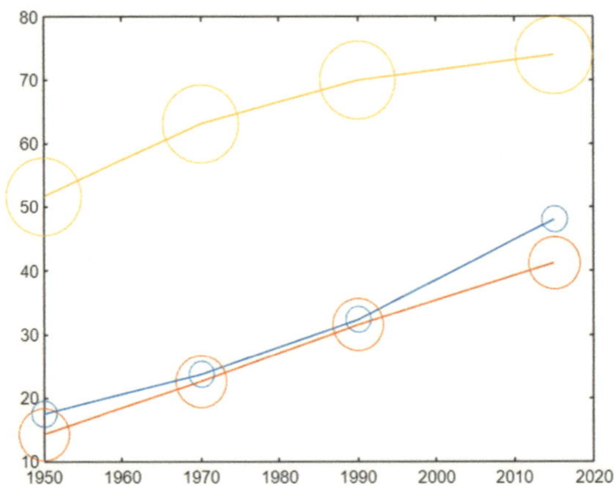

Fig. 5.30 Use of a table variable: Line plot using a for loop to adjust the sizes of circle markers

◀

5.5 Combining Optional Functions and the Plot Function

In the following program, using the **hold on** command, we overlay a line created by the **plot** function and a plot created by the **scatter** function on the same figure. The **scatter** function is explained in ▶ Chap. 3. Additionally, note that we include the **hold off** command at the end of the program to prevent subsequent plots from being overlaid on the existing figure. If you do not include the **hold off** command at the end, subsequent plots will be overlaid on the existing figure each time you plot. See ◘ Fig. 5.31.

```
plot(Year1,Global,"r-s");
hold on
scatter(Year1,Advanced,100,"k+");
hold off
```

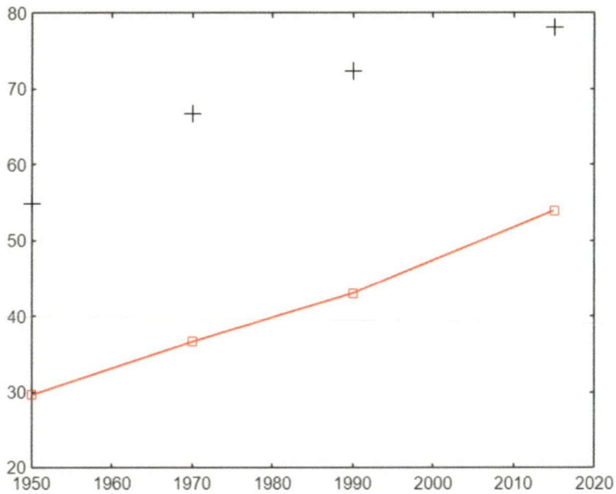

◘ **Fig. 5.31** Line plot with overlaid scatter plot

Here, using the **hold on** command, we overlay a bar graph created by the **bar** function on the same figure after creating a line using the **plot** function. The **bar** function is explained in ▶ Chap. 7. In the following program, we represent the dotted line with a blue color and square markers using ⌜b:s⌟, and then store in the variable ⌜pl⌟. Additionally, we set ⌜LineWidth⌟ to 3 and ⌜MarkerSize⌟ to 8, creating a thicker dotted line with large square markers.

After creating the bar graph, we store the properties of the **bar** function in the variable ⌜b⌟. Then, we obtain the x-axis and y-axis tick coordinates using ⌜XEndPoints⌟ and ⌜YEndPoints⌟ and store them in ⌜xtips⌟ and ⌜ytips⌟. Additionally, using the **string** function, we convert the ⌜Global⌟ data to a string type, and store it in ⌜labels⌟. Finally, using the **text** function, we place the values of the y-axis data given by ⌜labels⌟ at the coordinates obtained from ⌜xtips⌟ and ⌜ytips⌟. For more details of the **text** function, see ▶ Sect. 18.2. The resulting figure is shown in ◘ Fig. 5.32.

```
pl = plot(Year1,Global,"b:s");
    pl.LineWidth = 3;
    pl.MarkerSize = 8;
hold on
b = bar(Year1,Global,0.5);
    xtips = b.XEndPoints;
    ytips = b.YEndPoints;
labels = string(b.YData);
text(xtips,ytips,labels,...
    "HorizontalAlignment","center","VerticalAlignment","bottom");
hold off
```


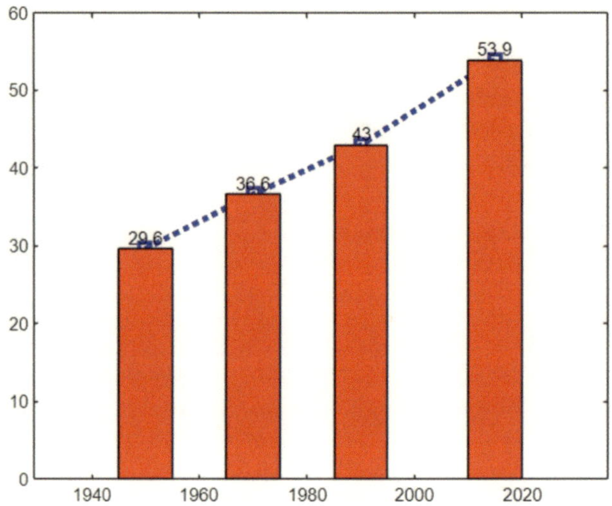

Fig. 5.32 Line plot with overlaid bar chart and data labels

Here, we use the **hold on** command. Unlike the previous program that created Fig. 5.32, after creating a bar graph using the **bar** function, we overlay a line using the **plot** function. Note that the line graph appears before the bar graph. Since the second argument of the **bar** function is 「[Global,Advanced])」, two bar graphs are placed side by side. We also use the **text** function. Refer to the previous program that created Fig. 5.32 for details on how to use the **text** function. For more details of the **text** function, see ► Sect. 18.2. The resulting figure is presented in Fig. 5.33.

```
b = bar(Year1,[Global,Advanced]);
    xtip1 = b(1).XEndPoints;
    ytip1 = b(1).YEndPoints;
 label1 = string(b(1).YData);
     xtip2 = b(2).XEndPoints;
     ytip2 = b(2).YEndPoints;
 label2 = string(b(2).YData);
text(xtip1,ytip1,label1,...
    "HorizontalAlignment","center","VerticalAlignment","bottom");
text(xtip2,ytip2,label2,...
    "HorizontalAlignment","center","VerticalAlignment","bottom");
hold on
pl = plot(Year1,[Global,Advanced],":");
    pl(1).LineWidth = 3;
    pl(1).Marker = "square";
    pl(1).MarkerSize = 8;
    pl(1).MarkerFaceColor = "yellow";
    pl(2).LineWidth = 3;
    pl(2).Marker = "x";
    pl(2).MarkerSize = 8;
    pl(2).Color= "black";
hold off
```

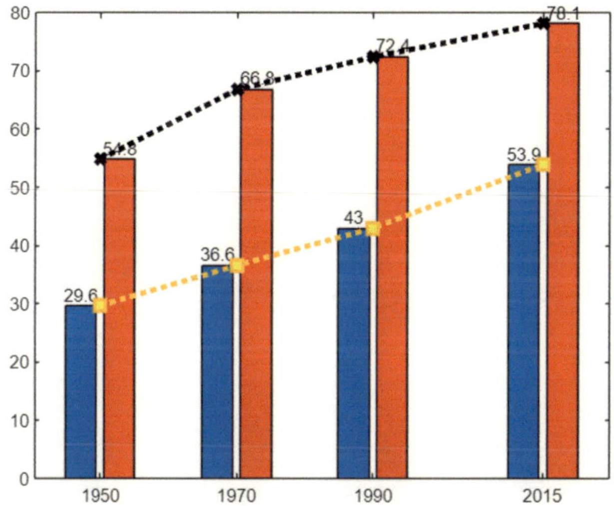

Fig. 5.33 Line plot with overlaid grouped bar chart and data labels

The previous program that created Fig. 5.33 and the following program in Fig. 5.34 differ mainly in two points. First, the **grid minor** command is used to display finer grid lines. Second, the **ylim** function is used to set the y-axis range from 0 to 150. The resulting figure is shown in Fig. 5.34.

```
b = bar(Year1,[Global,Advanced]);
    xtip1 = b(1).XEndPoints;
    ytip1 = b(1).YEndPoints;
 label1 = string(b(1).YData);
     xtip2 = b(2).XEndPoints;
     ytip2 = b(2).YEndPoints;
 label2 = string(b(2).YData);
text(xtip1,ytip1,label1,...
    "HorizontalAlignment","center","VerticalAlignment","bottom");
text(xtip2,ytip2,label2,...
    "HorizontalAlignment","center","VerticalAlignment","bottom");
hold on
pl = plot(Year1,[Global,Advanced],":");
    pl(1).LineWidth = 3;
    pl(1).Marker = "square";
    pl(1).MarkerSize = 8;
    pl(1).MarkerFaceColor = "blue";
    pl(2).LineWidth = 3;
    pl(2).Marker = "x";
    pl(2).MarkerSize = 8;
    pl(2).Color = "black";
hold off
ylim([0 150]);
grid minor
```


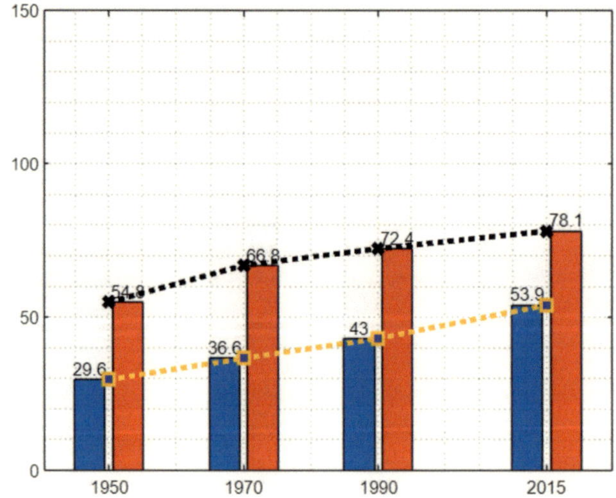

Fig. 5.34 Line plot with wide Y-axis limits and fine gridlines (cf. Fig. 5.33)

After using the **hold on** command, you can overlay multiple plots. Here, we first execute the **plot** function before the **hold on** command. Then, after the **hold on** command, we run two **plot** functions and one **scatter** function. These plots are all overlaid. See Fig. 5.35 for the created figure.

```
plot(Year1,Global,"r-s");
hold on
plot(Year1,Advanced,"k:+");
plot([1950,2015],[54,54],"b");
scatter([1900 1925 1975 2000],[30 40 50 60],[100 100 100 100],"filled");
grid on
hold off
```

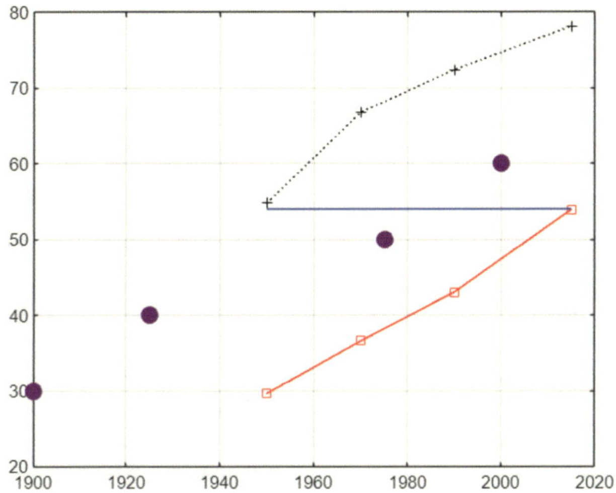

Fig. 5.35 Scatter and line plots with various line styles and marker types

Using the **annotation** function, we specify 「TextArrow」, to add a text arrow to the figure. The location of the text arrow is specified in coordinates, starting at the point 「(x,y)=(0.25,0.8)」 and ending at 「(x,y)=(0.35,0.75)」. By setting the 「String」 property, we include the text 「Advanced」. For more details on the **annotation** function, see ▶ Sect. 18.1. The resulting figure is shown in Fig. 5.36.

```
plot(Year1,Global,Year1,Advanced,":");
x = [0.25 0.35];
y = [0.8 0.75];
annotation("TextArrow",x,y,"String","Advanced");
```

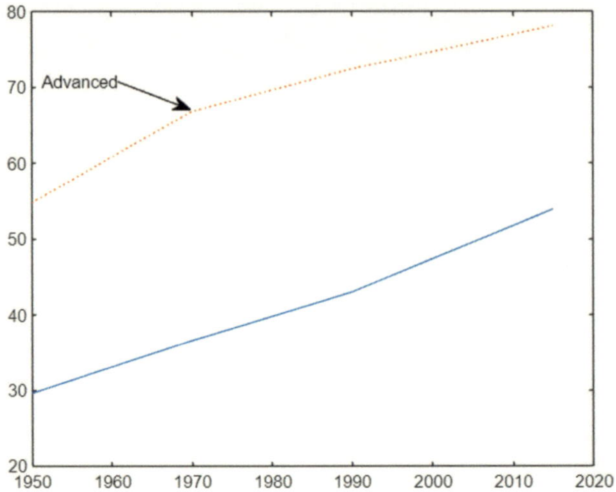

Fig. 5.36 Line plot with a text arrow

Comparing the previous program that created Fig. 5.36 with the following program, we add the following changes. First, we add two horizontal lines using the **yline** function. The first line is at y=50, styled as a blue dashed line with the text 「y=50」. The second line is at y=30, styled as a red dashed line with the text 「y=30」. For more details of the **yline** function, see ▶ Sect. 18.3. Second, using the **legend** function, we add a legend for the two horizontal lines based on the **yline** function's 「DisplayName」 property. By setting 「NumColumns」 to 2, a two-column legend is created. See Fig. 5.37.

```
plot(Year1,Global,Year1,Advanced,":");
x = [0.25 0.35];
y = [0.8 0.75];
annotation("TextArrow",x,y,"String","Advanced");
yl1 = yline(50,"--",{"y=50"});
    yl1.DisplayName = "y=50 line";
    yl1.Color = "blue";
yl2 = yline(30,"--",{"y=30"});
    yl2.DisplayName = "y=30 line";
    yl2.Color = "red";
lgd = legend([yl1,yl2]);
    lgd.Location = "southeast";
    lgd.NumColumns = 2;
```

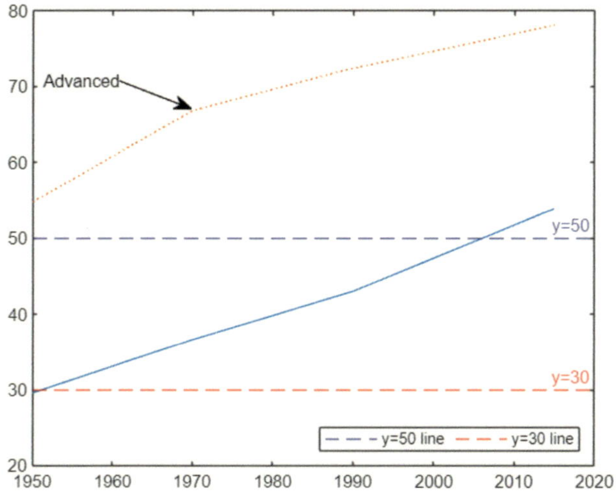

Fig. 5.37 Line plot with a text arrow, two horizontal lines, and a legend

Here, after creating a plot graph using the **plot** function, we use the **annotation** and **xregion** functions. The **annotation** function is used with line type specified as 「TextArrow」 and 「ellipse」 as shown in the first argument of each annotation, creating two arrows with text and a circle. Specifically, the outline of the circle is set to red and placed to enclose one of the points on the line graph. For more details on the **annotation** function, see ▶ Sect. 18.1.

Additionally, the **xregion** function is used to fill the area to the right of the rightmost point by providing 「xregion([Year1(end),Inf])」, and the area below the lowest point by providing 「yregion([-Inf Global(1)])」. Furthermore, the **legend** function displays 「filled regions」 in the legend. See Fig. 5.38 for the resulting figure.

```
plot(Year1,Global,"r-s",Year1,Advanced,"+k:");
x1 = [0.25 0.37];
y1 = [0.8 0.77];
ta = annotation("TextArrow",x1,y1,"String","Advanced");
    ta.Color = "blue";
x2 = [0.8 0.78];
y2 = [0.3 0.52];
ta = annotation("TextArrow",x2,y2,"String","Global");
    ta.Color = "red";
dim = [0.8 0.52 0.1 0.1];
annotation("ellipse",dim,"Color","red");
xg = xregion([Year1(end),Inf]);
    xg.DisplayName = "filled region";
yregion([-Inf Global(1)]);
legend(xg);
grid on
```

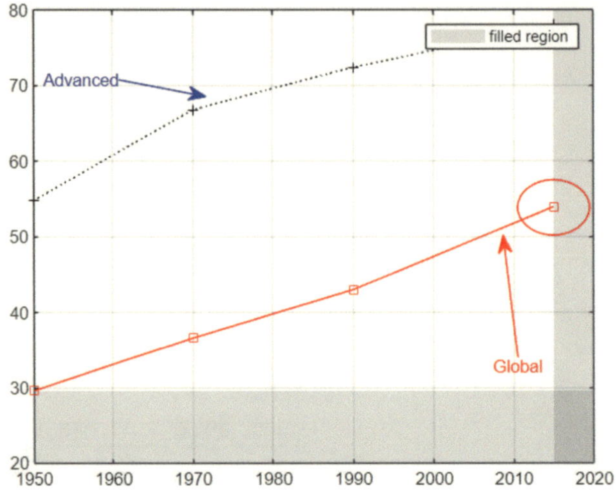

Fig. 5.38 Line plot with two shaded areas, a circle marker, two text arrows, and a legend

5

To create a text box, two ellipses, and a rectangle, we use the **annotation** function four times. Specifically, we specify ⌜dim1⌟, ⌜dim2⌟, ⌜dim3⌟, and ⌜dim4⌟ as a four-element vector of the form ⌜[x y w h]⌟ in the MATLAB documentation of the **annotation** function, where ⌜x⌟ and ⌜y⌟ elements determine the position, and ⌜w⌟ and ⌜h⌟ elements determine the size.

For instance, in the first **annotation** function, we set ⌜dim1 = [0.8 0.52 0.1 0.4]⌟, which positions and sizes the ellipse to enclose the two rightmost points. For more details on the **annotation** function, see ► Sect. 18.1. The resulting figure is shown in Fig. 5.39.

```
pl = plot(Year1,Global,Year1,Advanced);
    pl(1).Marker = ">";
    pl(1).MarkerFaceColor = "yellow";
    pl(1).MarkerEdgeColor = "black";
    pl(2).Marker = "hexagram";
    pl(2).MarkerFaceColor = [0.4940 0.1840 0.5560];
    pl(2).MarkerSize = 10;
dim1 = [0.8 0.52 0.1 0.4];
annotation("ellipse",dim1,"Color","magenta");
dim2 = [0.3 0.3 0.35 0.6];
ta2 = annotation("rectangle",dim2,"Color",[0.4660 0.6740 0.1880]);
    ta2.FaceColor = "yellow";
    ta2.FaceAlpha = 0.1;
dim3 = [0.08 0.22 0.1 0.4];
annotation("ellipse",dim3,"Color",[0.4660 0.6740 0.1880]);
dim4 = [0.3 0.15 0.2 0.08];
ta4 = annotation("TextBox",dim4,"String","1965-1995");
    ta4.FontSize = 14;
grid minor
```

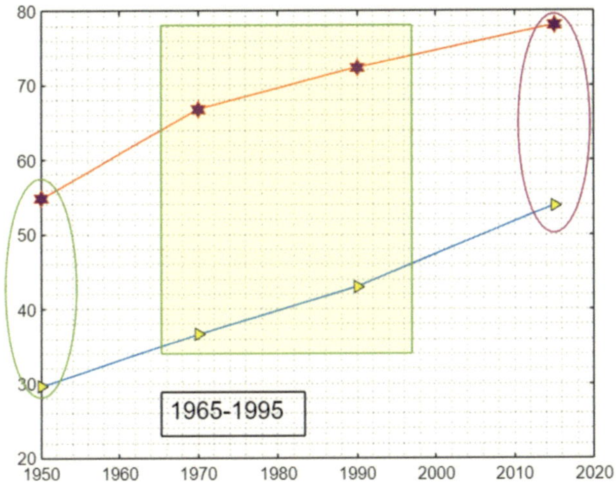

Fig. 5.39 Line plot with two ellipses, a rectangle, a textbox, and fine gridlines

Here, using the **annotation** function, we specify 「TextArrow」, 「doublearrow」, and 「ellipse」, to create two arrows with text, two double arrows, and an ellipse. When 「TextArrow」 and 「doublearrow」 are used, the start and end points are set in the second and third arguments of the **annotation** function. Alternatively, when using 「ellipse」, we set the position and size in the second argument of the **annotation** function. For more details on the **annotation** function, see ▶ Sect. 18.1. See Fig. 5.40.

```
plot(Year1,Global,"r-s",Year1,Advanced,"+k:");
x1 = [0.25 0.4];
y1 = [0.8 0.77];
ta1 = annotation("TextArrow",x1,y1,"String","Advanced");
    ta1.Color = "blue";
x2 = [0.8 0.78];
y2 = [0.3 0.5];
ta2 = annotation("TextArrow",x2,y2,"String","Global");
    ta2.Color = "red";
xlim([Year1(1) Year1(end)]);
dim = [0.85 0.52 0.1 0.1];
annotation("ellipse",dim,"Color","red");
ta3 = annotation("doublearrow",[0.37 0.37],[0.35 0.73]);
    ta3.Color = "magenta";
    ta3.LineStyle = "--";
ta4 = annotation("doublearrow",[0.61 0.61],[0.45 0.8]);
    ta4.Color = [0.4660 0.6740 0.1880];
    ta4.LineWidth = 3;
    ta4.LineStyle = "--";
grid on
```

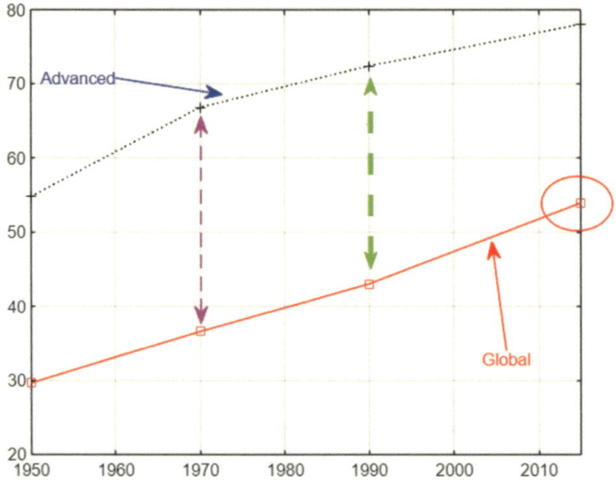

Fig. 5.40 Line plot with two text arrows, two vertical double arrows, and a circle marker

5

Here, using the **xlim** and **ylim** functions, we set the ranges for the x-axis and y-axis. Specifically, for the x-axis, we input the values 1900 and 2100 directly into the **xlim** function, setting the range from 1900 to 2100. For the y-axis, using the **ylim** function, we set the y-axis range from 「min(Global)」 to 「max(Global)」, where 「min(Global)」 and 「max(Global)」 are the minimum and maximum values of the variable 「Global」. The resulting figure is shown in Fig. 5.41.

```
plot(Year1,Global,"-");
xlim([1900,2100]);
ylim([min(Global),max(Global)]);
```

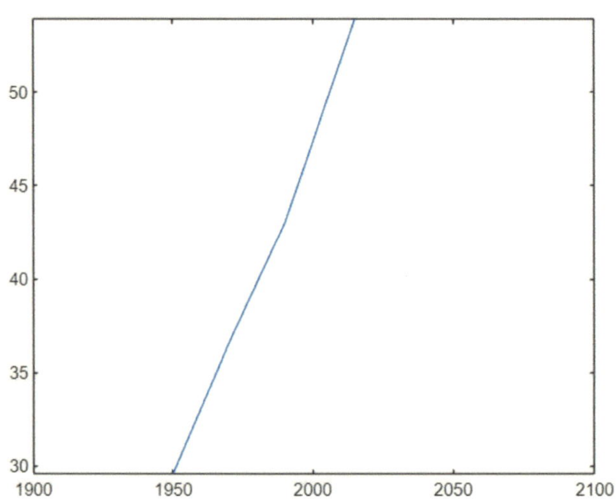

Fig. 5.41 Line plot with adjusted X- and Y-axis limits

In the following program, we concatenate the column vectors of 「Global」 and 「Advanced」 into a single column vector, 「combined_vector」. Then, we use the **max** and **min** functions to obtain the maximum and minimum values of 「combined_vector」 where the maximum and minimum values of 「combined_vector」 are respectively stored in the variables 「ymax」 and 「ymin」. Finally, these variables are used to set the y-axis range. See Fig. 5.42.

```
plot(Year1,Global,"-",Year1,Advanced,":");
combined_vector = [Advanced;Global];
ymax = max(combined_vector);
ymin = min(combined_vector);
ylim([ymin,ymax]);
xlim([min(Year1),max(Year1)]);
```

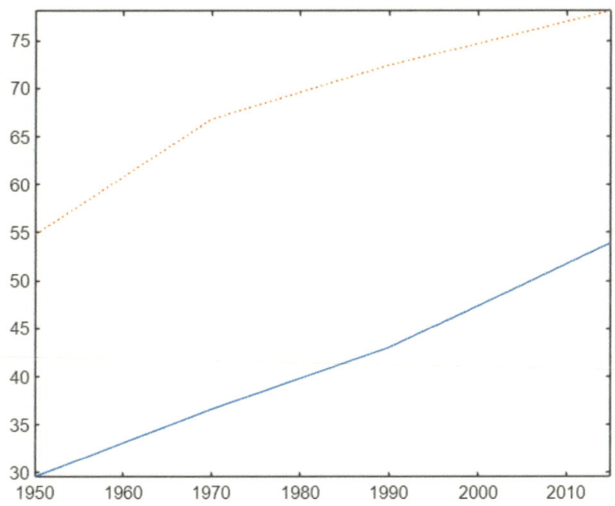

Fig. 5.42 Line plot with axis limits matching data range

First, using the **figure** command, we reset the figure settings to their defaults. Here, we use 「yyaxis」 to create two y-axes in the figure. Pay attention to the order of the following program. First, we set 「yyaxis」 to 「right」, and then generate the plot using the **plot** function. Then, the tick marks for the subsequently generated plot appear on the right. The resulting figure is shown in Fig. 5.43, demonstrating that the tick marks on the right y-axis are set from 50 to 80. This automatically corresponds to the data of 「Advanced」 on the y-axis.

Moreover, by using the **ylabel** function after 「yyaxis right」, we provide a label 「the urbanization rate」 on the right y-axis. On the other hand, the tick marks on the left y-axis are set to the default range of 0 to 1.

```
figure
yyaxis right;
plot(Year1,Advanced);
ylabel("the urbanization rate");
```

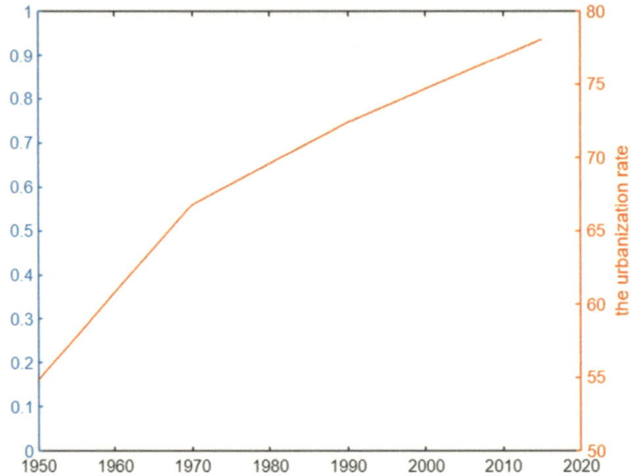

Fig. 5.43 Line plot with dual Y-axes (default left Y-axis)

Using the **figure** command, we reset the figure settings to their defaults. Here, we note the order of the **plot** function and the **yyaxis right** command. Compare the previous program that created Fig. 5.43 with the following program. First, after creating a new figure space using the **figure** command, we use the **plot** function to create a line graph. The tick marks for this line graph are automatically displayed on the left y-axis, adjusted to range from 25 to 55. Subsequently, after specifying 「yyaxis right」, we set 「YColor」 to 「green」, changing the axis color to green. Additionally, the axis label 「Default Y-axis Tick」 is also created on the right y-axis. Compare Fig. 5.43 with Fig. 5.44.

```
figure
plot(Year1,Global);
ylabel("Y-axis Tick for Global");
yyaxis right;
axright = gca;
    axright.YColor = "green";
yl = ylabel("Default Y-axis Tick");
    yl.Color = "blue";
    yl.FontSize = 20;
```

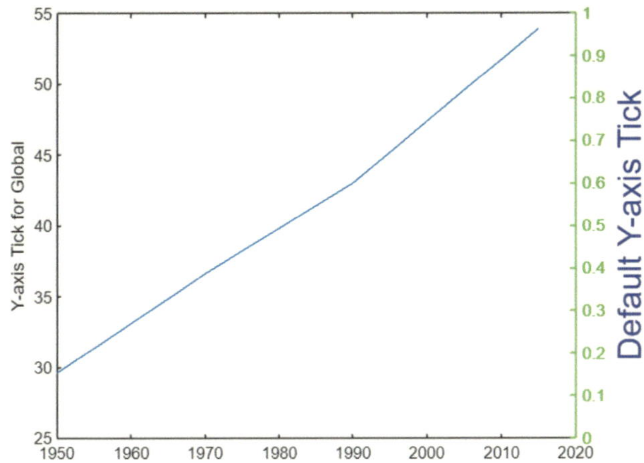

Fig. 5.44 Line plot with dual Y-axes (default right Y-axis ticks in green)

In the following program, we use the **plot** function after the **yyaxis right** command, so the tick marks for the variable 「Global」 correspond to the right y-axis. Additionally, since the **ylabel** function is used after 「yyaxis right」, the axis label 「the global urbanization rate」 is placed on the right y-axis. The color of the right y-axis is also changed to green by setting 「YColor」 to 「green」. On the other hand, after using the **yyaxis left** command, the left y-axis tick marks are removed by setting 「YTick」 to 「[ ]」. Moreover, the axis label 「No tick」 is placed on the left y-axis. The resulting figure is shown in Fig. 5.45.

```
figure
yyaxis right;
plot(Year1,Global);
axright = gca;
    axright.YColor = "green";
ylabel("the global urbanization rate");
yyaxis left;
axleft = gca;
    axleft.YTick = [];
yl = ylabel("No tick");
    yl.Color = "black";
    yl.FontSize = 20;
```

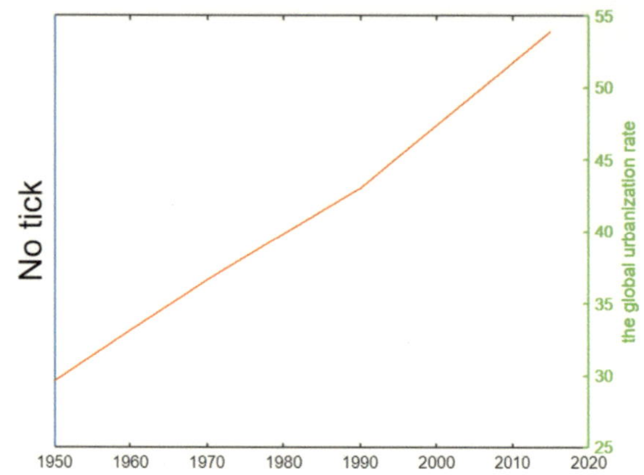

■ **Fig. 5.45** Line plot with dual Y-axes (no left ticks)

Here, using the **gca** command as the first argument of the **set** function, we change the settings of the x-axis and y-axis. Specifically, after using the **yyaxis left** command, we set 「YTick」 to an empty vector 「[ ]」 using the **set** function, thereby removing the tick marks on the left y-axis. Next, after using the **yyaxis right** command, we set 「YColor」 to 「black」, to change the color of the right y-axis to black. Finally, using the **set** function, we set 「XAxisLocation」 to 「top」, to position the x-axis at the top of the figure. The resulting figure is shown in ■ Fig. 5.46.

```
figure
yyaxis left;
set(gca,"YTick",[]);
yyaxis right;
set(gca,"YColor","black");
set(gca,"XAxisLocation","top");
plot(Year1,Global);
ylabel("the global urbanization rate");
title("Controlling axes using the set function");
```

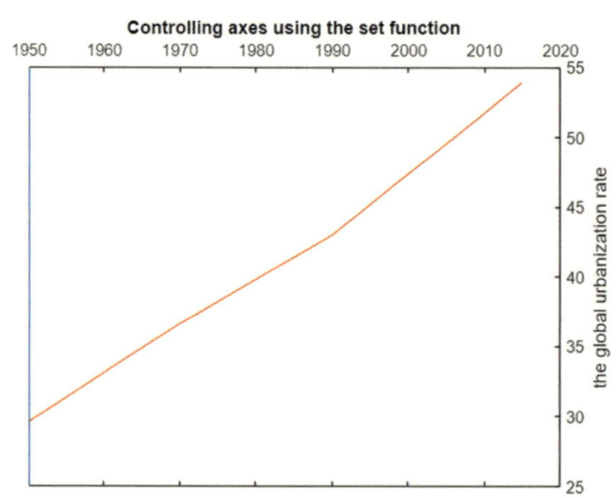

■ **Fig. 5.46** Line plot with a top-aligned X-axis via the set function

▶ Example of Table Variables

Here, we plot five line graphs 「(X,Y)=(「Year1」, 「urban_mat.Europe」)」, 「(X,Y)=(「Year1」, 「Advanced」)」, 「(X,Y)=(「Year1」, 「urban_mat.Asia」)」, 「(X,Y)=(「Year1」, 「urban_mat.Africa」)」, and 「(X,Y)=(「Year1」, 「Global」)」. Specifically, we use the **plot** function to create the first two graphs 「(X,Y)=(「Year1」, 「urban_mat. Europe」)」 and 「(X,Y)=(「Year1」, 「Advanced」)」, after the **yyaxis right** command. This results in the tick marks being displayed for the Y data on the right y-axis.

Next, after the **yyaxis left** command, we use the **plot** function to create the other graphs, 「(X,Y)=(「Year1」, 「urban_mat.Asia」)」, 「(X,Y)=(「Year1」, 「urban_mat.Africa」)」, and 「(X,Y)=(「Year1」, 「Global」)」, causing the tick marks for the Y data to correspond to the left y-axis. Additionally, using the **annotation** functions, we set 「TextArrow」 to create two text arrows labeled 「Advanced」 and 「Europe」. For more details of the **annotation** function, see ▶ Sect. 18.1. See ◘ Fig. 5.47.

```
figure
yyaxis right;
pl_high = plot(Year1,urban_mat.Europe,Year1,Advanced);
    pl_high(1).LineWidth = 2;
    pl_high(1).Color = "black";
    pl_high(2).LineWidth = 2;
    pl_high(2).Color = "black";
    pl_high(2).LineStyle = "--";
ax_high = gca;
    ax_high.YColor = "black";
    ax_high.YLim = [70 80];
ylabel("Y-axis for Europe and Advanced");
yyaxis left;
plot(Year1,[urban_mat.Asia,urban_mat.Africa,Global]);
ax_low = gca;
    ax_low.YColor = "blue";
ylabel("Y-axis for Asia, Africa and Global");
x1 = [0.45 0.5];
y1 = [0.2 0.2];
annotation("TextArrow",x1,y1,"String","Advanced");
x2 = [0.7 0.7];
y2 = [0.38 0.28];
annotation("TextArrow",x2,y2,"String","Europe");
```

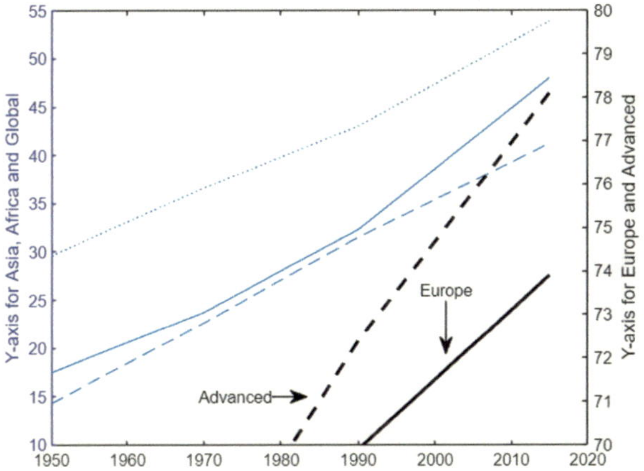

◘ **Fig. 5.47** Line plot with various line styles, text arrows, and dual Y-axes

◀

By executing the **figure** command, we reset the x-axis and y-axis settings in the previous program to their default values. After generating the plot using the **plot** function, we store the properties of the **xlabel** function in a variable 「xl」. Specifically, we set 「Color」 to 「blue」, 「FontSize」 to 30, and 「FontWeight」 to 「bold」, increasing the size of the font and making it blue and bold. See ◘ Fig. 5.48 for the created figure.

```
figure
plot(Year1,Global,"-");
xl = xlabel("Year");
    xl.Color = "blue";
    xl.FontSize = 30;
    xl.FontWeight = "bold";
```


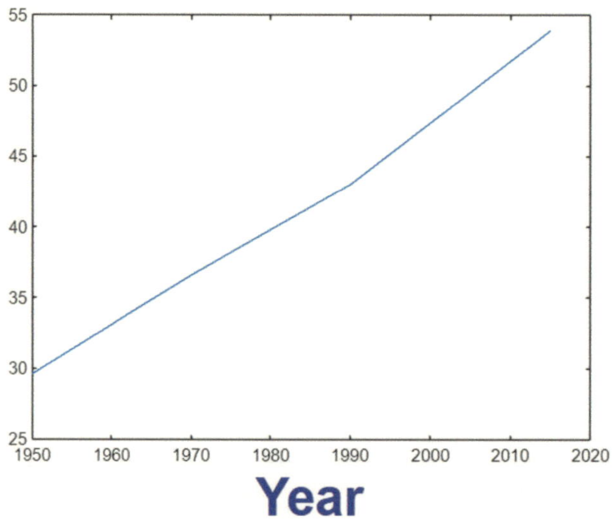

◘ **Fig. 5.48** Line plot with the X-axis labeled "Year" in bold blue

Here, we add a mathematical expression to the x-axis. Specifically, we can use the LaTeX format by setting 「Interpreter」 to 「latex」. See the following program that creates ◘ Fig. 5.49.

```
figure
plot(Year1,Global,"-");
xlabel({"Year","$¥int_1^{20} x^2 dx$"},"Interpreter","latex");
```

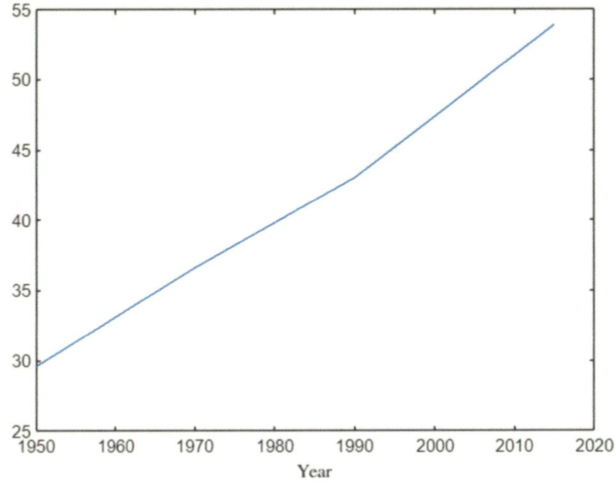

$\int_1^{20} x^2 dx$

Fig. 5.49 Line plot with a two-line X-axis label containing a text label and a LaTeX expression

Here, the **xlabel** function is used to label the x-axis as 「Year」 on the first line and a mathematical expression on the second line, with 「Interpreter」 set to 「latex」. Furthermore, 「FontSize」 is set to 15 and 「Color」 is set to 「red」, creating a red label with a large font size on the x-axis. See Fig. 5.50.

```
figure
plot(Year1,Global,"-");
xlabel({"Year","$¥lim_{n¥to ¥infty}¥sum_{t=0}^{n-1}f(x_{t})¥Delta x$"},...
    "FontSize",15,"Color","red","Interpreter","latex");
ylabel("Urban Population Rate");
```

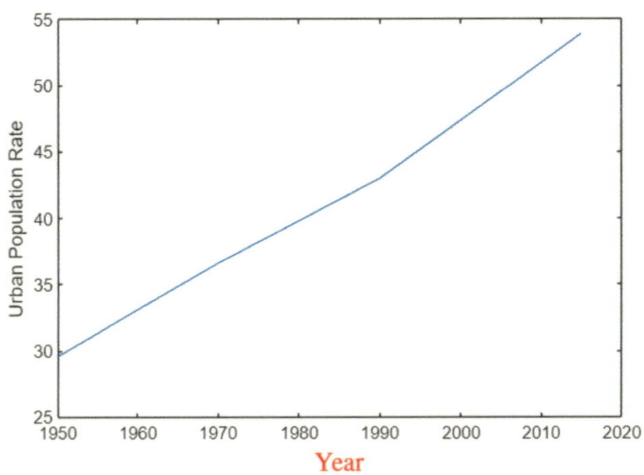

$\lim_{n\to\infty} \sum_{t=0}^{n-1} f(x_t)\Delta x$

Fig. 5.50 Line plot with a red two-line X-axis label and a Y-axis label

▶ Example of Table Variables

Here, using the **legend** function, we create a legend with labels for plot graphs. The legend includes five graphs, with each color properly represented.

Additionally, we set 「Location」 to 「southeastoutside」, to position the legend in the bottom-right corner outside the axes. The resulting figure is shown in ◘ Fig. 5.51.

```
figure
plot(urban_mat.Year,[urban_mat.Asia,urban_mat.Africa,urban_mat.Europe,...
    urban_mat.Global,urban_mat.Advanced]);
xlim([urban_mat.Year(1),urban_mat.Year(end)]);
title("Urbanization Rate");
lg = legend("Asia","Africa","Europe","Global","Advanced");
    lg.Location = "southeastoutside";
grid on
```


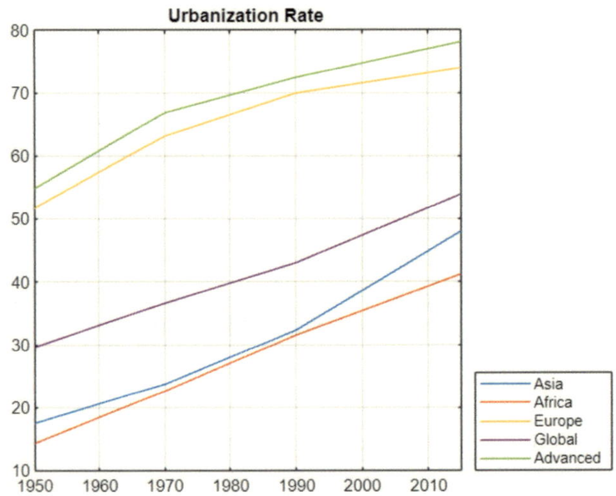

◘ **Fig. 5.51** Use of a table variable: Line plot with multiple lines and a legend (outside-bottom-right position)

◀

▶ Example of Table Variables

Here, we use the **title** and **legend** functions to create a title and legend. Specifically, after storing the properties of the **legend** function in 「lg」, we set 「Location」 to 「southeast」, to position the legend in the bottom-right corner inside the axes. Compare ◘ Fig. 5.51 with ◘ Fig. 5.52 to observe the difference in legend location. Additionally, we set 「boxoff」 in the **legend** function, to remove the box outline of the legend.

```
figure
plot(urban_mat.Year,[urban_mat.Asia,urban_mat.Africa,urban_mat.Europe,...
    urban_mat.Global,urban_mat.Advanced]);
xlim([urban_mat.Year(1),urban_mat.Year(end)]);
title("Urban Population Rate");
lg = legend("Asia","Africa","Europe","Global","Advanced");
    lg.Location = "southeast";
legend("boxoff");
grid on
```

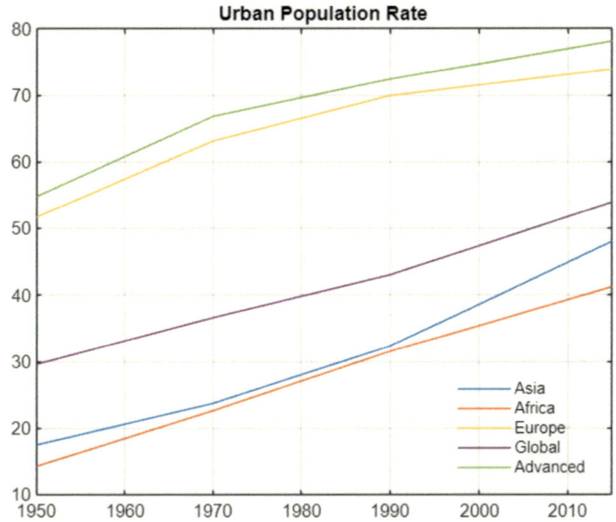

Fig. 5.52 Use of a table variable: Line plot with multiple lines and a frameless legend (bottom-right position)

◄

► Example of Table Variables

Here, using the **legend** function, we create a legend. Specifically, because we set 「LineStyle」 to 「--」 in the **plot** function, the legend represents the two dashed lines among the five plot graphs. Additionally, we set 「Location」 to 「southeast」, to locate the legend in the bottom-right corner inside the axes. Finally, we set 「NumColumns」 to 3, creating a legend with three columns. The resulting figure is shown in Fig. 5.53.

```
figure
pl = plot(urban_mat.Year,[urban_mat.Asia,urban_mat.Africa,...
    urban_mat.Europe,urban_mat.Global,urban_mat.Advanced]);
    pl(1).LineStyle = "--";
    pl(2).LineStyle = "--";
xlim([urban_mat.Year(1),urban_mat.Year(end)]);
title("Urban Population Rate");
lg = legend("Asia","Africa","Europe","Global","Advanced");
    lg.Location = "southeast";
    lg.NumColumns = 3;
grid on
```

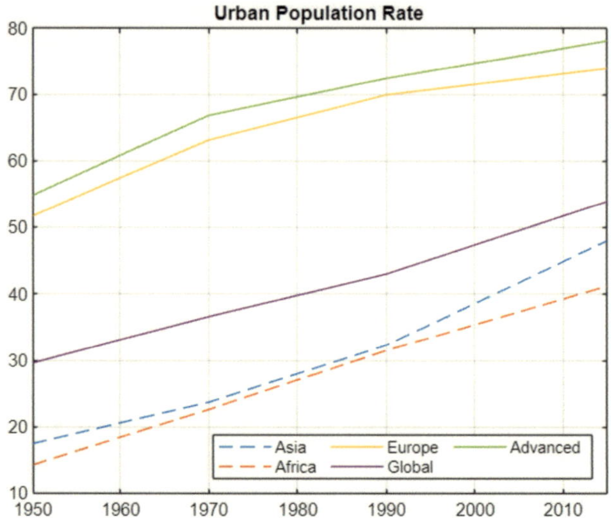

Fig. 5.53 Use of a table variable: Line plot with multiple lines and a 3-column legend (bottom-right position)

◄

► Example of Table Variables

Here, we create a 2-by-3 tiled layout using the **tiledlayout** function. Specifically, we can use「nexttile」to create the next layout each time. Note that by storing the properties of the **tiledlayout** function in the variable「t」, we can use「title(t,"Urbanization Rate")」to create a common title,「Urbanization Rate」, for all six plots. See Fig. 5.54.

```
figure
t = tiledlayout(2,3);
ti = title(t,"Urbanization Rate");
    ti.FontSize = 20;
nexttile;
plot(urban_mat.Year,urban_mat.Global);
title("Global countries");
grid minor
nexttile;
plot(urban_mat.Year,urban_mat.Advanced);
title("Advanced countries");
grid minor
nexttile;
plot(urban_mat.Year,urban_mat.Asia);
title("Asia");
grid minor
nexttile;
plot(urban_mat.Year,urban_mat.Africa);
title("Africa");
grid minor
nexttile;
plot(urban_mat.Year,urban_mat.Europe);
title("Europe");
grid minor
nexttile;
dim = [0.75 0.2 0.15 0.15];
ta = annotation("TextBox",dim,"String","Tile");
    ta.Color = "red";
```

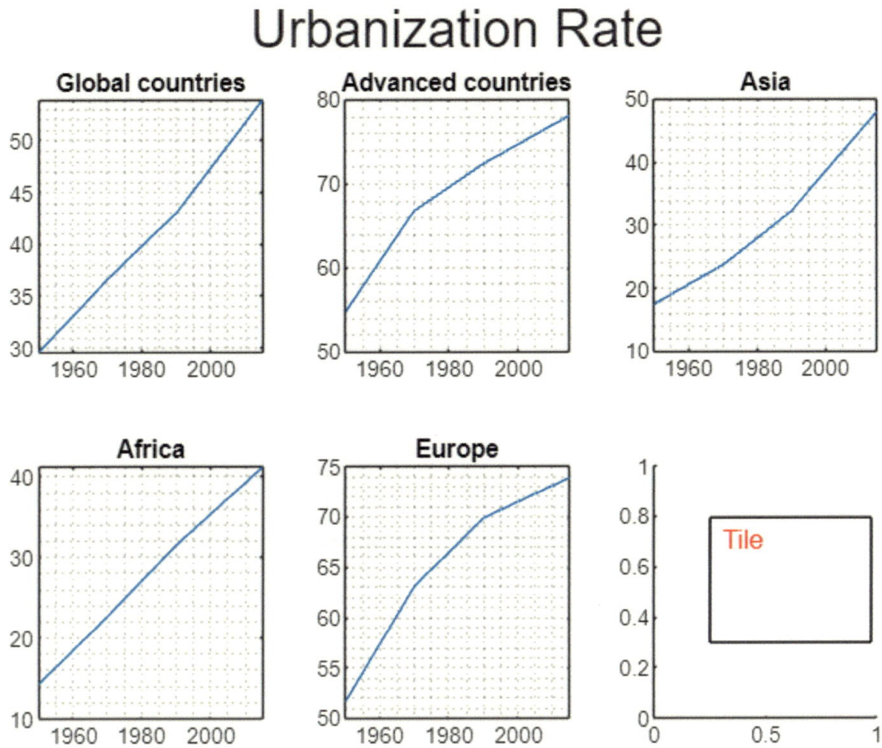

Fig. 5.54 Use of a table variable: Line plot (2-row, 3-column layout)

◄

► Example of Table Variables

Here, we use the **tiledlayout** function to create a 2-by-3 tiled layout. We show two different ways of using 「nexttile」. The first plot, specified with 「nexttile(2)」, selects the second tile in the layout, placed in the second position of a 1-by-2. Next, the second plot, specified with 「nexttile([1,3])」, is placed across the entire second row, occupying three columns. That is, 「nexttile([1,3])」 creates a 1-by-3 tile. Since there is no available space for a 1-by-3 tile in the first row, it is placed in the second row. The resulting figure is shown in Fig. 5.55.

```
figure
tiledlayout(2,3);
nexttile(2);
plot(urban_mat.Year,[urban_mat.Global,urban_mat.Advanced]);
lg1 = legend("Global","Advanced");
    lg1.Location = "westoutside";
nexttile([1,3]);
plot(urban_mat.Year,[urban_mat.Asia,urban_mat.Africa,urban_mat.Europe]);
lg2 = legend("Asia","Africa","Europe");
    lg2.Location = "north";
grid minor
```

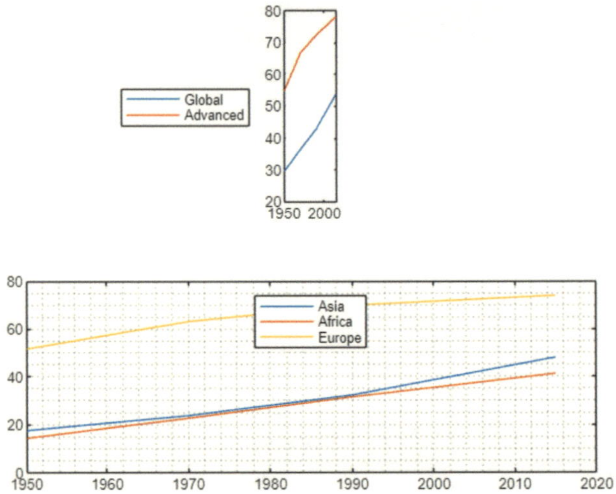

Fig. 5.55 Use of a table variable Line plot (2-row layout) with two legends

◄

5.6 Questions

Question 1

Create **Fig. 5.56**. Concretely, using the table variable 「urban_mat」, input 「urban_mat.Year」 as the first argument of the **plot** function and 「[urban_mat.Asia, urban_mat.Europe]」 as the second argument. Then, change two lines to dotted lines with green right-pointing triangles.

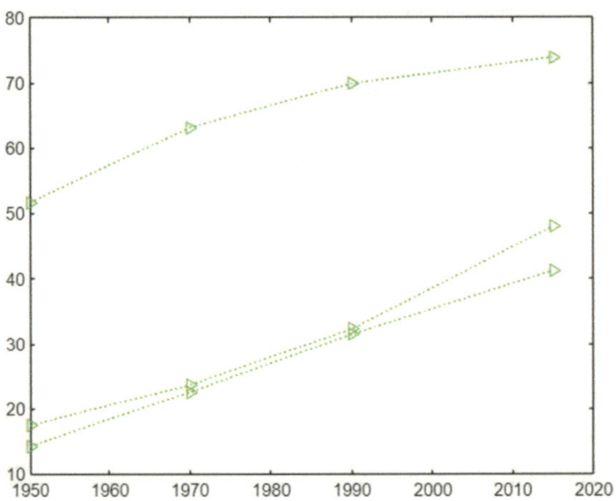

Fig. 5.56 Line plot (Question 1)

Answer 1

```
figure
plot(urban_mat.Year,[urban_mat.Asia,urban_mat.Africa,urban_mat.Europe],">:g");
```

Question 2

Using the **bar** function, create Fig. 5.57 with the first argument as 「Year1」 and the second argument as 「Global」. Specifically, to place the data values on top of the bars, use 「XEndPoints」, 「YEndPoints」, and 「YData」 of the bar graph. Adjust the figure so that the numbers do not overlap with the bars.

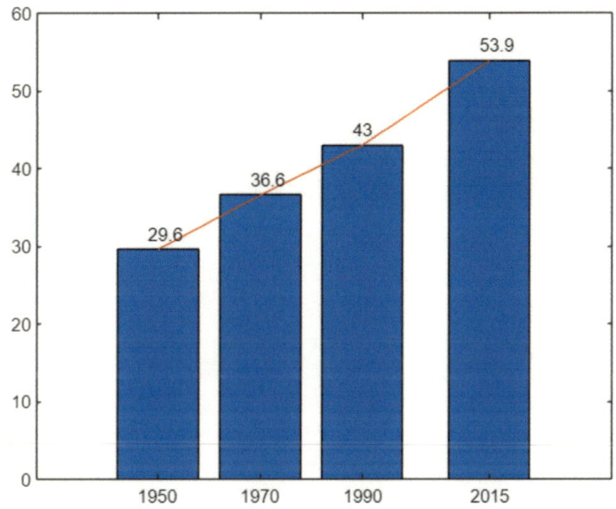

Fig. 5.57 Line plot (Question 2)

Answer 2

```
figure
Ex2 = bar(Year1,Global);
    xtip = Ex2.XEndPoints;
    ytip = Ex2.YEndPoints;
    label = string(Ex2.YData);
hold on
    plot(xtip,ytip);
hold off
text(xtip-2,ytip+2,label);
```

Question 3

Using the **plot** function, create Fig. 5.58. Use a solid black line with circular markers for the points. Additionally, use the **xlim** function, **yyaxis left**, and **yyaxis right** commands.

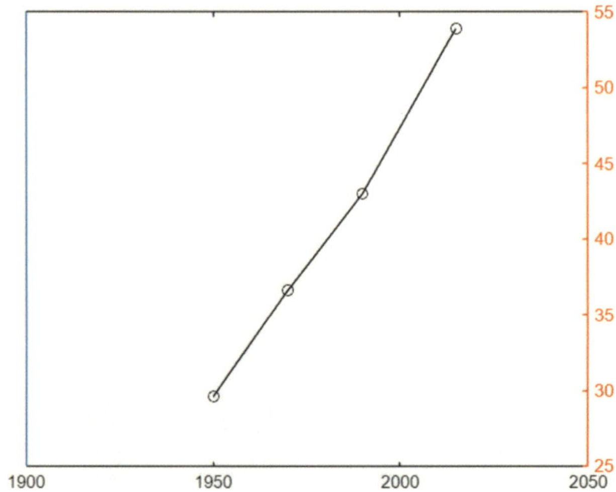

Fig. 5.58 Line plot (Question 3)

Answer 3

```
figure
yyaxis right;
plot(Year1,Global,"k-o");
xlim([1900 2050]);
yyaxis left;
set(gca,"YTick",[]);
```

Plot3 Function

Contents

6.1 Details of the Data Used in this Chapter – 176

6.2 Excel Data of ▶ Section 6.1 – 177

6.3 Basic Syntax of the Plot3 Function – 178

6.4 Setting and Controlling the Plot3 Function – 185

6.5 Combining Optional Functions and the Plot3 Function – 188

6.6 Questions – 198

Supplementary Information The online version contains supplementary material available at https://doi.org/10.1007/978-981-96-6213-5_6.

Y. Nakamoto, *A Comprehensive Guide to Creating Graphics with MATLAB Programming*, Classroom Companion: Economics, https://doi.org/10.1007/978-981-96-6213-5_6

Learning Objectives

This chapter creates 3D line plots using the **plot3** function. By reading this chapter, you will learn how to use the **plot3** function and understand the similarities and differences in both the created figures and syntax among the **plot3** function, the **scatter3** function (covered in ▶ Chap. 4), and the **plot** function (covered in ▶ Chap. 5).

Overview

The **plot3** function is useful for visualizing 3D data. The **scatter3** function, introduced in ▶ Chap. 4, also visualizes 3D data. Unlike the **scatter3** function, the **plot3** function allows us to plot curves or continuous data in 3D space. This has the advantage of visualizing complex data sets, such as time-varying data, which can be difficult to understand in 2D plots. On the other hand, similar to the **scatter3** function, a disadvantage is that we may not be able to interpret figures easily due to the 3D space. In this case, it is helpful to use the optional function introduced in ▶ Sect. 6.5.

6.1 Details of the Data Used in this Chapter

In this chapter, we use data from OECD (Organization for Economic Cooperation and Development) countries and SSA (Sub-Saharan Africa) countries. Specifically, 「Year」 represents the Year. 「Internet_OECD」 and 「Internet_SSA」 indicate the percentage of individuals using the Internet in OECD and SSA countries, respectively. Next, 「School_enrollment_OECD」 and 「School_enrollment_SSA」 represent the gross enrollment ratio in secondary education for OECD and SSA countries, respectively. In particular, the reason why the gross enrollment ratio in secondary education exceeds 100 is due to the nature of the gross enrollment ratio. For instance, students outside the typical age range and those who repeat or return to school can lead to an increase in the numerator (total number of students enrolled). As a result, the numerator may be larger than the denominator (population of the relevant age group), resulting in values greater than 100 percent. Finally, 「Young_unemployment_OECD」 and 「Young_unemployment_SSA」 represent the unemployment rate for the youth population aged 15–24.

```
Year = [2000,2001,2002,2003,2004,2005,2006,2007,...
    2008,2009,2010,2011,2012,2013,2014,2015,...
    2016,2017,2018,2019,2020,2021]';

Internet_OECD = [26.98,32.33,40.23,44.37,49.07,52.63,...
    55.18,60.50,62.62,64.15,66.56,67.54,70.19,71.53,...
    73.14,75.68,79.86,80.51,82.39,84.78,86.20,87.39]';

Internet_SSA = [0.52,0.66,0.84,1.11,1.50,2.02,2.66,3.37,...
    3.67,4.31,6.14,7.60,9.28,11.16,13.37,16.01,18.64,...
    21.27,24.83,27.76,31.82,35.86]';

School_enrollment_OECD = [94.80,96.12,97.02,97.73,97.06,...
    97.69,98.07,98.49,98.85,98.38,99.51,99.78,99.49,...
    103.78,104.77,105.42,105.76,106.69,106.46,106.65,...
    106.19,105.88]';

School_enrollment_SSA = [25.56,26.86,27.85,28.96,30.48,...
    31.39,32.43,33.36,35.30,37.08,38.99,40.19,41.43,...
    44.04,42.85,43.37,42.85,43.09,43.28,43.81,44.38,44.78]';

Young_unemployment_OECD = [14.19,13.89,14.74,15.19,15.27,...
    14.73,13.88,13.12,13.82,17.61,17.69,17.20,17.34,...
    17.24,16.22,15.06,14.23,13.22,12.46,12.64,15.82,13.66]';

Young_unemployment_SSA = [12.09,12.05,11.95,11.93,11.75,...
    11.66,11.53,11.46,11.44,11.59,12.39,12.01,11.85,...
    11.49,11.48,11.49,11.47,11.46,11.33,11.48,12.03,11.44]';

OECD_SSA  =  table(Year,Internet_OECD,Internet_SSA,...
    School_enrollment_OECD,School_enrollment_SSA,...
    Young_unemployment_OECD,Young_unemployment_SSA);
```

6.2 Excel Data of ▶ Section 6.1

▶ Section 6.2 contains the data used in ▶ Sects. 6.3, 6.4, 6.5, and 6.6, formatted as an Excel file. First, the file name (「alldata.xlsx」) and sheet name (「Chap6_OECD_SSA」) are identified. For more information about importing data from Excel, see ▶ Chap. 2. Next, we extract data from the table variable 「OECD_SSA」 and store it in the corresponding variables.

```
OECD_SSA = readtable("alldata.xlsx","Sheet","Chap6_OECD_SSA");
OECD_SSA = renamevars(OECD_SSA,"YEAR","Year");
Year = OECD_SSA.Year;
Internet_OECD = OECD_SSA.Internet_OECD;
Internet_SSA = OECD_SSA.Internet_SSA;
School_enrollment_OECD = OECD_SSA.School_enrollment_OECD;
School_enrollment_SSA = OECD_SSA.School_enrollment_SSA;
Young_unemployment_OECD = OECD_SSA.Young_unemployment_OECD;
Young_unemployment_SSA = OECD_SSA.Young_unemployment_SSA;
```

6.3 Basic Syntax of the Plot3 Function

Basic Syntax:

Since we create a 3D plot figure, we need three types of data. Referring to the first syntax 「plot3(X,Y,Z)」, enclosed in the black box in Figure 6.1, we need to specify 「X」, 「Y」, and 「Z」 to plot a 3D line. Note that these variables must be the same length.

Next, 「plot3(X,Y,Z,LineSpec)」 adds a fourth argument, 「LineSpec」, which allows us to specify the line style. For example, we can set the line style, marker, and color.

Moreover, the syntax 「plot3(X1,Y1,Z1,...,Xn,Yn,Zn)」 allows us to use multiple datasets simultaneously.

For instance, 「X1」, 「Y1」, and 「Z1」 could represent OECD data, while 「X2」, 「Y2」, and 「Z2」 could represent SSA data.

Finally, 「plot3(X1,Y1,Z1,LineSpec1,...,Xn,Yn,Zn,LineSpecn)」 allows us to adjust the line style, markers, and color for each 3D plot created with the different datasets.

Syntax

```
plot3(X,Y,Z)
plot3(X,Y,Z,LineSpec)
plot3(X1,Y1,Z1,...,Xn,Yn,Zn)
plot3(X1,Y1,Z1,LineSpec1,...,Xn,Yn,Zn,LineSpecn)

plot3(tbl,xvar,yvar,zvar)

plot3(ax, __ )
plot3( __ ,Name,Value)
p = plot3( __ )
```

Fig. 6.1 The syntax of the plot3 function

Here, we enter the numbers 1, 3, and 5 in the first, second, and third arguments of the **plot3** function. However, since the **plot3** function is used to draw lines, at least two points are required to connect a line between them. If only one point is provided as in the following program, you must specify a marker. Consequently, the following program, which does not specify any marker, does not generate any plot. See Fig. 6.2.

```
plot3(1,3,5);
```

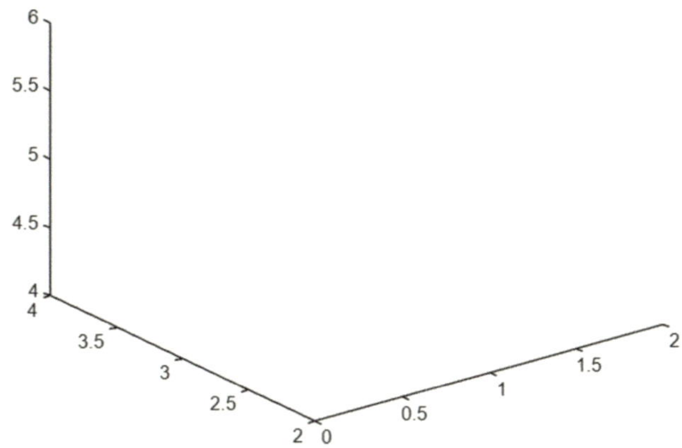

Fig. 6.2 3D plot not generated due to a single data point

The difference from the previous program that created Fig. 6.2 is that the fourth argument is specified with the marker 「o」. As a result, a circle marker is created at the point 「(X,Y,Z)=(1,3,5)」. See Fig. 6.3.

```
plot3(1,3,5,"o");
```

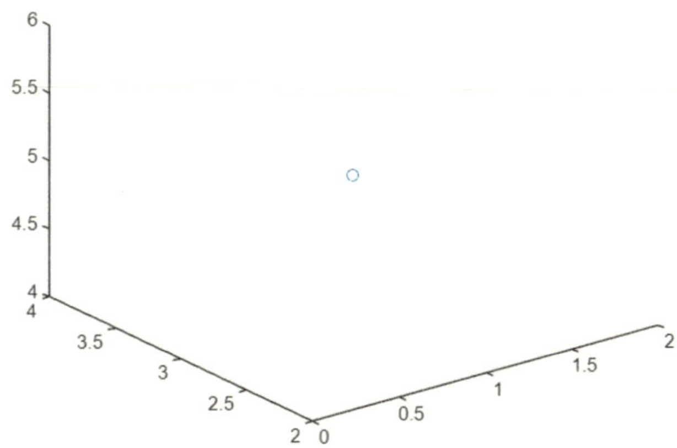

Fig. 6.3 3D plot generated due to a marker for a single data point

Here, we enter 1-by-2 vectors for each argument. Consequently, two points are specified: (X,Y,Z)=(1,3,5) and (X,Y,Z)=(2,4,6), and a line is drawn connecting these two points. See Fig. 6.4

```
plot3([1 2],[3 4],[5 6]);
```

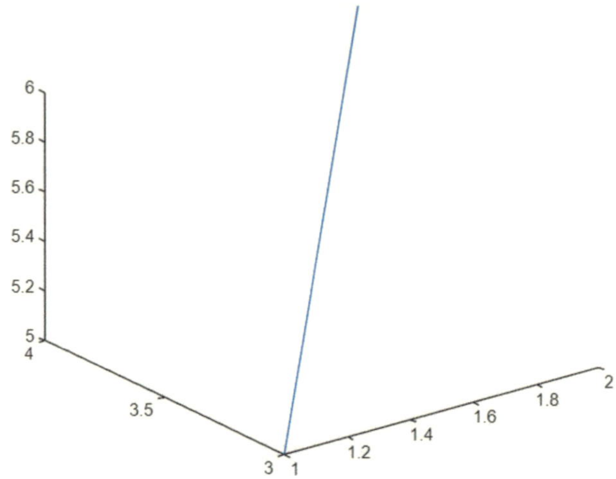

Fig. 6.4 3D plot with one line

Here, unlike the previous program that created Fig. 6.4, we enter a 2-by-2 matrix for the third argument. Consequently, we create not only a line connecting the two points, (X,Y,Z)=(1,3,5) and (X,Y,Z)=(2,4,6), but also an additional line connecting the points, (X,Y,Z)=(1,3,10) and (X,Y,Z)=(2,4,20). See Fig. 6.5.

```
plot3([1 2],[3 4],[5 6;10 20]);
```

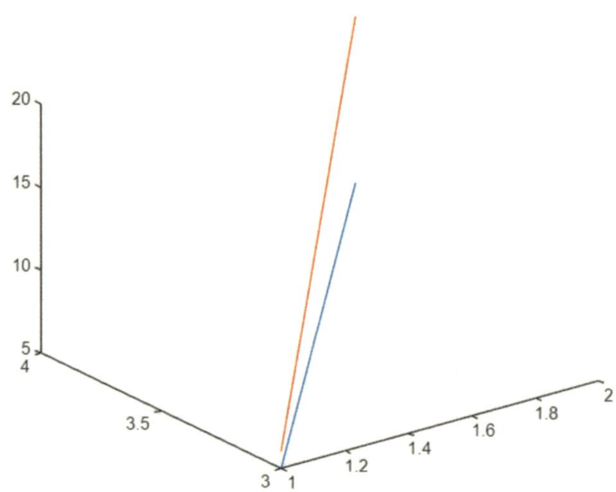

Fig. 6.5 3D plot with two lines

The following program is the same as the previous one that generated ◘ Fig. 6.5, except for the fourth argument. In this program, the fourth argument specifies the marker as 「o」, creating circle markers at each of the specified coordinates. See ◘ Fig. 6.6.

```
plot3([1 2],[3 4],[5 6;10 20],"o");
```

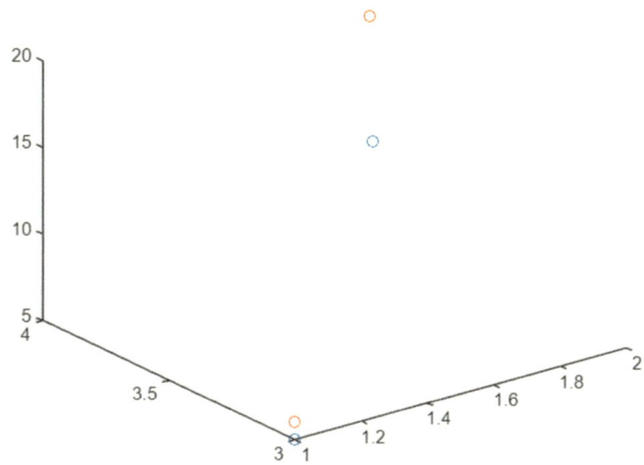

◘ **Fig. 6.6** 3D plot with two groups of circle markers

In the following program, we use the basic syntax, 「plot3(X1,Y1,Z1,LineSpec1,X2,Y2,Z2,LineSpec2)」. In the first dataset, we connect the two points, (X,Y,Z)=(1,3,5) and (X,Y,Z)=(2,4,6), with a dashed line where the marker is set to 「x」. In the second dataset, we connect the two points, (X,Y,Z)=(20,5,1) and (X,Y,Z)=(10,10,2), with a dash-dotted line, where the marker is set to 「o」. The resulting figure is presented in ◘ Fig. 6.7.

```
plot3([1 2],[3 4],[5 6],"--x",[20,10],[5,10],[1,2],"-.o");
```

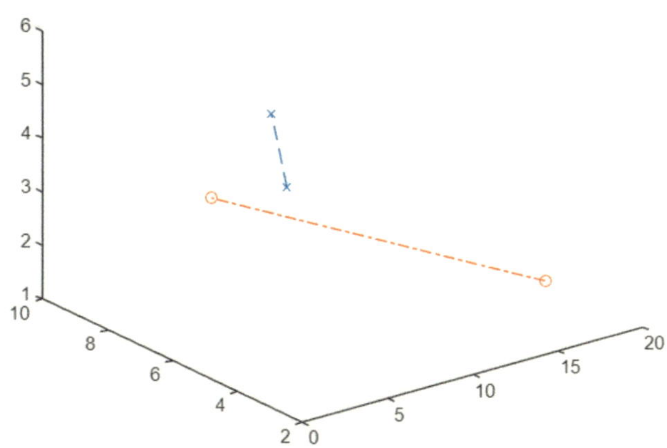

◘ **Fig. 6.7** 3D plot with dashed and dash-dotted lines and different markers

Here, we create a 3D plot using 「Year」 as the first argument, 「Young_unemployment_OECD」 as the second argument, and 「School_enrollment_OECD」 as the third argument. Since no marker is specified, a line is drawn connecting each of the points. See ◘ Fig. 6.8.

```
plot3(Year,Young_unemployment_OECD,School_enrollment_OECD);
```

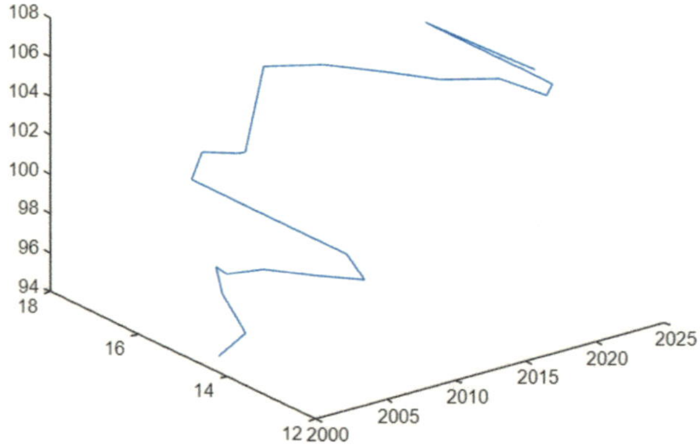

◘ **Fig. 6.8** 3D plot with a line from three variables

Here, we change the order of the variables in the previous program that generated ◘ Fig. 6.8. Specifically, we use the **plot3** function to create a 3D plot with 「Young_unemployment_OECD」 as the first argument, 「School_enrollment_OECD」 as the second argument, and 「Internet_OECD」 as the third argument. See ◘ Fig. 6.9.

```
plot3(Young_unemployment_OECD,School_enrollment_OECD,Internet_OECD);
```

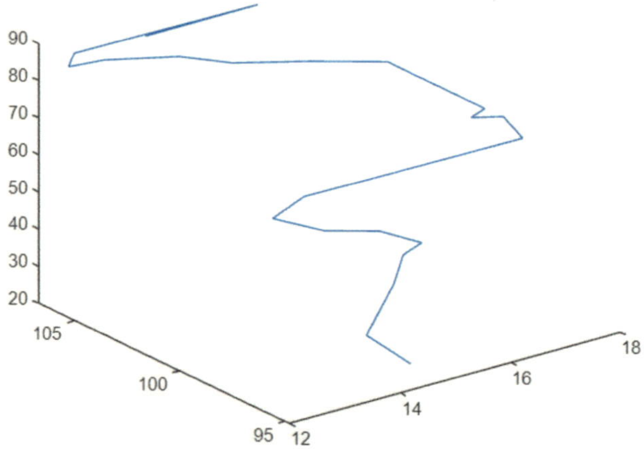

◘ **Fig. 6.9** 3D plot with a line from different three variables (cf. ◘ Fig. 6.8)

Here, we enter 「[Year(1:11),Year(12:22)]」 as the first argument, 「[Internet_OECD(1:11), Internet_OECD(12:22)]」 as the second argument, and 「[Young_unemployment_OECD(1:11),Young_unemployment_OECD(12:22)]」 as the third argument. This creates two lines, with each column representing one group. See ◘ Fig. 6.10 for the created figure.

```
plot3([Year(1:11),Year(12:22)],...
    [Internet_OECD(1:11),Internet_OECD(12:22)],...
    [Young_unemployment_OECD(1:11),Young_unemployment_OECD(12:22)]);
```

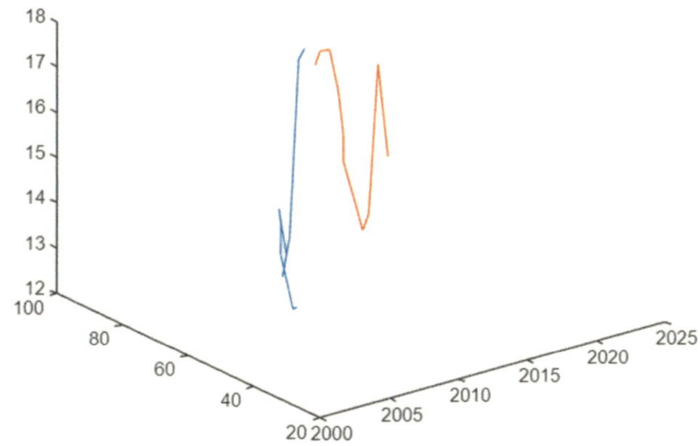

◘ **Fig. 6.10** 3D plot with two lines created by two subsets of variables

Here, we create the same figure as ◘ Fig. 6.10. However, the following program is different from the previous one. Specifically, while the previous program used three arguments, the following program uses six arguments. Compare the previous program in ◘ Fig. 6.10 with that in ◘ Fig. 6.11.

```
plot3(Year(1:11),Internet_OECD(1:11),Young_unemployment_OECD(1:11),...
    Year(12:22),Internet_OECD(12:22),Young_unemployment_OECD(12:22));
```

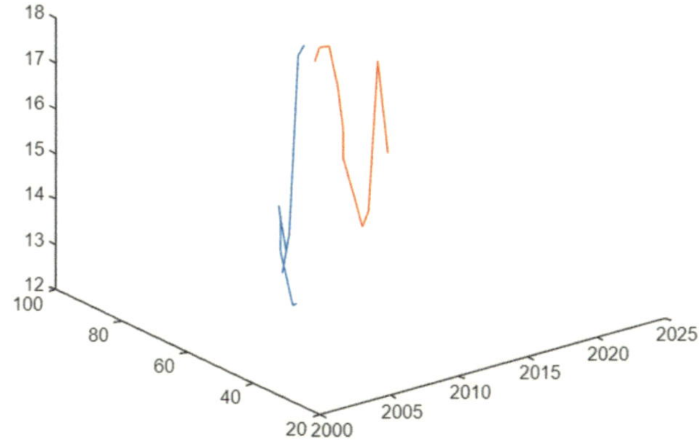

◘ **Fig. 6.11** 3D plot using repeated X-, Y-, and Z-axis data for two subsets (cf. ◘ Fig. 6.10)

Here, we enter 「Year」 as the first argument, 「Young_unemployment_SSA」 as the second argument, and 「School_enrollment_SSA」 as the third argument to create a 3D plot. In this respect, the following program is similar to the previous one that created ◘ Fig. 6.8. However, in the following program, we enter 「o」 as the fourth argument to create only the marker with no line. See ◘ Fig. 6.12.

```
plot3(Year,Young_unemployment_SSA,School_enrollment_SSA,"o");
```

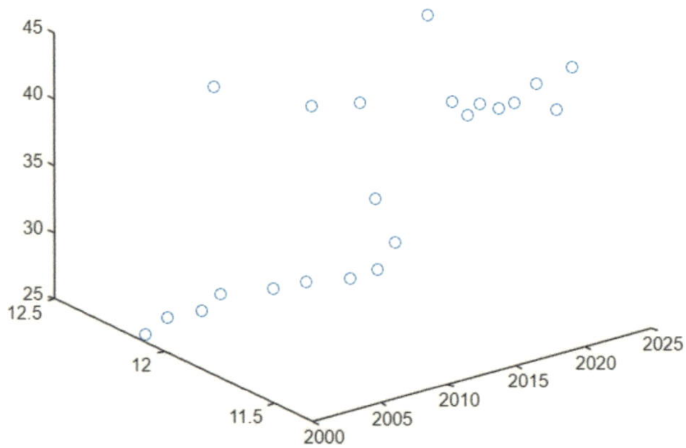

◘ **Fig. 6.12** 3D plot with circle markers

Here, we input 「:x」 as the fourth argument. As a result, we commonly create the dotted lines with cross markers in two groups. The resulting figure is shown in ◘ Fig. 6.13.

```
plot3([Year(1:11),Year(12:22)],[Internet_SSA(1:11),Internet_SSA(12:22)],...
    [Young_unemployment_SSA(1:11),Young_unemployment_SSA(12:22)],":x");
```

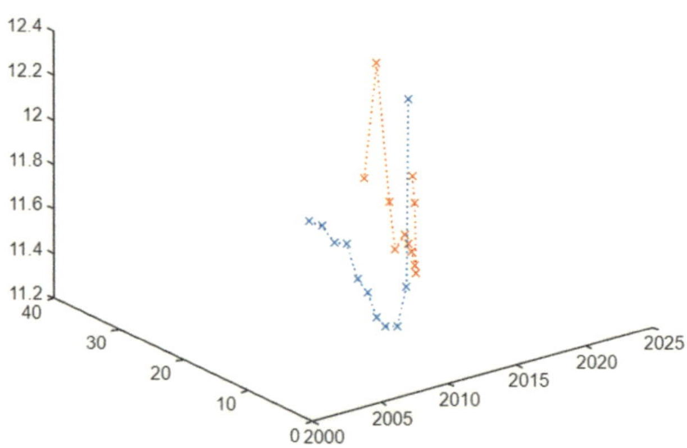

◘ **Fig. 6.13** 3D plot with two dotted lines and cross markers

Here, we enter 「-.^」 as the fourth argument, creating a dash-dotted line with upward-pointing triangle markers for the first group's data. Next, we enter 「--hexagram」 as the eighth argument, which creates a dashed line with hexagram markers for the second group's data. See ◘ Fig. 6.14 for the created figure.

```
plot3(Year,Internet_OECD,Young_unemployment_OECD,"-.^",...
    Year,Internet_SSA,Young_unemployment_SSA,"--hexagram");
```

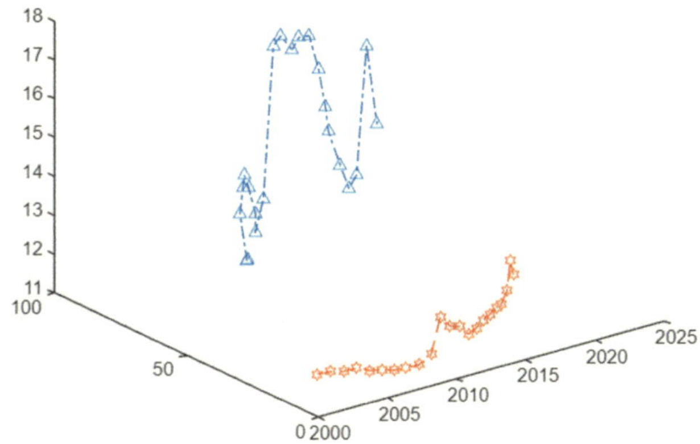

◘ **Fig. 6.14** 3D plot with dash-dotted triangle and dashed hexagram lines

6.4 Setting and Controlling the Plot3 Function

Here, we store the properties of the **plot3** function in the variable 「pl3」. Then, we set 「LineStyle」 to 「--」, 「Color」 to 「red」, 「MarkerFaceColor」 to 「green」, and 「Marker」 to 「^」, creating a red dashed line with green upward-pointing triangle markers. See ◘ Fig. 6.15.

```
pl3 = plot3(Year,Young_unemployment_OECD,School_enrollment_OECD);
    pl3.LineStyle = "--";
    pl3.Color = "red";
    pl3.MarkerFaceColor = "green";
    pl3.Marker = "^";
```

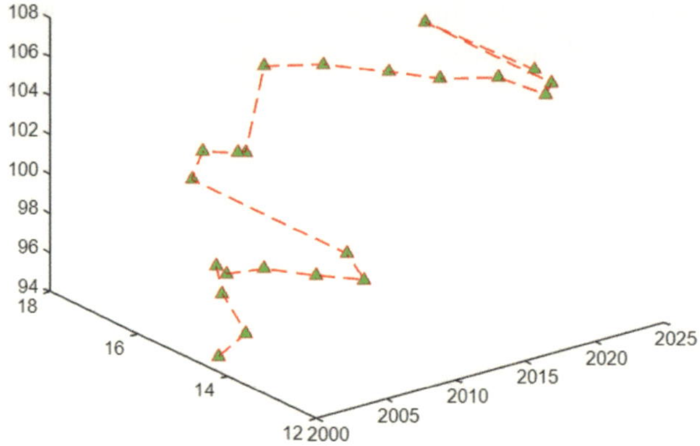

Fig. 6.15 3D plot with a red dashed line and green-filled triangle markers

Here, in addition to the previous program used to create Fig. 6.15, we set「LineWidth」to 2 and「MarkerSize」to 10, creating a thick red dashed line with large green upward-pointing triangle markers. See Fig. 6.16 for the created figure.

```
pl3 = plot3(Year,Young_unemployment_OECD,School_enrollment_OECD);
    pl3.LineStyle = "--";
    pl3.Color = "red";
    pl3.MarkerFaceColor = "green";
    pl3.Marker = "^";
    pl3.LineWidth = 2;
    pl3.MarkerSize = 10;
```

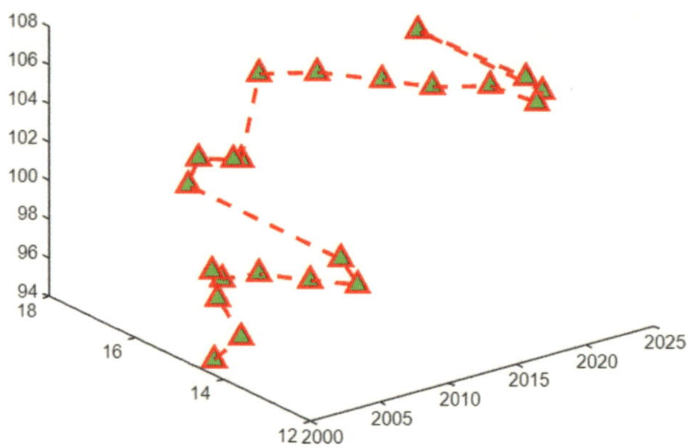

Fig. 6.16 3D plot with a thick line and large markers (cf. Fig. 6.15)

We set「LineStyle」to「-」,「Marker」to「x」,「MarkerEdgeColor」to「black」, and「MarkerSize」to 8, creating a solid line with large black cross markers. The resulting figure is shown in Fig. 6.17.

```
pl3 = plot3(Year,School_enrollment_SSA,Internet_SSA);
    pl3.LineStyle = "-";
    pl3.Marker = "x";
    pl3.MarkerEdgeColor = "black";
    pl3.MarkerSize = 8;
```

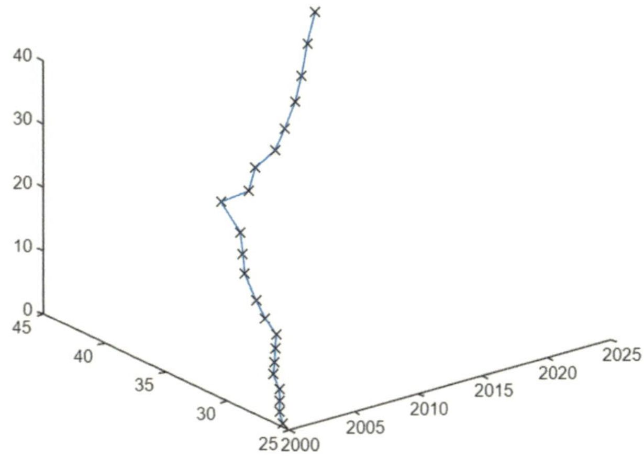

Fig. 6.17 3D plot with one line and cross black markers

Here, we set 「LineStyle」 to 「-」, 「MarkerEdgeColor」 to 「black」, 「Marker」 to 「x」, 「LineWidth」 to 4, and 「MarkerSize」 to 12, creating a thick solid line with large black cross markers. See Fig. 6.18 for the created figure.

```
pl3 = plot3(Year,School_enrollment_SSA,Internet_SSA);
    pl3.LineStyle = "-";
    pl3.MarkerEdgeColor = "black";
    pl3.Marker = "x";
    pl3.LineWidth = 4;
    pl3.MarkerSize = 12;
```

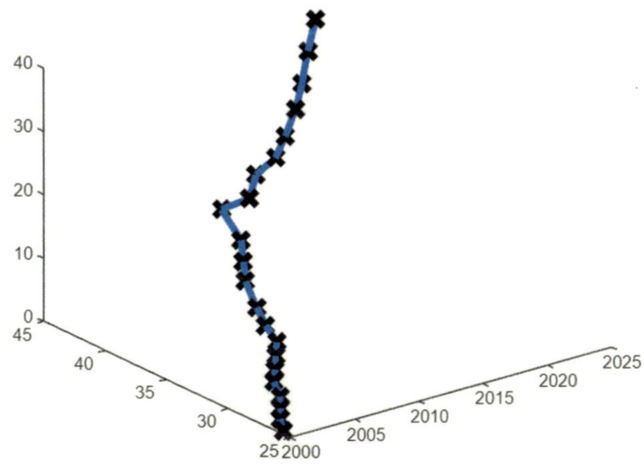

Fig. 6.18 3D plot with a thick line and large markers (cf. Fig. 6.17)

6.5 Combining Optional Functions and the Plot3 Function

Here, we use the **xlabel**, **ylabel**, and **zlabel** functions. Additionally, the **grid on** and **box on** commands are used to make the position of the plot lines clearer. See ◘ Fig. 6.19.

```
plot3(Year,School_enrollment_OECD,Internet_OECD,":o",...
    Year,School_enrollment_SSA,Internet_SSA,"--x");
xlabel("Year");
ylabel("Enrollment");
zlabel("Internet");
grid on
box on
```


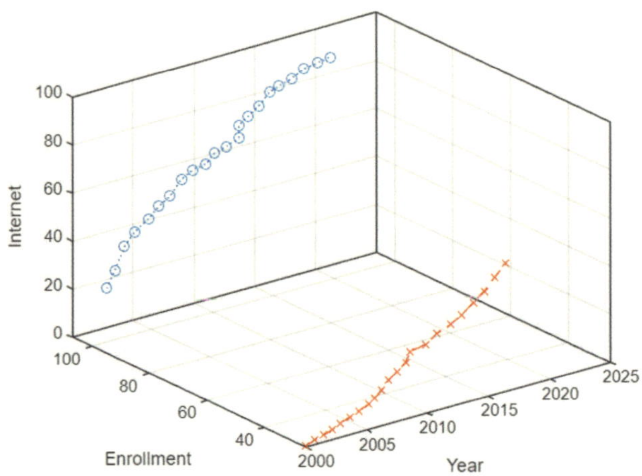

◘ **Fig. 6.19** 3D plot with black frame, gridlines, and axis labels

Here, we set「:o」and「--x」in the fourth and eighth arguments of the **plot3** function. This creates a dotted line with circle markers for the first group's plot and a dashed line with cross markers for the second group's plot.

Next, we store the properties of the **plot3** function in the variable「pl3」. Then, for「pl3(1)」, we set「MarkerSize」to 5 and「MarkerFaceColor」to「black」, creat ing a plot line with larger, filled black markers for the first group. Additionally, we adjust the viewing angle by setting「view([10 20])」. See ◘ Fig. 6.20.

```
pl3 = plot3(Year,School_enrollment_OECD,Internet_OECD,":o",...
    Year,School_enrollment_SSA,Internet_SSA,"--x");
    pl3(1).MarkerSize = 5;
    pl3(1).MarkerFaceColor = "black";
xlabel("Year");
ylabel("Enrollment");
zlabel("Internet");
grid on
box on
view([10 20]);
```

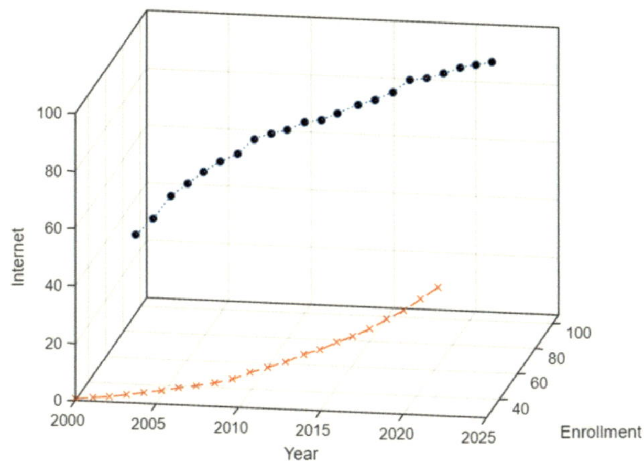

Fig. 6.20 3D plot from a different angle, with black frame, gridlines, and axis labels

The following program is almost the same as those shown in Figs. 6.19 and 6.20. Additionally, we use the **hold on** command and a **for** loop to create descent lines from each marker. Specifically, within the **plot3** function of the **for** loop, we use 「[Internet_OECD(i),0]」 in the third argument to draw descent lines along the z-axis. See ▶ Figs. 4.31 through ▶ 4.34 for more details on creating descent lines. The resulting figure is displayed in Fig. 6.21.

```
pl3 = plot3(Year,School_enrollment_OECD,Internet_OECD,"--xr");
    pl3.MarkerSize = 10;
    pl3.LineWidth = 1;
xlabel("Year");
ylabel("Enrollment");
zlabel("Internet");
grid on
box on
view([10 20]);
hold on
for i = 1:length(Year)
    plot3([Year(i),Year(i)],...
        [School_enrollment_OECD(i),School_enrollment_OECD(i)],...
        [Internet_OECD(i),0],"k:","LineWidth",0.5);
end
hold off
```

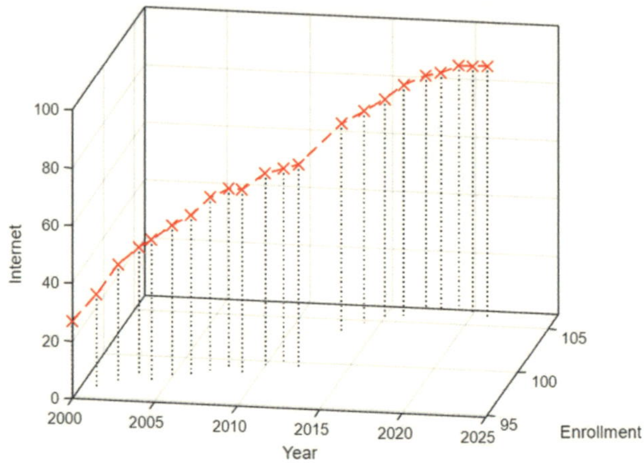

6

Fig. 6.21 3D plot using a for loop to create vertical dotted droplines

Here, as in the previous program that created Fig. 6.21, we create descent lines. Additionally, using the **text** function, we place the value of 「Internet_OECD(i)」 next to each marker. Specifically, note that in the second argument of the **text** function, we use 「School_enrollment_OECD(i)-1」 to shift the text along the y-axis by -1, which prevents the generated text from overlapping with the markers. For more details about creating the text in the 3D figure, see the explanation of ▶ Fig. 4.26. For more details on the **text** function, see ▶ Sect. 18.2. The resulting figure is presented in Fig. 6.22.

```
pl3 = plot3(Year,School_enrollment_OECD,Internet_OECD,"--x");
    pl3.MarkerSize = 10;
    pl3.LineWidth = 1;
grid on
hold on
for i = 1:length(Year)
    plot3([Year(i),Year(i)],...
        [School_enrollment_OECD(i),School_enrollment_OECD(i)],...
        [Internet_OECD(i),0],"k:","LineWidth",0.5);
    text(Year(i),School_enrollment_OECD(i)-1,Internet_OECD(i),...
        num2str(Internet_OECD(i),"%.2f"),"FontSize",8,"Color","blue");
end
hold off
```

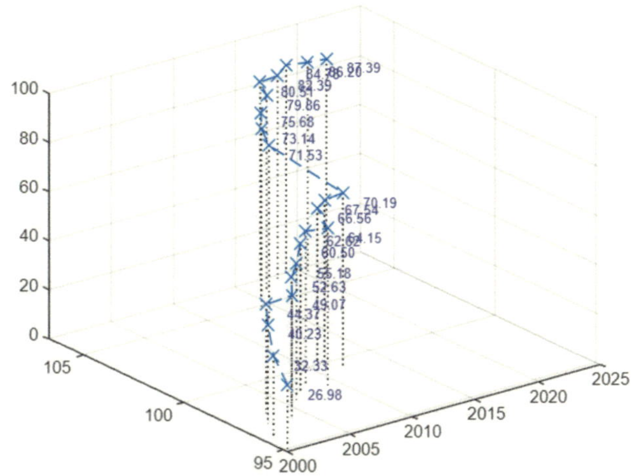

Fig. 6.22 3D plot using a for loop to create vertical dotted droplines and data labels

Here, as in the program that created Fig. 6.22, we create descent lines and place the text showing the value of 「Internet_SSA(i)」 next to each marker. Specifically, note that in the **for** loop, we set 「for i=1:4:length(Year)」. This means that instead of placing the text next to every marker, we place it next to every fourth marker. In the **text** function, we set 「HorizontalAlignment」 to 「right」 to adjust the position of the text. For more details on the **text** function, see ▶ Sect. 18.2. We also use the **grid minor** command to create finer grid lines. See Fig. 6.23.

```
pl3 = plot3(Year,School_enrollment_SSA,Internet_SSA,"--x");
    pl3.MarkerSize = 10;
    pl3.LineWidth = 1;
hold on
for i = 1:4:length(Year)
    plot3([Year(i),Year(i)],...
        [School_enrollment_SSA(i),School_enrollment_SSA(i)],...
        [Internet_SSA(i),0],"k:","LineWidth",1.5);
    text(Year(i),School_enrollment_SSA(i),Internet_SSA(i),...
        num2str(Internet_SSA(i),"%.2f"),...
        "HorizontalAlignment","right","FontSize",8,"Color","blue");
end
grid minor
view([10 20]);
hold off
box on
```

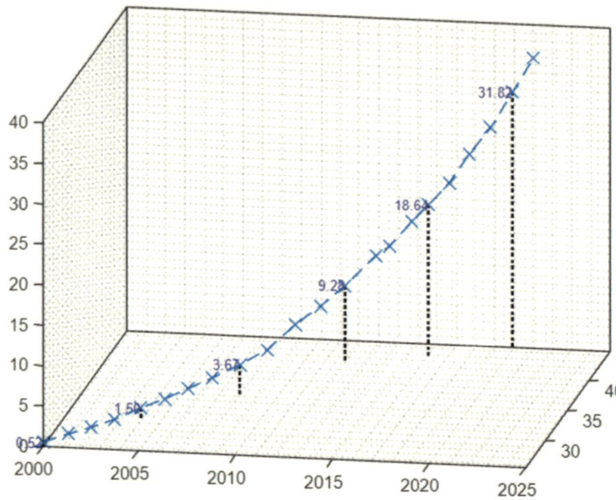

Fig. 6.23 3D plot using a for loop to create vertical dotted droplines and data labels for selected points

6

Here, as in the previous program that created Fig. 6.23, we use a **for** loop to place text next to every fourth marker. Additionally, by setting 「view([0 90])」, we adjust the viewing angle of a 2D plot. The **xlim** function sets the lower and upper limits of the x-axis. The resulting figure is shown in Fig. 6.24.

```
pl3 = plot3(Year,School_enrollment_SSA,Internet_SSA,"--x");
    pl3.MarkerSize = 10;
    pl3.LineWidth = 1;
grid minor
view([0 90]);
hold on
for i = 1:4:length(Year)
    text(Year(i),School_enrollment_SSA(i),Internet_SSA(i),...
        num2str(Internet_SSA(i),"%.2f"),...
        "HorizontalAlignment","right","FontSize",8,"Color","blue");
end
hold off
xlim([2000 2022]);
xlabel("Year");
```

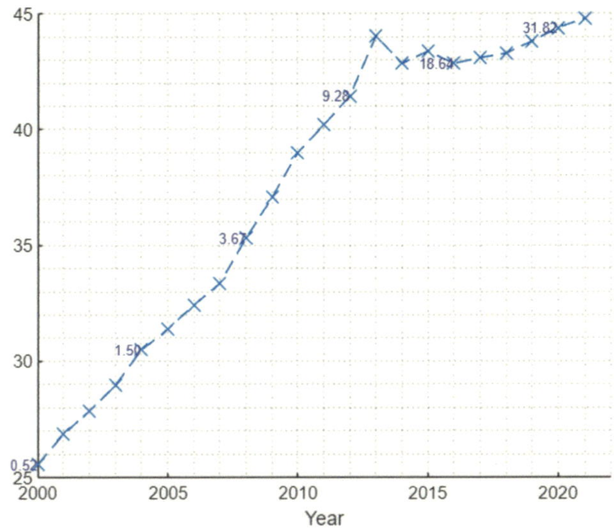

Fig. 6.24 Flat 3D plot using a for loop to create data labels for selected points

Here, as in the previous program that created Fig. 6.24, we adjust the viewing angle of a 2D plot; however, we use 「view(2)」. Additionally, within the **for** loop, we place text next to each marker. Specifically, by setting the second argument of the **text** function to 「School_enrollment_OECD(i)+1」 and 「HorizontalAlignment」 to 「center」, we ensure that the text does not overlap the markers. For more details on the **text** function, see ▶ Sect. 18.2.

Moreover, we use the **annotation** function twice. In the first **annotation** function, we set the first argument to 「Arrow」, creating an arrow. In the second **annotation** function, we set the first argument to 「TextBox」, creating a text box with the label 「Large increase」 enclosed in a rectangle. For more details on the **annotation** function, see ▶ Sect. 18.1. See Fig. 6.25.

```
pl3 = plot3(Year,School_enrollment_OECD,Internet_OECD,"o");
    pl3.MarkerSize = 10;
    pl3.MarkerFaceColor = "red";
    pl3.LineWidth = 1;
grid minor
view(2);
hold on
for i = 1:length(Year)
    text(Year(i),School_enrollment_OECD(i)+1,Internet_OECD(i),...
        num2str(Internet_OECD(i),"%.2f"),...
        "HorizontalAlignment","center","FontSize",6,"Color","red");
end
x = [0.55 0.59];
y = [0.46 0.66];
dim = [0.6 0.5 0.1 0.1];
annotation("Arrow",x,y);
annotation("TextBox",dim,"String","Large increase");
hold off
xlim([2000 2022]);
xlabel("Year");
```

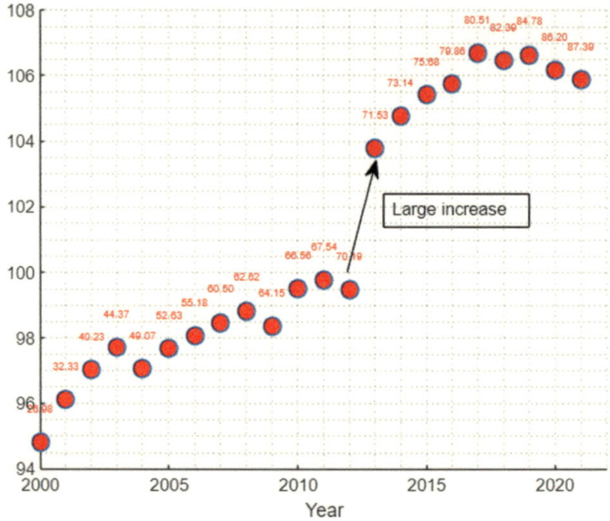

Fig. 6.25 Flat 3D plot with data labels, a text arrow, and a textbox

Here, as in Fig. 6.25, we adjust the viewing angle of a 2D plot by using 「view(2)」. Additionally, using the **hold on** command, we overlay plot lines created by the **plot3** function. Specifically, the second **plot3** function draws a dotted line at y=100, and the third **plot3** function draws a dotted line at y=103.5. Note that in both **plot3** functions, the third argument is set to 「[0, 0]」 to correspond to the 2D viewing angle. Although the **view** function adjusts the viewing angle of a 2D plot, note that the **xline** and **yline** functions cannot be used in this case. For more details on the **xline** and **yline** functions, see ► Sect. 18.3. See Fig. 6.26.

```
pl3 = plot3(Year,School_enrollment_OECD,Internet_OECD,"o");
    pl3.MarkerSize = 10;
    pl3.MarkerFaceColor = "red";
    pl3.LineWidth = 1;
grid minor
view(2);
hold on
    plot3([Year(1),2022],[100,100],[0,0],"b:","LineWidth",2);
    plot3([Year(1),2022],[103.5,103.5],[0,0],"b:","LineWidth",2);
hold off
xlim([2000 2022]);
xlabel("Year");
```

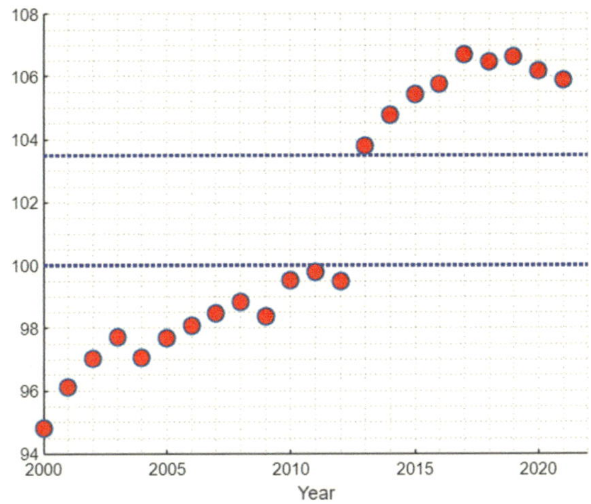

Fig. 6.26 Flat 3D plot with two horizontal dotted lines

Here, for the point at 「Year=2021」 (「Year(end)」), we create not only a descent line but also lines parallel to the x-axis and y-axis. Specifically, we use the **plot3** function four times.

The first **plot3** function creates a black dotted line with cross markers. In the second **plot3** function, the first argument is 「[Year(end), Year(end)]」, and the second argument is 「[School_enrollment_OECD(end),School_enrollment_OECD(end)]」, keeping the x-axis and y-axis fixed. Then, the third argument is 「[Internet_OECD(end),0]」, which draws a descent line along the z-axis. In the third **plot3** function, as the first argument is 「[Year(1),Year(end)]」 and the third argument is 「[0,0]」, we draw a line parallel to the x-axis at z=0. Moreover, in the fourth **plot3** function, as the second argument is 「[0,School_enrollment_OECD(end)]」 and the third argument is 「[0,0]」, we draw a line parallel to the y-axis at z=0. See Fig. 6.27 for the created figure. The program that created Fig. 4.34 made use of a similar approach, so please refer to it as well.

Additionally, we use the **text** function. For more details on the **text** function, see ► Sect. 18.2.

```
pl3 = plot3(Year,School_enrollment_OECD,Internet_OECD,":xk");
    pl3.MarkerSize = 5;
    pl3.LineWidth = 1;
grid on
hold on
    plot3([Year(end),Year(end)],...
        [School_enrollment_OECD(end),School_enrollment_OECD(end)],...
        [Internet_OECD(end),0],...
        "r:","LineWidth",2);
    plot3([Year(1),Year(end)],...
        [School_enrollment_OECD(end),School_enrollment_OECD(end)],...
        [0, 0],"r:","LineWidth",2);
    plot3([Year(end),Year(end)],...
        [0,School_enrollment_OECD(end)],...
        [0,0],"r:","LineWidth",2);
    text(Year(end),School_enrollment_OECD(end),Internet_OECD(end)+3,...
        num2str(Internet_OECD(end),"%.2f"),...
        "HorizontalAlignment","left","FontSize",8,"Color","blue");
hold off
```

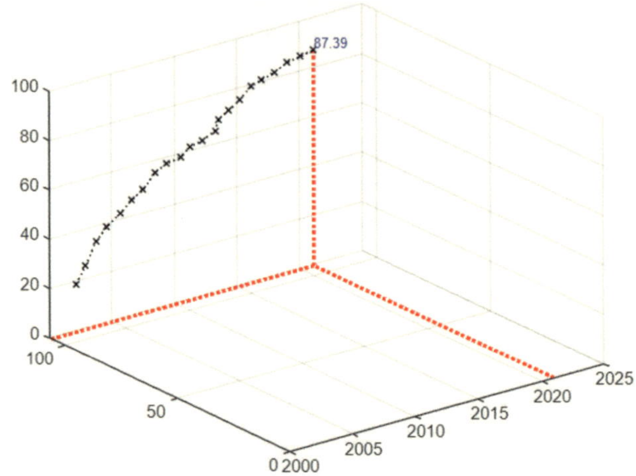

Fig. 6.27 3D plot with three-directional dotted droplines

Here, by using 「tiledlayout(1,2)」, we create a 1-by-2 tiled layout. Then, we create a 3D figure in each of these spaces using the **plot3** function. Specifically, the data for OECD is used in the 1-by-1 tile, while the data for SSA is used in the 1-by-2 tile. See Fig. 6.28 for the resulting figure.

```
figure
ti = tiledlayout(1,2);
title(ti,"School enrollment and young unemployment")
xlabel(ti,"Year");
nexttile;
plot3(Year,School_enrollment_OECD,Young_unemployment_OECD,"o-");
title("OECD");
view([10 20]);
grid minor
box on
nexttile;
plot3(Year,School_enrollment_SSA,Young_unemployment_SSA,"x-");
title("SSA");
view([10 20]);
grid minor
box on
```

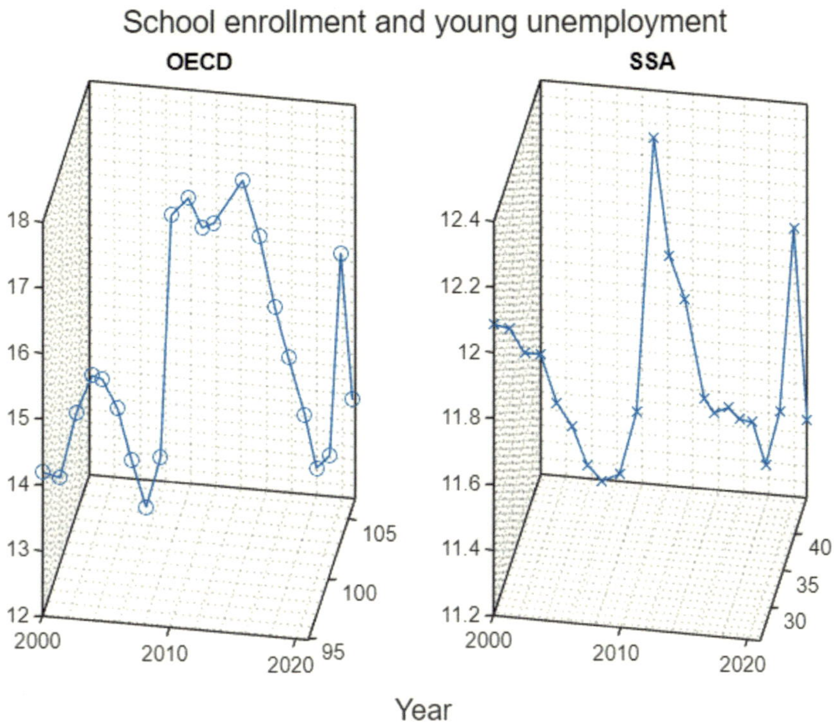

Fig. 6.28 3D plot (2-column layout)

Here, we use 「tiledlayout(2,2)」 to create a 2-by-2 tiled layout. Then, we create a 3D plot in each of these tiles using the **plot3** function. The resulting figure is given in Fig. 6.29.

```
figure
ti = tiledlayout(2,2);
title(ti,"OECD and SSA:School enrollment");
xlabel(ti,"Year");
nexttile;
plot3(Year,School_enrollment_OECD,Young_unemployment_OECD,"o-");
title("OECD: Young unemployment");
view([10 20]);
grid minor
box on
nexttile;
plot3(Year,School_enrollment_OECD,Internet_OECD,"o-");
title("OECD: Use of Internet");
grid minor
nexttile;
plot3(Year,School_enrollment_SSA,Young_unemployment_SSA,"x-");
title("SSA: Young unemployment");
view([10 20]);
grid minor
box on
nexttile;
plot3(Year,School_enrollment_SSA,Internet_SSA,"x-");
title("SSA: Use of Internet");
grid minor
```

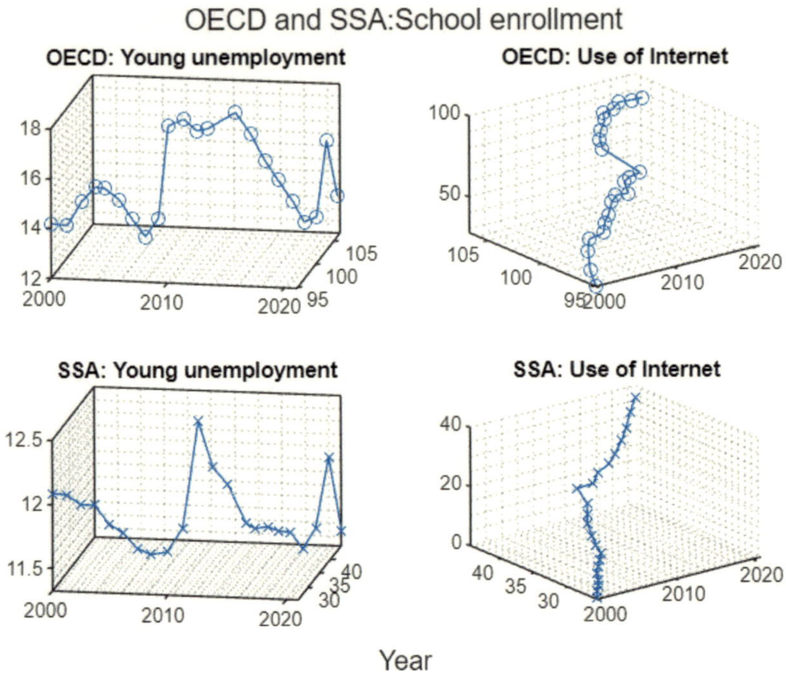

Fig. 6.29 3D plot (2-row, 2-column layout)

6

6.6 Questions

Question 1

Using the table variable 「OECD_SSA」, create Fig. 6.30. Use 「Year」 as the first argument, 「School_enrollment_SSA」 as the second argument, and 「Internet_SSA」 as the third argument. Then, set 「LineStyle」 to dashed, 「Color」 to cyan, 「Marker」 to a left-pointing triangle, and 「MarkerFaceColor」 to blue. Additionally, add grid lines.

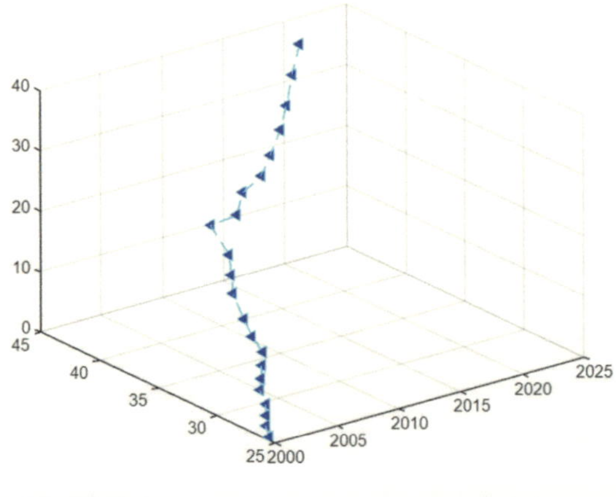

Fig. 6.30 3D plot (Question 1)

Answer 1

```
figure
Ex1 = plot3(OECD_SSA.Year,OECD_SSA.School_enrollment_SSA,...
    OECD_SSA.Internet_SSA);
    Ex1.LineStyle = "--";
    Ex1.Color = "cyan";
    Ex1.MarkerFaceColor = "blue";
    Ex1.Marker = "<";
grid on
```

Question 2

Using the **plot3** function, create Fig. 6.31 with 「Year」 as the first argument, 「School_enrollment_SSA」 as the second argument, and 「Internet_SSA」 as the third argument, and set the marker to a cross. Moreover, use the **text** function to place the values of 「Internet_SSA」.

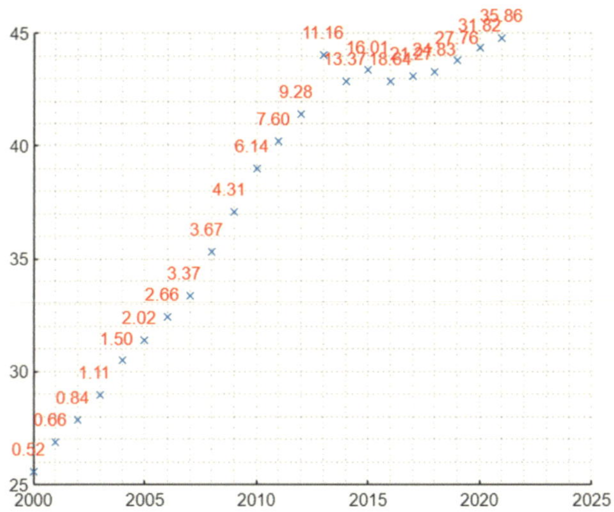

Fig. 6.31 3D plot (Question 2)

Answer 2

```
figure
plot3(Year,School_enrollment_SSA,Internet_SSA,"x");
view([0 90]);
grid minor
hold on
for i = 1:length(Year)
    text(Year(i)-1,School_enrollment_SSA(i)+1,Internet_SSA(i),...
        num2str(Internet_SSA(i), "%.2f"),"Color", "red");
end
hold off
```

Question 3

Using the **hold on** command, create descent lines as shown in Fig. 6.32. First, create a 3D plot using the **plot3** function, with the first argument as 「Year」, the second argument as 「School_enrollment_SSA」, and the third argument as 「Internet_SSA」. After that, use the **plot3** function to draw a thick solid green descending line.

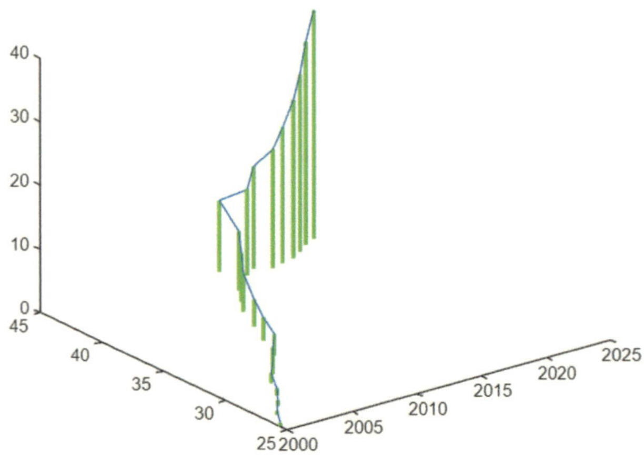

Fig. 6.32 3D plot (Question 3)

Answer 3

```
figure
plot3(Year,School_enrollment_SSA,Internet_SSA);
hold on
for i = 1:length(Year)
    plot3([Year(i),Year(i)],...
        [School_enrollment_SSA(i),School_enrollment_SSA(i)],...
        [Internet_SSA(i),0],"green","LineWidth",2);
end
hold off
```

Bar Function

Contents

7.1 Details of the Data Used in this Chapter – 202

7.2 Excel Data of ► Section 7.1 – 202

7.3 Basic Syntax of the Bar Function (◘ Fig. 7.1) – 203

7.4 Setting and Controlling the Bar Function – 214

7.5 Combining Optional Functions and the Bar Function – 226

7.6 Questions – 247

Supplementary Information The online version contains supplementary material available at https://doi.org/10.1007/978-981-96-6213-5_7.

Y. Nakamoto, *A Comprehensive Guide to Creating Graphics with MATLAB Programming*, Classroom Companion: Economics, https://doi.org/10.1007/978-981-96-6213-5_7

Learning Objectives

This chapter creates bar charts using the **bar** function. By reading this chapter, you will learn how to use the **bar** function. Moreover, you will create not only standard bar charts but also stacked bar charts. In addition, you will learn the similarities and differences in the created figures, syntax, and the type of data used among the **bar** function, the **scatter** function (covered in ▶ Chap. 3), and the **plot** function (covered in ▶ Chap. 5). In addition to this chapter, it is beneficial to study horizontal bar charts in ▶ Chap. 9.

Overview

In MATLAB, the **bar** function creates bar graphs, which represent quantities or values by the vertical length of bars. Alternatively, ▶ Chap. 9 introduces the **barh** function to create horizontal bars. Longer bars represent larger values, while shorter bars represent smaller values, making it easier to interpret data. In addition, bar graphs allow you to display different data sets side by side, making it easy to compare different items or groups.

7.1 Details of the Data Used in this Chapter

In this chapter, we use data on the proportion of the population aged 65 and over in Japan, Germany, the United States, Italy, and France. We use data from 2000 to 2050, with simulated data for 2030, 2040, and 2050. 「Year」 represents the year. Additionally, 「Japan」, 「Germany」, 「US」, 「Italy」, and 「France」 are the proportions of the population aged 65 and over in each country. Using these variables, we generate a table variable called 「population65」.

```
Year = [2000,2010,2020,2030,2040,2050]';

Japan = [17.8,23.6,29.6,31.4,35.2,37.5]';

Germany = [16.4,20.5,22,26.4,29.5,30.5]';

US = [12.3,13,16.2,20.5,22.4,23.6]';

Italy = [18.3,20.4,23.4,28.3,34.5,37.1]';

France = [16.2,17,21,24.4,27.2,28.5]';

population65 = table(Year,Japan,Germany,US,Italy,France);
```

7.2 Excel Data of ▶ Section 7.1

In ▶ Sect. 7.2, we import the data used in ▶ Sects. 7.3, 7.4, 7.5, and 7.6 in Excel format. First, we specify the file name and sheet name. Next, we use the **renamevars** function to change the column name 「Period」 to 「Year」. For more details on how to import Excel data, refer to ▶ Chap. 2.

```
file = "alldata.xlsx";
sheet = "Chap7_population_aged_65";
population65 = readtable(file,"Sheet",sheet);
population65 = renamevars(population65,"Period","Year");
```

We will extract each column from the table variable 「population65」.

```
Year = population65.Year;
Japan = population65.Japan;
Germany = population65.Germany;
US = population65.US;
Italy = population65.Italy;
France = population65.France;
```

7.3 Basic Syntax of the Bar Function (◘ Fig. 7.1)

Basic Syntax:

The **bar** function in MATLAB is used to create vertical bars. By typing 「doc bar」 in the Command Window, you can see the syntaxes for the **bar** function, as shown in Figure 7.1. The first syntax inside the black box, 「bar(y)」, is the basic syntax that creates a bar graph with one bar for each element in data 「y」. This data 「y」 can be specified as either a vector or a matrix. Additionally, the second syntax, 「bar(x,y)」, creates a bar graph with one bar for each element in data 「y」 at the locations specified by data 「x」

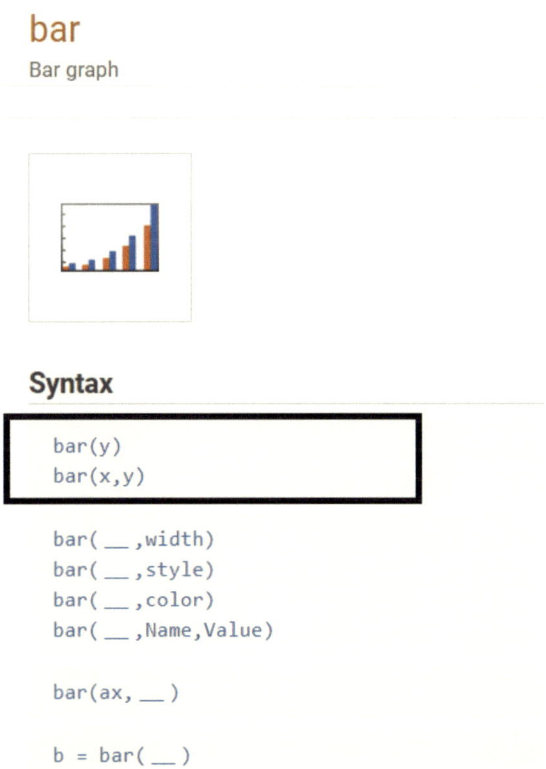

◘ **Fig. 7.1** The syntax of the bar function

The basic syntax of the **bar** function is 「bar(y)」, which takes only one argument. When using 「US」 as 「y」 in the basic syntax, we run the following program. The y-axis data represents the percentage of Americans who are 65 and older. Note that the x-axis tick values are automatically assigned as 1, 2, 3, 4, 5, and 6. See Fig. 7.2.

```
bar(US);
```

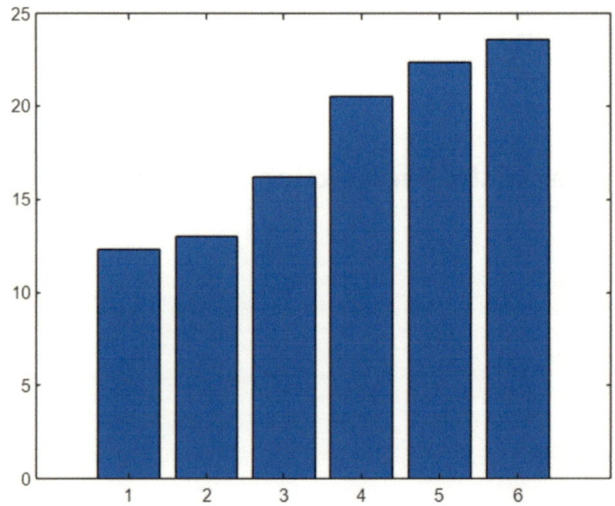

Fig. 7.2 Bar chart (X-axis: Default index)

The following program is an application of the basic syntax 「bar(y)」. It uses five column vectors as the first argument. As a result, 「[Japan,Germany,US,Italy,France]」 becomes a 6-by-5 matrix. In this case, the data for 「Japan」, 「Germany」, 「US」, 「Italy」, and 「France」 are displayed vertically, side by side, at positions 1, 2, 3, 4, 5, and 6 on the x-axis. See Fig. 7.3.

```
bar([Japan,Germany,US,Italy,France]);
```

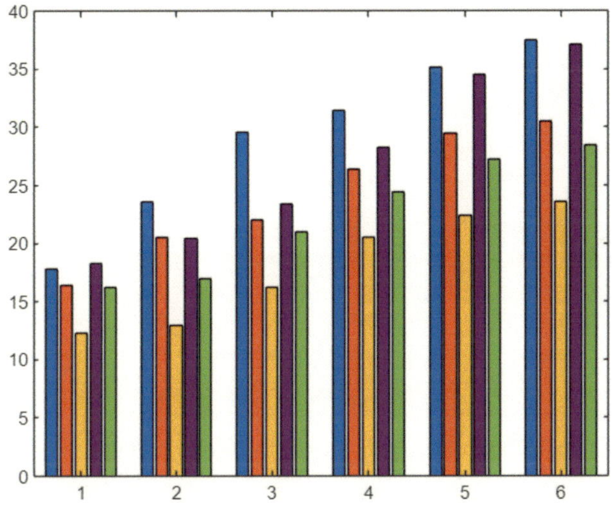

■ **Fig. 7.3** Grouped bar chart

We use the basic syntax, 「bar(x,y)」. Here, using the **bar** function, we enter 「Year」 as the first argument and 「US」 as the second argument. Therefore, the ticks on the x-axis are labeled 2000, 2010,...2050, while the height of the bars represents the values of 「US」. See ■ Fig. 7.4.

```
bar(Year,US);
```

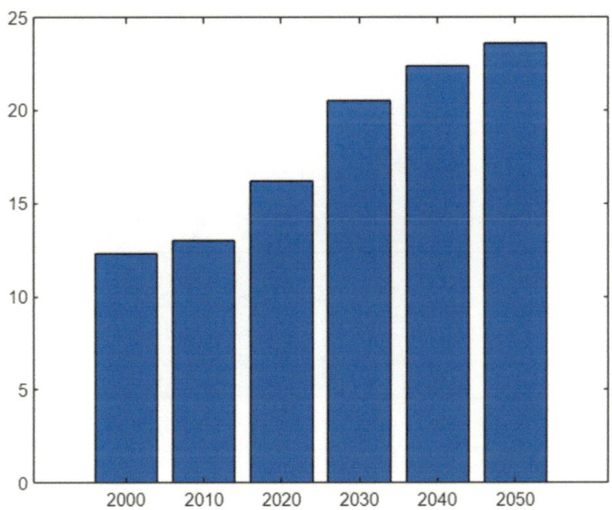

■ **Fig. 7.4** Bar chart of yearly data (X-axis: Year)

Attention

Error Program: Here, we use a column vector 「x」 with five elements as the first argument. However, for the second argument, we use the variable 「Germany」, which is a column vector with six elements. The number of elements in the first argument does not match the number of elements in the second argument. Consequently, an error occurs. The error message is presented in ◘ Fig. 7.5.

```
%%% Error program
%x = [1;2;3;4;5];
%bar(x,Germany);
```

```
Error using bar
X must be same length as Y.
```

◘ **Fig. 7.5** Error output of the bar function: Inconsistent vector lengths

Here, using the **bar** function, we enter data directly. Note that the number of elements in the vectors should be the same to avoid errors. In the following program, the first argument vector contains three elements, and the second argument vector also contains three elements. The created figure is presented in ◘ Fig. 7.6.

```
bar([3,5,7],[100,50,80]);
```

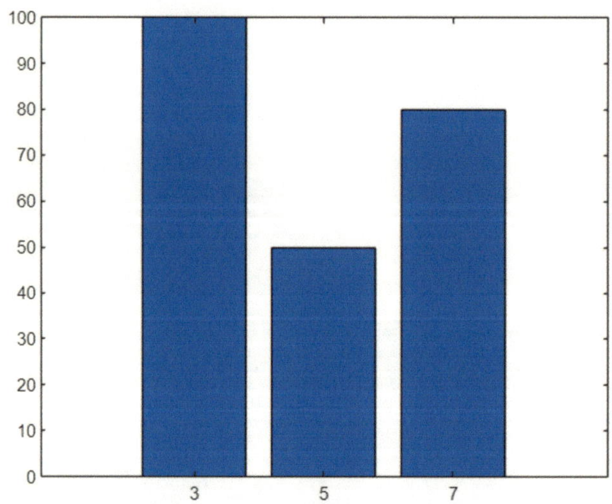

◘ **Fig. 7.6** Bar chart created from manually input data

Attention

Error Program: In the following program, the first argument vector contains three elements, but the second argument vector contains only two elements. Because the number of elements is not the same, an error occurs. See ◘ Fig. 7.7 for the error message.

```
%%% Error program
% bar([3,5,7],[100,50]);
```

```
Error using bar
X must be same length as Y.
```

◘ **Fig. 7.7** Error output of the bar function: Inconsistent vector lengths from manually input data

Here, the elements of the first argument vector, 「[10,1,5]」, are not arranged in ascending order of their numerical values. Alternatively, the plot displays the elements of the first argument vector on the x-axis in ascending order of their numerical values. In other words, the bar graph for (x,y)=(1,80) is created on the left. Next, the bar graph for (x,y)=(5,-20) is created in the middle. Finally, the bar graph for (x,y)=(10,80) is created on the right. ◘ Fig. 7.8 shows the resulting figure.

```
bar([10,1,5],[80,10,-20]);
```

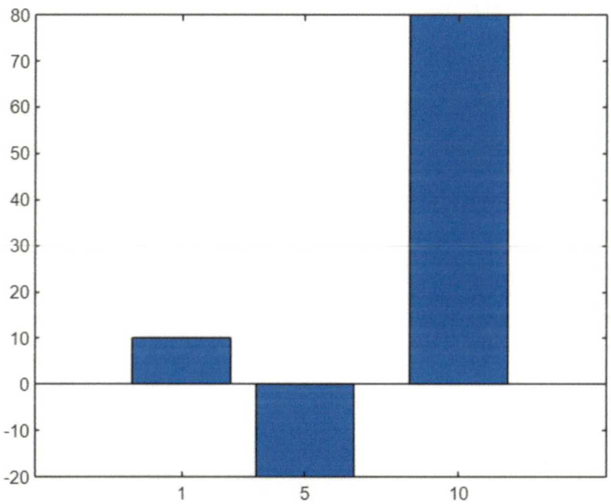

◘ **Fig. 7.8** Bar chart with a negative value

Here, using the **bar** function, we enter 「Year」 as the first argument and 「[Japan,Germany,France]」 as the second argument. Then, the program 「bar(Year, [Japan,Germany,France])」 creates three bar graphs (「Japan」, 「Germany」, and 「France」) side by side for each value of 「Year」. See ◘ Fig. 7.9.

```
bar(Year,[Japan,Germany,France]);
```

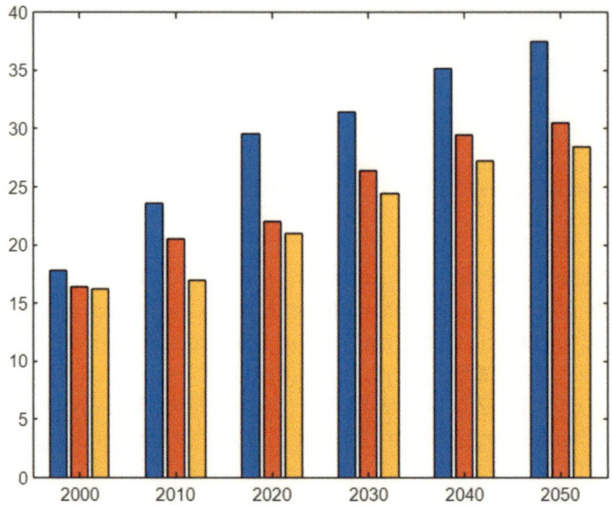

Fig. 7.9 Grouped bar chart of yearly data (X-axis: Year)

▶ Example of Table Variables

The table variable「population65」is used to create bar graphs. Specifically, for each「population65.Year」, the bar graphs for「population65.Japan」and「population65.Germany」are created side by side. The resulting figure is displayed in Fig. 7.10.

```
bar(population65.Year,[population65.Japan,population65.Germany]);
```

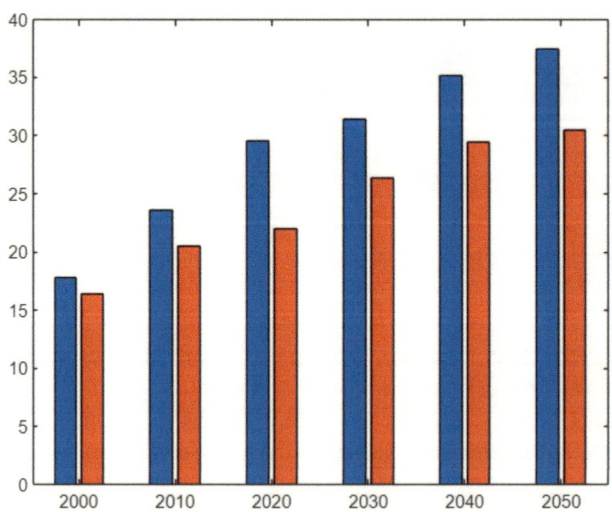

Fig. 7.10 Use of a table variable: Grouped bar chart

◀

▶ Example of Table Variables

Here, we create a figure by specifying elements of each variable in 「population65」. Specifically, when we see 「population65.Year(2:4)」, 「population65.Japan(2:4)」, and 「population65.Germany(2:4)」, we create a bar graph using the second, third, and fourth elements of each variable. The resulting figure is shown in ◘ Fig. 7.11.

```
bar(population65.Year(2:4),...
    [population65.Japan(2:4),population65.Germany(2:4)]);
```

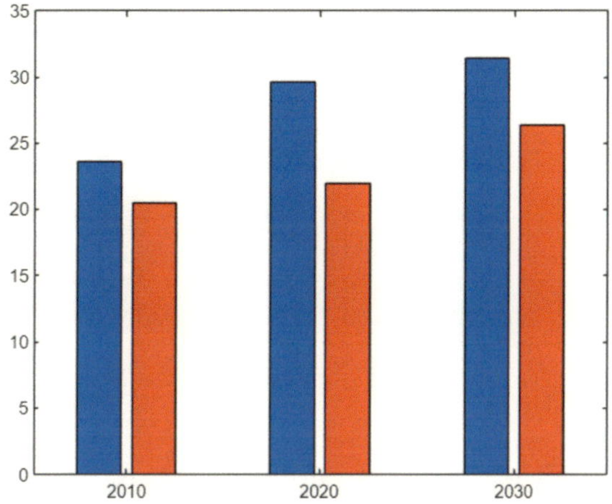

◘ **Fig. 7.11** Use of a subset of a table variable: Grouped bar chart

◀

In the following program, the first argument is a scalar, and the second argument is a vector. In this case, all elements of the vector are displayed as bar graphs at a single coordinate on the x-axis. Concretely, the following program shows that at x=100, four bar graphs with heights of 「[50,80,20,100]」 are created. See ◘ Fig. 7.12.

```
bar(100,[50,80,20,100]);
```

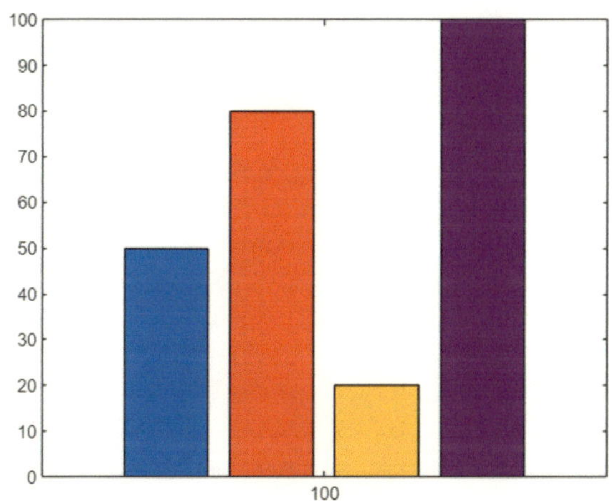

◘ **Fig. 7.12** Grouped bar chart at a single X-point

Here, three bars (「Japan(3)」, 「Germany(3)」, and 「France(3)」) are created at x=2020, which corresponds to 「Year(3)」. See Fig. 7.13.

```
bar(Year(3),[Japan(3),Germany(3),France(3)]);
```

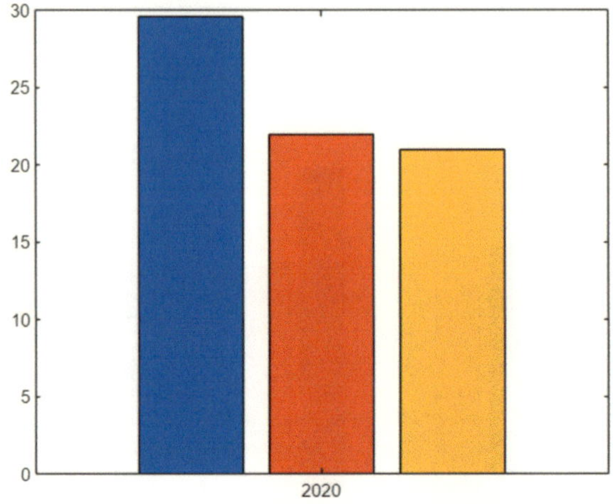

Fig. 7.13 Bar chart from the elements of variables

We can use a text in the first argument of the **bar** function. Here, by placing the text 「the population ratio of those aged 65 and over」 as the first argument, it is displayed on the x-axis of the created figure. See Fig. 7.14.

```
bar("the population ratio of those aged 65 and over",...
    [Japan(1),Germany(1),France(1)]);
```

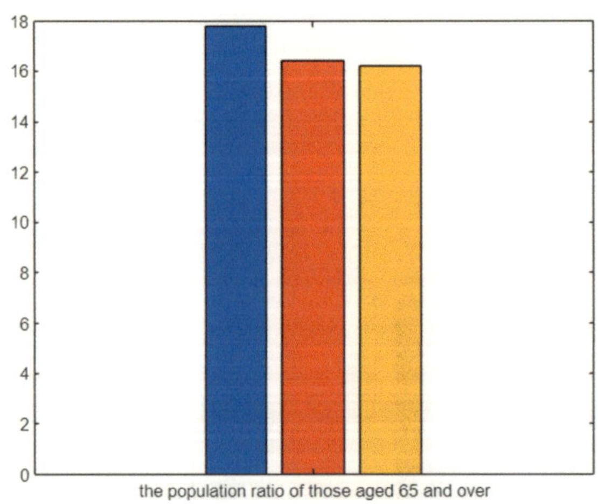

Fig. 7.14 Bar chart with text label at a single X-point

In the following program, the first argument is a vector with two elements, and the second argument is a vector with six elements. Normally, an error would occur because the number of elements is not the same. In this case, however, no error occurs. This is because the variable 「y1」 is defined with line breaks without 「...」 after 3, 10, and 15, and then continues with 20, 5, and 1. The variable 「y1」 is a 2-by-3 double matrix as shown in ◘ Fig. 7.15. Therefore, for the first group with an x-coordinate of 100, bar graphs with heights of 「[3,10,15]」 are created. Next, for the second group with an x-coordinate of 200, bar graphs with heights of 「[20,5,1]」 are created. The created figure is presented in ◘ Fig. 7.16.

2x3 double

	1	2	3	4
1	3	10	15	
2	20	5	1	
3				

◘ **Fig. 7.15** Contents of the matrix "y1"

```
x1 = [100,200];
y1 = [3,10,15,
    20,5,1];
bar(x1,y1);
```

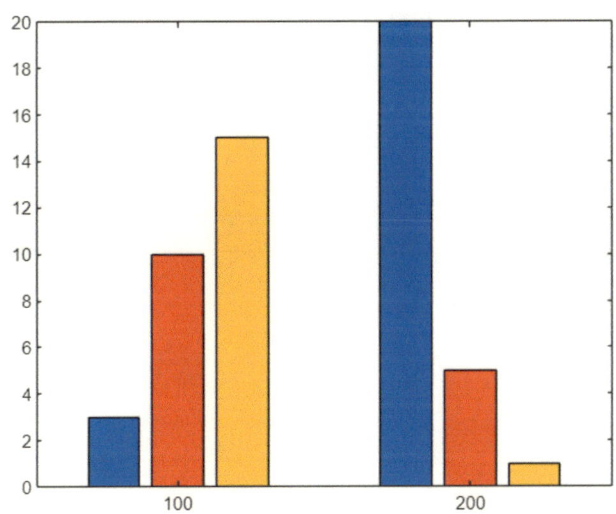

◘ **Fig. 7.16** Grouped bar chart at two X-points

Attention

Error Program: Here, the variable 「y2」 is not broken into multiple lines. As a result, 「y2」 becomes a 1-by-6 row vector, as shown in Fig. 7.17. Therefore, an error occurs when using the **bar** function. Refer to Fig. 7.18 for the error message. Also, compare the following program with the program that created Fig. 7.16.

1x6 double

	1	2	3	4	5	6
1	3	10	15	20	5	1
2						

Fig. 7.17 Contents of the vector "y2"

7

```
%%% Error program
%x1 = [100,200];
%y2 = [3,10,15,20,5,1];
%bar(x1,y1);
```

```
Error using bar (line 202)
X must be same length as Y.
```

Fig. 7.18 Error output of the bar function: Mismatch between first and second arguments

Here, the figure is similar to Fig. 7.16; however, it differs in whether the variable of the second argument is a 2-by-3 matrix or a 3-by-2 matrix. In sum, in Fig. 7.16, the second argument was a 2-by-3 matrix, while here the second argument uses a 3-by-2 matrix called the variable 「y1_3_2」. Additionally, we use the text 「["1st group","2nd group"]」 in the first argument. See Fig. 7.19.

```
x1 = ["1st group","2nd group"];
y1_3_2 = [3,20;10,5;15,1];
bar(x1,y1_3_2);
```

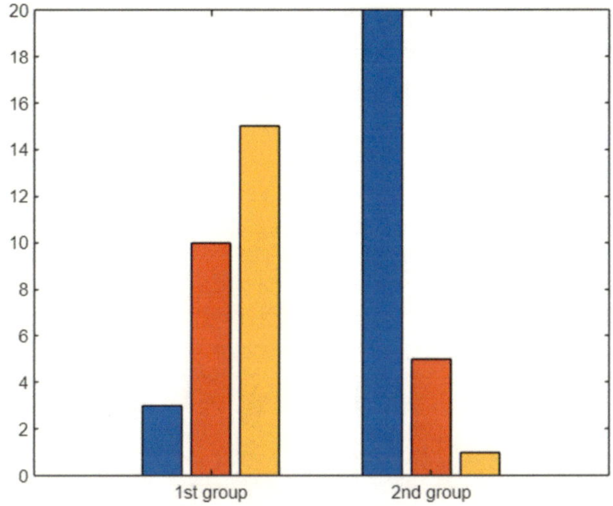

Fig. 7.19 Bar chart using manually entered Y-data, with text labels at two X-points

Here, 「y_pop」 is a 6-by-2 matrix. Consequently, we create six bars above the text labels 「Germany 2000-2050」 and 「US 2000-2050」, respectively. See Fig. 7.20.

```
countries = ["Germany 2000-2050","US 2000-2050"];
y_pop = [Germany,US];
bar(countries,y_pop);
```

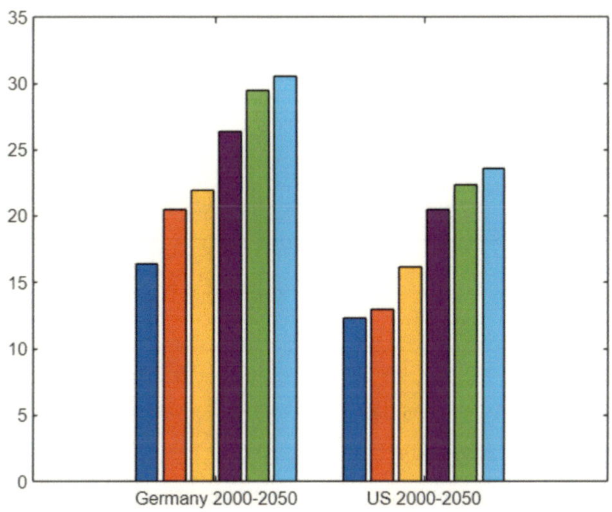

Fig. 7.20 Bar chart using variables for Y-data, with text labels at two X-points

7.4 Setting and Controlling the Bar Function

Here, we use the third argument to change the color of the bar graph to red. See Fig. 7.21.

```
bar(Year,US,"r");
```

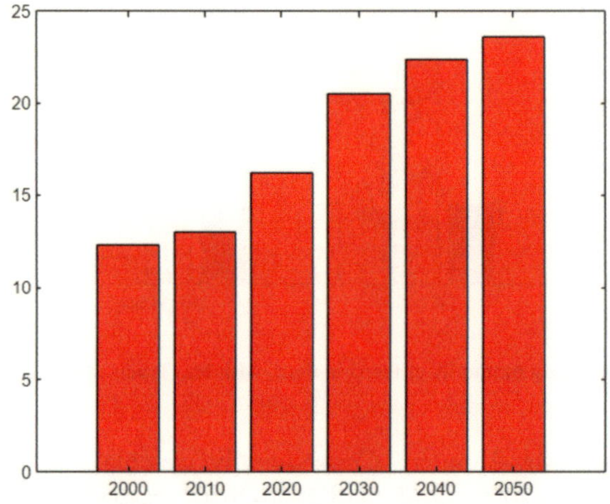

Fig. 7.21 Bar chart in red

In the following program, we store the properties of the **bar** function in the variable 「b」. Then, we set 「EdgeColor」 to 「blue」, 「LineWidth」 to 4, and 「LineStyle」 to 「--」. As a result, the outline of the bar graph becomes a thick, blue dashed line. The resulting figure is presented in Fig. 7.22.

```
b = bar(Year,US,"r");
    b.EdgeColor = "blue";
    b.LineWidth = 4;
    b.LineStyle = "--";
```

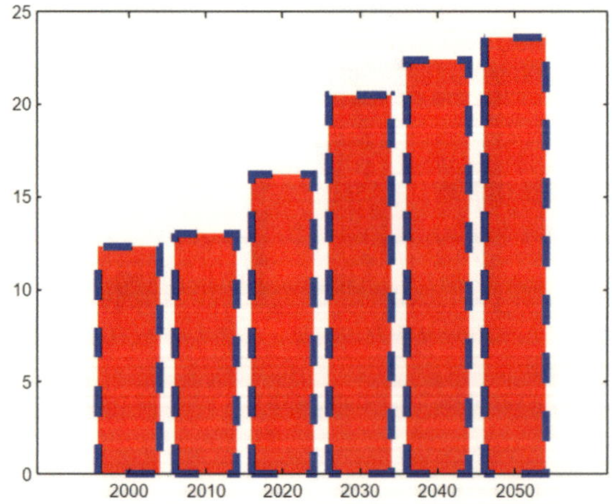

Fig. 7.22 Bar chart in red with blue dashed edges

Attention

Error Program: Suppose that you want to change 「LineWidth」 to 2 for only the second bar graph. When you run the following program, an error occurs. The error message is shown in ◘ Fig. 7.23. The reason is that the variable 「b」, which stores the properties of the **bar** function, is a 1-by-1 Bar object. Therefore, 「b(1)」 exists; however, as 「b(2)」 does not exist, you cannot specify 「b(2)」. To change the properties of only the second bar graph, you need to use 「CData()」. See ◘ Fig. 7.39 for a description of 「CData()」.

```
%%% Error program
% b = bar(Year,US);
% b(2).LineWidth = 2;
```

```
Unrecognized property 'LineWidth' for class 'matlab.graphics.GraphicsPlaceholder'.
```

◘ **Fig. 7.23** Error output of the bar function: Index beyond single bar object

In the following program, the variable 「b」 contains the properties of the bar graphs for the variables 「Japan」, 「Germany」, and 「US」. That is, it is a 1-by-3 bar object. Specifically, the properties of 「Japan」, 「Germany」, and 「US」 correspond to 「b(1)」, 「b(2)」, and 「b(3)」, respectively. Here, we change 「b(2).LineWidth」 to 2, adjusting the width of its outline for the 「Germany」 bar graph. See ◘ Fig. 7.24.

```
b = bar([Japan,Germany,US]);
    b(2).LineWidth = 3;
```

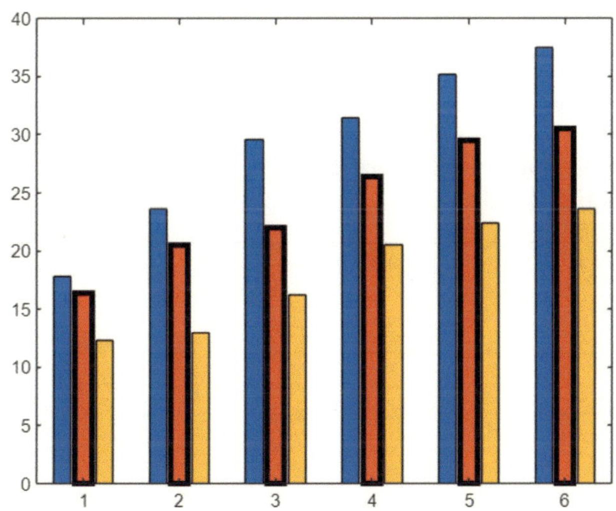

◘ **Fig. 7.24** Grouped bar chart with one thick-edged group

Here, for the first bar graph 「b(1)」, we set 「FaceColor」 to 「magenta」, 「LineStyle」 to 「--」, and 「LineWidth」 to 3, creating a magenta-filled bar with a dashed and thicker outline. Furthermore, for the third bar graph 「b(3)」, we set 「EdgeColor」 to 「blue」 and 「LineWidth」 to 3, making the outline thicker and blue. See Fig. 7.25 for the created figure.

```
b = bar([Japan,Germany,US]);
    b(1).FaceColor = "magenta";
    b(1).LineStyle = "--";
    b(1).LineWidth = 3;
    b(3).EdgeColor = "blue";
    b(3).LineWidth = 3;
```

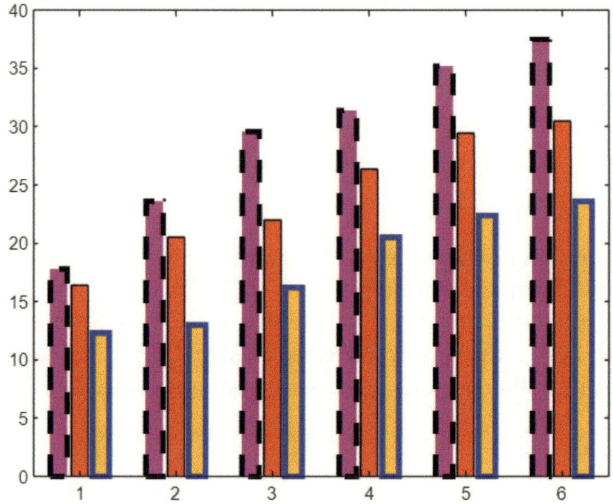

Fig. 7.25 Grouped bar chart with various edge styles by group

Here, we change the 「BaseValue」 property, which is specific to bar graphs. When 「BaseValue」 is set to 30, a reference line is drawn on y=30. Bar values greater than this reference line are displayed above it, while smaller values are displayed below it. See Fig. 7.26.

```
bar([Japan,Germany,US],"BaseValue",30);
```

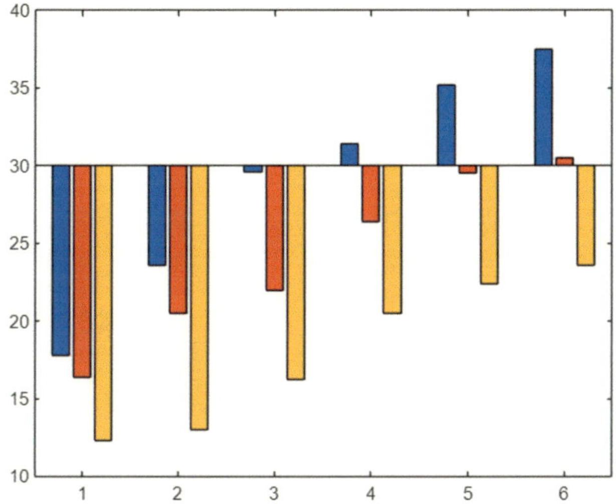

Fig. 7.26 Grouped bar chart with a reference line

Using the 「BaseLine」 property of the **bar** function, we modify the reference line. Since the variable 「b」 is a 1-by-3 Bar object, we specify 「b(1).BaseLine」 here; instead, if we specify 「b(2).BaseLine」 or 「b(3).BaseLine」, we will obtain the same result. We then change the color of the reference line using 「b(1).BaseLine.Color」 and adjust its width using 「b(1).BaseLine.LineWidth」. See Fig. 7.27 for the created figure.

```
b = bar([Japan,Germany,US]);
    b(1).BaseLine.Color = "red";
    b(1).BaseLine.LineWidth = 5;
```

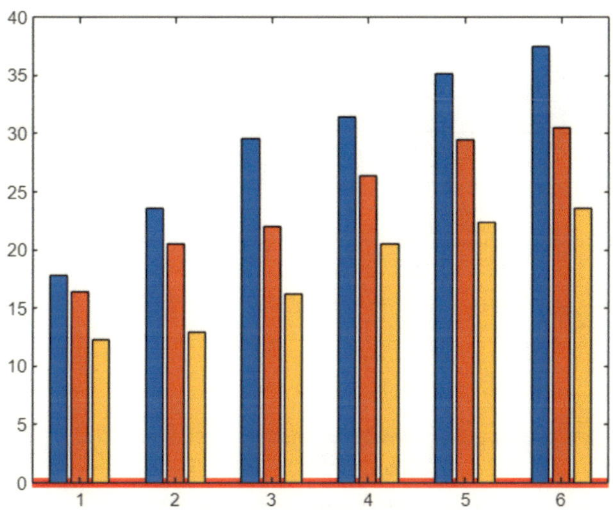

Fig. 7.27 Grouped bar chart with a thick red reference line

Attention

Error Program: As explained in the program that created Fig. 7.27, because the variable 「b」 is a 1-by-3 bar object, we cannot use 「b.BaseLine」 directly. Therefore, if you try to use 「b.BaseLine」 as shown below, an error will occur. See Fig. 7.28 for details of the error.

```
%%% Error program
%b = bar([Japan,Germany,US]);
%b.BaseLine.Color = "red";
```

```
Intermediate dot '.' indexing produced a comma-separated list with 3
values, but it must produce a single value when followed by subsequent
indexing operations.
```

Related documentation

Fig. 7.28 Error output of the bar function: Invalid use of "BaseLine" property

Here, we modify the reference line of 「BaseLine」 as follows. At first, we set 「BaseValue」 to 35, drawing the reference line at y=35. Then, we set 「LineStyle」 to 「--」, 「Color」 to 「red」, and 「LineWidth」 to 3, creating a thicker red dashed reference line. The resulting figure is displayed in Fig. 7.29.

```
br = bar(Year,[Japan,Germany,US,Italy,France]);
    br(2).BaseValue = 35;
    br(2).BaseLine.LineStyle = "--";
    br(2).BaseLine.Color = "red";
    br(2).BaseLine.LineWidth = 3;
```

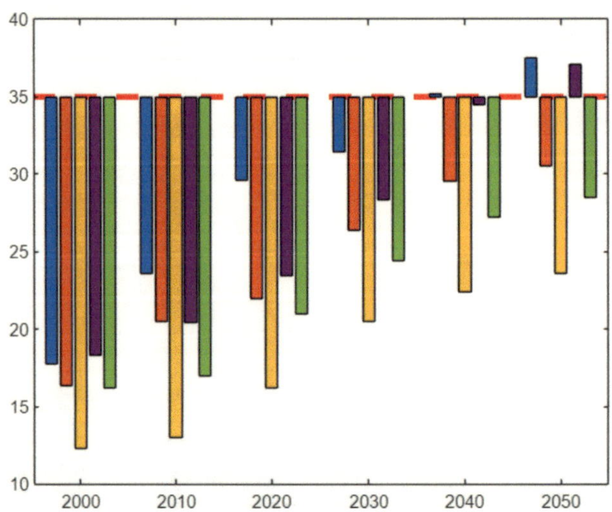

Fig. 7.29 Grouped bar chart with a thick dashed red reference line

Here, in the **bar** function, we set 「BaseValue」 to 30. Additionally, we store the properties of 「BaseLine」 in the variable 「b2BL」, which differs from the method used in the previous programs that created ◘ Figs. 7.27 and ◘ 7.29. Then, we set 「Color」 to 「red」, 「LineStyle」 to 「-.」, and 「LineWidth」 to 2, creating a thick red dash-dotted reference line. See ◘ Fig. 7.30.

```
b = bar([Japan,Germany,US],"BaseValue",30);
    b2BL = b(2).BaseLine;
    b2BL.Color = "red";
    b2BL.LineStyle = "-.";
    b2BL.LineWidth = 2;
```

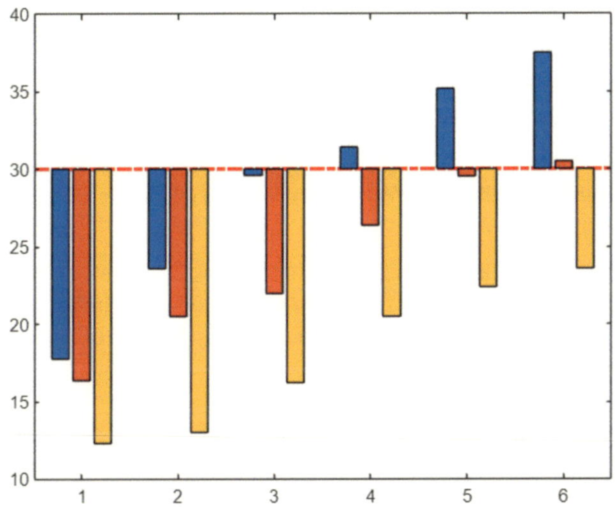

◘ **Fig. 7.30** Grouped bar chart with a thick dash-dotted red reference line

Here, we set 「BaseValue」 to 20 and store the properties of the **bar** function in the variable 「b」. For 「b(2)」, we set 「FaceAlpha」 to 0.2 and 「LineWidth」 to 3, creating the second bar with a thick outline and high transparency. In addition, for 「b(5)」, we set 「EdgeColor」 to 「green」 and 「LineWidth」 to 3, creating the fifth bar with a thick green outline.

We also set 「b(1).BaseLine.LineWidth=3」, 「b(1).BaseLine.LineStyle = ":"」, and 「b(1).BaseLine.Color="magenta"」, creating a thick dotted magenta reference line. The same change of the reference line can be applied when using 「b(2)」 or 「b(3)」. But, using 「b.BaseLine.LineWidth」 will result in an error. See ◘ Fig. 7.31.

```
b = bar([Japan,Germany,US,Italy,France],"BaseValue",20);
    b(2).FaceAlpha = 0.2;
    b(2).LineWidth = 3;
    b(5).EdgeColor = "green";
    b(5).LineWidth = 3;
    b(1).BaseLine.LineWidth = 3;
    b(1).BaseLine.LineStyle = ":";
    b(1).BaseLine.Color = "magenta";
```

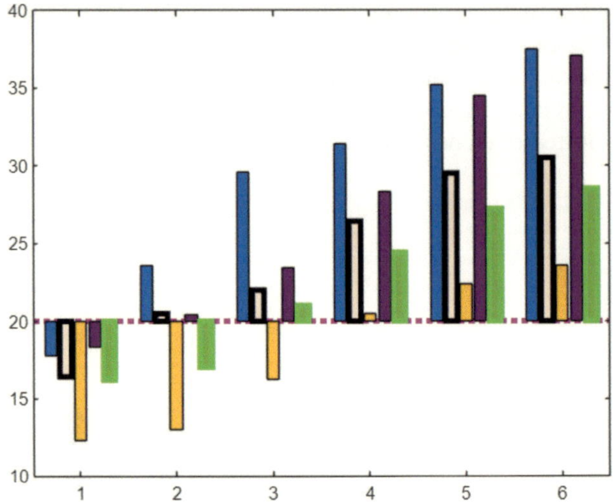

Fig. 7.31 Grouped bar chart with various edge styles and a magenta reference line

Here, we create a stacked bar chart using the **bar** function. By using 「stacked」 for the third argument of the **bar** function, we can display the stacked bar chart. In the following program, we stack the elements 「(3,10,15)」 of 「y_s」 at x=10. The resulting figure is presented in Fig. 7.32.

```
x_s = 10;
y_s = [3,10,15];
bar(x_s,y_s,"stacked");
```

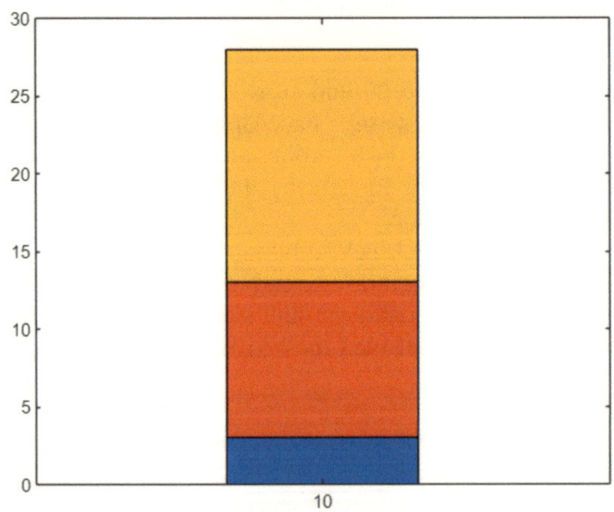

Fig. 7.32 Stacked bar chart at a single X-point

Here, we use 「stacked」 to display each group as one multicolored bar. Additionally, we store the properties of the **bar** function in the variable 「bs」. For the first stack, 「bs(1)」, we set 「FaceColor」 to 「green」 and 「LineWidth」 to 2, creating a green bar with a thick outline. See ◘ Fig. 7.33.

```
bs = bar(10,[3,10,15],"stacked");
    bs(1).FaceColor = "green";
    bs(1).LineWidth = 2;
```

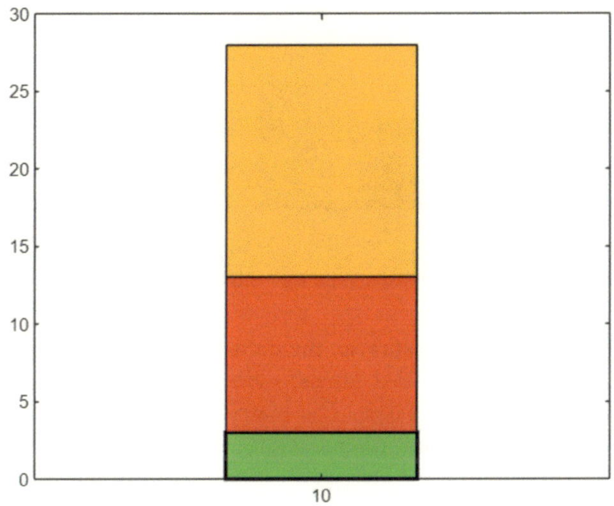

◘ **Fig. 7.33** Stacked bar chart with a styled bottom segment

Here, we create a stacked bar chart by using 「stacked」 as the second argument of the **bar** function where x-axis coordinates are not specified. The variable 「y_st」 is a 5-by-3 matrix. Therefore, five bars are created, with each bar stacking three elements. For 「bs(2)」, we set 「FaceColor」 to 「green」, 「EdgeColor」 to 「blue」, 「LineWidth」 to 2, and 「LineStyle」 to 「--」, creating a green stacked bar chart with a thick dashed blue outline. See ◘ Fig. 7.34.

```
y_st = [10,20,30; 10,1,3; 1,50,0; 1,1,50; 5,5,5];
bs = bar(y_st,"stacked");
    bs(2).FaceColor = "green";
    bs(2).EdgeColor = "blue";
    bs(2).LineWidth = 2;
    bs(2).LineStyle = "--";
```

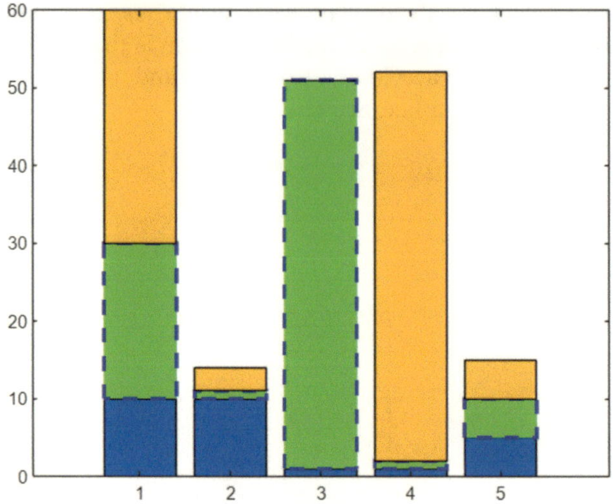

Fig. 7.34 Stacked bar chart with styled middle segments

Here, we use the matrix 「y_sta」 as the data to create a stacked bar chart. The variable 「x_sta」 contains two text labels: 「No minus」 and 「Minus」, which are displayed as categories on the x-axis. The matrix 「y_sta」 has two rows and three columns, so two bars are created, each stacking three elements. Specifically, for the first bar labeled 「No minus」, the elements 「(5,20,10)」 are stacked. For the second bar labeled 「Minus」, the elements 「(-50,-20,10)」 are stacked, with negative values appearing below the baseline. See Fig. 7.35.

```
x_sta = ["No minus","Minus"];
y_sta = [5,20,10; -50,-20,10];
bar(x_sta,y_sta,"stacked");
```

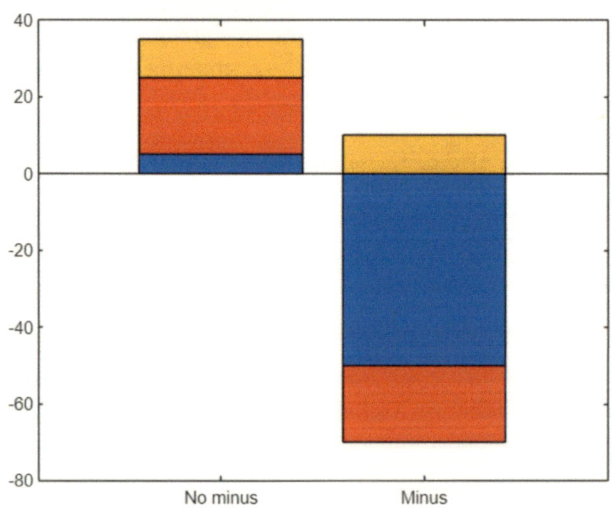

Fig. 7.35 Stacked bar chart with negative values

Here, we create a stacked bar chart using the 5-by-3 matrix 「y_st」. This creates five bars, each with three stacked elements. By setting 「bs(1).BaseLine.LineStyle=":" 」, we create a dotted reference line. Additionally, setting 「bs(2).BaseValue = 20」, we change the base value of the reference line to 20. Finally, with 「bs(3).FaceColor = "black"」, we change the color of the third stack to black. The resulting figure is presented in ◘ Fig. 7.36.

```
y_st = [10,20,30; 10,1,3; 1,50,0; 1,1,50; 5,5,5];
bs = bar(y_st,"stacked");
    bs(1).BaseLine.LineStyle = ":";
    bs(2).BaseValue = 20;
    bs(3).FaceColor = "black";
```

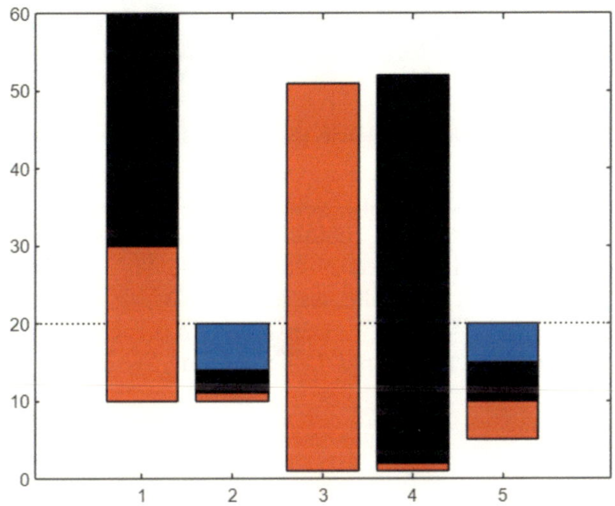

◘ **Fig. 7.36** Stacked bar chart with various segment styles and a reference line

Here, we create a stacked bar chart using the variables 「US」 and 「Japan」 in the second argument of the **bar** function. The data for 「US」 is represented by a blue bar, and a green bar representing the data for 「Japan」 is stacked on top of it. See ◘ Fig. 7.37.

```
b_3 = bar(Year,[US,Japan],"stacked");
    b_3(2).FaceColor = "green";
```

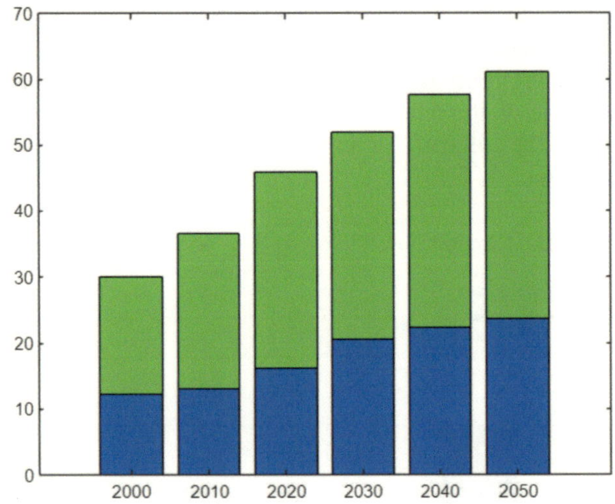

Fig. 7.37 Stacked bar chart from two variables (X-axis: Year)

The following program creates a stacked bar chart using the variables 「US」 and 「Japan」, with the text variable 「countries」 as the x-axis labels. The properties of the stacked bar chart are stored in the variable 「b_3」. For the fifth stack, we modify its properties. In particular, we set 「LineWidth」 to 1.5, 「EdgeColor」 to 「red」, 「LineStyle」 to 「--」, and 「FaceAlpha」 to 0.1, creating a semi-transparent stacked bar chart with a thicker red dashed outline. See Fig. 7.38.

```
countries = ["US","Japan"];
b_3 = bar(countries,[US,Japan],"stacked");
    b_3(5).LineWidth = 1.5;
    b_3(5).EdgeColor = "red";
    b_3(5).LineStyle = "--";
    b_3(5).FaceAlpha = 0.1;
```

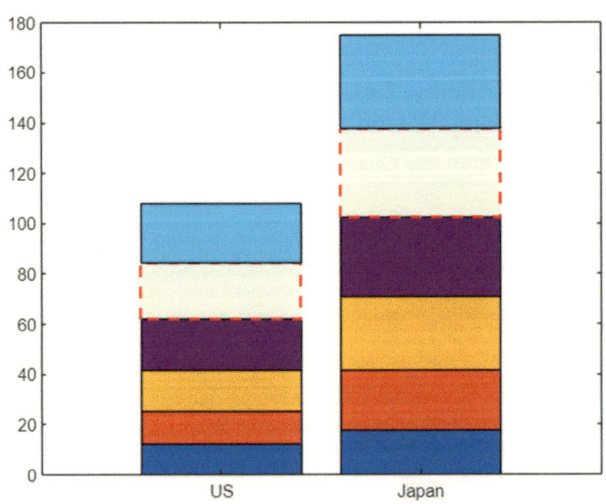

Fig. 7.38 Stacked bar chart from two variables (X-axis: Countries)

In the following program, we set 「FaceColor」 to 「Flat」. This option allows us to use 「CData (= Color Data)」 to assign different colors to each bar. Normally, setting 「FaceColor」 to a specific color such as 「red」 or 「blue」 in the **bar** function would make all bars the same color; however, in the following program, we set 「CData(4,:)=[1 0.5 0.5]」, which specifies the color of the fourth bar as 「[1 0.5 0.5]」. The created figure is shown in ◘ Fig. 7.39.

```
bh1 = bar(Year,US,"FaceColor","Flat");
    bh1.CData(4,:) = [1 0.5 0.5];
```

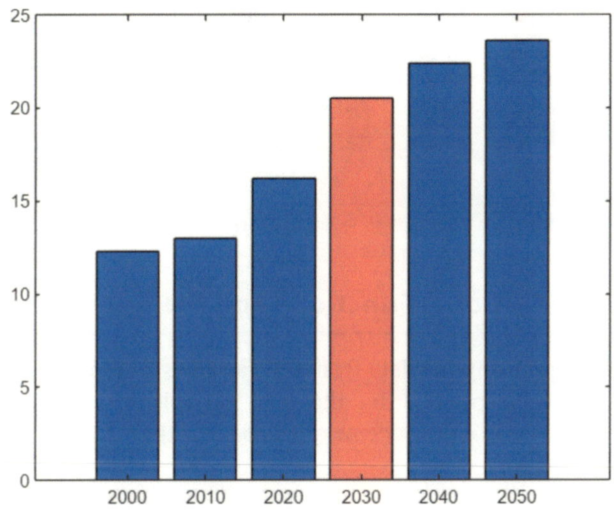

◘ **Fig. 7.39** Bar chart with one pink-colored bar (X-axis: Year)

Here, we set 「FaceColor」 to 「Flat」 where the second argument of the **bar** function contains two types of data: 「US」 and 「Germany」. Based on the order of the second argument, 「[US, Germany]」, in the **bar** function, 「bh2(1)」 corresponds to 「US」 and 「bh2(2)」 corresponds to 「Germany」.

Therefore, 「bh2(1).CData(4,:)=[1 1 1]」 specifies that the fourth bar for 「US」 is set to white. Additionally, 「bh2(2).CData(2,:)=[0 0 0]」 and 「bh2(2).CData(6,:)=[0 1 1]」 specify that the second and sixth bars for 「Germany」 are set to black and cyan, respectively. See ◘ Fig. 7.40 for the resulting figure.

```
bh2 = bar(Year,[US,Germany],"FaceColor","Flat");
    bh2(1).CData(4,:) = [1 1 1];
    bh2(2).CData(2,:) = [0 0 0];
    bh2(2).CData(6,:) = [0 1 1];
```

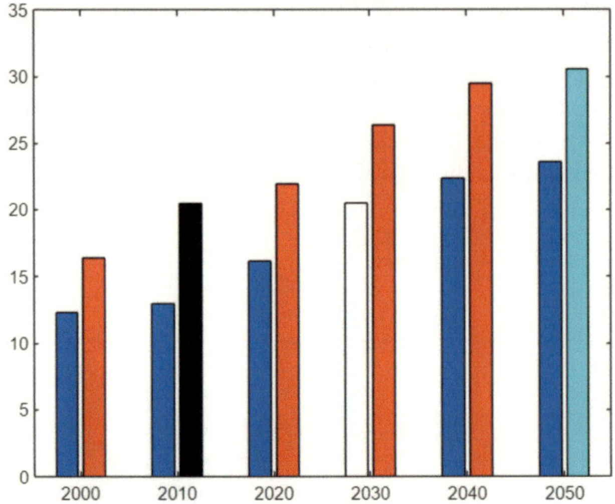

Fig. 7.40 Grouped bar chart with bars in various colors (X-axis: Year)

7.5 Combining Optional Functions and the Bar Function

We create a bar graph using the data from the variables 「Germany」 and 「Japan」, with the x-axis labels specified by the text variable 「country」. Using 「colororder("earth")」, we set the color scheme of the bars to the 「earth」 color palette. See Fig. 7.41.

```
country = ["Japan","Germany"];
bar(country,[Japan,Germany]);
colororder("earth");
```

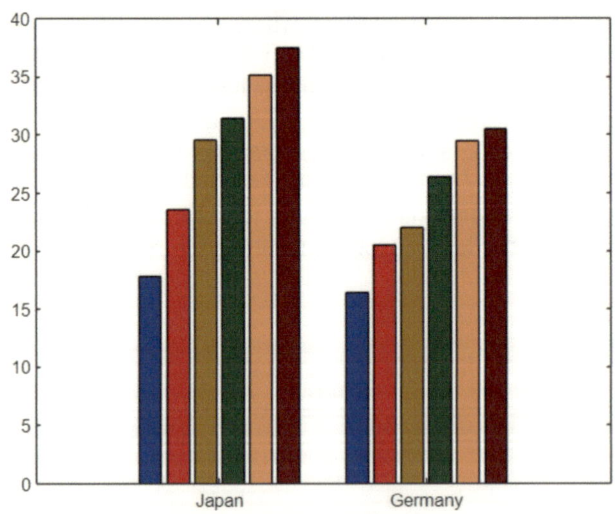

Fig. 7.41 Grouped bar chart with predefined color order

Here, we set the y-axis range from 0 to 100 using the **ylim** function. Additionally, we use 「colororder("sail")」 to change the color palette to the 「sail」 scheme. The resulting figure is shown in ◘ Fig. 7.42.

```
country = ["Japan","Germany"];
bar(country,[Japan,Germany]);
ylim([0,100]);
colororder("sail");
```

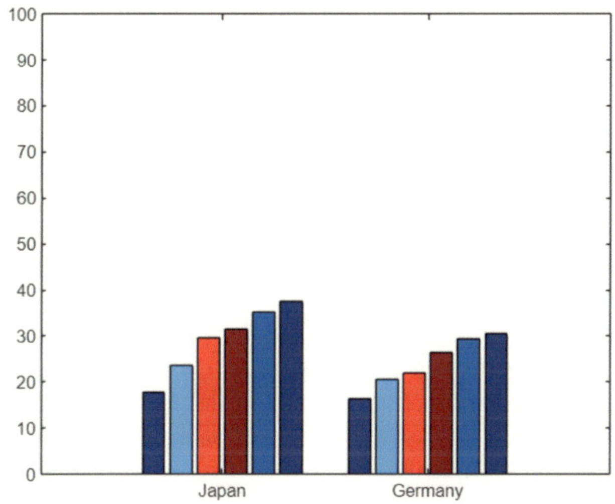

◘ **Fig. 7.42** Grouped bar chart with adjusted Y-axis limits and predefined color order

In the previous program that created ◘ Fig. 7.42, we used 「colororder("sail")」. Therefore, to reset the color palette, we use the **figure** command at the beginning of the program. Next, we set the y-axis range using the **ylim** function. Additionally, we set the tick marks using the **yticks** function. Concretely, by using 「yticks(0:2:50)」, the tick marks on the y-axis range from 0 to 50 with increments of 2. See ◘ Fig. 7.43.

```
figure
bar(["France","US","Italy"],[France,US,Italy]);
ylim([0,50]);
yticks(0:2:50);
```

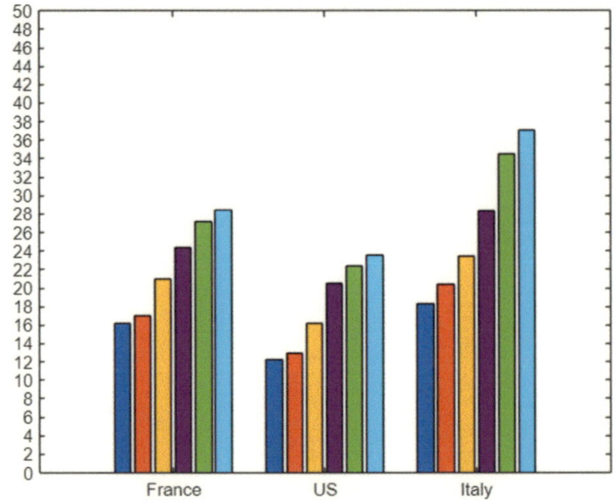

Fig. 7.43 Grouped bar chart with specified Y-axis ticks and predefined color order

To make the tick marks point outward, we set 「TickDir」 to 「out」. In addition, we set 「TickLength」 to 「[0.05 1]」, where the first element specifies the 2D tick length as 0.05. Since the **bar** function does not create any 3D plot, the second element of 「TickLength」 (which is 1) is irrelevant and can be any positive number. See Fig. 7.44 for the resulting figure.

```
bar([France,US,Italy]);
ylim([0,50]);
yticks(0:2:50);
set(gca,"TickDir","out");
set(gca,"TickLength",[0.05 1]);
```

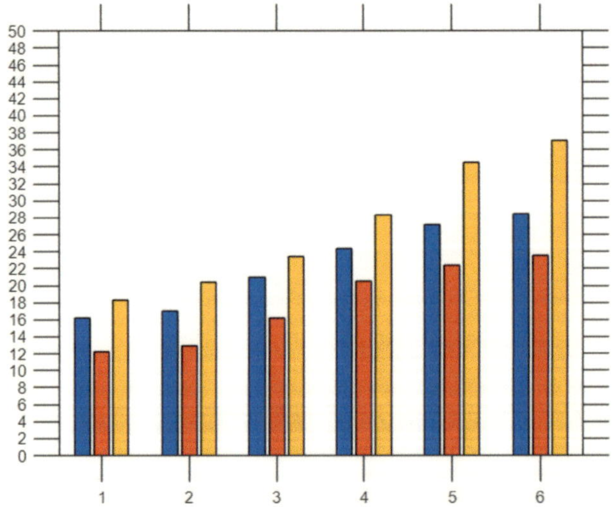

Fig. 7.44 Use of gca: Grouped bar chart with large outward tick marks

In the following program, we use the **legend** function to create a legend for the bar graph. The **legend** function labels the bars with the corresponding country names: 「France」, 「US」, and 「Italy」. See Fig. 7.45.

```
bar(Year,[France,US,Italy]);
legend(["France","US","Italy"]);
```

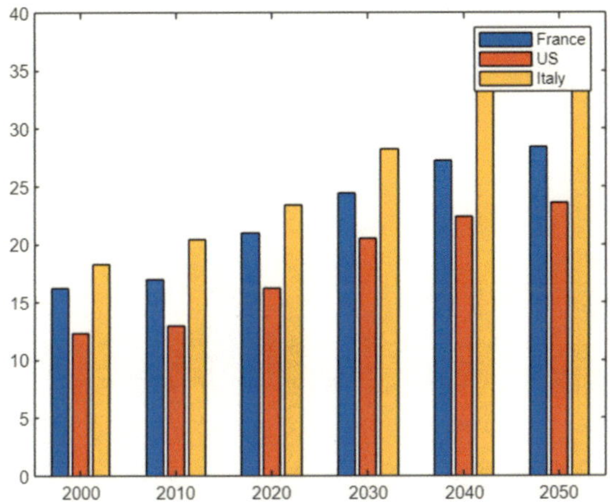

Fig. 7.45 Grouped bar chart with a legend

Here, the **legend** function is used to create a legend. Specifically, by setting 「NumColumns」 to 3, a 3-column legend is created. Additionally, 「Location」 is set to 「north」, positioning the legend at the top center inside the figure. The resulting figure is presented in Fig. 7.46.

```
bar(Year,[France,US,Italy]);
lg = legend(["France","US","Italy"]);
    lg.NumColumns = 3;
    lg.Location = "north";
```

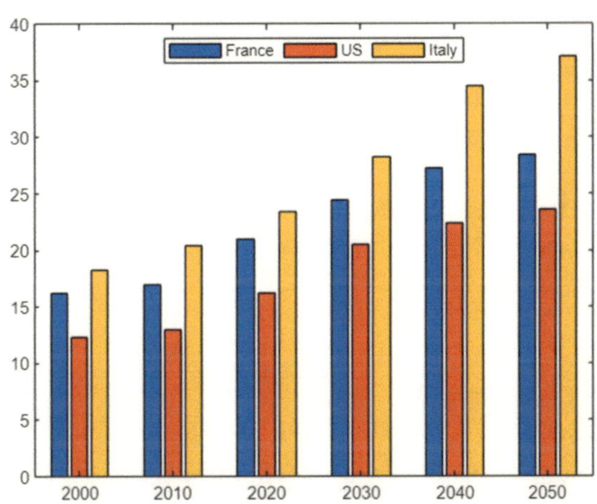

Fig. 7.46 Grouped bar chart with a horizontal legend (top-center position)

Here, the **legend** function is used to create a legend. By setting 「Location」 to 「eastoutside」, we position the legend outside the figure on the right. In addition, the **title**, **xlabel**, and **ylabel** functions are used to add text for the title, x-axis label, and y-axis label, respectively. See Fig. 7.47.

```
bar(Year,[France,US,Italy,Japan,Germany]);
lg = legend(["France","US","Italy"," "]);
    lg.Location = "eastoutside";
title("The population ratio of people aged 65 and over");
xlabel("Year");
ylabel("Percent");
```

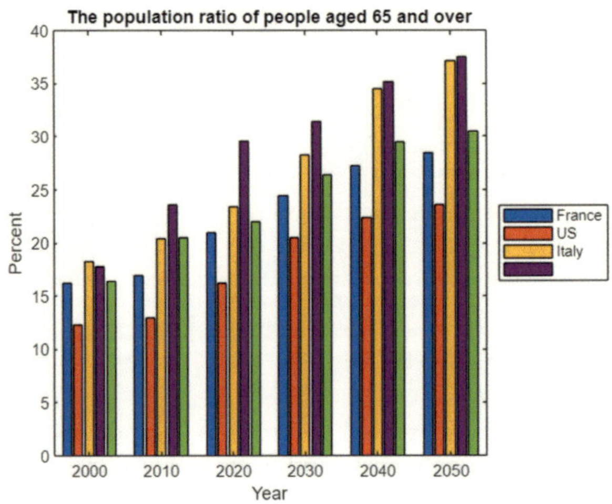

Fig. 7.47 Grouped bar chart with a blank legend entry (outside-right position)

Here, the **legend** function is used to create a legend, with 「NumColumns」 set to 5 and 「Location」 set to 「north」, placing a 5-column legend at the top inside the figure. Next, the **title** function is used to create a title, with 「Color」 set to 「blue」 and 「FontSize」 set to 15, creating the title with large blue text. For the **xlabel** function, 「FontWeight」 is set to 「bold」. Lastly, for the **ylabel** function, we set 「Color」 to 「red」 and 「Rotation」 to 0, displaying horizontally aligned red text on the y-axis. See Fig. 7.48 for the created figure.

```
bar(Year,[France,US,Italy,Japan,Germany]);
lg = legend(["France","US","Italy","Japan","Germany"]);
    lg.NumColumns = 5;
    lg.Location = "north";
tl = title("The population ratio of people aged 65 and over");
    tl.Color = "blue";
    tl.FontSize = 15;
xl = xlabel({"Year","(2000-2050)"});
    xl.FontWeight = "bold";
yl = ylabel("Percent");
    yl.Rotation = 0;
    yl.Color = "red";
grid minor
```

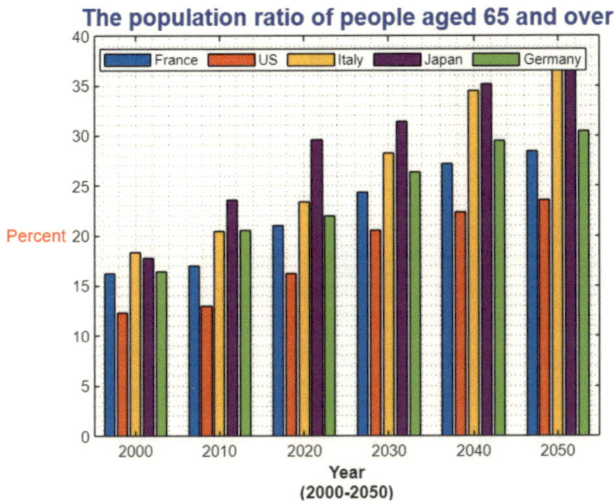

Fig. 7.48 Grouped bar chart with a title, axis labels, and a horizontal legend

Here, we use the **annotation** function to create three annotations. In detail, using 「an_j=annotation("textarrow",x1,y1,"String",{"Japan","2000"})」, we specify 「textarrow」 to create a text arrow. Additionally, we specify the start and end points of the arrow as 「x1=[0.2 0.25]」 and 「y1=[0.85 0.7]」. This means that the start point of the arrow is at (x,y)=(0.2,0.85), and the end point is at (x,y)=(0.25,0.7).

We add text at (x,y)=(0.2,0.85). 「String」 is set to 「{"Japan","2000"}」, indicating that 「Japan」 is on the first line and 「2000」 is on the second line. The creation of the other two annotations follows a similar process. For more details of the **annotation** function, see ▶ Sect. 18.1. Fig. 7.49 shows the created figure.

```
Japan2 = Japan(1:2);
France2 = France(1:2);
US2 = US(1:2);
bar(["Japan","France","US"],[Japan2,France2,US2]);
x1 = [0.2 0.25];
y1 = [0.85 0.7];
an_j = annotation("textarrow",x1,y1,"String",{"Japan","2000"});
    an_j.Color = "red";
    an_j.FontSize = 5;
x2 = [0.5 0.5];
y2 = [0.8 0.65];
an_f = annotation("textarrow",x2,y2,"String",{"France","2000"});
    an_f.LineWidth = 6;
x3 = [0.8 0.7];
y3 = [0.3 0.3];
an_u = annotation("textarrow",x3,y3,"String",{"US 2000"});
    an_u.FontSize = 20;
```

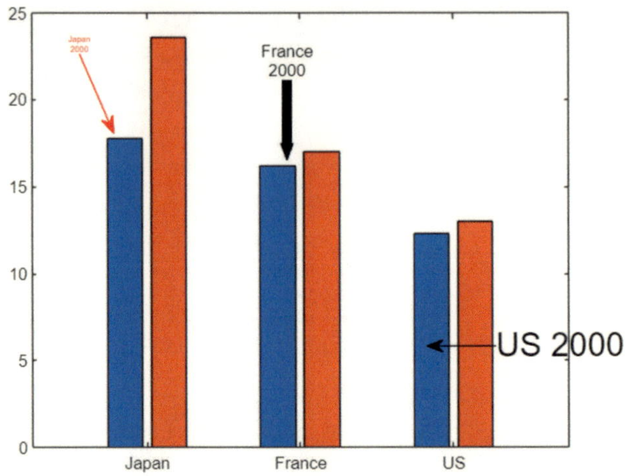

7

Fig. 7.49 Grouped bar chart with various text arrows

Here, we use the **annotation** function. In the first **annotation** function, we set the first argument to 「arrow」. Then, we set 「Color」 to 「red」. With 「x1=[0.2, 0.5]」 and 「y1=[0.55, 0.8]」, the starting point of the arrow is at (x,y)=(0.2,0.55), and the ending point is at (x,y)=(0.5,0.8). In the next **annotation** function, we use 「arrow」 again. Lastly, in the third **annotation** function, we set the first argument to 「textbox」. We specify the position and size with 「dim=[0.15 0.7 0.15 0.15]」, where the position is determined by 0.15 and 0.7, and the size is determined by the elements 0.15 and 0.15. The box is created larger than the text inside it. For more details on the **annotation** function, see ▶ Sect. 18.1. See Fig. 7.50.

```
bar(Year,US);
x1 = [0.2 0.5];
y1 = [0.55 0.8];
annotation("arrow",x1,y1,"Color","red");
x2 = [0.55 0.8];
y2 = [0.82 0.92];
an = annotation("arrow",x2,y2);
    an.Color = "magenta";
    an.LineStyle = "--";
    an.LineWidth = 2;
dim = [0.15 0.7 0.15 0.15];
str = "Annotation Application";
annotation("textbox",dim,"String",str);
```

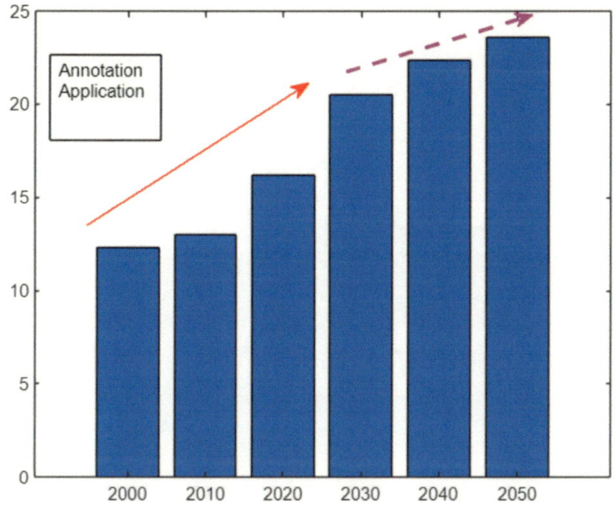

Fig. 7.50 Bar chart with two text arrows and a textbox

Here, we use the **annotation** function twice. We use ⌜textbox⌟ in the first **annotation** function. This is similar to the use of the previous **annotation** function in Fig. 7.50; however, we now set ⌜FitBoxToText⌟ to ⌜on⌟. Consequently, the text box on the left side of the figure automatically adjusts the specified width and height to fit the text characters ⌜Annotation Application (Good Size)⌟. The second example adds an annotation using ⌜rectangle⌟ as the first argument in the **annotation** function, creating a rectangular annotation. ⌜FaceColor⌟ is set to ⌜blue⌟, filling the rectangle with blue. Additionally, ⌜FaceAlpha⌟ is used to make the blue-filled rectangle semi-transparent. For more details on the **annotation** function, see ▶ Sect. 18.1. The created figure is shown in Fig. 7.51.

```
bar(Year,US);
dim = [0.15 0.7 0.15 0.15];
dim2 = [0.6 0.6 0.3 0.3];
str = "Annotation Application (Good Size)";
annotation("textbox",dim,"String",str,...
    "Color","red","FitBoxToText","on");
annotation("rectangle",dim2,"FaceColor","blue","FaceAlpha",0.2);
```

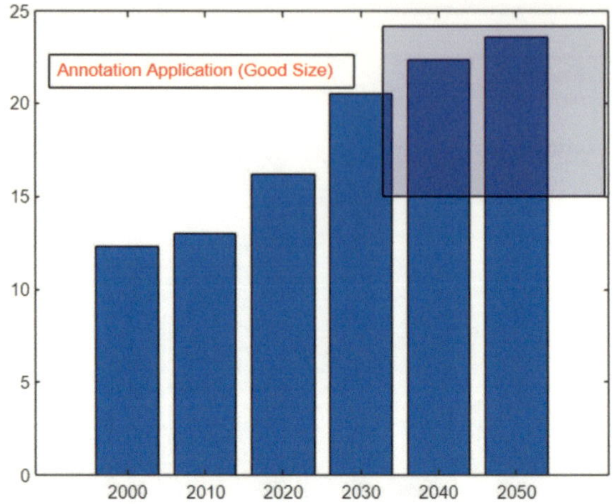

7

■ **Fig. 7.51** Bar chart with a textbox and a rectangle

Here, we use the **yline** function to draw four horizontal lines. First, we store the properties of the **bar** function in the variable「bar1」. The variable「bar1」is a 1-by-3 bar object. For example,「bar1(1)」contains「Japan(1)」and「Germany(1)」. Next, the variable「ym1」contains the y-axis values of the first group of bars using「YEndPoints」. Similarly, the variable「ym2」contains the y-axis values of the second group of bars.

We use the **yline** function to draw four horizontal lines at the positions specified by「ym1」and「ym2」. For more details of the **yline** function, see ▶ Sect. 18.3. The resulting figure is displayed in ■ Fig. 7.52.

```
bar1 = bar(["Japan","Germany"],[Japan(1:3),Germany(1:3)]);
    ym1 = bar1(1).YEndPoints;
    ym2 = bar1(2).YEndPoints;
yline(ym1,"LineStyle","--","Color","black");
yline(ym2,"LineWidth",3,"Color","blue");
```

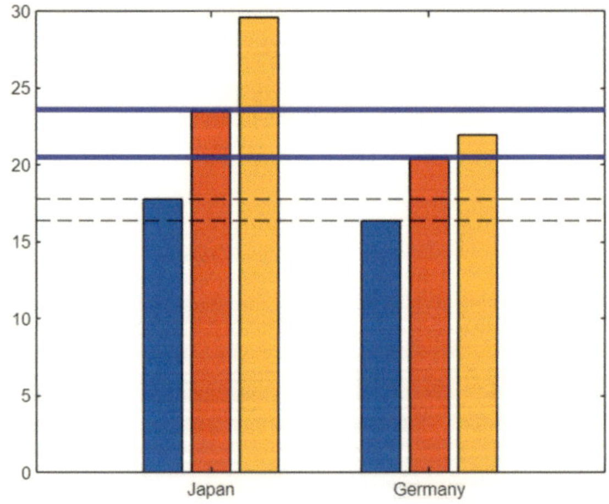

■ **Fig. 7.52** Grouped bar chart with horizontal lines of different styles and colors

Here, we store the properties of the bar graph (⌜bar(Year,US)⌟) in the variable ⌜b_us⌟. We then use the ⌜YEndPoints⌟ property to obtain the y-axis end-points of the bars and store them in ⌜y_us⌟. Next, ⌜yregion(y_us(end),Inf)⌟ creates a filled region on the y-axis. Note that ⌜Inf⌟ is assigned to approximately the highest value on the y-axis. Finally, ⌜yg=yregion(y_us(1),y_us(end))⌟ fills the area between ⌜y_us(1)⌟ and ⌜y_us(end)⌟. See ◘ Fig. 7.53.

```
b_us = bar(Year,US);
    y_us = b_us.YEndPoints;
yregion(y_us(end),Inf);
yg = yregion(y_us(1),y_us(end));
    yg.FaceColor = "green";
```

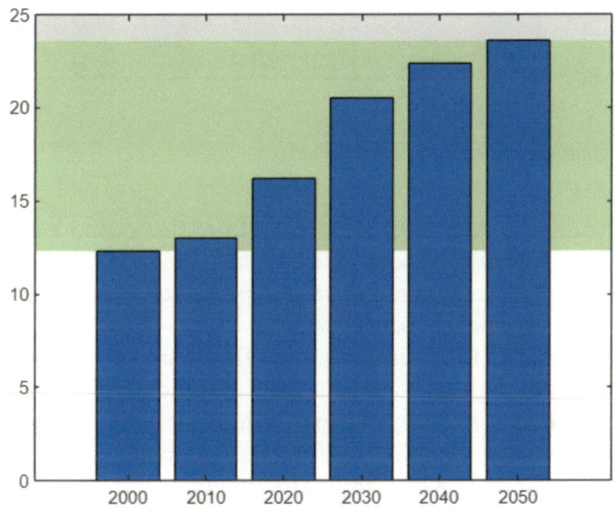

◘ **Fig. 7.53** Bar chart with two horizontal shaded areas

Here, we use the **xregion** function three times. For example, we create a filled region on the x-axis from 1995 to 2015 with ⌜x1 = xregion(1995, 2015)⌟. We store the properties of this region in the variable ⌜x1⌟. Next, we set ⌜FaceColor⌟ to ⌜red⌟, ⌜LineWidth⌟ to 2, and ⌜EdgeColor⌟ to ⌜black⌟, creating a red filled region with a thicker black outline. See ◘ Fig. 7.54.

```
bar(Year,US,0.5);
x1 = xregion(1995,2015);
    x1.FaceColor = "red";
    x1.LineWidth = 2;
    x1.EdgeColor = "black";
x2 = xregion(2015,2045);
    x2.FaceColor = "blue";
x3 = xregion(2045,2055);
    x3.FaceColor = "yellow";
```

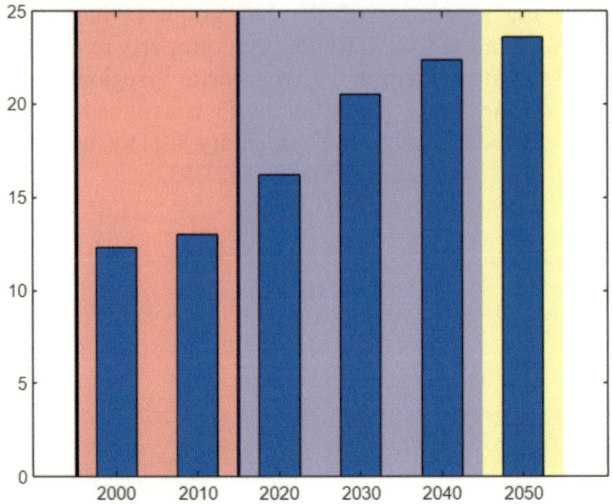

Fig. 7.54 Bar chart with three vertical shaded areas

Here, we retrieve the values (23.6 and 20.5) from the second bar graphs for each category (Japan and Germany) and place these values on top of the second bar graphs. In detail, we obtain the coordinates for placing these numbers using 「xp=bh(2).XEndPoints」 and 「yp=bh(2).YEndPoints」.

Specifically, 「bh(2).XEndPoints」 returns the x-coordinates of the second bar graph. Note that since 「bh(2)」 contains second data for 「Japan」 and 「Germany」 respectively, 「xp」 becomes a 1-by-2 array. In the same way, 「YEndPoints」 returns the y-coordinates. Next, 「string(bh(2).YData)」 returns the 「YData」 values (23.6 and 20.5) for 「Japan」 and 「Germany」 as string type, respectively.

Finally, in 「text(xp,yp,labels)」, we place the label text at the positions specified by 「xp」 and 「yp」. For more details on the **text** function, see ▶ Sect. 18.2. In Fig. 7.55, note that the numbers 23.6 and 20.5 are slightly misaligned. This will be corrected in the next program of creating Fig. 7.56.

```
countries = ["Japan","Germany"];
bh = bar(countries,[Japan,Germany]);
    xp = bh(2).XEndPoints;
    yp = bh(2).YEndPoints;
labels = string(bh(2).YData);
text(xp,yp,labels);
```

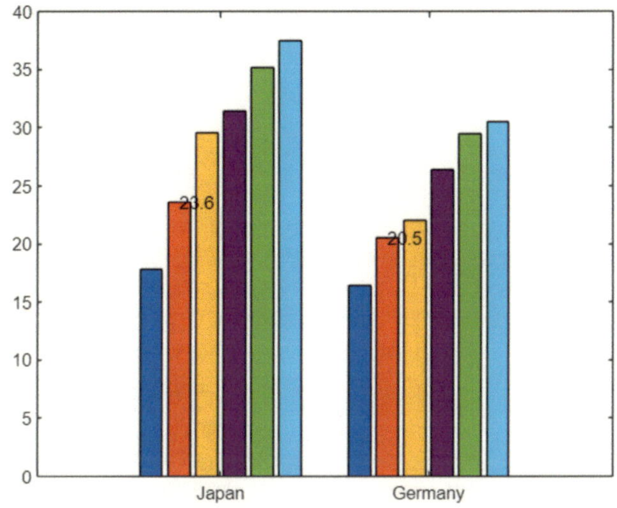

Fig. 7.55 Grouped bar chart with data labels

Except for the use of the **text** function in the last line, the following program is the same as the previous one that created Fig. 7.55. We only explain the **text** function. The **text** function is used to specify the location of the text string (「labels」) with the first and second arguments 「xp」 and 「yp」. Additionally, 「HorizontalAlignment」 is set as 「center」 to center the text. Next, 「VerticalAlignment」 is set to 「bottom」 to align the text at the bottom. For more details on the **text** function, see ▶ Sect. 18.2. The resulting figure is presented in Fig. 7.56.

```
countries = ["Japan","Germany"];
bh = bar(countries,[Japan,Germany]);
    xp = bh(2).XEndPoints;
    yp = bh(2).YEndPoints;
labels = string(bh(2).YData);
text(xp,yp,labels,"HorizontalAlignment","center",...
    "VerticalAlignment","bottom","FontSize",8);
```

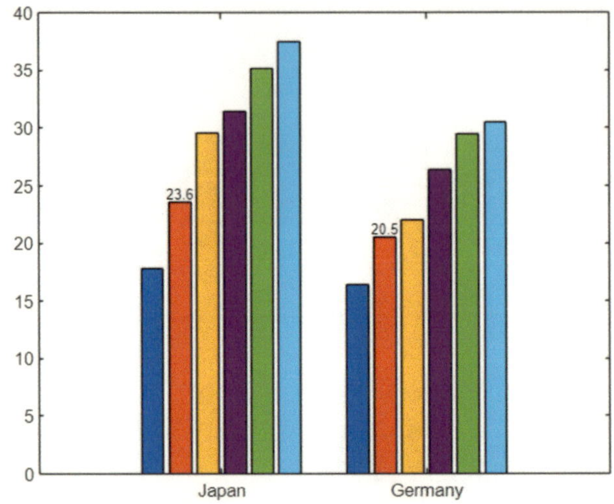

Fig. 7.56 Grouped bar chart with position-adjusted data labels (cf. Fig. 7.55)

We place the text characters given by 「label_txt=["1","2","3 \beta","4","5","6"]」 above the y-coordinates of six bar graphs, where 「\beta」 represents the Greek character. For more details on the **text** function, see ▶ Sect. 18.2. See Fig. 7.57.

```
b = bar(Year,US);
    xEnd = b.XEndPoints;
    yEnd = b.YEndPoints;
label_txt = ["1","2","3 ¥beta","4","5","6"];
    tx = text(xEnd,yEnd,label_txt,...
    "HorizontalAlignment","center","VerticalAlignment","bottom");
    tx(3).Color = "red";
    tx(3).FontSize = 20;
```

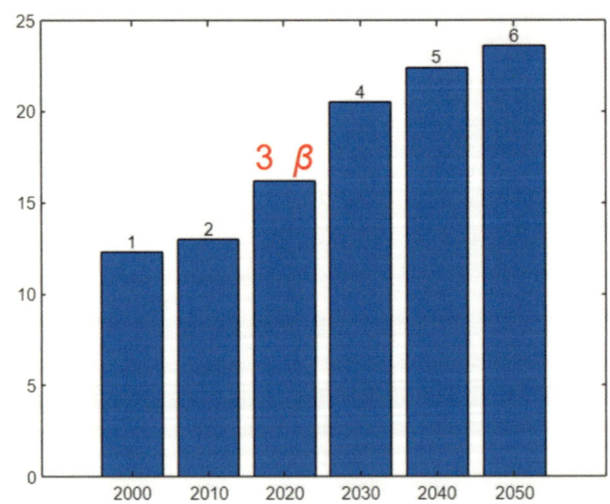

Fig. 7.57 Bar chart with Greek and numeric data labels

Here, we place text characters on top of multiple bar graphs. In the **text** function, the first, second, and third arguments are ｢xp1｣, ｢yp1｣, and ｢labels111｣. Note that the text variable ｢labels111｣ is a 1-by-1 string, while ｢xp1｣ and ｢yp1｣ are 1-by-3 doubles. In this case, the same text is placed on all bar graphs. Specifically, we use ｢["Japan"+newline+"2000"]｣, where the first line is ｢Japan｣ and the second line is ｢2000｣. We then store the properties of the **text** function in the variable ｢tx｣. For ｢tx(1)｣, ｢Color｣ is set to blue and ｢FontSize｣ is set to 20, making the left-most text blue and large. For more details on the **text** function, see ▶ Sect. 18.2. See ◘ Fig. 7.58.

```
Year3 = Year(1:3);
Japan3 = Japan(1:3);
bh = bar(Year3,Japan3,"FaceColor","red","EdgeColor","cyan");
    xp1 = bh.XEndPoints;
    yp1 = bh.YEndPoints;
labels111 = ["Japan" + newline + "2000-2020"];
tx = text(xp1,yp1,labels111,...
    "HorizontalAlignment","center","VerticalAlignment","bottom");
    tx(1).Color = "blue";
    tx(1).FontSize = 20;
```

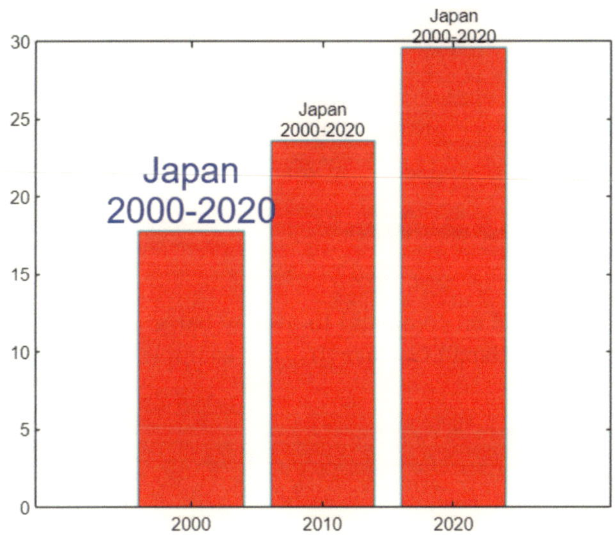

◘ **Fig. 7.58** Bar chart with two-line and colored text labels

Here, we specify the positions for each x-axis coordinate using ｢Year3=Year(1:3)｣, as in the previous program that created ◘ Fig. 7.58. Additionally, ｢labels222｣ provides three different text labels for the three bar graphs. Specifically, the first element of ｢labels222｣, ｢["Japan" + newline + "2000"]｣, is placed above the first bar graph. Similarly, ｢["Japan" + newline + "2010"]｣ is placed above the second bar graph, and ｢["Japan" + newline + "2020"]｣ is placed above the third bar graph. See ◘ Fig. 7.59 for the created figure.

```
Year3 = Year(1:3);
Japan3 = Japan(1:3);
bh = bar(Year3,Japan3,"FaceColor","red","EdgeColor","cyan");
    xp = bh.XEndPoints;
    yp = bh.YEndPoints;
labels222 = [["Japan" + newline + "2000"],...
    ["Japan"+newline+"2010"],["Japan"+newline+"2020"]];
text(xp,yp,labels222,...
    "HorizontalAlignment","center","VerticalAlignment","bottom");
```

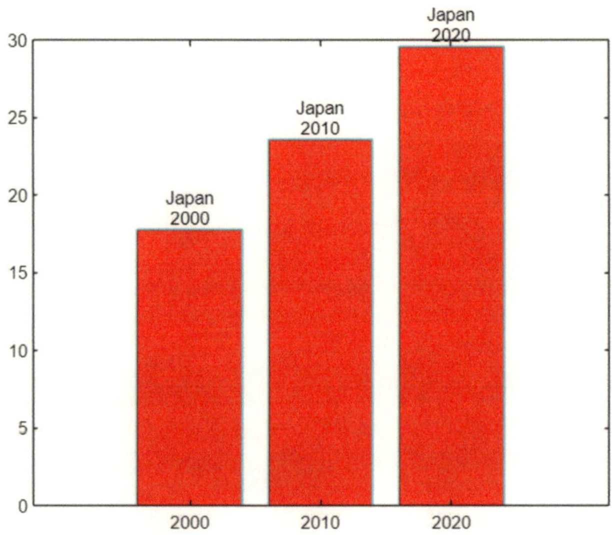

Fig. 7.59 Bar chart with two-line text labels

Here, we use the **hold on** command to overlay bar graphs. Note that this differs from using 「bar(Year, [Japan,Germany])」. For example, compare Fig. 7.9 with Fig. 7.60.

```
bar(Year,Japan);
hold on
bar(Year,Germany);
hold off
```

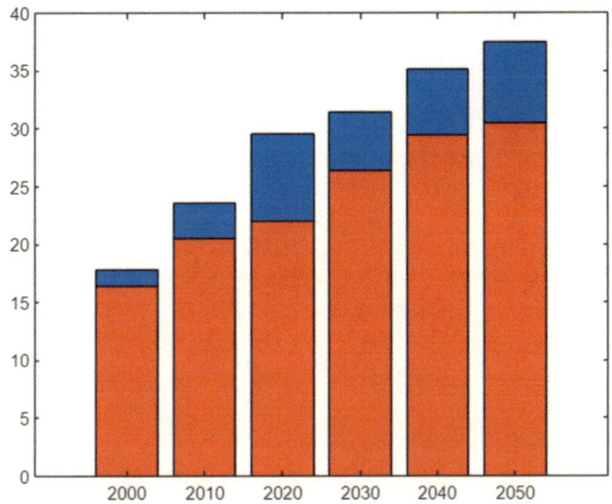

Fig. 7.60 Bar chart with overlaid bars (non-stacked)

Here, we use the **hold on** command to overlay two bar graphs. For the first bar graph, we set 「FaceAlpha」 to 0.2, which increases the transparency. For the second bar graph, we set 「FaceAlpha」 to 0.5, 「LineWidth」 to 2, 「EdgeColor」 to 「blue」, and 「LineStyle」 to 「-.」, creating a bar graph with a thick blue dash-dot outline and increased transparency. See Fig. 7.61.

```
bj = bar(Year,Japan);
    bj.FaceAlpha = 0.2;
hold on
bg = bar(Year,Germany);
    bg.FaceAlpha = 0.5;
    bg.LineWidth = 2;
    bg.EdgeColor = "blue";
    bg.LineStyle = "-.";
hold off
```

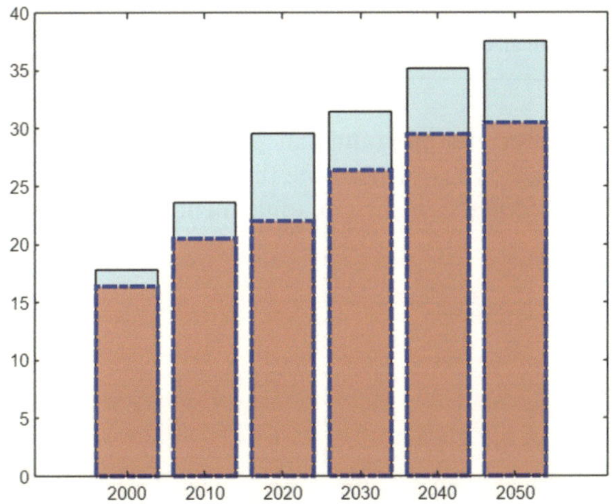

Fig. 7.61 Bar chart with overlaid transparent bars (non-stacked)

Here, using the **hold on** command, we overlay a bar graph created by the **bar** function with a line graph created by the **plot** function. After storing the properties of the **bar** function in ｢bh｣, we retrieve the x-axis and y-axis endpoints using ｢XEndpoints｣ and ｢YEndPoints｣, and store them in ｢xp｣ and ｢yp｣. The variables ｢xp｣ and ｢yp｣ are then used in the **plot** function. The resulting figure is presented in ◘ Fig. 7.62.

```
bh = bar(Year,Japan);
hold on
    xp = bh.XEndPoints;
    yp = bh.YEndPoints;
pl = plot(xp,yp,":x");
    pl.MarkerSize = 10;
    pl.Color = "black";
    pl.LineWidth = 2.5;
hold off
```


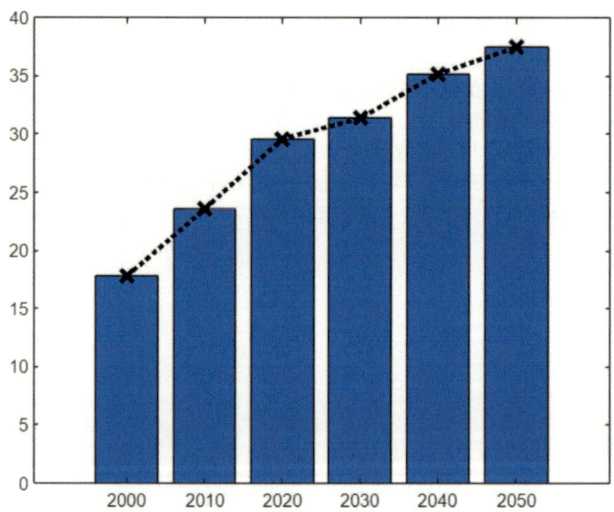

◘ **Fig. 7.62** Bar chart with an overlaid line connecting data points

Here, unlike the previous program that generated ◘ Fig. 7.62, we adjust the y-axis coordinates using ｢yp2｣, where note that ｢yp2=yp+2｣. Additionally, we use the **grid on** command to make the height adjustments clearer. Compare ◘ Fig. 7.62 with ◘ Fig. 7.63.

```
bh = bar(Year,Japan);
hold on
    xp = bh.XEndPoints;
    yp = bh.YEndPoints;
yp2 = yp+2;
pl = plot(xp,yp2,":o");
    pl.Color = "green";
    pl.LineWidth = 2.5;
    pl.MarkerSize = 10;
hold off
grid on
```

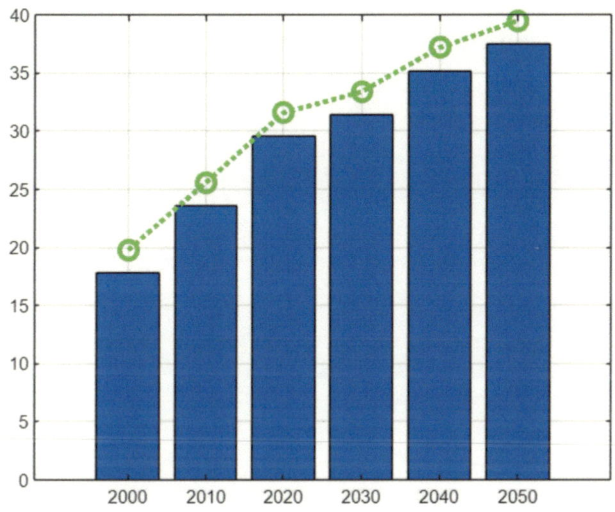

Fig. 7.63 Bar chart with an overlaid line connecting adjusted data points (cf. Fig. 7.62)

Here, we use the **tiledlayout** function to create a 1-by-3 tiled layout. For each space, we overlay bar and plot graphs using the **hold on** command. For more details on the overlaying program, see the program that created Figs. 7.62 and 7.63. The resulting figure is presented in Fig. 7.64.

```
figure
ti = tiledlayout(1,3);
title(ti,"The population ratio of people aged 65 and over");
nexttile;
bh1 = bar(Year,Japan);
hold on
    xp_j = bh1.XEndPoints;
    yp_j = bh1.YEndPoints;
    yp_j2 = yp_j+1;
pl_j = plot(xp_j,yp_j2,":o");
    pl_j.LineWidth = 2.5;
hold off
title("Japan");
xlabel("Year");
yl = ylabel("Percent");
    yl.Rotation = 0;
grid on
nexttile;
bh2 = bar(Year,US);
hold on
    xp_u = bh2.XEndPoints;
    yp_u = bh2.YEndPoints;
yp_u2 = yp_u+1;
pl = plot(xp_u,yp_u2,"-.s");
    pl.LineWidth = 2.5;
hold off
title("US");
xlabel("Year");
grid on
nexttile;
bh3 = bar(Year,Italy);
hold on
    xp_i = bh3.XEndPoints;
    yp_i = bh3.YEndPoints;
yp_i2 = yp_i+1;
pl = plot(xp_i,yp_i2,"-.*");
    pl.LineWidth = 2.5;
    pl.MarkerSize = 10;
hold off
title("Italy");
xlabel("Year");
grid on
```

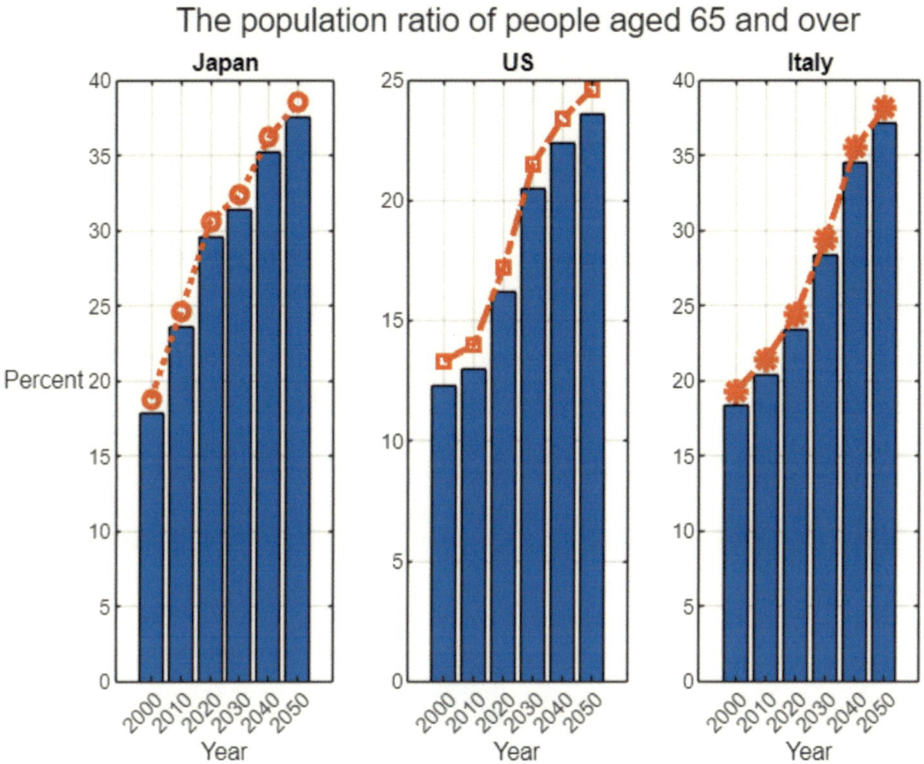

Fig. 7.64 Bar chart with overlaid lines connecting data points (1-row, 3-column layout)

Here, we use the **bar** function to create a stacked bar chart. Then, by using the **hold on** command, we overlay the **plot** function, which connects the points at each coordinate with a line. For example, in the command 「plot(xs1,ys1,"k:o","LineWi dth",2)」, we use the horizontal and vertical coordinate data 「xs1」 and 「ys1」 corresponding to 「US(1)」, 「Japan(1)」, and 「France(1)」. Note that because the **tiled-layout** function was used in the previous program that created Fig. 7.64, we reset the figure space to the default by using the **figure** command at the beginning of the program. See Fig. 7.65.

```
figure
countries = ["US","Japan","France"];
bs = bar(countries,[US,Japan,France],"stacked");
hold on
    xs1 = bs(1).XEndPoints;
    ys1 = bs(1).YEndPoints;
plot(xs1,ys1,"k:o","LineWidth",2);
    xs2 = bs(2).XEndPoints;
    ys2 = bs(2).YEndPoints;
plot(xs2,ys2,"k:o","LineWidth",2);
    xs3 = bs(3).XEndPoints;
    ys3 = bs(3).YEndPoints;
plot(xs3,ys3,"k:o","LineWidth",2);
    xs4 = bs(4).XEndPoints;
    ys4 = bs(4).YEndPoints;
plot(xs4,ys4,"k:o","LineWidth",2);
    xs5 = bs(5).XEndPoints;
    ys5 = bs(5).YEndPoints;
plot(xs5,ys5,"k:o","LineWidth",2);
    xs6 = bs(6).XEndPoints;
    ys6 = bs(6).YEndPoints;
plot(xs6,ys6,"k:o","LineWidth",2);
hold off
grid minor
```

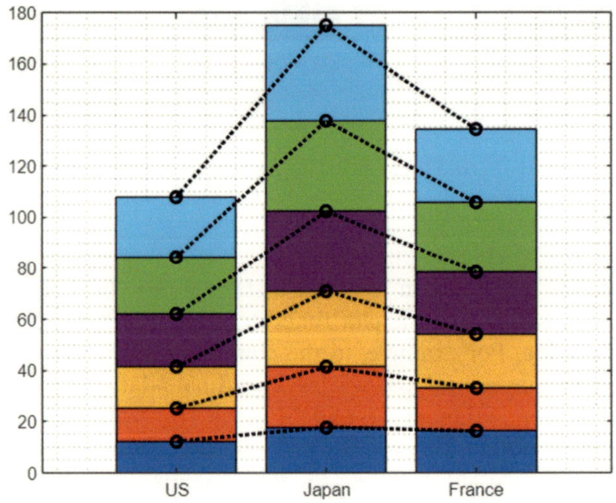

Fig. 7.65 Stacked bar chart with overlaid lines connecting data points

Here, using the **bar** and **plot** functions, we overlay them by executing the **hold on** command. Specifically, we create a stacked bar chart using the **bar** function. See Fig. 7.66 for the created figure.

```
bs = bar(Year,[Italy,Germany],"stacked");
hold on
    bsx1 = bs(1).XEndPoints;
    bsy1 = bs(1).YEndPoints;
pl1 = plot(bsx1,bsy1,"m:s");
    pl1.LineWidth = 2;
    pl1.MarkerSize = 10;
    bsx2 = bs(2).XEndPoints;
    bsy2 = bs(2).YEndPoints;
pl2 = plot(bsx2,bsy2,"k-.x");
    pl2.LineWidth = 2;
    pl2.MarkerSize = 10;
hold off
grid minor
```

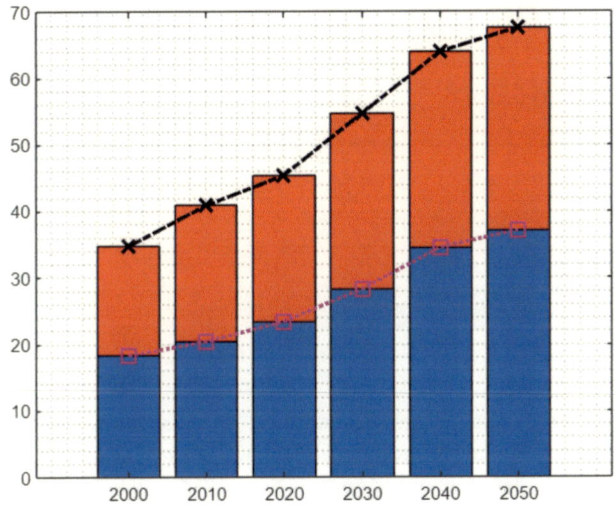

Fig. 7.66 Stacked bar chart with overlaid lines connecting data points (X-axis: Year)

7.6 Questions

Question 1

By entering 「Year」 in the first argument of the **bar** function and 「[US,Germany]」 in the second argument, create Fig. 7.67. In detail, store the properties of the **bar** function in the variable 「Ex1」. Use 「CData」 to change only the first, third, and fifth bars of the US bar graph to red ([1 0 0]).

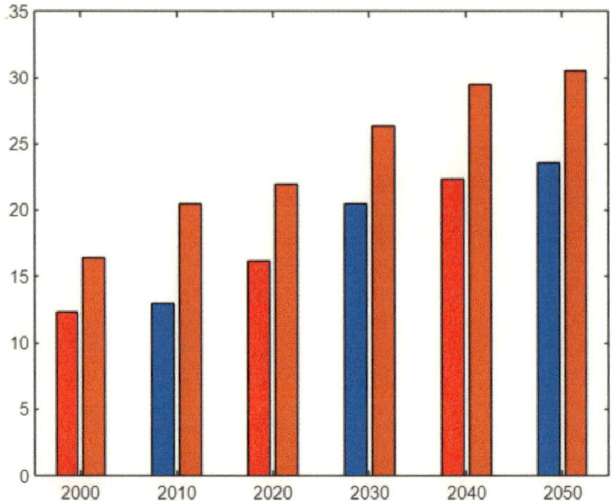

7

Fig. 7.67 Bar chart (Question 1)

Answer 1

```
figure
Ex1 = bar(Year,[US,Germany],"FaceColor","Flat");
    Ex1(1).CData(1,:) = [1 0 0];
    Ex1(1).CData(3,:) = [1 0 0];
    Ex1(1).CData(5,:) = [1 0 0];
```

Question 2

Using the **hold on** command and the **bar** function three times, create Fig. 7.68. Concretely, overlay three bar graphs (「Japan」, 「Germany」, 「US」), where the first argument in each **bar** function is 「Year」. Additionally, for the 「Japan」 bar graph, set 「FaceAlpha」 to 0.2 to make the bar graph more transparent. For the 「Germany」 bar graph, set 「LineStyle」 to a dash-dotted line and 「LineWidth」 to 2. For the 「US」 bar graph, set 「EdgeColor」 to blue.

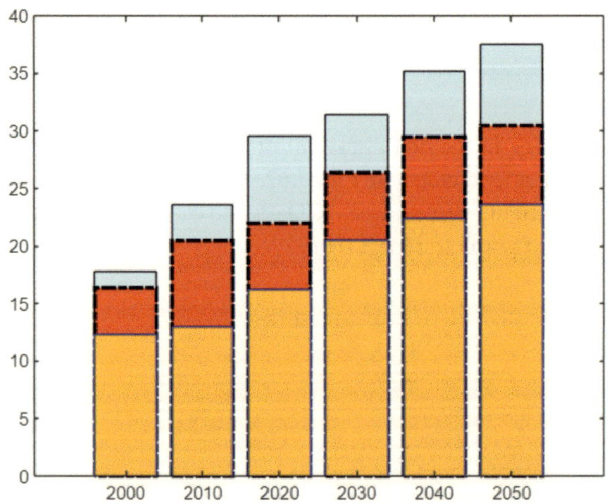

Fig. 7.68 Bar chart (Question 2)

Answer 2

```
figure
bar(Year,Japan,"FaceAlpha",0.2);
hold on
bar(Year,Germany,"LineStyle","-.","LineWidth",2);
bar(Year,US,"EdgeColor","blue");
hold off
```

Question 3

Using the data for 「Japan」, create Fig. 7.69. It is useful to use a **for** loop to obtain the x-axis data, y-axis data, and the values of 「Japan」 for each of the six bars and place the text using the **text** function.

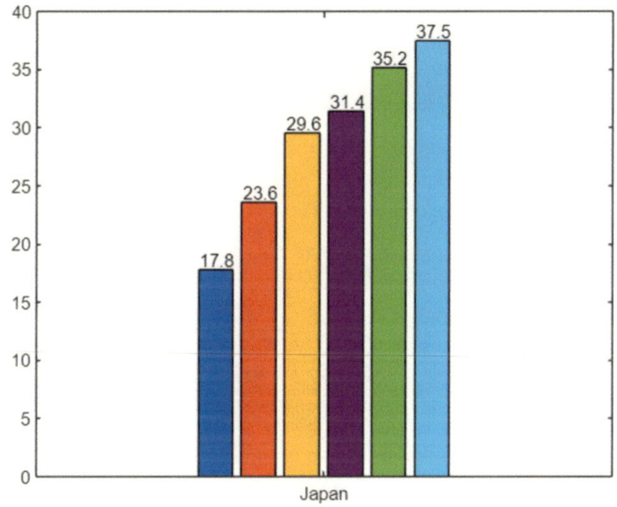

Fig. 7.69 Bar chart (Question 3)

Answer 3

```
figure
Ex2 = bar("Japan",Japan);
for i = 1:6
    xp = Ex2(i).XEndPoints;
    yp = Ex2(i).YEndPoints;
    labels = string(Ex2(i).YData);
    text(xp-0.05,yp+0.8,labels);
end
```

Bar3 Function

Contents

8.1 Details of the Data Used in this Chapter – 252

8.2 Excel Data of ► Section 8.1 – 253

8.3 Basic Syntax of the Bar3 Function – 253

8.4 Setting and Controlling the Bar3 Function – 259

8.5 Combining Optional Functions and the Bar3 Function – 269

8.6 Questions – 279

Supplementary Information The online version contains supplementary material available at https://doi.org/10.1007/978-981-96-6213-5_8.

Y. Nakamoto, *A Comprehensive Guide to Creating Graphics with MATLAB Programming*, Classroom Companion: Economics, https://doi.org/10.1007/978-981-96-6213-5_8

Learning Objectives

This chapter creates 3D bar charts using the **bar3** function. By reading this chapter, you will learn how to use the **bar3** function. Moreover, you will create not only standard 3D bar charts but also stacked 3D bar charts. You will understand the similarities and differences in the created figures, syntax, and the type of data between the **bar3** function and the **bar** function (covered in ► Chap. 7).

Overview

In this chapter, we create a 3D bar graph using the **bar3** function. Unlike the 2D bar graphs created with the **bar** function in ► Chap. 7, which plotted bars on two axes (x-axis and y-axis), the **bar3** function plots bars on three axes (x-axis, y-axis, and z-axis). The advantage of using the **bar3** function is that adding a third axis may reveal relationships that are difficult to understand in a 2D graph. However, as mentioned in ► Chap. 4 (the **scatter3** function) and ► Chap. 6 (the **plot3** function), a potential drawback is that as the data becomes more complex, it may become more difficult to grasp the relationships in a 3D plot. In such cases, it is useful to use the optional function, as described in ► Sect. 8.5.

8

8.1 Details of the Data Used in this Chapter

In this chapter, we use time-series data on crude oil prices (price per barrel in U.S. dollars). Crude oil is an important resource in modern society, widely used for gasoline and as a raw material for plastics. Regarding the variables used in this chapter, 「Year」 represents the year, 「Ave」 represents the average price per barrel of crude oil, 「Brent」 represents the price per barrel of Brent crude oil in U.S. dollars, 「Dubai」 represents the price per barrel of Dubai crude oil in U.S. dollars, and 「WTI」 represents the price per barrel of West Texas Intermediate (WTI) crude oil in U.S. dollars.

```
Year = [1990,1995,2000,2005,2010,2015,2020,2023]';

Ave = [23,17,28,53,79,51,41,81]';

Brent = [24,17,28,54,80,52,42,83]';

Dubai = [20,16,26,49,78,51,42,82]';

WTI = [24,18,30,56,79,49,39,78]';

Price_oil = table(Year,Ave,Brent,Dubai,WTI);
```

8.2 Excel Data of ▶ Section 8.1

In ▶ Sect. 8.2, we import the Excel data used in ▶ Sects. 8.3, 8.4, 8.5, and 8.6. We then specify the file name and sheet name. For more details on how to import Excel data, see ▶ Chap. 2.

```
file = "alldata.xlsx";
sheet = "Chap8_Price_CrudeOil";
Price_oil = readtable(file,"Sheet",sheet);
```

```
Warning: Column headers from the file were modified to make them valid MATLAB identifiers before creating
variable names for the table. The original column headers are saved in the VariableDescriptions property.
Set 'VariableNamingRule' to 'preserve' to use the original column headers as table variable names.
```

Using the **round** and **renamevars** functions, we round the data to the nearest integer and change the column names from 「CrudeOil_Average___bbl_」 to 「Ave」, 「CrudeOil_Brent___bbl_」 to 「Brent」, and 「CrudeOil_WTI___bbl_」 to 「WTI」. Additionally, we extract each column from the table variable.

```
Price_oil = round(Price_oil);
Price_oil = renamevars(Price_oil,
["CrudeOil_Average___bbl_","CrudeOil_Brent___bbl_",...
    "CrudeOil_Dubai___bbl_","CrudeOil_WTI___bbl_"],...
    ["Ave","Brent","Dubai","WTI"]);
Year = Price_oil.Year;
Ave = Price_oil.Ave;
Brent = Price_oil.Brent;
Dubai = Price_oil.Dubai;
WTI = Price_oil.WTI;
```

8.3 Basic Syntax of the Bar3 Function

Basic Syntax:

Here, we explain the two basic syntaxes of the **bar3** function highlighted in the black rectangle of Figure 8.1. First, as for 「bar3(z)」, we enter 「z」 in the first argument and create a 3D bar graph. Based on this syntax, the values of 「z」 correspond to the heights along the z-axis. Next, as for 「bar3(y,z)」, we enter 「y」 in the first argument and 「z」 in the second argument. In this case, the values of 「y」 correspond to the y-axis of the bar graph, while the values of 「z」 represent the height of the bars along the z-axis.

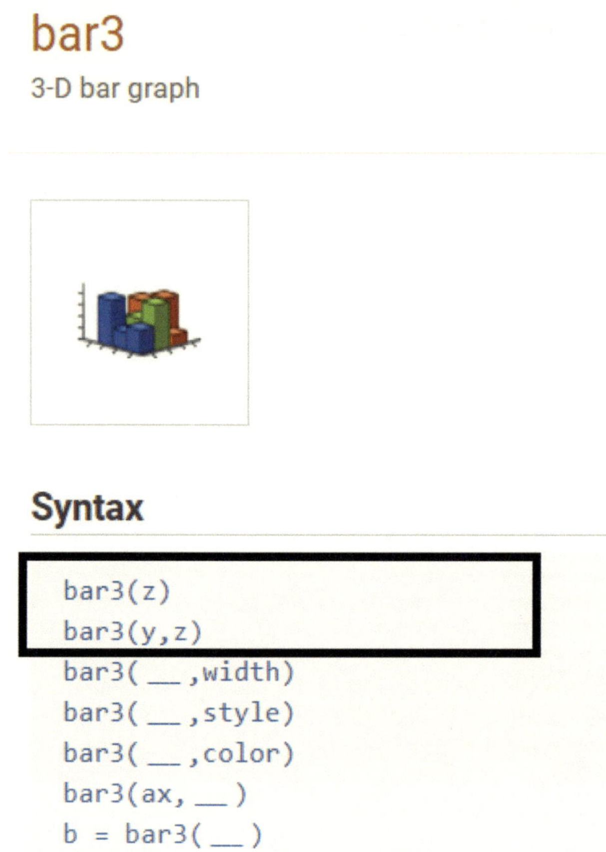

Fig. 8.1 The syntax of the bar3 function

Here, using the **bar3** function, we enter 「Ave」 in the first argument. As a result, the values of 「Ave」 correspond to the heights along the z-axis. The y-axis ticks are automatically assigned to 1, 2, ..., and 8. See Fig. 8.2.

```
bar3(Ave);
```

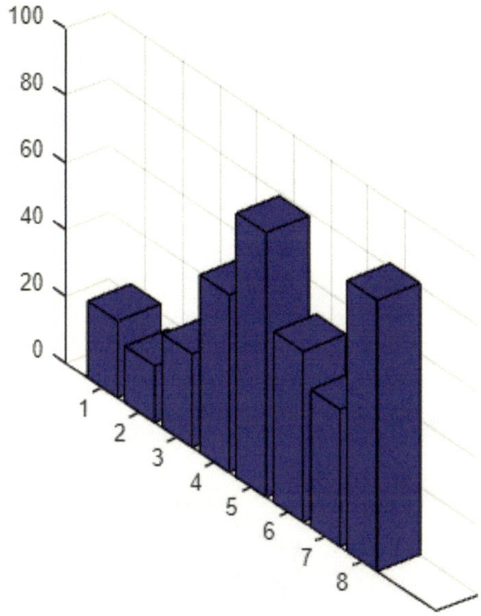

Fig. 8.2 3D bar chart (Y-axis: Default index)

Here, using the **bar3** function, we enter 「Year」 in the first argument and 「Ave」 in the second argument. Consequently, the y-axis in the previous figure (Fig. 8.2) was labeled 1, 2, 3, 4, 5, 6, 7, and 8. In Fig. 8.3, however, the y-axis is changed to reflect the values of the variable 「Year」. See Fig. 8.3.

```
bar3(Year,Ave);
```

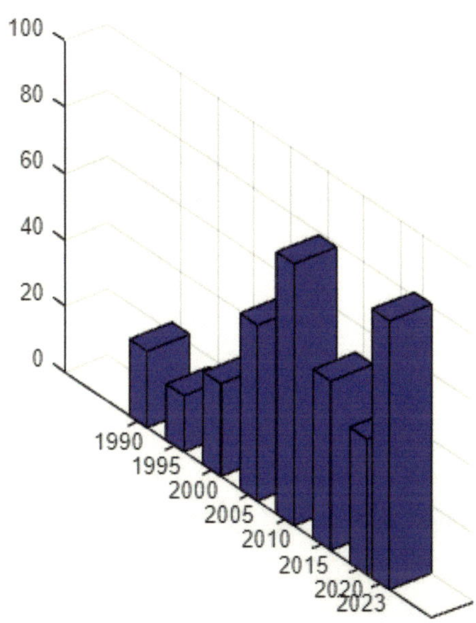

Fig. 8.3 3D bar chart (Y-axis: Year)

Here, we enter an 8-by-4 matrix 「[Ave,Brent,Dubai,WTI]」 in the second argument. As a result, unlike Fig. 8.3, four different 3D bar graphs are created along the x-axis. The x-axis is automatically labeled 1, 2, 3, and 4 for each group. In addition, each group is distinguished by different colors. See Fig. 8.4.

```
bar3(Year,[Ave,Brent,Dubai,WTI]);
```

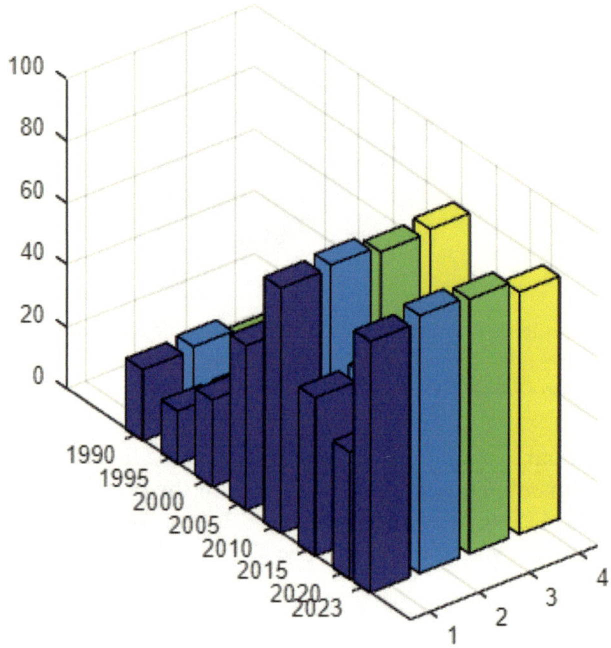

Fig. 8.4 3D bar chart with multiple groups (X-axis: Default index; Y-axis: Year)

The variables 「Year」, 「Ave」, 「Brent」, 「Dubai」, and 「WTI」 are each an 8-by-1 column vector. Here, except for the eighth data point for the year 2023, we use 7-by-1 column vectors in each argument of the **bar3** function. See Fig. 8.5 for the created figure.

```
bar3(Year(1:7),[Ave(1:7),Brent(1:7),Dubai(1:7),WTI(1:7)]);
```

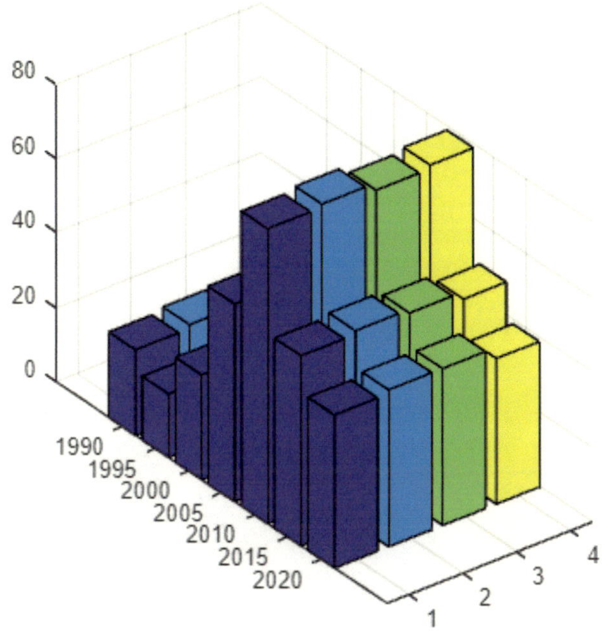

Fig. 8.5 3D bar chart with multiple groups using a subset of variables (X-axis: Default index; Y-axis: Year)

Here, using the data from the second through fifth data points of each variable, we create a 3D bar graph. See Fig. 8.6.

```
bar3(Year(2:5),[Ave(2:5),Brent(2:5),Dubai(2:5),WTI(2:5)]);
```

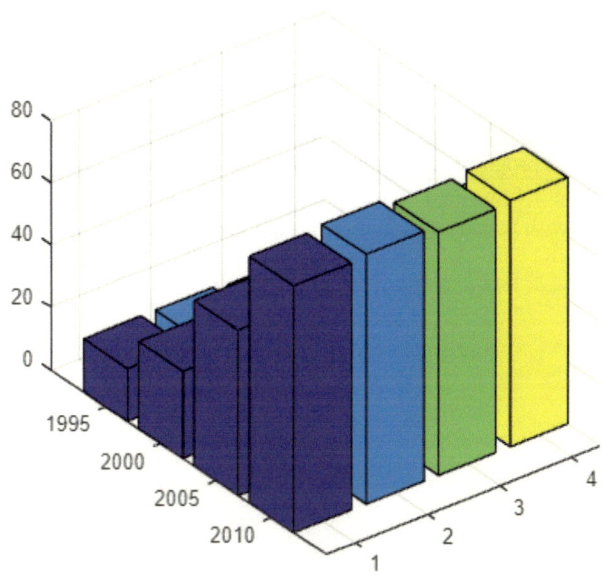

Fig. 8.6 3D bar chart with multiple groups using a different subset of variables (cf. Fig. 8.5)

▶ Example of Table Variables

Here, using the table variable 「Price_oil」, we create a 3D bar graph similar to ◘ Fig. 8.3. See ◘ Fig. 8.7.

```
bar3(Price_oil.Year,Price_oil.WTI);
```

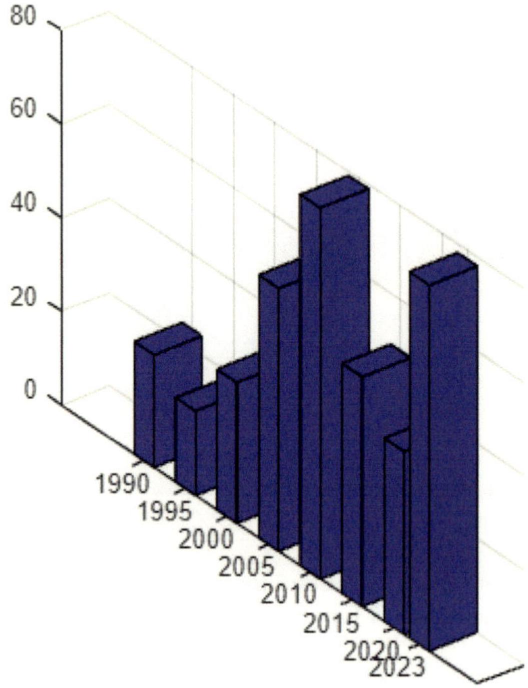

◘ **Fig. 8.7** Use of a table variable: 3D bar chart

◀

▶ Example of Table Variables

As in the previous program that created ◘ Fig. 8.7, we use the table variable 「Price_oil」. Here, by entering 「[Price_oil.Dubai, Price_oil.Brent]」 in the second argument, we create a 3D bar graph with two groups. The resulting figure is presented in ◘ Fig. 8.8.

```
bar3(Price_oil.Year,[Price_oil.Dubai,Price_oil.Brent]);
```

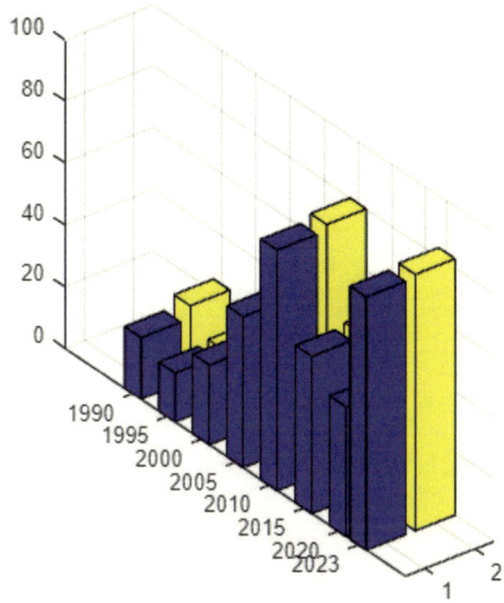

Fig. 8.8 Use of a table variable: 3D bar chart with two groups

◀

8.4 Setting and Controlling the Bar3 Function

Here, by entering 0.5 in the third argument of the **bar3** function, we create bars that are narrower than the default width. See Fig. 8.9.

```
bar3(Year,Ave,0.5);
```

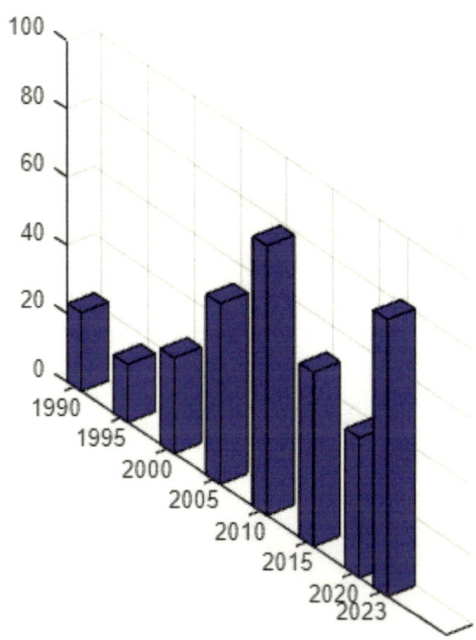

Fig. 8.9 3D bar chart with narrow bars

Here, in addition to the previous program that created ◘ Fig. 8.9, we specify 「red」 as the fourth argument, changing the color of the bars to red. See ◘ Fig. 8.10.

```
bar3(Year,Ave,0.5,"red");
```

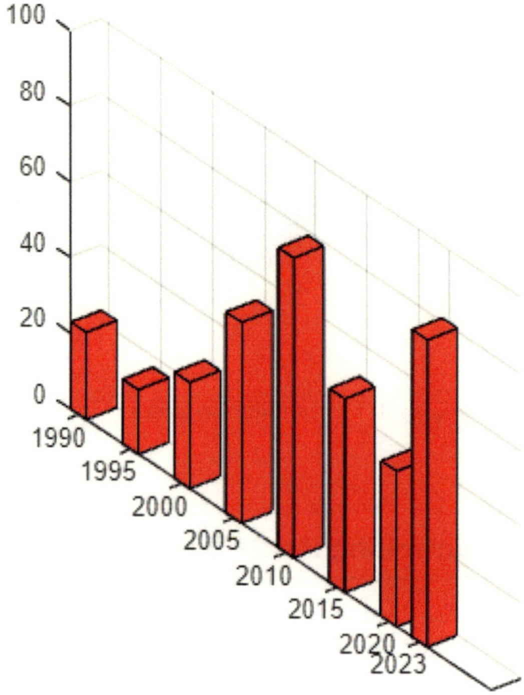

◘ **Fig. 8.10** 3D bar chart with narrow red bars

Here, we enter 「[Ave,Brent]」 in the second argument of the **bar3** function and store the properties in the variable 「b3」. Consequently, 「b3(1)」 corresponds to the variable 「Ave」 (the first group), and 「b3(2)」 corresponds to the variable 「Brent」 (the second group).

Then, for the first group, we set 「FaceColor」 to 「none」 and 「LineStyle」 to 「--」, creating a transparent 3D bar graph with dashed outlines. Next, for the second group, we set 「FaceColor」 to 「yellow」, 「LineWidth」 to 2, and 「EdgeColor」 to 「green」, creating a yellow filled 3D bar graph with thick green outlines. The resulting figure is shown in ◘ Fig. 8.11.

```
b3 = bar3(Year,[Ave,Brent]);
    b3(1).FaceColor = "none";
    b3(1).LineStyle = "--";
    b3(2).FaceColor = "yellow";
    b3(2).LineWidth = 2;
    b3(2).EdgeColor = "green";
```

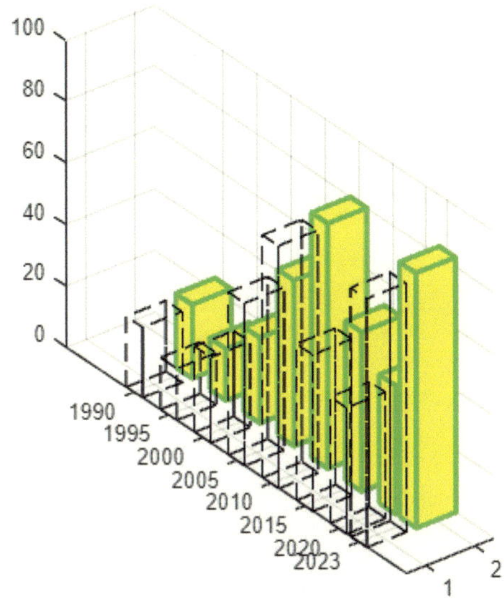

Fig. 8.11 3D bar chart with one group of transparent bars with dashed edges and another group of yellow bars with green edges

Here, we store the properties of the **bar3** function in the variable 「b3」. Then, we set 「FaceAlpha」 to 0.2, creating a 3D bar graph with high transparency. See Fig. 8.12.

```
b3 = bar3(Year,Ave,0.5,"red");
    b3.FaceAlpha = 0.2;
```

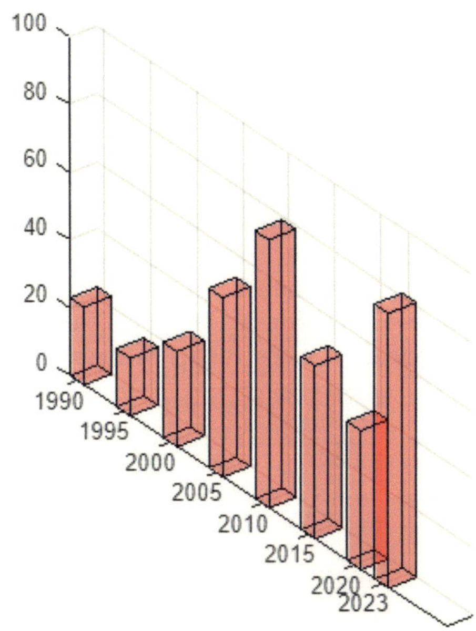

Fig. 8.12 3D bar chart with semi-transparent red bars

Here, using 「stacked」 in the **bar3** function, we create a stacked bar graph. However, note that the following program is incorrect in its use of matrices. First, 「a1」 is a row vector with a size of 1-by-3. Similarly, 「b1」 is also a row vector of size 1-by-3. Looking at the first argument of the **bar3** function, it is written as 「[a1,b1]」. Essentially, 「[a1,b1]」 is a row vector 「[1,3,5,2,4,6]」. Therefore, even when using 「stacked」, the resulting graph is not stacked as expected. See Fig. 8.13.

```
a1 = [1,3,5];
b1 = [2,4,6];
bar3([a1,b1],1,"stacked");
```

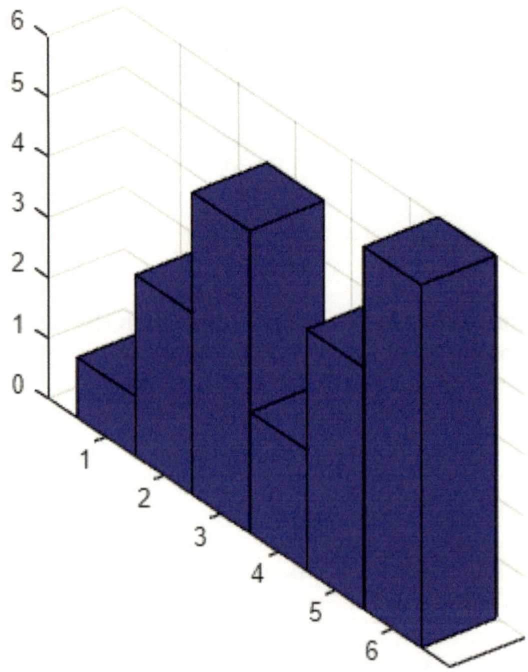

Fig. 8.13 3D bar chart with "stacked" option (not stacked)

The following program differs from the previous one that created Fig. 8.13. The first argument of the **bar3** function is 「[a1;b1]」. Therefore, 「[a1;b1]」 becomes a 2-by-3 matrix. Using 「stacked」, 「a1=[1,3,5]」 is stacked at y=1, and 「b1=[2,4,6]」 is stacked at y=2. See Fig. 8.14.

```
a1 = [1,3,5];
b1 = [2,4,6];
bar3([a1;b1],1,"stacked");
```

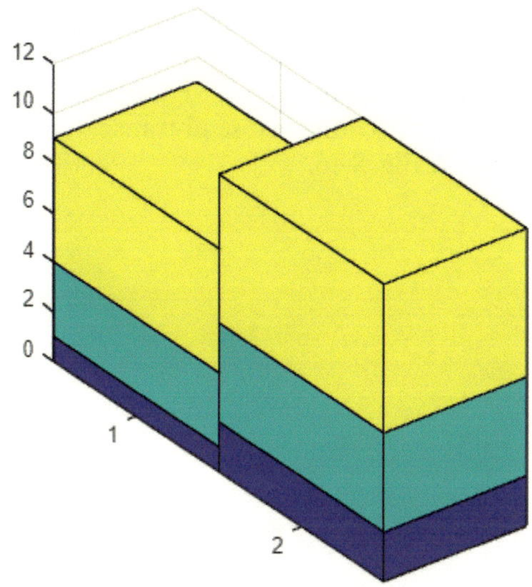

Fig. 8.14 Stacked 3D bar chart at two Y-points

Here, we enter an 8-by-2 matrix 「[Ave, Brent]」 in the second argument of the **bar3** function. As a result, the values of 「Brent」 are stacked on top of the corresponding values of 「Ave」 in the bar graph. See Fig. 8.15.

```
bar3(Year,[Ave,Brent],"stacked");
```

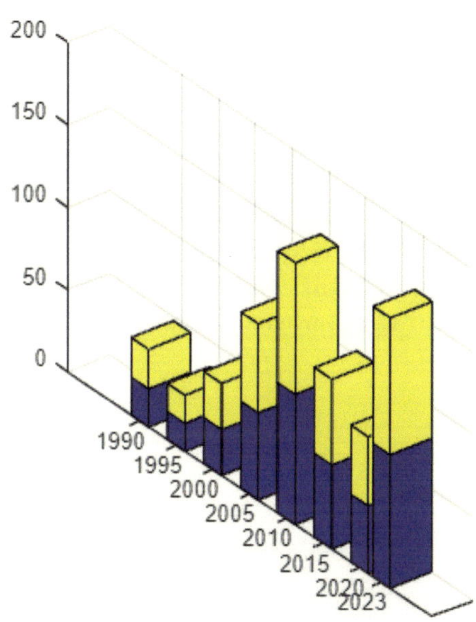

Fig. 8.15 Stacked 3D bar chart (Y-axis: Year)

Here, we set the third argument of the **bar3** function to 1.5, creating a slightly wider stacked bar graph and storing its properties in the variable 「b3」. For 「b3(1)」, we set 「FaceColor」 to 「red」, 「FaceAlpha」 to 0.5, 「LineWidth」 to 2, and 「LineStyle」 to 「:」, resulting in a semi-transparent red 3D bar graph with a thick dotted outline. See ◘ Fig. 8.16.

```
b3 = bar3(Year,[Ave,Brent],1.5,"stacked");
    b3(1).FaceColor = "red";
    b3(1).FaceAlpha = 0.5;
    b3(1).LineWidth = 2;
    b3(1).LineStyle = ":";
```

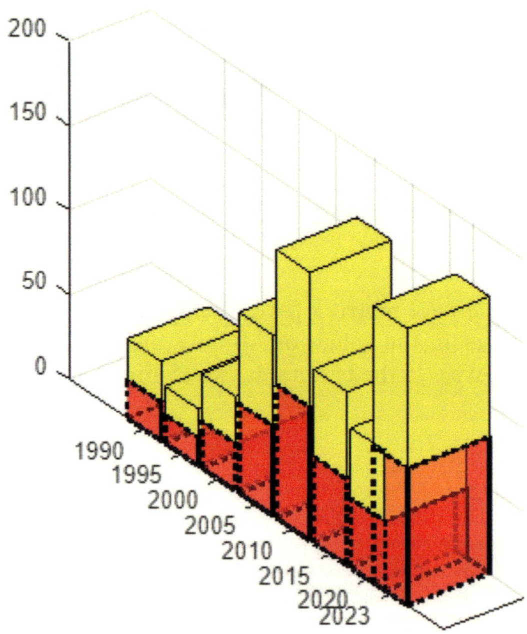

◘ **Fig. 8.16** Stacked 3D bar chart with a semi-transparent red bottom segment

Here, we enter 「grouped」 in the third argument. Consequently, the four different groups (「Ave」, 「Brent」, 「Dubai」, and 「WTI」) are displayed side by side. Compare ◘ Fig. 8.17 with ◘ Fig. 8.4 where the previous program in ◘ Fig. 8.4 did not use 「grouped」.

```
bar3(Year,[Ave,Brent,Dubai,WTI],"grouped");
```

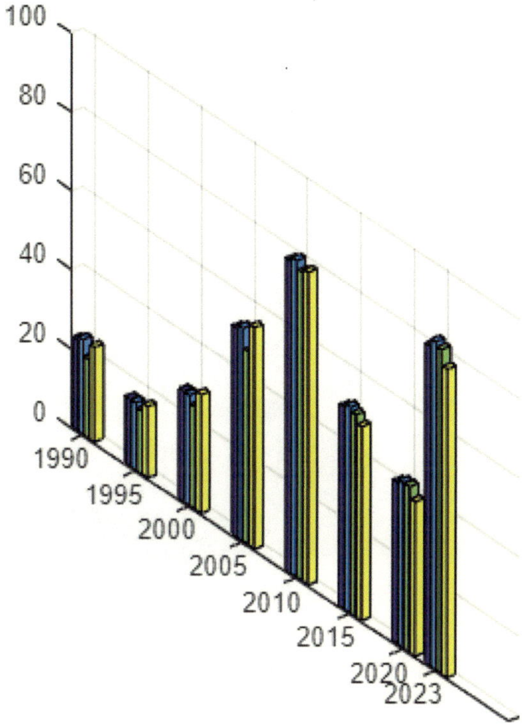

Fig. 8.17 Grouped 3D bar chart

Here, using 「grouped」 in the fourth argument of the **bar3** function, we store its properties in the variable 「b3」. Additionally, we modify the properties of the fourth bar (「b3(4)」). Specifically, we set 「FaceColor」 to 「none」, 「LineWidth」 to 2, and 「LineStyle」 to 「-.」, creating a transparent 3D bar graph with a thick dash-dotted outline. The resulting figure is shown in Fig. 8.18.

```
b3 = bar3(Year,[Ave,Brent,Dubai,WTI],2,"grouped");
    b3(4).FaceColor = "none";
    b3(4).LineWidth = 2;
    b3(4).LineStyle = "-.";
```

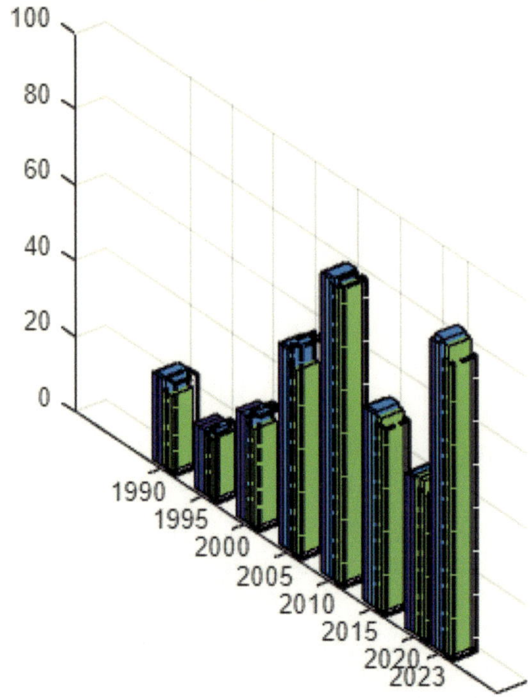

Fig. 8.18 Grouped 3D bar chart (one bar with no fill and dash-dotted outline)

Here, we obtain the z-axis data of the first bar graph using 「b3(1).ZData」 and store it in the variable 「zdata」. Then, by setting 「b3(1).CData=zdata」, we replace the color data (「CData」) with 「zdata」. As a consequence, the color of each bar is adjusted based on the values in 「zdata」. Finally, we set the surface color of the bars to 「interp」 using 「b3(1).FaceColor="interp"」. See the 3D bar for the first group in Fig. 8.19.

```
b3 = bar3(Year,[Ave,Dubai],0.5);
    zdata = b3(1).ZData;
    b3(1).CData = zdata;
    b3(1).FaceColor = "interp";
```

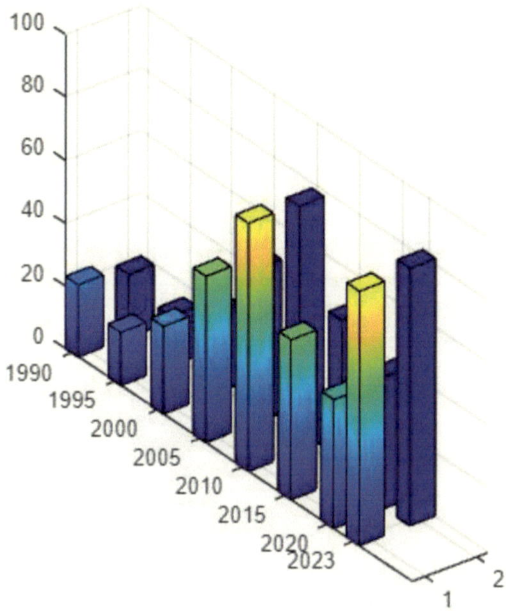

Fig. 8.19 3D bar chart with one group having interpolated face colors

Unlike the previous program that created Fig. 8.19, here, we obtain the x-axis data of the first group using「xdata=b3(1).XData」. Then, as in the previous program in Fig. 8.19, we set the「CData」and「FaceColor」properties accordingly. Compare Fig. 8.19 with Fig. 8.20.

```
b3 = bar3(Year,[Ave,Dubai],0.5);
    xdata = b3(1).XData;
    b3(1).CData = xdata;
    b3(1).FaceColor = "interp";
```

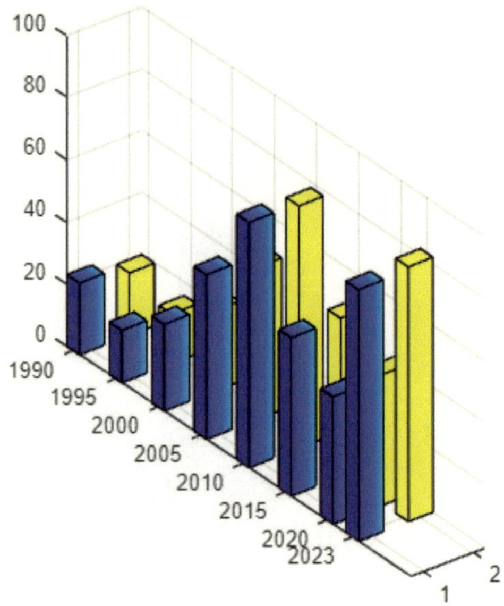

Fig. 8.20 3D bar chart with one group having interpolated face colors (cf. Fig. 8.19)

Here, we hide the bars of the first group by setting 「b3(1).Visible」 to 「off」. See Fig. 8.21.

```
b3 = bar3(Year,[Ave,Dubai],0.5);
    b3(1).Visible = "off";
```

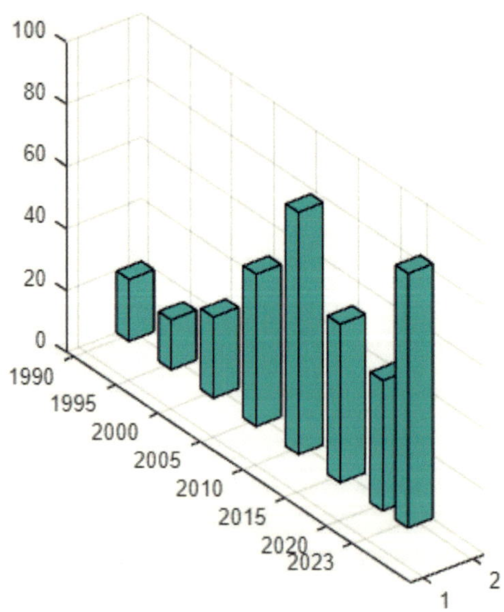

Fig. 8.21 3D bar chart with one group being invisible

8.5 Combining Optional Functions and the Bar3 Function

Here, we use the **xticklabels** function to label the x-axis as 「Ave」, 「Brent」, 「Dubai」, and 「WTI」. These differ from the default x-axis labels in ◘ Fig. 8.4. Additionally, we use the **yticks** function to display only the ticks for 1990, 2000, 2010, and 2020. See ◘ Fig. 8.22.

```
bar3(Year,[Ave,Brent,Dubai,WTI]);
xticklabels({"Ave","Brent","Dubai","WTI"});
yticks(1990:10:2020);
```

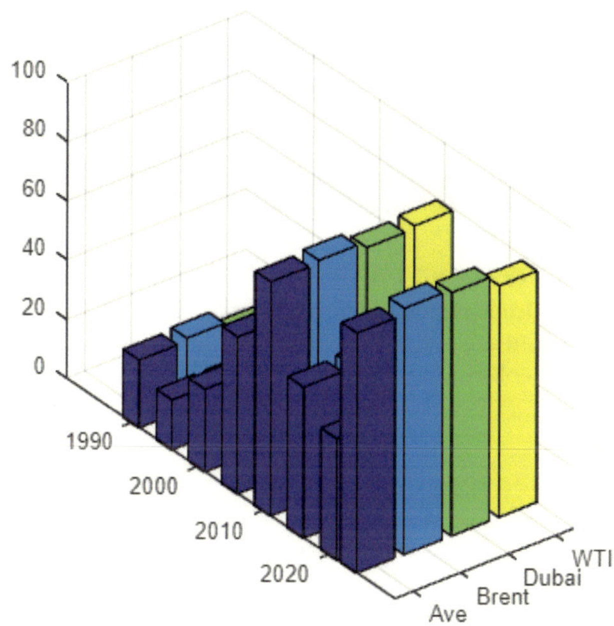

◘ **Fig. 8.22** 3D bar chart with specified X-tick labels

Here, after storing the properties of the **bar3** function in the variable 「b3」, we use the 「FaceAlpha」 property. Specifically, we utilize a **for** loop to gradually reduce the transparency.

For example, when 「i=1」 is set in the **for** loop, 「b3(1).FaceAlpha=0.2」 makes the 3D bar graph of the first group highly transparent. On the other hand, for 「i=2」, 「b3(2).FaceAlpha=0.4」 makes the 3D bar graph of the second group less transparent than that of the first group. Similarly, for 「i=3」 and 「i=4」, the transparency is further reduced. The resulting figure is given in ◘ Fig. 8.23.

```
b3 = bar3(Year,[Ave,Brent,Dubai,WTI]);
for i = 1:3
    b3(i).FaceAlpha = i*0.2;
end
```

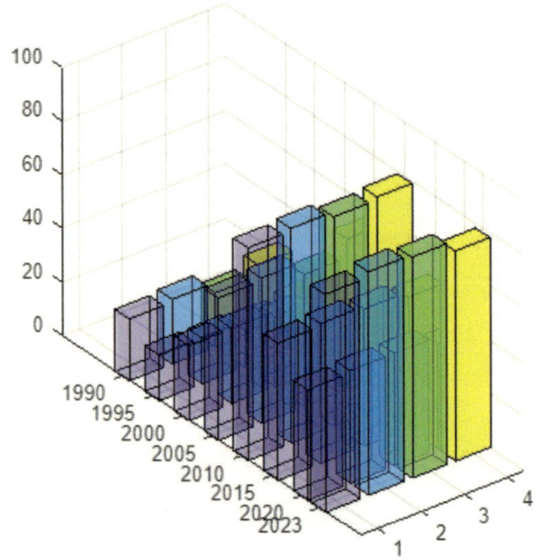

Fig. 8.23 3D bar chart using a for loop to vary transparency

As in the previous Fig. 8.23, we use the 「FaceAlpha」 property within the **for** loop to adjust the transparency. The difference from Fig. 8.23 is that we enter 「red」 in the third argument of the **bar3** function. Consequently, all the bars in the 3D bar graph are colored red. Compare Fig. 8.23 with Fig. 8.24.

```
b3 = bar3(Year,[Ave,Brent,Dubai,WTI],"red");
for i = 1:3
    b3(i).FaceAlpha = i*0.2;
end
```

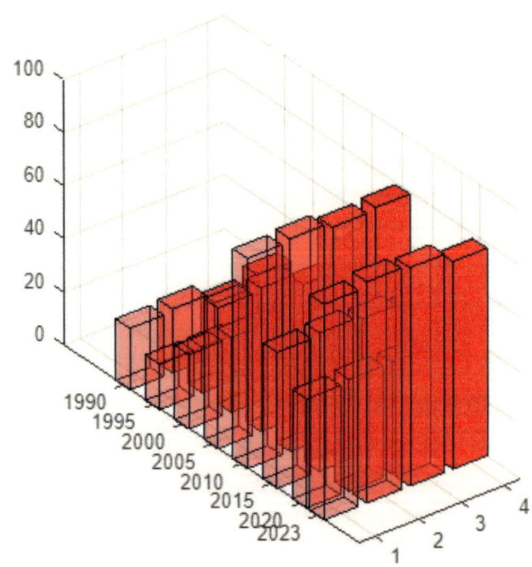

Fig. 8.24 3D bar chart using a for loop to vary red transparency

Here, we use the **xticklabels** and **annotation** functions simultaneously. Specifically, by using 「xticklabels({})」, we remove the labels from the x-axis. Next, we use the **annotation** function to create an ellipse by specifying 「ellipse」 in the first argument of the **annotation** function. As a consequence, we draw a red ellipse around the area. For more details on the **annotation** function, see ▶ Sect. 18.1. The resulting figure is given in ◘ Fig. 8.25.

```
bar3(Year,[Ave,Brent,Dubai,WTI]);
xticklabels({});
dim = [0.6 0.05 0.25 0.15];
annotation("ellipse",dim,"Color","red");
```

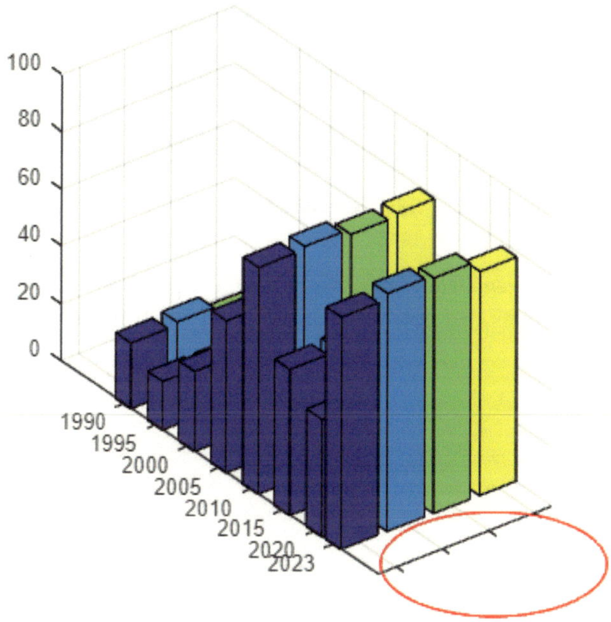

◘ **Fig. 8.25** 3D bar chart with no X-tick labels and a red ellipse

Here, as in ◘ Fig. 8.25, we use 「xticklabels({})」 to remove the labels from the x-axis. In the following program, we use the **legend** function to add a legend. Compare ◘ Fig. 8.22 with ◘ Fig. 8.26 regarding the use of labels on the x-axis.

```
bar3(Year,[Ave,Brent,Dubai,WTI]);
xticklabels({});
legend(["Average","Brent","Dubai","WTI"]);
```

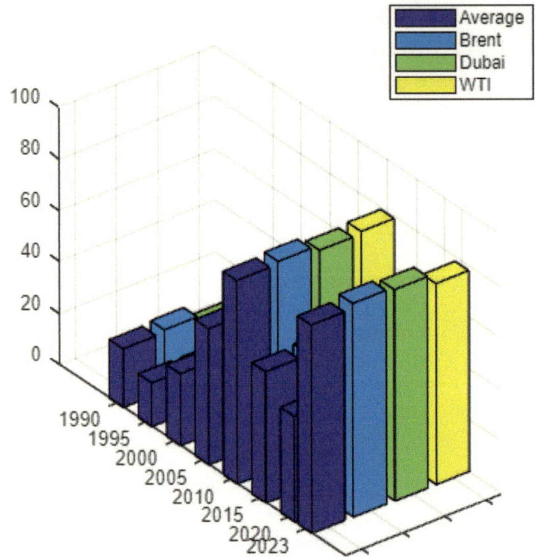

Fig. 8.26 3D bar chart with no X-tick labels and a legend

Here, using the **legend** function, we create a legend. Additionally, by setting 「Location」 to 「north」, we position the legend at the top center of the figure. See Fig. 8.27.

```
bar3(Year,[Ave,Brent,Dubai,WTI]);
xticklabels({});
lg = legend(["Average","Brent","Dubai","WTI"]);
    lg.Location = "north";
```

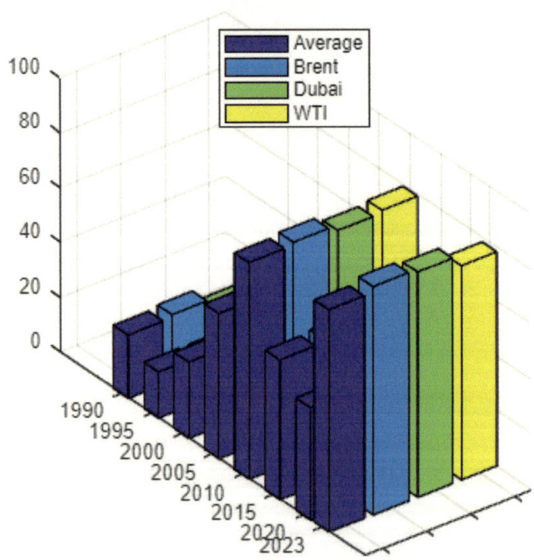

Fig. 8.27 3D bar chart with no X-tick labels and a legend (top-center position)

Here, using the **view** function, which is useful for 3D graphs, we change the viewing angle. See ◘ Fig. 8.28.

```
bar3(Year,[Ave,Brent,Dubai,WTI]);
xticklabels({"Ave","Brent","Dubai","WTI"});
view(-135,30);
```

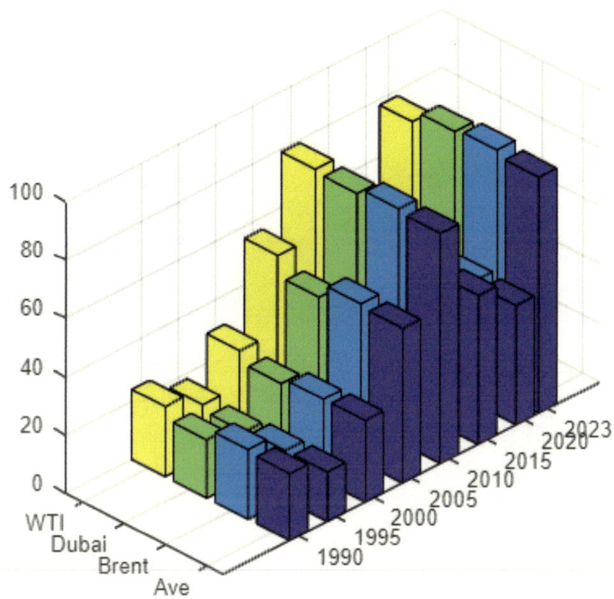

◘ **Fig. 8.28** 3D bar chart from a different angle

As in ◘ Fig. 8.28, we use the **view** function. Specifically, we set 「view(170,0)」 to change the viewing angle. Compare ◘ Fig. 8.28 with ◘ Fig. 8.29.

```
bar3(Year,[Ave,Brent,Dubai,WTI]);
xticklabels({"Ave","Brent","Dubai","WTI"});
view(170,0);
```

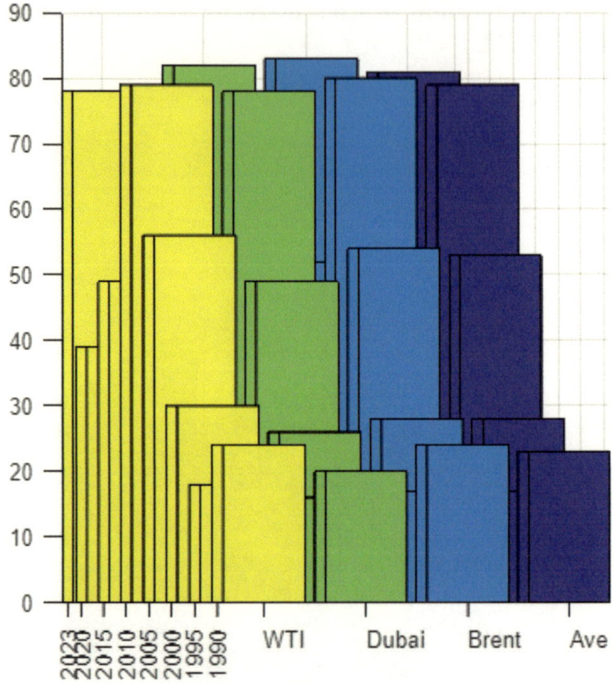

Fig. 8.29 3D bar chart from a different angle (cf. Fig. 8.28)

Here, we use「view(2)」to change the visualization to a 2D top-down perspective. See Fig. 8.30.

```
bar3(Year,[Ave,Brent,Dubai,WTI]);
xticklabels({"Ave","Brent","Dubai","WTI"});
view(2);
```

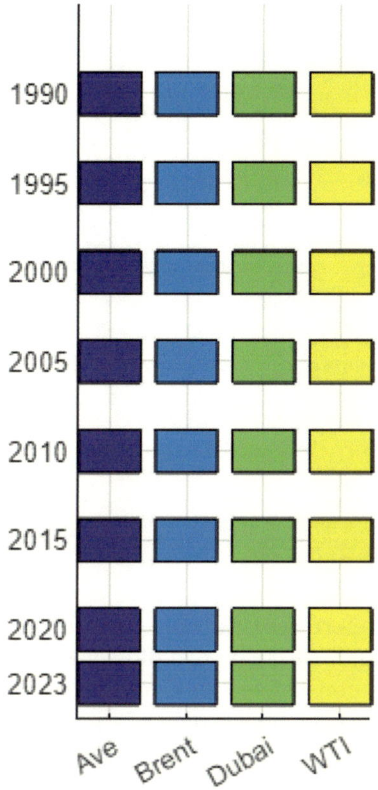

Fig. 8.30 Flat 3D bar chart from 2D top-down perspective

Here, as in Figs. 8.19 and 8.20, we use a **for** loop with 「CData」 and 「FaceColor」 to color the 3D bar graph according to the values on the z-axis. Specifically, by setting 「b3(i).FaceColor="interp"」, we adjust the bar colors. In addition, using the **colorbar** command, we add a colorbar. Finally, we label the axes using the **ylabel** and **zlabel** functions. See Fig. 8.31.

```
b3 = bar3(Year,[Ave,Dubai,WTI,Brent],0.5);
for i = 1:4
    zdata = b3(i).ZData;
        b3(i).CData = zdata;
        b3(i).FaceColor = "interp";
end
colorbar;
view(-135,30);
xticklabels({"Ave","Dubai","WTI","Brent"});
yl = ylabel("Year");
    yl.Color = "red";
zl = zlabel("Dollars per barrel");
    zl.Color = "red";
box on
```

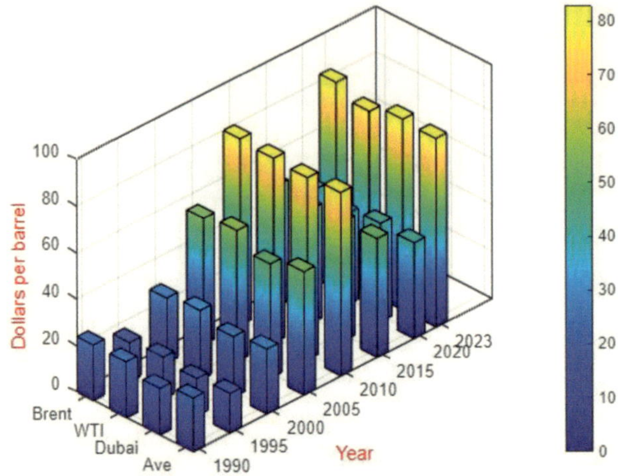

Fig. 8.31 3D bar chart using a for loop to apply interpolated colors, with labeled Y- and Z-axes and a color bar

Here, we store the properties of the **bar3** function in the variable 「b3」. Then, using the **hold on** command, we create two plots with the **plot3** function and overlay them. Both plots are constructed similarly. Specifically, 「XData」, 「YData」, and 「ZData」 contain the coordinates data of the vertices of the bar graph. From these data, we extract the coordinates of the vertices of the 3D bar graph created with the **bar3** function. Since 「(X,Y,Z)=(xp2,yp2,zp2)」 represents the coordinates of a corner of each 3D bar, using these coordinates in the **plot3** function connects their points and generates a 3D plot. Similarly, using 「xp3」, 「yp3」, and 「zp3」 connects the 8 points with the **plot3** function. See Fig. 8.32 for the created figure.

```
figure
b3 = bar3(Year,Ave,0.5);
    b3.FaceAlpha = 0.4;
hold on
    xp1 = b3(1).XData(:,1);
    yp1 = b3(1).YData(:,1);
    zp1 = b3(1).ZData(:,2);
xp2 = xp1(2:6:48);
yp2 = yp1(2:6:48);
zp2 = zp1(2:6:48);
pl = plot3(xp2,yp2,zp2,"o:r");
    pl.LineWidth = 2;
xp3 = xp1(3:6:48);
yp3 = yp1(3:6:48);
zp3 = zp1(3:6:48);
pl2 = plot3(xp3,yp3,zp3,"x--k");
    pl2.LineWidth = 2;
hold off
```

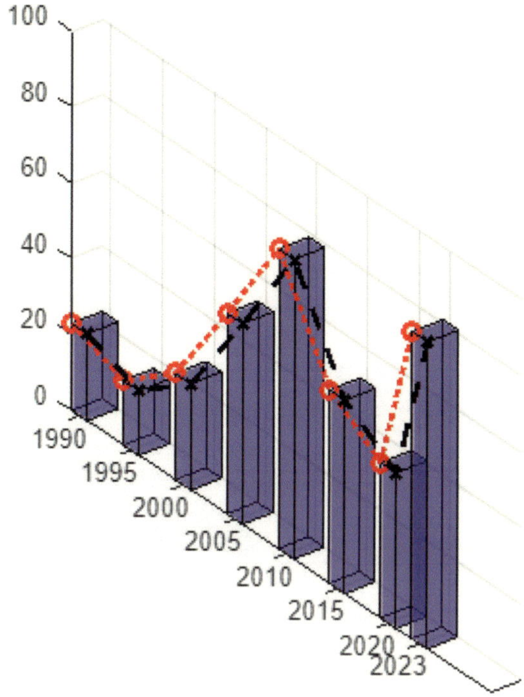

Fig. 8.32 3D bar chart with overlaid lines connecting data points

Here, using the **hold on** command, we overlay two 3D bar graphs created by the **bar3** function. However, since both bar graphs are placed in the same location, it is difficult to distinguish between the two overlapping 3D bar graphs. See Fig. 8.33.

```
figure
bar3(Year,Ave,0.5);
hold on
bar3(Year,WTI,0.5);
hold off
box on
```

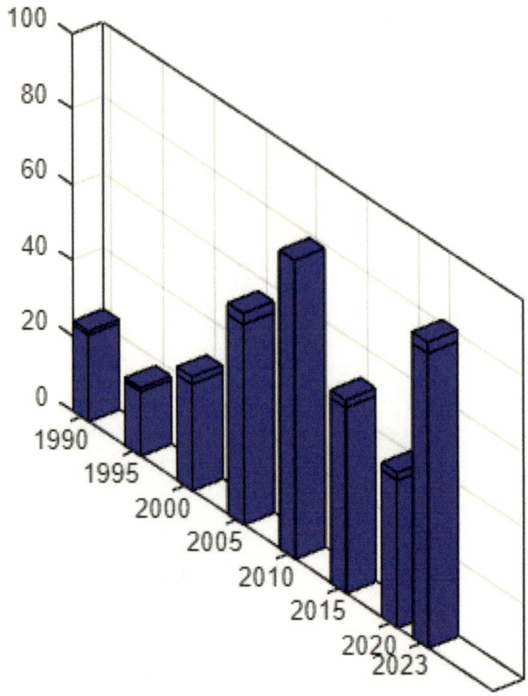

Fig. 8.33 3D bar chart with overlaid bars (hard to distinguish due to identical color)

As in Fig. 8.33, we use the **hold on** command to overlay the 3D bar graphs. Here, we use 「FaceAlpha」, 「LineStyle」, 「LineWidth」, and 「EdgeColor」, to make the distinction between the two overlapping bar graphs clearer. Additionally, we adjust the viewing angle using the **view** function. Compare Fig. 8.33 with Fig. 8.34.

```
figure
b1 = bar3(Year,Ave,0.5);
    b1.FaceColor = "red";
    b1.FaceAlpha = 0.4;
    b1.LineStyle = ":";
    b1.LineWidth = 2;
hold on
b2 = bar3(Year,WTI,0.5);
    b2.FaceColor = "none";
    b2.LineStyle = "-.";
    b2.LineWidth = 1;
    b2.EdgeColor = "blue";
hold off
view(-100,20);
box on
```

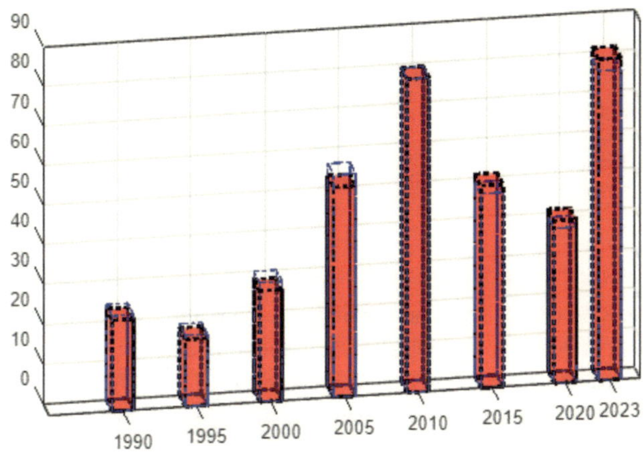

Fig. 8.34 3D bar chart with overlaid bars (cf. Fig. 8.33)

8.6 Questions

Question 1

Create Fig. 8.35 with 「Year」 as the first argument and 「[Brent,Dubai,WTI]」 as the second argument. Specifically, for the bar graph representing the data of 「Brent」, set 「FaceColor」 to transparent and make the outline a thick dashed line.

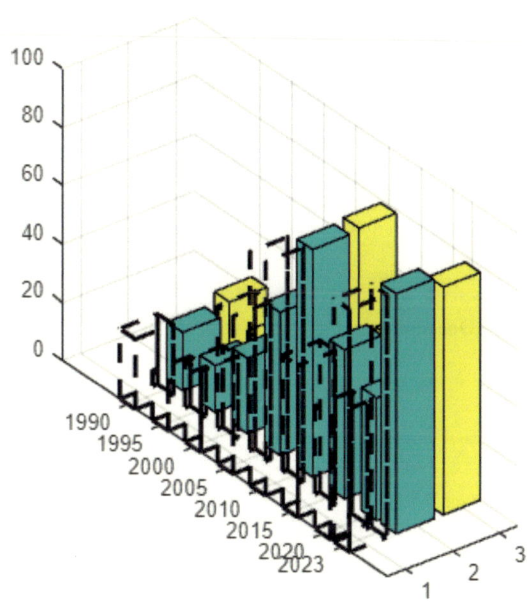

Fig. 8.35 3D bar chart (Question 1)

Answer 1

```
figure
Ex1 = bar3(Year,[Brent,Dubai,WTI]);
    Ex1(1).FaceColor = "none";
    Ex1(1).LineStyle = "--";
    Ex1(1).LineWidth = 2;
```

Question 2

Create Fig. 8.36 with 「Year」 as the first argument and 「[Ave, Brent, Dubai, WTI]」 as the second argument. Specifically, stack the four data sets 「[Ave, Brent, Dubai, WTI]」. Then use the **legend** function to name each of the four datasets (Ave, Brent, Dubai, WTI) and add a title (「Oil Price」) to the legend.

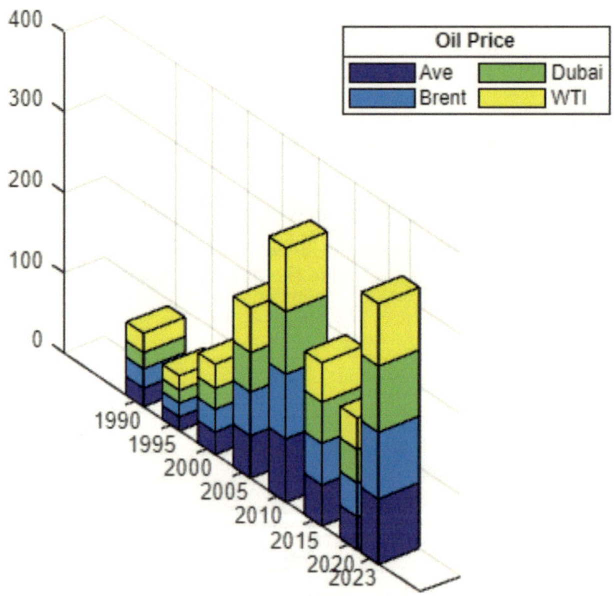

Fig. 8.36 3D bar chart (Question 2)

Answer 2

```
figure
bar3(Year,[Ave,Brent,Dubai,WTI],"stacked");
lgd = legend("Ave","Brent","Dubai","WTI");
    lgd.NumColumns = 2;
    lgd.Title.String = "Oil Price";
```

Question 3

Using the data for 「Year」, 「Ave」, 「Brent」, 「Dubai」, and 「WTI」, create Fig. 8.37.

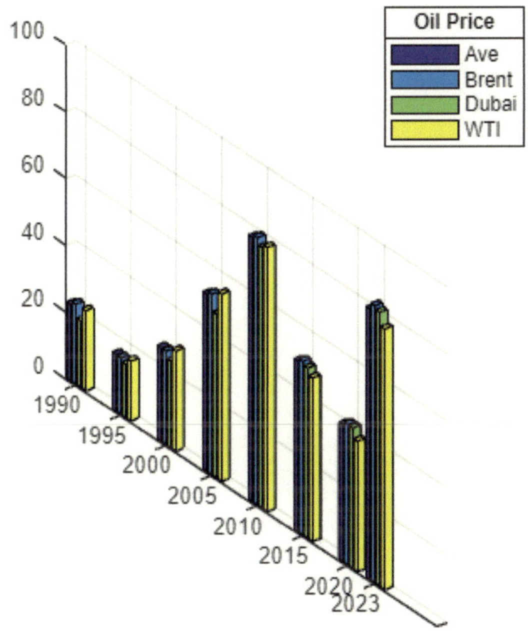

Fig. 8.37 3D bar chart (Question 3)

Answer 3

```
figure
bar3(Year,[Ave,Brent,Dubai,WTI],"grouped");
    lgd = legend("Ave","Brent","Dubai","WTI");
    lgd.Title.String = "Oil Price";
```

Barh Function

Contents

9.1 Details of the Data Used in this Chapter – 284

9.2 Excel Data of ▶ Section 9.1 – 285

9.3 Basic Syntax of the Barh Function – 286

9.4 Setting and Controlling the Barh Function – 291

9.5 Combining Optional Functions and the Barh Function – 298

9.6 Questions – 311

Supplementary Information The online version contains supplementary material available at https://doi.org/10.1007/978-981-96-6213-5_9.

Y. Nakamoto, *A Comprehensive Guide to Creating Graphics with MATLAB Programming*, Classroom Companion: Economics, https://doi.org/10.1007/978-981-96-6213-5_9

Learning Objectives

This chapter creates horizontal bar charts using the **barh** function. By reading this chapter, you will learn how to use the **barh** function and understand the similarities and differences in the created figures, syntax, and the type of data between the **barh** function and the **bar** function (covered in ▶ Chap. 7). As a result, you will be able to determine whether to use vertical or horizontal bar charts.

Overview

The **barh** function in MATLAB is used to create horizontal bar graphs for visualizing data. In a horizontal bar graph, the bars are arranged horizontally, with categories on the vertical axis and values on the horizontal axis. Alternatively, the **bar** function in MATLAB creates vertical bar graphs, in which the bars are arranged vertically, with categories on the horizontal axis and values on the vertical axis. Refer to ▶ Chap. 7 (the **bar** function) for more details. The choice between these two formats depends on the nature of the data and what you want to highlight. The use of the **barh** function in this chapter is essentially similar to that of the **bar** function. Therefore, it may be efficient to read ▶ Chap. 7.

9

9.1 Details of the Data Used in this Chapter

In this chapter, we use data on the percentages of male and female employment in the original member countries of ASEAN (Association of South East Asian Nations). ASEAN, which is composed of 10 Southeast Asian countries, was established by the "Bangkok Declaration" in 1967. The original member countries were Thailand, Indonesia, Singapore, the Philippines, and Malaysia. In this chapter, we use the data from these countries. 「Indonesia_f」 represents the percentage of female employees (wage or salaried workers) in Indonesia. Similarly, 「Malaysia_f」, 「Philippines_f」, 「Singapore_f」, and 「Thailand_f」 represent the percentage of female employees in Malaysia, the Philippines, Singapore, and Thailand, respectively. On the other hand, 「Indonesia_m」, 「Malaysia_m」, 「Philippines_m」, 「Singapore_m」, and 「Thailand_m」 represent the percentage of male employees in those countries. 「Year」 indicates the year. The table variable 「employment」 is created.

```
Year = [2000,2010,2020]';

Indonesia_f = [28.16,34.46,40.01]';

Malaysia_f = [76.43,78.59,73.82]';

Philippines_f = [50.79,52.84,58.96]';

Singapore_f = [90.59,90.6,91.33]';

Thailand_f = [34.96,43.26,49.37]';

Indonesia_m = [35.69,43.05,51.59]';

Malaysia_m = [73.05,72.82,75.66]';

Philippines_m = [50.68,55.0,64.87]';

Singapore_m = [80.79,82.07,82.68]';

Thailand_m = [40.93,45.63,49.56]';

employment = table(Year,Indonesia_f,Malaysia_f,Philippines_f,Singapore_f,...
    Thailand_f,Indonesia_m,Malaysia_m,Philippines_m,Singapore_m,Thailand_m);
```

9.2 Excel Data of ▶ Section 9.1

To import the data used in ▶ Sects. 9.3, 9.4, 9.5, and 9.6 in Excel format, we specify the file name and sheet name. For more details on how to import Excel data, refer to ▶ Chap. 2.

```
file_chap9 = "alldata.xlsx";
sheet_chap9 = "Chap9_Employment";
employment = readtable(file_chap9,"Sheet",sheet_chap9);
```

```
Warning: Column headers from the file were modified to make them valid MATLAB identifiers before creating
variable names for the table. The original column headers are saved in the VariableDescriptions property.
Set 'VariableNamingRule' to 'preserve' to use the original column headers as table variable names.
```

```
employment = renamevars(employment,...
    ["IndonesiaFemale","MalaysiaFemale","PhilippinesFemale","SingaporeFemale",...
    "ThailandFemale","IndonesiaMale","MalaysiaMale","PhilippinesMale",...
    "SingaporeMale","ThailandMale"],...
    ["Indonesia_f","Malaysia_f","Philippines_f","Singapore_f",...
    "Thailand_f","Indonesia_m","Malaysia_m","Philippines_m",...
    "Singapore_m","Thailand_m"]);
```

We will extract each column from the table variable 「employment」.

```
Year = employment.Year;
Indonesia_f = employment.Indonesia_f;
Malaysia_f = employment.Malaysia_f;
Philippines_f = employment.Philippines_f;
Singapore_f = employment.Singapore_f;
Thailand_f = employment.Thailand_f;
Indonesia_m = employment.Indonesia_m;
Malaysia_m = employment.Malaysia_m;
Philippines_m = employment.Philippines_m;
Singapore_m = employment.Singapore_m;
Thailand_m = employment.Thailand_m;
```

9.3 Basic Syntax of the Barh Function

Basic Syntax:

The **barh** function is used to create a horizontal bar graph. Various syntaxes can be viewed by typing 「doc barh」 in the Command Window. Figure 9.1 shows the syntaxes for the **barh** function. The first syntax inside the black rectangle is the basic syntax of the **barh** function, 「barh(y)」. Based on the basic syntax, the **barh** function requires one type of data, 「y」. The second syntax, 「barh(x,y)」, requires two types of data: 「x」 and 「y」 where 「x」 represents the locations along the vertical axis and 「y」 is the lengths of the bars.

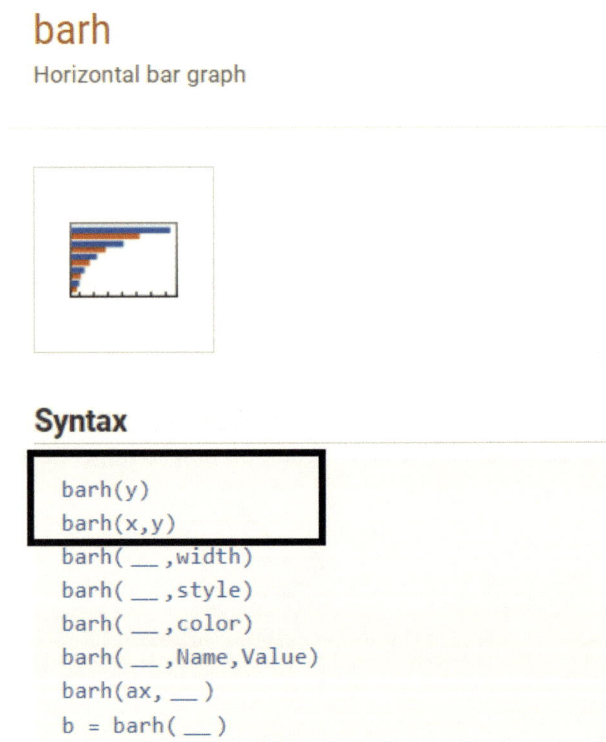

Fig. 9.1 The syntax of the barh function

Here, we use the **barh** function to create a horizontal bar graph. We only specify the variable 「Indonesia_f」 as the first argument. This creates a horizontal bar for each element of the vector 「Indonesia_f」, with the vertical axis automatically numbered. See Fig. 9.2.

```
barh(Indonesia_f);
```

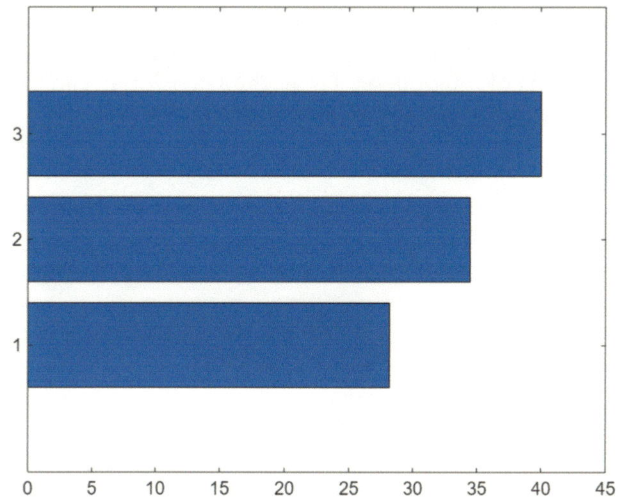

Fig. 9.2 Horizontal bar chart (Y-axis: Default index)

Here, we use a 3-by-2 matrix as the first argument of the **barh** function. This creates grouped horizontal bar graphs, with two bars for each category. In this example, the matrix 「Indonesia=[Indonesia_f,Indonesia_m]」 represents the female and male employment data for each category side by side. The **barh** function then creates two horizontal bars for each category, one for female data and one for male data. For instance, the first group corresponds to the values of 「Indonesia_f(1)」 and 「Indonesia_m(1)」 at y-axis position 1. See Fig. 9.3 for the resulting figure.

```
Indonesia = [Indonesia_f,Indonesia_m];
barh(Indonesia);
```

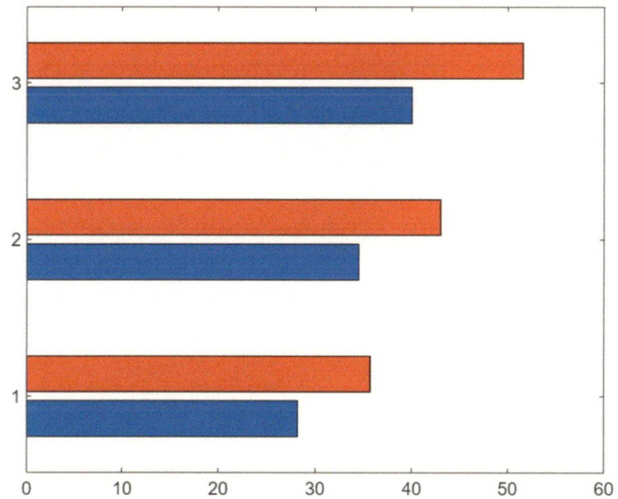

Fig. 9.3 Horizontal bar chart with two groups (Y-axis: Default index)

In the following program, as confirmed in 「barh(Indonesia')」, we transpose 「Indonesia」 using 「'」, which changes the 2-by-3 matrix into a 3-by-2 matrix. Then, the first group consists of 「Indonesia_f」 on the y-axis of 1, and the second group consists of 「Indonesia_m」 on the y-axis of 2. See ◘ Fig. 9.4.

```
Indonesia = [Indonesia_f,Indonesia_m];
barh(Indonesia');
```

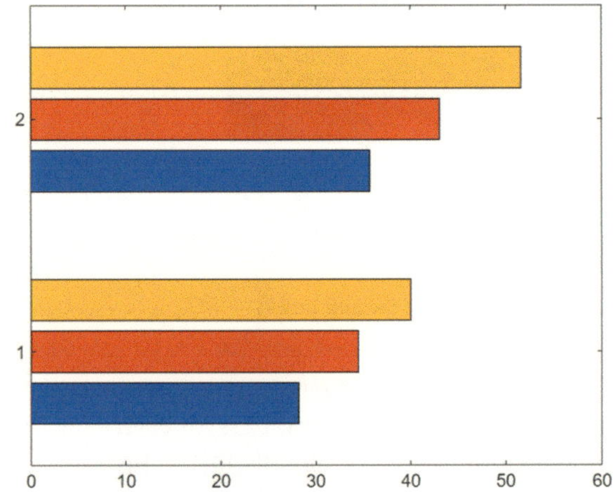

◘ **Fig. 9.4** Horizontal bar chart with multiple groups (cf. ◘ Fig. 9.3)

9

Here, by specifying 「barh(Year, Malaysia)」, we create a horizontal bar graph with 「Year」 on the y-axis. Each year has two bars: 「Malaysia_f」 and 「Malaysia_m」. See ◘ Fig. 9.5.

```
Malaysia = [Malaysia_f,Malaysia_m];
barh(Year,Malaysia);
```

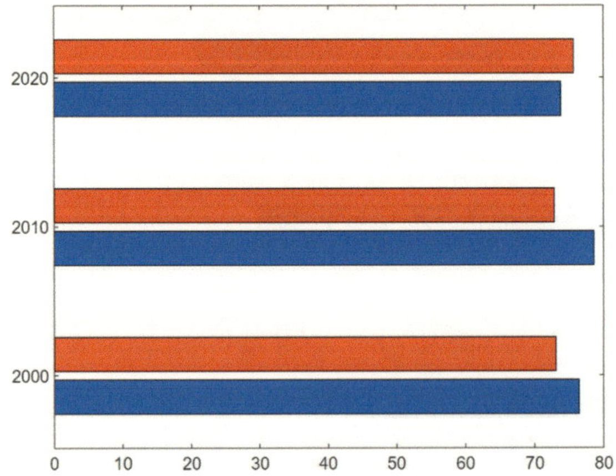

◘ **Fig. 9.5** Horizontal bar chart with two groups (Y-axis: Year)

▶ Example of Table Variables

Here, we use the table variable to create a horizontal bar graph. The resulting figure (◘ Fig. 9.6) is the same as ◘ Fig. 9.5.

```
barh(employment.Year,[employment.Malaysia_f,employment.Malaysia_m]);
```

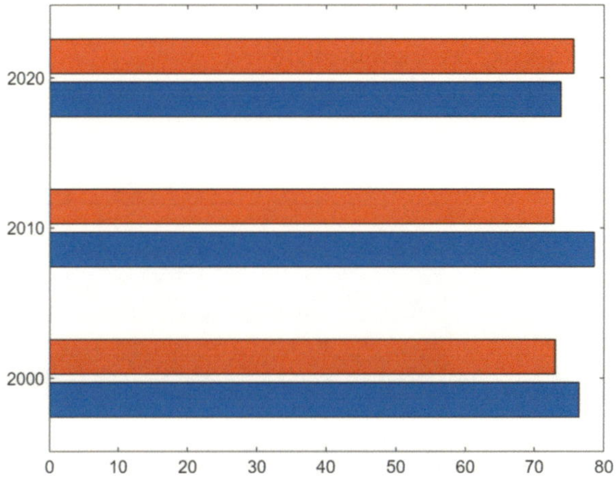

◘ **Fig. 9.6** Use of a table variable: Horizontal bar chart

◀

Here, we use a 3-by-4 matrix as the second argument in the **barh** function. For example, in the group for 「Year(1) (= 2000)」, four horizontal bar graphs are created for 「Indonesia_f(1)」, 「Indonesia_m(1)」, 「Malaysia_f(1)」, and 「Malaysia_m(1)」. See ◘ Fig. 9.7.

```
Indonesia = [Indonesia_f,Indonesia_m];
Malaysia = [Malaysia_f,Malaysia_m];
barh(Year,[Indonesia,Malaysia]);
```

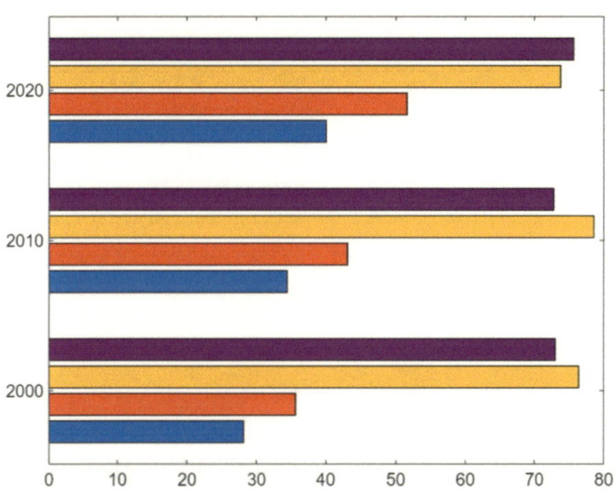

◘ **Fig. 9.7** Horizontal bar chart with multiple groups (Y-axis: Year)

Here, the first argument of the **barh** function consists of four text strings, namely: 「Indonesia and Malaysia」, 「Male and Female」, 「Year」, and 「2000」. Additionally, we use the data for 「Indonesia_f(1)」, 「Indonesia_m(1)」, 「Malaysia_f(1)」, and 「Malaysia_m(1)」 as the second argument. As a consequence, the four text labels are displayed on the vertical axis, and corresponding horizontal bars are given. See ◘ Fig. 9.8.

```
Indonesia_2000 = [Indonesia_f(1),Indonesia_m(1)];
Malaysia_2000 = [Malaysia_f(1),Malaysia_m(1)];
barh(["Indonesia and Malaysia","Male and Female","Year","2000"],...
    [Indonesia_2000,Malaysia_2000]);
```

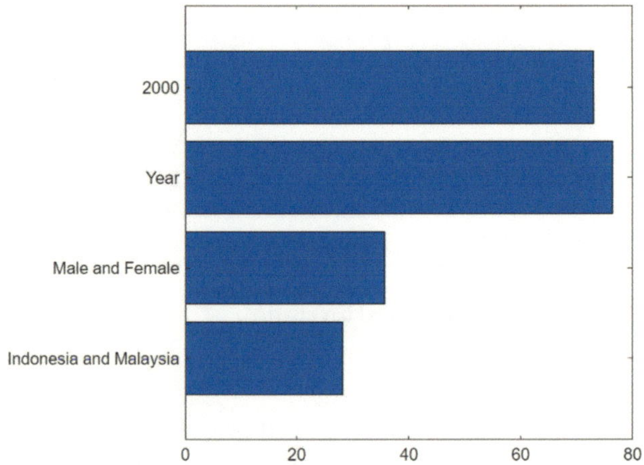

◘ **Fig. 9.8** Horizontal bar chart with Y-axis text labels

Attention

Error Program: Here, the first argument consists of three text strings, namely: 「Indonesia and Malaysia」, 「Male and Female」, and 「Year 2000」. However, as the second argument has four elements, the first and second arguments do not have the same length, which results in an error. See ◘ Fig. 9.9 for the error message.

```
%%% Error program
% Indonesia_2000=[Indonesia_f(1),Indonesia_m(1)];
% Malaysia_2000=[Malaysia_f(1),Malaysia_m(1)];
% barh(["Indonesia and Malaysia","Male and Female","Year 2000"],...
% [Indonesia_2000,Malaysia_2000]);
```

```
Error using bar (line 202)
X must be same length as Y.

Error in barh (line 46)
h = bar(varargin{:});
```

◘ **Fig. 9.9** Error output of the barh function: Inconsistent vector lengths

Here, the first argument consists of four text strings, namely: 「Indonesia female」, 「Indonesia male」, 「Malaysia female」, and 「Malaysia male」. Additionally, we use a 3-by-4 matrix as the second argument. See ◘ Fig. 9.10.

```
Indonesia = [Indonesia_f,Indonesia_m];
Malaysia = [Malaysia_f,Malaysia_m];
barh(["Indonesia female","Indonesia male",...
    "Malaysia female","Malaysia male"],[Indonesia,Malaysia]);
```

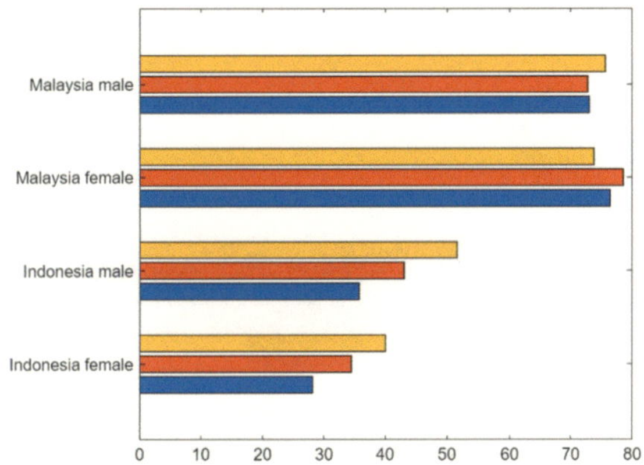

◘ **Fig. 9.10** Grouped horizontal bar chart with Y-axis text labels

9.4 Setting and Controlling the Barh Function

Here, the **barh** function specifies 0.25 as the third argument to adjust the width of the horizontal bars. See ◘ Fig. 9.11.

```
Philippines = [Philippines_f,Philippines_m];
Singapore = [Singapore_f,Singapore_m];
barh(Year,[Philippines,Singapore],0.25);
```

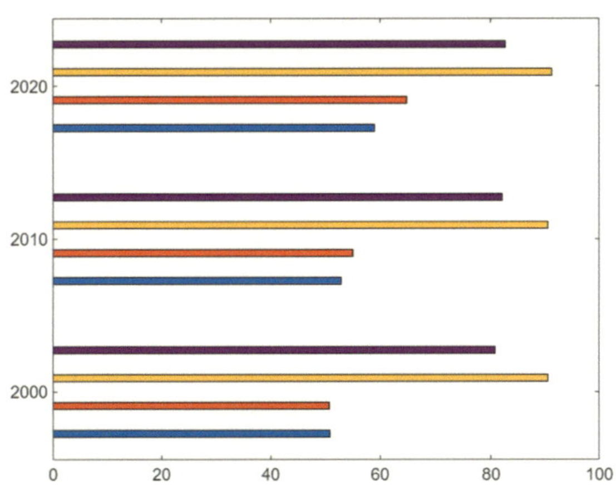

◘ **Fig. 9.11** Grouped horizontal bar chart with narrow bars

Here, the third argument of the **barh** function is set to 1.25, making the bars wider. The properties of the **barh** function are then stored in the variable 「bh」, a 1-by-4 bar object.

By using 「bh(3)」, the third horizontal bar graph is customized. Specifically, we set 「FaceColor」 to 「green」, 「LineWidth」 to 1.5, and 「LineStyle」 to 「--」, creating a green bar with a thick dashed outline. The created figure is illustrated in ◘ Fig. 9.12.

```
Philippines = [Philippines_f,Philippines_m];
Singapore = [Singapore_f,Singapore_m];
bh = barh(Year,[Philippines,Singapore],1.25);
    bh(3).FaceColor = "green";
    bh(3).LineWidth = 1.5;
    bh(3).LineStyle = "--";
```

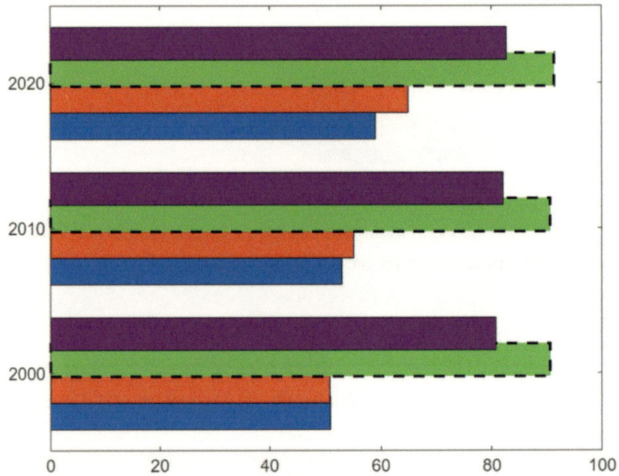

◘ **Fig. 9.12** Grouped horizontal bar chart with one group of dashed-outlined green bars

In addition to the previous program that created ◘ Fig. 9.12, we use 「bh(4)」 to customize the fourth bar graph, setting 「FaceColor」 to 「magenta」. The resulting figure is presented in ◘ Fig. 9.13.

```
Philippines = [Philippines_f,Philippines_m];
Singapore = [Singapore_f,Singapore_m];
bh = barh(Year,[Philippines,Singapore],1.25);
    bh(3).FaceColor = "green";
    bh(3).LineWidth = 1.5;
    bh(3).LineStyle = "--";
    bh(4).FaceColor = "magenta";
```

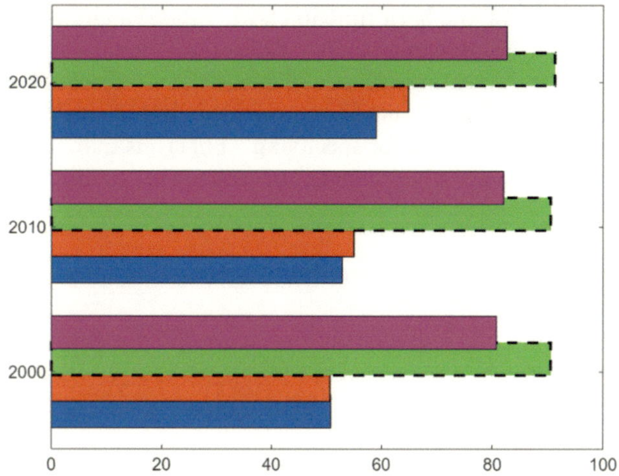

■ **Fig. 9.13** Grouped horizontal bar chart with one group of dashed-outlined green bars and another group of magenta bars

Here, we set 「BaseValue」 to 70 in the **barh** function. As a result, the vertical line (baseline) is displayed at x=70. Values less than 70 are displayed to the left of the baseline, while values greater than 70 are displayed to the right of the baseline. See ■ Fig. 9.14.

```
Philippines = [Philippines_f,Philippines_m];
Singapore = [Singapore_f,Singapore_m];
barh(Year,[Philippines,Singapore],"BaseValue",70);
```

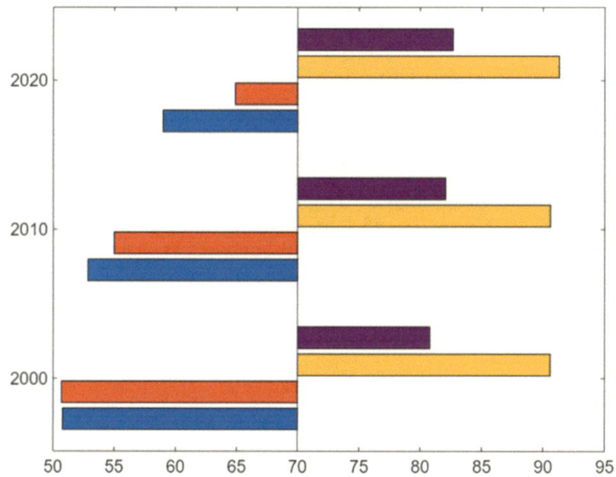

■ **Fig. 9.14** Grouped horizontal bar chart with a reference line

Here, the properties of the **barh** function with 「BaseValue」 set to 90 are stored in 「bh」. Then, we set the reference line's 「Color」 to 「red」, 「LineStyle」 to 「-.」, and 「LineWidth」 to 3, creating a thick red dash-dotted reference line.

For modifying 「BaseLine」, 「bh(1)」 is specified, but it could also be 「bh(2)」, 「bh(3)」, or 「bh(4)」. The reason for specifying 「bh(1)」 is that, as 「bh」 is a 1-by-4 bar object, we have to specify which element of 「bh」. The resulting figure is shown in ◘ Fig. 9.15.

```
Philippines = [Philippines_f,Philippines_m];
Singapore = [Singapore_f,Singapore_m];
bh = barh(Year,[Philippines,Singapore],"BaseValue",90);
    bh(1).BaseLine.Color = "red";
    bh(1).BaseLine.LineStyle = "-.";
    bh(1).BaseLine.LineWidth = 3;
```

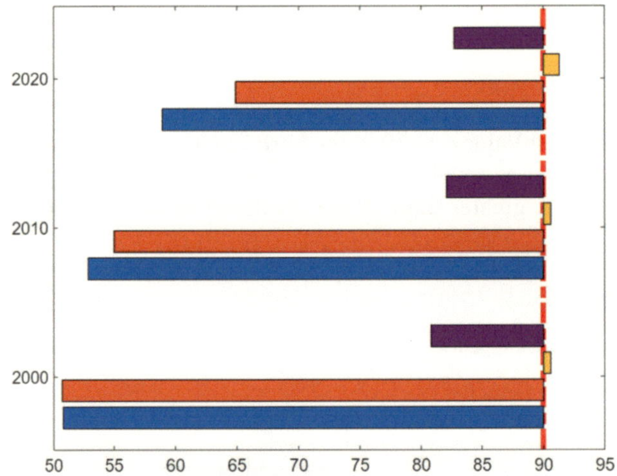

◘ **Fig. 9.15** Grouped horizontal bar chart with a dash-dotted red reference line

Here, the second argument 「stacked」 is used in the **barh** function to create a stacked bar chart. For example, at y=1, 「Philippines_f(1)」, 「Singapore_f(1)」, and 「Thailand_f(1)」 are stacked. See ◘ Fig. 9.16.

```
barh([Philippines_f,Singapore_f,Thailand_f],"stacked");
```

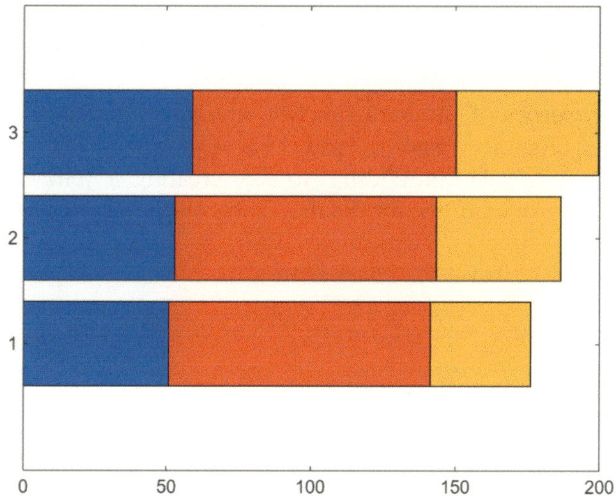

Fig. 9.16 Horizontal stacked bar chart

In the previous program that created Fig. 9.16, the vertical axis was labeled 1, 2, and 3. Here, text strings 「["Philippines female","Singapore female","Thailand female"]」 are used as the first argument for the vertical axis labels. See Fig. 9.17.

```
barh(["Philippines female","Singapore female","Thailand female"],...
    [Philippines_f,Singapore_f,Thailand_f],"stacked");
```

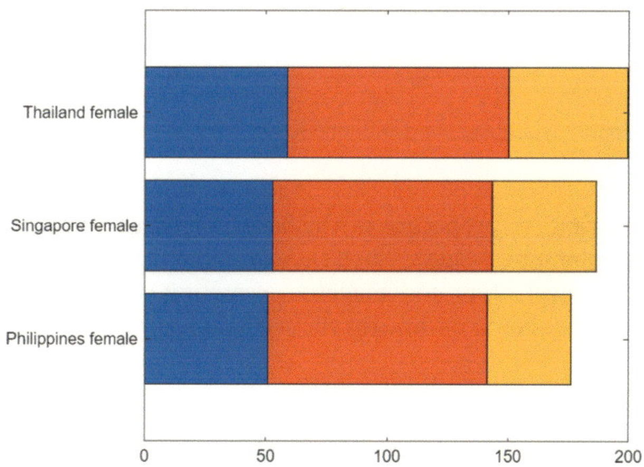

Fig. 9.17 Horizontal stacked bar chart with Y-axis text labels

Here, when using a string array with four elements in the first argument of the **barh** function and a 3-by-4 matrix in the second argument, we create a stacked bar chart by setting the third argument to 「stacked」.

When the properties of the **barh** function are stored in the variable 「pst」, we set 「pst(1).EdgeColor」 to 「black」, 「pst(1).FaceColor」 to 「yellow」, 「pst(1).Line Style」 to 「:」, and 「pst(1).LineWidth」 to 2.5, creating a yellow horizontal bar with a thick dotted black outline for the first category. See Fig. 9.18.

```
pst = barh(["Philippines female","Singapore female",...
    "Thailand female","Malaysia female"],...
    [Philippines_f,Singapore_f,Thailand_f,Malaysia_f],"stacked");
    pst(1).EdgeColor = "black";
    pst(1).FaceColor = "yellow";
    pst(1).LineStyle = ":";
    pst(1).LineWidth = 2.5;
```

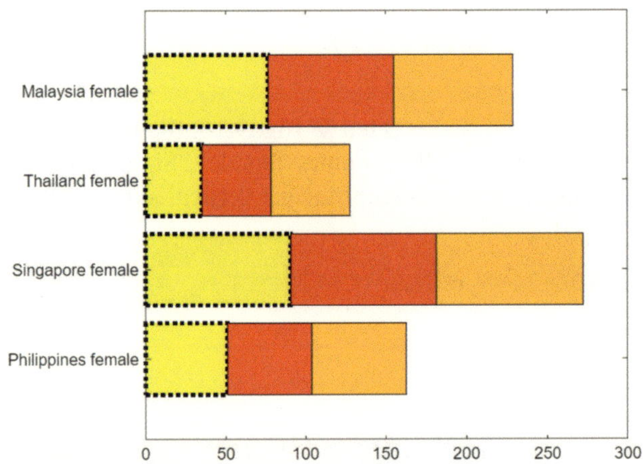

Fig. 9.18 Horizontal stacked bar chart with one yellow segment outlined in thick dotted black

Here, after storing the properties of the **barh** function in the variable 「b_t」, we set 「b_t(2).FaceColor」 to 「blue」, 「b_t(2).FaceAlpha」 to 0.2, 「b_t(2).LineWidth」 to 2, and 「b_t(2).EdgeColor」 to 「magenta」, creating a blue bar with high transparency and a thick magenta outline for the second category. See Fig. 9.19 for the created figure.

```
Thailand = [Thailand_f,Thailand_m];
b_t = barh(Year,Thailand,"stacked");
    b_t(2).FaceColor = "blue";
    b_t(2).FaceAlpha = 0.2;
    b_t(2).LineWidth = 2;
    b_t(2).EdgeColor = "magenta";
```

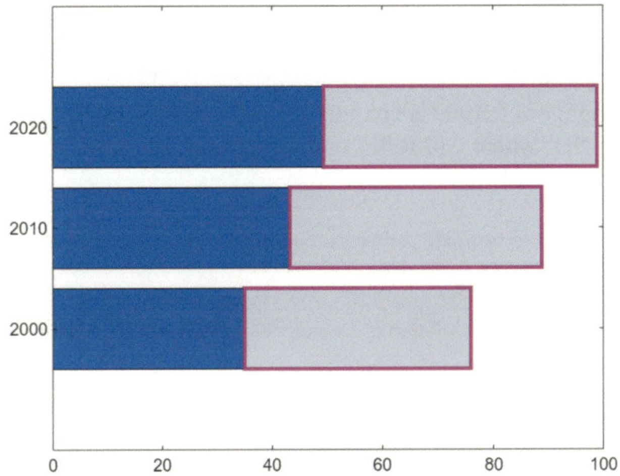

Fig. 9.19 Horizontal stacked bar chart with one pale segment outlined in thick magenta

Here, we use the 「BaseValue」 option for a stacked bar chart. Specifically, we set 「BaseValue」 to 「Thailand_f(2)」. See Fig. 9.20.

```
Thailand = [Thailand_f,Thailand_m];
b_t = barh(Year,Thailand,"stacked");
    b_t(1).BaseValue = Thailand_f(2);
```

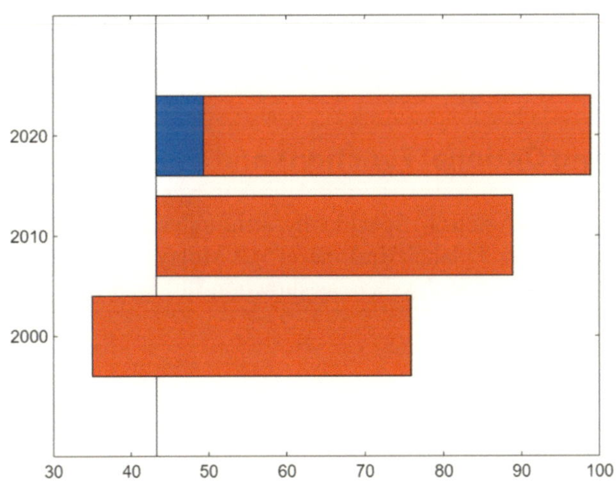

Fig. 9.20 Horizontal stacked bar chart with a reference line

After setting 「FaceColor」 to 「flat」, we use 「CData」 to change the color of the specified bars. For example, in 「bhf(2).CData(2,:) = [1 1 1]」, 「bhf(2)」 corresponds to the Singapore female data, and by modifying 「CData(2,:)」, the second data point for the Singapore female is changed to white. The same is applied to 「bhf(3). CData(3,:)=[0 0 0]」, where 「[0 0 0]」 corresponds to black. The resulting figure is illustrated in ◘ Fig. 9.21.

```
bhf = barh(["Philippines female","Singapore female","Thailand female"],...
    [Philippines_f,Singapore_f,Thailand_f],"FaceColor","flat");
    bhf(2).CData(2,:) = [1 1 1];
    bhf(3).CData(3,:) = [0 0 0];
```

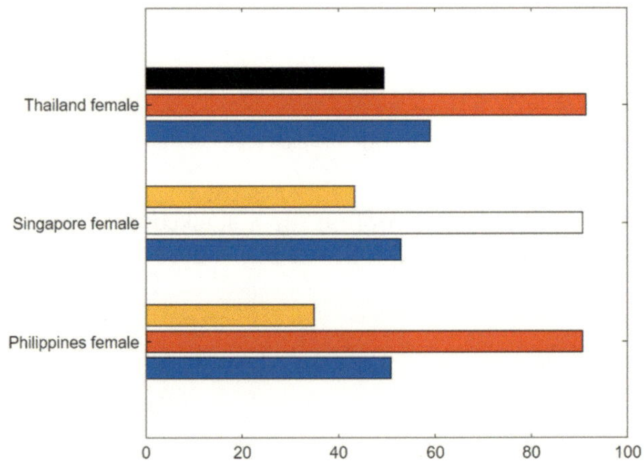

◘ **Fig. 9.21** Use of the RGB vector: Horizontal bar chart

9.5 Combining Optional Functions and the Barh Function

Here, we use the **ylabel** function. Specifically, although it is somewhat classical, effective use of spaces such as 「ylabel("Indonesia   Malaysia")」 can create the categories of 「Indonesia」 and 「Malaysia」 as shown in ◘ Fig. 9.22.

```
Indonesia = [Indonesia_f,Indonesia_m];
Malaysia = [Malaysia_f,Malaysia_m];
barh(["Indonesia female","Indonesia male",...
    "Malaysia female","Malaysia male"],[Indonesia,Malaysia]);
ylabel("Indonesia        Malaysia");
```

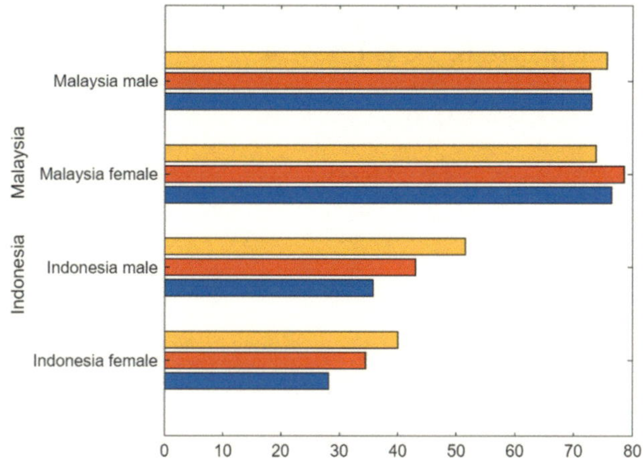

Fig. 9.22 Grouped horizontal bar chart with manually spaced Y-axis labels

Unlike the previous program that created Fig. 9.22, the **ylabel** function is used differently to set the label for the y-axis. Specifically, the label is set as a cell array of strings, 「{"Malaysia","","","","","","","","Indonesia"}」. This array contains 「Malaysia」 and 「Indonesia」, with seven empty strings, creating space between 「Malaysia」 and 「Indonesia」 on the y-axis. Additionally, 「yl.Rotation=0」 sets the rotation of the y-axis label to 0 degrees, which displays the labels horizontally instead of the default vertical orientation. The 「Malaysia」 and 「Indonesia」 labels are separated by blank spaces, displayed horizontally. See Fig. 9.23.

```
Indonesia = [Indonesia_f,Indonesia_m];
Malaysia = [Malaysia_f,Malaysia_m];
barh(["Indonesia female","Indonesia male",...
    "Malaysia female","Malaysia male"],[Indonesia,Malaysia]);
yl = ylabel({"Malaysia","","","","","","","","Indonesia"});
    yl.Rotation = 0;
```

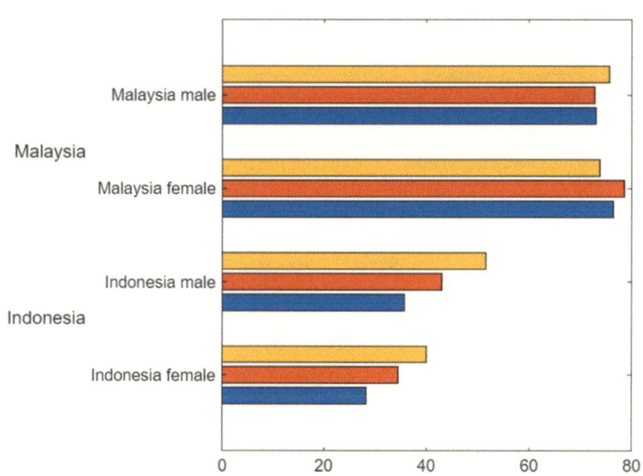

Fig. 9.23 Grouped horizontal bar chart with manually spaced Y-axis labels using text rotation (cf. Fig. 9.22)

Here, the **yticklabels** function is used to place 「Indonesia」 and 「Malaysia」 at specific locations on the y-axis. See ◘ Fig. 9.24.

```
Indonesia = [Indonesia_f,Indonesia_m];
Malaysia = [Malaysia_f,Malaysia_m];
barh(["Indonesia female","Indonesia male",...
    "Malaysia female","Malaysia male"],[Indonesia,Malaysia]);
yticklabels({"Indonesia","","Malaysia",""});
```

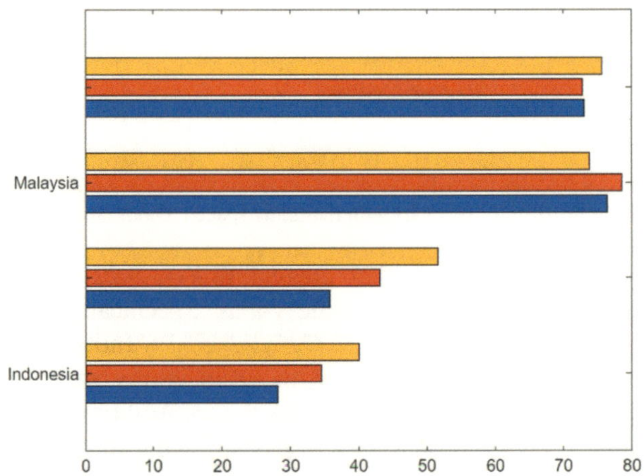

◘ **Fig. 9.24** Grouped horizontal bar chart with partially labeled Y-axis

Similar to the program in ◘ Fig. 9.23, this program uses the **ylabel** function. Additionally, the **xlabel** function is used to place the text 「Percent of employment」 on the x-axis. The **xlim** function is used to set the lower and upper limits of the x-axis to 0 and 100, respectively. Lastly, the **grid on** command is used. See ◘ Fig. 9.25.

```
Indonesia = [Indonesia_f,Indonesia_m];
Malaysia = [Malaysia_f,Malaysia_m];
barh(["Indonesia female","Indonesia male",...
    "Malaysia female","Malaysia male"],[Indonesia,Malaysia]);
yl = ylabel({"Malaysia","","","","","","","","Indonesia"});
    yl.Rotation = 0;
xlabel("Percent of employment");
xlim([0 100]);
grid on
```

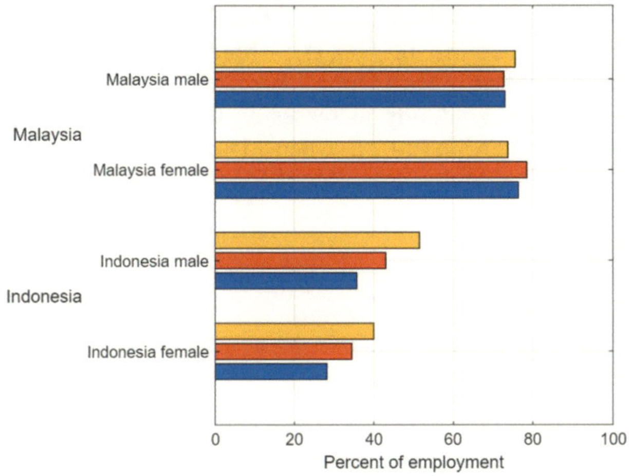

Fig. 9.25 Grouped horizontal bar chart with X- and Y-axis labels and X-axis limits

Here, the **xlim** function is used to set the x-axis limits from 0 to 100. Additionally, the **xregion** function is used to fill the area from 「Thailand_m(3)」 to 100 (= 「Inf」 in the following program) with yellow. Finally, the **legend** function is used to create a legend. See Fig. 9.26.

```
Thailand = [Thailand_f,Thailand_m];
barh(["Thailand female","Thailand male"],Thailand);
xlim([0 100]);
 xr = xregion(Thailand_m(3),Inf);
     xr.FaceColor = "yellow";
 legend(["2000","2010","2020"]);
```

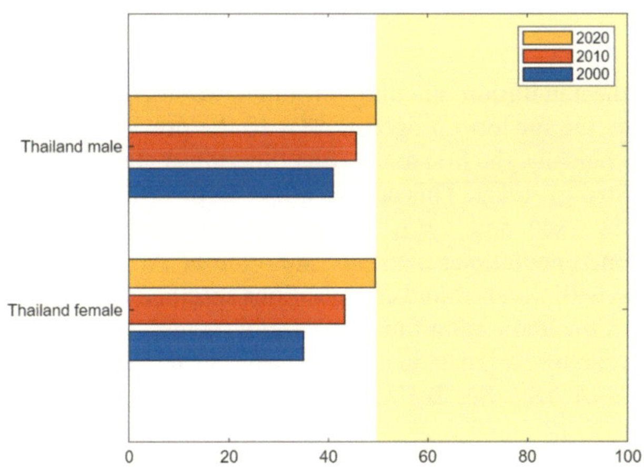

Fig. 9.26 Grouped horizontal bar chart with a yellow shaded area and a legend

Here, the **for** loop is used to specify 「FaceColor」 and 「FaceAlpha」 iterati vely. Specifically, 「FaceColor」 of 「bh(1)」, 「bh(2)」, and 「bh(3)」 is set to 「red」, making the bars red. Next, 「FaceAlpha」 is set to 「1-0.3 * i」, so that the trans parency of the red bars gradually increases as the **for** loop iterates. Specifically, when 「i=1」, 「bh(1).FaceAlpha=0.7」; when 「i=2」, 「bh(2).FaceAlpha=0.4」; and when 「i=3」, 「bh(3).FaceAlpha=0.1」. See ◘ Fig. 9.27.

```
Thailand = [Thailand_f,Thailand_m];
bh = barh(["Thailand female","Thailand male"],Thailand,"FaceColor","flat");
for i = 1:3
    bh(i).FaceColor = "red";
    bh(i).FaceAlpha = 1-0.3*i;
end
```

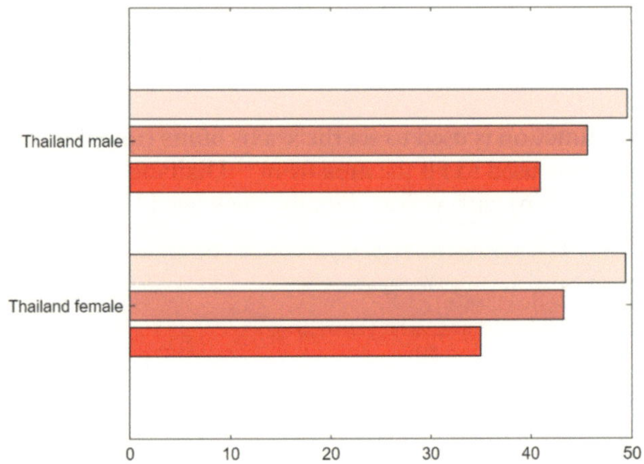

◘ **Fig. 9.27** Grouped horizontal bar chart using a for loop to vary red transparency

Here, using the **annotation** function four times, we create two text arrows and two arrows. First, for the **for**-loop program, refer to the previous program that created ◘ Fig. 9.27. Specifically, the first **annotation** function is given by 「annotation("Text Arrow",x1,y1,"String",str1)」. This creates a text arrow, with the start and end positions specified by 「x1」 and 「y1」, and the text specified by 「String」 as 「str1」. Next, in the second **annotation** function, 「annotation("Arrow",x2,y2)」 is used. This creates an arrow, with the start and end positions specified by 「x2」 and 「y2」. Note that 「Arrow」 in the **annotation** function cannot include text. The remaining **annotation** functions can be explained in a similar way. For more details on the **annotation** function, see ► Sect. 18.1. See ◘ Fig. 9.28.

```
Thailand = [Thailand_f,Thailand_m];
bh = barh(["Thailand female","Thailand male"],Thailand,"FaceColor","flat");
for i = 1:3
    bh(i).FaceColor = "red";
    bh(i).FaceAlpha = 1-0.3*i;
end
xlim([0 100]);
x1 = [0.7 0.55];
y1 = [0.75 0.75];
str1 = "Year 2020";
annotation("TextArrow",x1,y1,"String",str1);
x2 = [0.7 0.55];
y2 = [0.75 0.45];
annotation("Arrow",x2,y2);
x3 = [0.7 0.52];
y3 = [0.55 0.68];
str3 = "Year 2010";
an3 = annotation("TextArrow",x3,y3,"String",str3);
    an3.TextColor = "blue";
    an3.FontSize = 15;
    an3.Color = "green";
    an3.LineWidth = 2;
    an3.HeadStyle = "vback3";
x4 = [0.7 0.52];
y4 = [0.55 0.35];
an4 = annotation("Arrow",x4,y4);
    an4.Color = "green";
    an4.HeadStyle = "plain";
grid minor
```

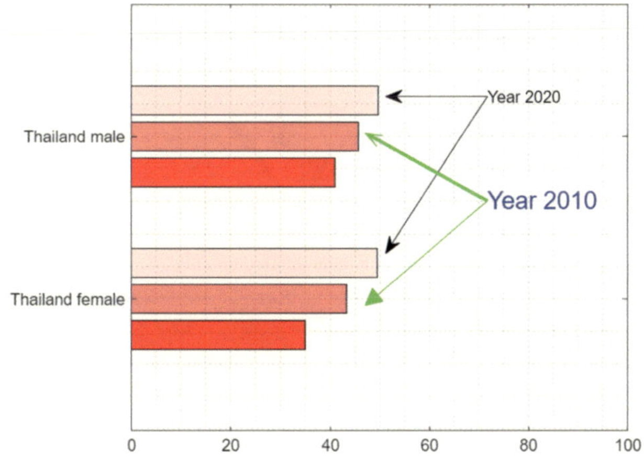

Fig. 9.28 Grouped horizontal bar chart with arrows and text arrows

Here, we use a **for** loop together with the **xlim**, **annotation**, and **xline** functions. For details on the **annotation** function, see the explanation of the previous program that created Fig. 9.28. Similarly, for the **for**-loop explanation, refer to the program that created Fig. 9.27. The **xlim** function is used to set the lower and upper limits to 0 and 100. Additionally, 「xline(Thailand_m(3), ":", "Male 2020")」 draws a dotted vertical line at x=Thailand_m(3), and places the text 「Male 2020」 next to the line. For more details on the **annotation** and **xline** functions, see ► Sects. 18.1 and 18.3. The resulting figure is illustrated in ◘ Fig. 9.29.

```
Thailand = [Thailand_f,Thailand_m];
bh = barh(["Thailand female","Thailand male"],Thailand,"FaceColor","flat");
for i = 1:3
    bh(i).FaceColor = "blue";
    bh(i).FaceAlpha = 1-0.3*i;
end
xlim([0 100]);
x1 = [0.5 0.6];
y1 = [0.58 0.75];
an1 = annotation("Arrow",x1,y1);
    an1.LineWidth = 2;
    an1.HeadWidth = 25;
    an1.Color = "red";
x2 = [0.48 0.6];
y2 = [0.25 0.45];
str2 = "Thailand female";
an2 = annotation("TextArrow",x2,y2,"String",str2);
    an2.LineWidth = 4;
    an2.HeadStyle = "plain";
xline(Thailand_m(3),":","Male 2020");
```

9

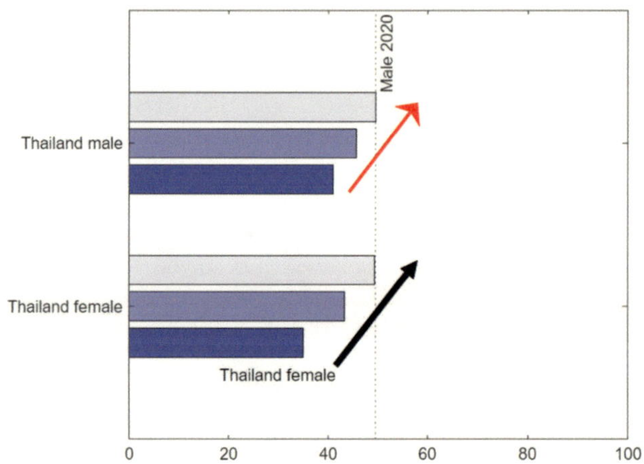

◘ **Fig. 9.29** Grouped horizontal bar chart using a for loop to vary blue transparency, with an arrow, a text arrow, and a vertical dotted line

Here, we use a **for** loop along with the **xlim**, **xline**, **xticks**, **xticklabels**, and **xregion** functions. For details on the **for** loop, refer to the explanation of the program that created ◘ Fig. 9.27. The **xlim** function sets the lower and upper limits to 0 and 100. Additionally, the **xline** function, with the first argument as 「[Thailand_m(1),Thailand_m(2),Thailand_m(3)]」, draws vertical dotted lines at these three values. As confirmed in the third argument of the **xline** function, each line is labeled with 「2000」, 「2010」, and 「2020」, respectively.

Next, the **xticks** and **xticklabels** functions are used to place the text labels 「{"0%","25%","50%","75%","100%"}」 on the x-axis ticks. Finally, the **xregion** function is used to fill the areas with yellow and the default color (gray). For more details on the **xline** function, see ▶ Sect. 18.3. See ◻ Fig. 9.30.

```
bh = barh("Thailand male",Thailand_m,"FaceColor","flat");
for i = 1:3
    bh(i).FaceColor = "blue";
    bh(i).FaceAlpha = 1-0.3*i;
end
xlim([0 100]);
labels = {"2000","2010","2020"};
xl = xline([Thailand_m(1),Thailand_m(2),Thailand_m(3)],":",labels);
    xl(2).Color = "red";
    xl(2).LineWidth = 2;
xticks([0 25 50 75 100]);
xticklabels({"0%","25%","50%","75%","100%"});
xr1 = xregion([50 75]);
    xr1.FaceColor ="yellow";
xregion([75 100]);
```

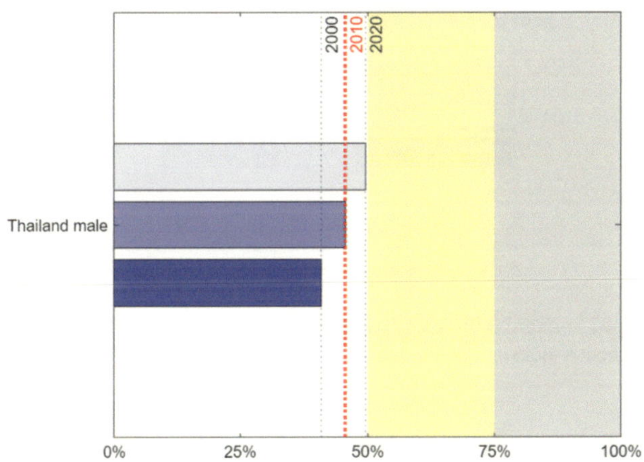

◻ **Fig. 9.30** Grouped horizontal bar chart using a for loop to vary blue transparency, with vertical reference lines and shaded areas

Here, we use a **for** loop along with the **legend**, **xticks**, and **set** functions. For details on the **for** loop, see the explanation of the program that created ◻ Fig. 9.27. Next, the **legend** function creates a legend, with 「NumColumns」 set to 3 and 「Location」 set to 「south」, which places a 3-column legend at the bottom center of the figure. Although the **xticks** function in the program specifies the range from 0 to 100, the x-axis ticks are created at intervals of 2, from 0 to 50.

The **set** function is used with 「gca」 as the first argument to change the tick settings. Specifically, 「TickLength」 is set to 「[0.02 1]」, which specifies the length for both 2D and 3D figures. Since the **barh** function creates a 2D figure, only the first value, 0.02, is used. See ◻ Fig. 9.31 for the resulting figure.

```
Thailand = [Thailand_f,Thailand_m];
bh = barh(["Thailand female","Thailand male"],Thailand,...
    "FaceColor","flat");
for i = 1:3
    bh(i).FaceColor = "blue";
    bh(i).FaceAlpha = 1-0.3*i;
end
lg = legend(["2000","2010","2020"]);
    lg.NumColumns = 3;
    lg.Location = "south";
xticks(0:2:100);
set(gca,"TickLength",[0.02 1]);
```

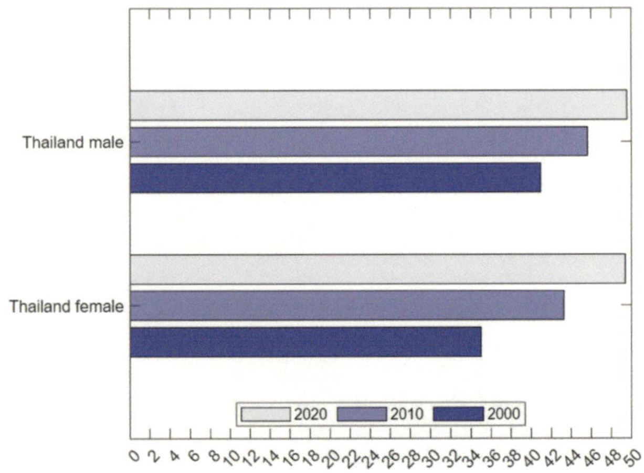

Fig. 9.31 Grouped horizontal bar chart using a for loop to vary blue transparency, with a horizontal legend and specified X-ticks

9

Here, the **hold on** command overlays the figures created by the **barh** and **plot** functions. Specifically, the properties of the **barh** function are stored in the variable 「hb」. Additionally, 「XEndPoints」 and 「YEndPoints」 provide the coordinate data, stored in 「xp1」, 「yp1」, 「xp2」, and 「yp2」. The **plot** function uses these variables to connect the points. In 「plot(yp1,xp1,":x")」, the points with coordinates (x,y)=(yp1,xp1) are marked with 「x」 and connected with a dotted line. Similarly, in 「plot(yp2,xp2,"--o")」, the points with coordinates (x,y)=(yp2,xp2) are marked with 「o」 and connected with a dashed line. The resulting figure is displayed in Fig. 9.32.

```
Malaysia=[Malaysia_f,Malaysia_m];
hb = barh(Year,Malaysia);
      xp1 = hb(1).XEndPoints;
      yp1 = hb(1).YEndPoints;
      xp2 = hb(2).XEndPoints;
      yp2 = hb(2).YEndPoints;
hold on
pl1 = plot(yp1,xp1,":x");
      pl1.LineWidth = 2;
pl2 = plot(yp2,xp2,"--o");
      pl2.Color = "red";
      pl2.LineWidth = 2;
hold off
xlim([50 80]);
title("% of Employment: Malaysia");
legend(["female","male"],"Location","eastoutside");
```

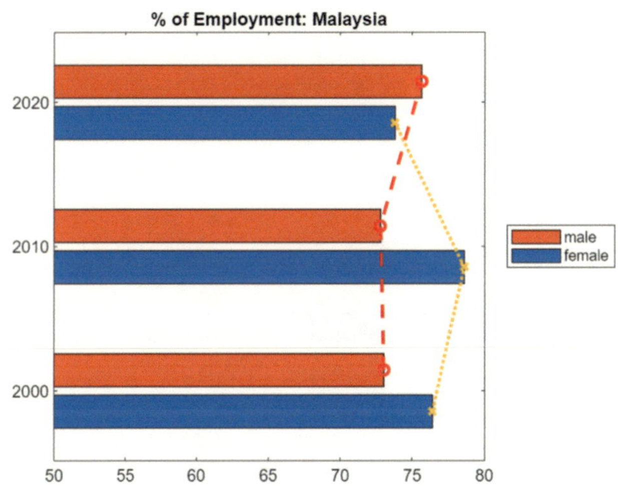

Fig. 9.32 Grouped horizontal bar chart with lines connecting data points and a legend (outside-right position)

Here, similar to the previous program that created Fig. 9.32, we use the **hold on** command. Since 「hb」 is a 1-by-3 bar object, the coordinate data is retrieved using 「XEndPoints」 and 「YEndPoints」 for 「hb(1)」, 「hb(2)」, and 「hb(3)」. The **plot** function then connects the retrieved coordinate data with lines. See Fig. 9.33 for the created figure.

```
figure
Malaysia = [Malaysia_f,Malaysia_m];
hb = barh(["Malaysia female","Malaysia male"],Malaysia);
    xp1 = hb(1).XEndPoints;
    yp1 = hb(1).YEndPoints;
    xp2 = hb(2).XEndPoints;
    yp2 = hb(2).YEndPoints;
    xp3 = hb(3).XEndPoints;
    yp3 = hb(3).YEndPoints;
hold on
pl1 = plot(yp1,xp1,":");
    pl1.Color = "blue";
    pl1.LineWidth = 2;
pl2 = plot(yp2,xp2,"--");
    pl2.Color = "red";
    pl2.LineWidth = 2;
pl3 = plot(yp3,xp3,":");
    pl3.Color = "magenta";
    pl3.LineWidth = 4;
hold off
lg = legend(["2000","2010","2020"]);
    lg.NumColumns = 3;
    lg.Location ="south";
```


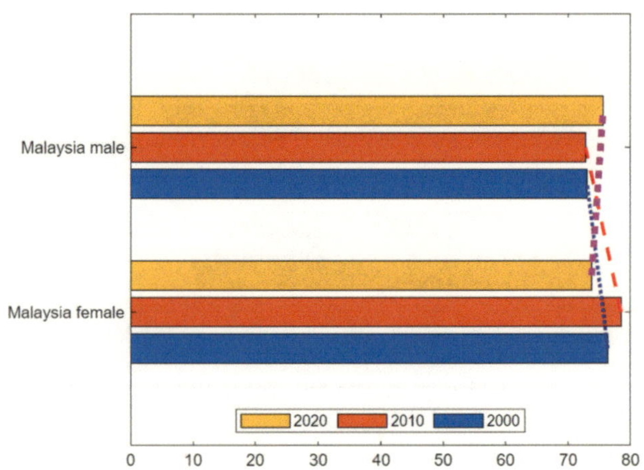

Fig. 9.33 Grouped horizontal bar chart with lines connecting data points and a horizontal legend (cf. Fig. 9.32)

Here, the **hold on** command is used to overlay horizontal bar graphs, similar to the overlay of vertical bar graphs with the **bar** function described in ► Figs. 7.60 and 7.61. In the first **barh** function, we set 「FaceAlpha」 to 0.2, 「LineStyle」 to 「:」, 「EdgeColor」 to 「magenta」, and 「LineWidth」 to 2, resulting in a semi-transparent bar graph with a thick dotted magenta outline. In the second **barh** function, we set 「FaceAlpha」 to 0.2, 「LineStyle」 to 「--」, and 「LineWidth」 to 2, creating a semi-transparent bar graph with a thick dashed outline. See Fig. 9.34.

```
figure
bf = barh(Year,Malaysia_f);
     bf.FaceAlpha = 0.2;
     bf.LineStyle = ":";
     bf.EdgeColor = "magenta";
     bf.LineWidth = 2;
hold on
bm = barh(Year,Malaysia_m);
     bm.FaceAlpha = 0.2;
     bm.LineStyle = "--";
     bm.LineWidth = 2;
hold off
title("% of Employment (Malaysia)");
xlabel("Percentage");
```

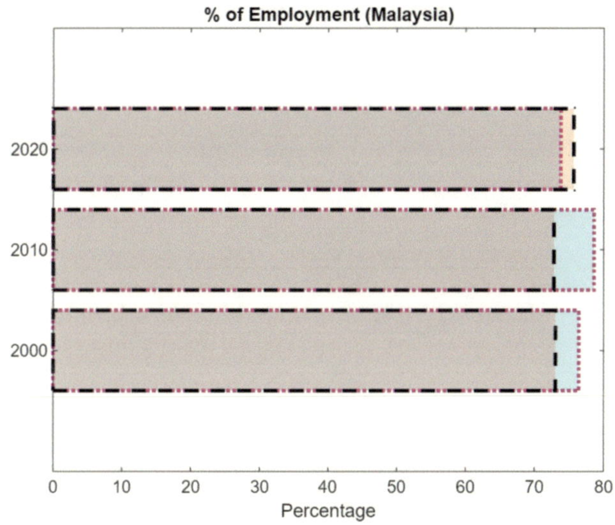

Fig. 9.34 Horizontal bar chart with overlaid bars and customized edges (no stacked)

Here, the properties created by the first **barh** function are stored in the variable 「bt」. We then modify the 「FaceAlpha」, 「LineWidth」, 「LineStyle」, and 「EdgeColor」 properties in a **for** loop. Next, we use the **hold on** command to overlay the figure created by the second **barh** function. In addition, using 「colororder("earth")」, the color palette is set to 「earth」. For the legend created by the **legend** function, 「NumColumns」 is set to 3, creating a 3-column legend. See Fig. 9.35.

```
figure
Thailand = [Thailand_f,Thailand_m];
Singapore = [Singapore_f,Singapore_m];
bt = barh(["Female","Male"],Thailand);
for i = 1:3
    bt(i).FaceAlpha = 0.4;
    bt(i).LineWidth = 2;
    bt(i).LineStyle = "--";
    bt(i).EdgeColor = "blue";
end
hold on
bs = barh(["Female","Male"],Singapore);
for i = 1:3
    bs(i).FaceAlpha = 0.1;
end
hold off
colororder("earth");
lg = legend(["Thailand 2000","Thailand 2010","Thailand 2020",...
    "Singapore 2000","Singapore 2010","Singapore 2020"]);
    lg.NumColumns = 3;
```

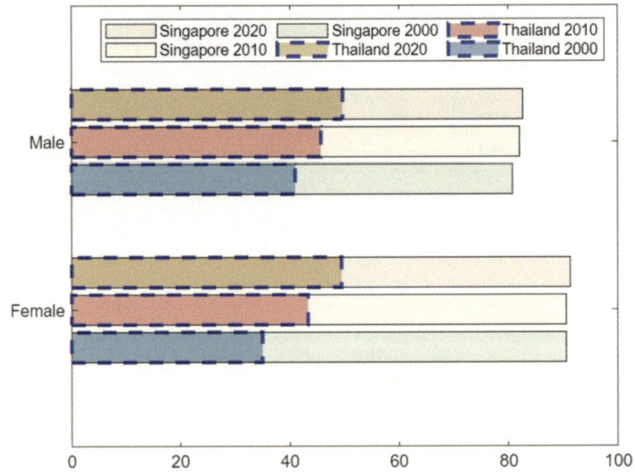

Fig. 9.35 Horizontal bar chart with blue thick dashed edges and a two-line legend (no stacked)

9

Here, ⌜tiledlayout(2,1)⌟ creates a 2-by-1 tiled layout. ⌜nexttile(1)⌟ uses the **barh** function to create a horizontal bar graph within the 1-by-1 space. The x-axis ticks are removed with ⌜xticks([])⌟. Then, ⌜set(gca,"TickDir","out")⌟ directs the tick marks outward, and ⌜set(gca,"TickLength",[0.02,0.02])⌟ makes the tick lines longer. Additionally, ⌜set(gca,"YAxisLocation","right")⌟ places the y-axis labels and tick marks on the right side of the plot. As a result, the text ⌜Female⌟ from ⌜ylabel("Female")⌟ appears on the right side of the y-axis. In ⌜nexttile(2)⌟, ⌜YAxisLocation⌟ is not specified, so the y-axis labels and ticks are placed on the default left side. See Fig. 9.36.

```
figure
ti = tiledlayout(2,1);
title(ti,"% of Employment");
nexttile(1);
barh(["Malaysia","Singapore","Indonesia"],...
    [Malaysia_f,Singapore_f,Indonesia_f]);
xticks([]);
set(gca,"TickDir","out");
set(gca,"TickLength",[0.02,0.02]);
set(gca,"YAxisLocation","right");
yl1 = ylabel("Female");
    yl1.FontSize = 20;
    yl1.Color = "red";
    yl1.Rotation = 0;
nexttile(2);
barh(["Malaysia","Singapore","Indonesia"],...
    [Malaysia_m,Singapore_m,Indonesia_m]);
set(gca,"TickDir","out");
set(gca,"TickLength",[0.02,0.02]);
    yl2 = ylabel("Male");
    yl2.FontSize = 20;
    yl2.Color = "blue";
    yl2.Rotation = 0;
```

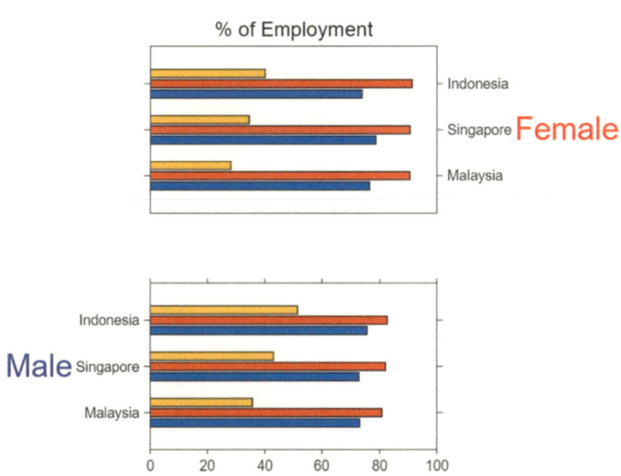

Fig. 9.36 Horizontal bar charts (2-row layout with outward ticks and right Y-axis in top chart)

9.6 Questions

Question 1

Create Fig. 9.37 using 「Year」 as the first argument and 「[Philippines_m, Singapore_m]」 as the second argument. Additionally, use 「grid minor」 to create fine grid lines. Additionally, use the **xlim** function to extend the range of the horizontal axis from 0 to 200.

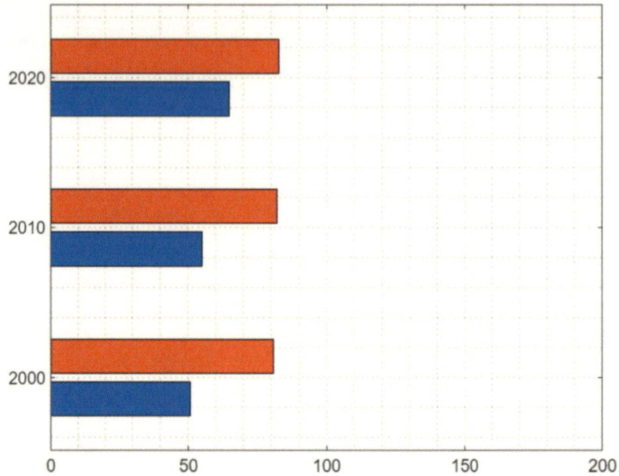

Fig. 9.37 Horizontal bar chart (Question 1)

Answer 1


```
figure
barh(Year,[Philippines_m,Singapore_m]);
grid minor
xlim([0 200]);
```

Question 2

Create Fig. 9.38 using「Year」as the first argument and「[Singapore_f,Singapore_m]」as the second argument. Additionally, increase the transparency of the two stacked horizontal bars. Finally, set the title of the figure to「Percent of Employment in Singapore」and add a legend with「Female」and「Male」.

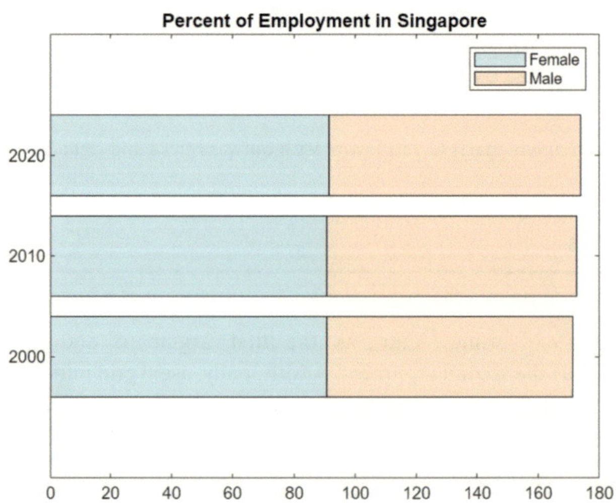

Fig. 9.38 Horizontal bar chart (Question 2)

Answer 2

```
figure
Ex2 = barh(Year,[Singapore_f Singapore_m],"stacked");
    Ex2(1).FaceAlpha = 0.2;
    Ex2(2).FaceAlpha = 0.2;
title("Percent of Employment in Singapore");
legend("Female","Male");
```

Question 3

Using the variables 「Indonesia_f」, 「Malaysia_f」, 「Philippines_f」, and 「Singapore_f」, create Fig. 9.39 with data for 2000 and 2010. Note that these variables contain data for three years: 2000, 2010, and 2020.

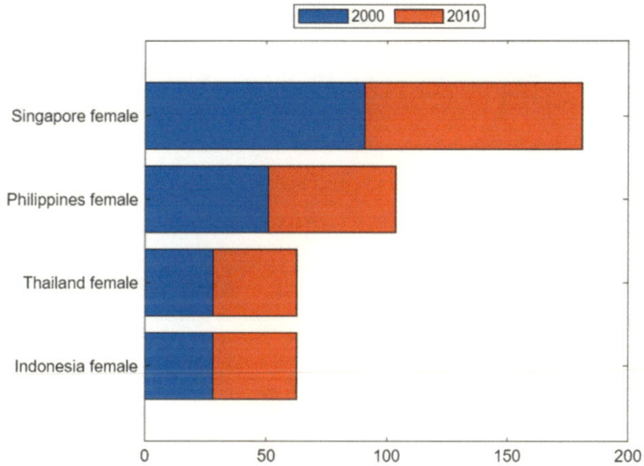

Fig. 9.39 Horizontal bar chart (Question 3)

Answer 3

```
figure
Indonesia_f2 = Indonesia_f(1:2);
Malaysia_f2 = Malaysia_f(1:2);
Philippines_f2 = Philippines_f(1:2);
Singapore_f2 = Singapore_f(1:2);
barh(["Indonesia female","Thailand female",...
    "Philippines female","Singapore female"],...
    [Indonesia_f2,Indonesia_f2,Philippines_f2,Singapore_f2],...
    "stacked");
lgd = legend("2000","2010");
    lgd.NumColumns = 2;
    lgd.Location = "northoutside";
```

Bar3h Function

Contents

10.1 Details of the Data Used in this Chapter – 316

10.2 Excel Data of ► Section 10.1 – 316

10.3 Basic Syntax of the Bar3h Function – 317

10.4 Setting and Controlling the Bar3h Function – 321

10.5 Combining Optional Functions and the Bar3h Function – 330

10.6 Questions – 335

Supplementary Information The online version contains supplementary material available at https://doi.org/10.1007/978-981-96-6213-5_10.

Y. Nakamoto, *A Comprehensive Guide to Creating Graphics with MATLAB Programming*, Classroom Companion: Economics, https://doi.org/10.1007/978-981-96-6213-5_10

Learning Objectives

This chapter creates 3D horizontal bar charts using the **bar3h** function. By reading this chapter, you will learn how to use the **bar3h** function and understand the similarities and differences in the created figures, syntax, and the type of data between the **bar3h** function and the **bar3** function (covered in ▶ Chap. 8). As a result, you will be able to determine whether to use 3D horizontal or 3D vertical bar charts.

Overview

The **bar3h** function in MATLAB creates 3D horizontal bar graphs. The bars in the graph extend horizontally (along the x-axis), while the y-axis and z-axis are used to represent the data. In contrast, the **bar3** function in ▶ Chap. 8 creates 3D vertical bar graphs. Since the only difference is the direction in which the bars extend, the advantages and disadvantages of the **bar3h** function are similar to those of the **bar3** function. See ▶ Chap. 8.

10.1 Details of the Data Used in this Chapter

10 In this chapter, we use the time series data for the prices of gold, platinum, and silver (measured in $/troy ounce). The troy ounce is commonly used for precious metals. The variable 「Year」 represents the year, and the variables 「Gold」, 「Platinum」, and 「Silver」 represent the prices of gold, platinum, and silver, respectively.

```
Year = [1960,1970,1980,1990,2000,2010,2020]';

Gold = [35,36,608,383,279,1225,1770]';

Platinum = [84,133,679,472, 545, 1610, 883]';

Silver = [1,2,21,5,5,20,21]';

Price_metals = table(Year,Gold,Platinum,Silver);
```

10.2 Excel Data of ▶ Section 10.1

In ▶ Sect. 10.2, we import the Excel formatted data used in ▶ Sects. 10.3, 10.4, 10.5, and 10.6. First, the file and sheet names are specified. See ▶ Chap. 2 for details on importing Excel data. Additionally, as in Sect. 8.2, we use the **round** function, to round the data. Moreover, we use the **renamevars** function to change the column names 「Year」, 「Gold」, 「Platinum」, and 「Silver」 from 「Var1」, 「Gold___troyOz_」, 「Platinum___troyOz_」, and 「Silver___troyOz_」.

```
file = "alldata.xlsx";
sheet = "Chap10_Price_Precious_Metals";
Price_metals = readtable(file,"Sheet",sheet);
```

```
Warning: Column headers from the file were modified to make them valid MATLAB identifiers before creating
variable names for the table. The original column headers are saved in the VariableDescriptions property.
Set 'VariableNamingRule' to 'preserve' to use the original column headers as table variable names.
```

```
Price_metals = round(Price_metals);
Price_metals = renamevars(Price_metals,["Var1","Gold___troyOz_",...
    "Platinum___troyOz_","Silver___troyOz_"],...
    ["Year","Gold","Platinum","Silver"]);
```

We will extract each column from the table variable 「Price_metals」.

```
Year = Price_metals.Year;
Gold = Price_metals.Gold;
Platinum = Price_metals.Platinum;
Silver = Price_metals.Silver;
```

10.3 Basic Syntax of the Bar3h Function

Basic Syntax:

We explain the two basic syntaxes of the **bar3h** function, enclosed in black. See Figure 10.1. The syntax 「bar3h(y)」 represents that 「y」 contains the data values, and the corresponding bars extend horizontally (along the x-axis). Next, the syntax 「bar3h(z,y)」 uses the data in 「z」 to distinguish between different groups or categories. 「y」 also contains the data values. Along the z-axis, the different groups or categories are displayed, and the corresponding bars extend horizontally.

Fig. 10.1 The syntax of the bar3h function

Here, using the syntax 「bar3h(y)」, we create a 3D horizontal bar graph for the variable 「Silver」. See Fig. 10.2.

```
bar3h(Silver);
```

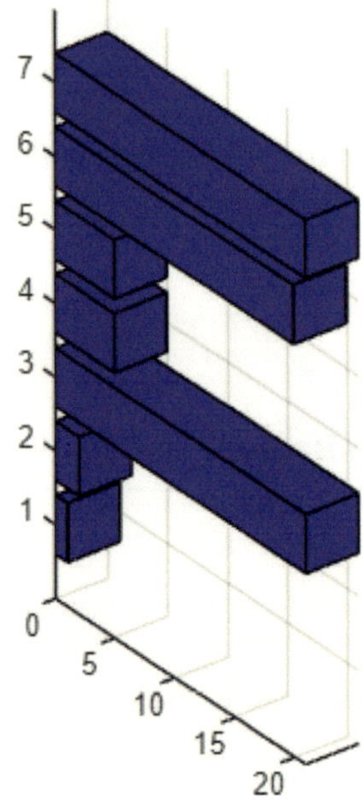

Fig. 10.2 Horizontal 3D bar chart (Z-axis: Default index)

Here, we use the syntax 「bar3h(z,y)」, where 「z」 and 「y」 correspond to 「Year」 and 「Silver」, respectively. As a result, the z-axis displays numbers from 1960 to 2020. Compare Fig. 10.2 with Fig. 10.3.

```
bar3h(Year,Silver);
```

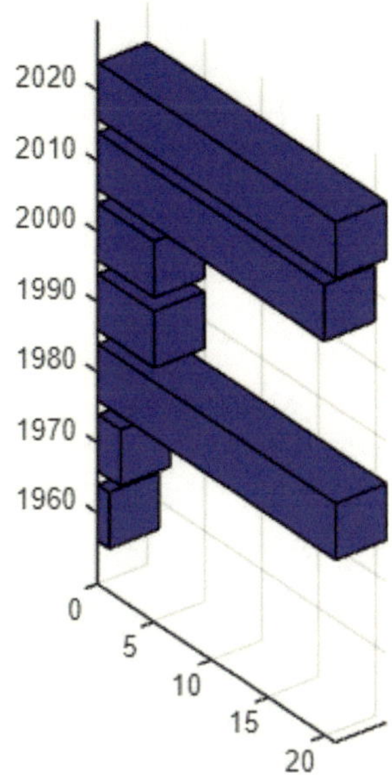

Fig. 10.3 Horizontal 3D bar chart (Z-axis: Year)

Here, we enter a 7-by-3 matrix of 「[Silver, Gold, Platinum]」 as the second argument. As a result, three groups are created, and each color is automatically assigned. The resulting figure is presented in **Fig. 10.4**.

```
bar3h(Year,[Silver,Gold,Platinum]);
```

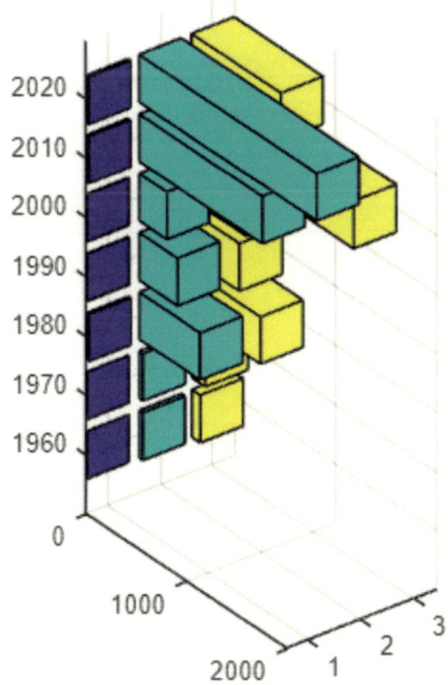

Fig. 10.4 Horizontal 3D bar chart with multiple groups (Z-axis: Year)

Here, instead of using all the elements of each vector, only the second, third, and fourth elements are used. For example, 「Year(2:4)」 corresponds to the data for 「Year(2) (=1970)」, 「Year(3) (=1980)」, and 「Year(4) (=1990)」. See ◘ Fig. 10.5.

```
bar3h(Year(2:4),[Silver(2:4),Gold(2:4),Platinum(2:4)]);
```

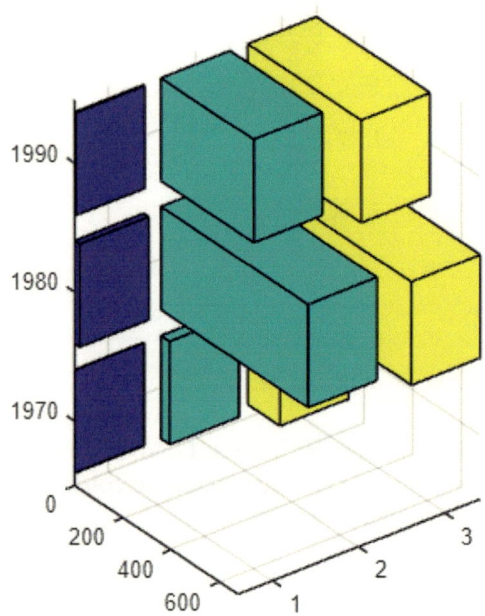

◘ **Fig. 10.5** Horizontal 3D bar chart with multiple groups using a subset of variables

▶ Example of Table Variables

Here, we use the table variable 「Price_metals」 to create a 3D horizontal bar graph. See ◘ Fig. 10.6.

```
bar3h(Price_metals.Year,...
    [Price_metals.Silver,Price_metals.Gold,Price_metals.Platinum]);
```

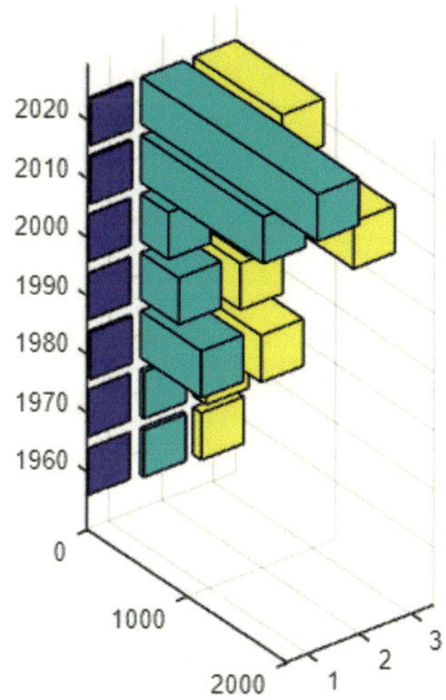

Fig. 10.6 Use of a table variable: Horizontal 3D bar chart

◀

10.4 Setting and Controlling the Bar3h Function

Here, we enter 0.5 as the third argument. As a result, we create a 3D horizontal bar graph with narrower bars. The resulting figure is presented in Fig. 10.7. This uses the same syntax as ▶ Fig. 8.9 generated by the **bar3** function.

```
bar3h(Year,Silver,0.5);
```

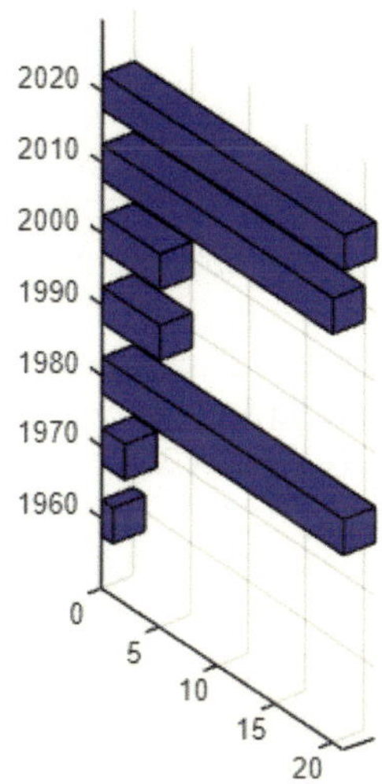

Fig. 10.7 Horizontal 3D bar chart with narrow bars

Here, we set the third argument of the **bar3h** function to 0.2 and the fourth argument to 「blue」, creating a narrow blue 3D horizontal bar graph. The properties of the graph are stored in the variable 「bh3」. Then, we set 「EdgeColor」 to 「green」 and 「LineWidth」 to 2, resulting in the thicker green outline. See Fig. 10.8.

```
bh3 = bar3h(Year,Silver,0.2,"blue");
    bh3.EdgeColor = "green";
    bh3.LineWidth = 2;
```

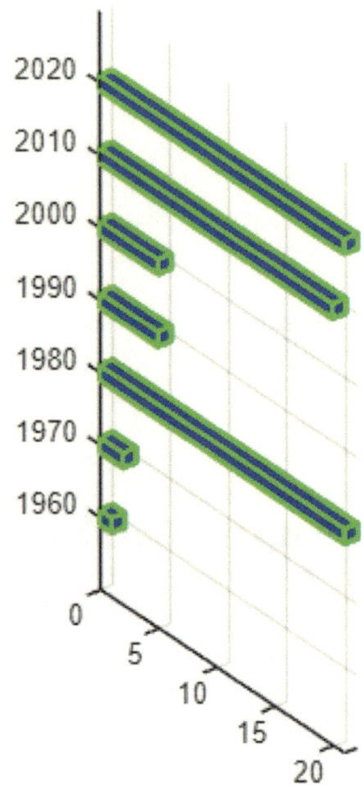

Fig. 10.8 Horizontal 3D bar chart with narrow bars having thick green edges

Here, the properties of the **bar3h** function are stored in the variable 「b3h」. In the Workspace, note that 「b3h」 is shown as a 1-by-2 Surface. This means that 「b3h(1)」 contains the properties of the 3D horizontal bar graph created using 「Platinum」, and 「b3h(2)」 contains the properties of the 3D horizontal bar graph created using 「Gold」.

Specifically, for 「b3h(1)」, we set 「FaceColor」 to 「red」, 「FaceAlpha」 to 0.2, 「LineStyle」 to 「--」, and 「LineWidth」 to 2, creating a highly transparent red bar with a thick dashed outline. See Fig. 10.9.

```
b3h = bar3h(Year,[Platinum,Gold]);
    b3h(1).FaceColor = "red";
    b3h(1).FaceAlpha = 0.2;
    b3h(1).LineStyle = "--";
    b3h(1).LineWidth = 2;
```

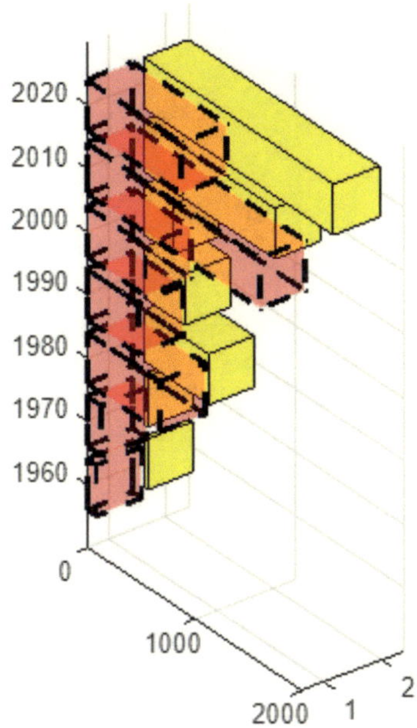

Fig. 10.9 Horizontal 3D bar chart with one group of semi-transparent red bars having dashed black edges

Here, we set the third argument of the **bar3h** function to 「yellow」, making the bars yellow. The properties of the graph are stored in the variable 「b3h」. Then, using a **for** loop, we adjust the transparency using 「FaceAlpha」.

Specifically, when 「i=1」 in the first iteration, we set 「b3h(1). FaceAlpha=0.2」, resulting in a highly transparent yellow 3D horizontal bar graph for the first group. Next, when 「i=2」, we set 「b3h(2).FaceAlpha=0.4」, making the transparency lower. When 「i=3」, the transparency becomes further lower. Additionally, for all bar graphs, we set 「LineWidth」 to 2 and 「LineStyle」 to 「:」, creating a thick dotted outline for the bars. See Fig. 10.10.

```
b3h = bar3h(Year,[Silver,Gold,Platinum],"yellow");
for i = 1:3
    b3h(i).FaceAlpha = i*0.2;
    b3h(i).LineWidth = 2;
    b3h(i).LineStyle = ":";
end
```

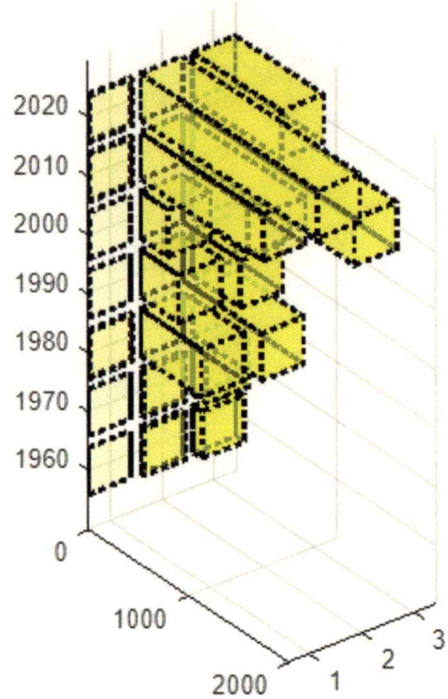

Fig. 10.10 Horizontal 3D bar chart using a for loop to vary yellow transparency, with dotted black edges

Here, we create a stacked bar graph using 「stacked」 in the third argument of the **bar3h** function. However, the next program is incorrect in vector concatenation. First, 「a1」 is a 1-by-3 row vector. Similarly, 「b1」 is also a 1-by-3 row vector. Looking at the first argument of the **bar3h** function, it is written as 「[a1,b1]」, leading to a row vector 「[1,3,5,2,4,6]」. Therefore, even when using 「stacked」, the resulting graph is not stacked as expected. See Fig. 10.11.

```
a1 = [1,3,5];
b1 = [2,4,6];
bar3h([a1,b1],1,"stacked","red");
```

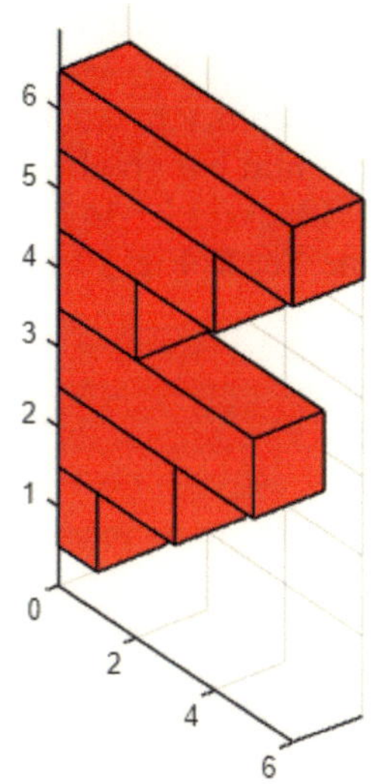

■ **Fig. 10.11** Horizontal 3D bar chart with "stacked" option (not stacked)

The following program differs from the previous one that created ■ Fig. 10.11. The first argument of the **bar3h** function is 「[a1;b1]」, with a semicolon inserted to create a new line. Thus, 「[a1;b1]」 becomes a matrix with 2 rows and 3 columns. Using 「stacked」, 「a1=[1,3,5]」 is stacked at z=1. In the same way, 「b1=[2,4,6]」 is stacked at z=2. See ■ Fig. 10.12.

```
a1 = [1,3,5];
b1 = [2,4,6];
bar3h([a1;b1],1,"stacked");
```

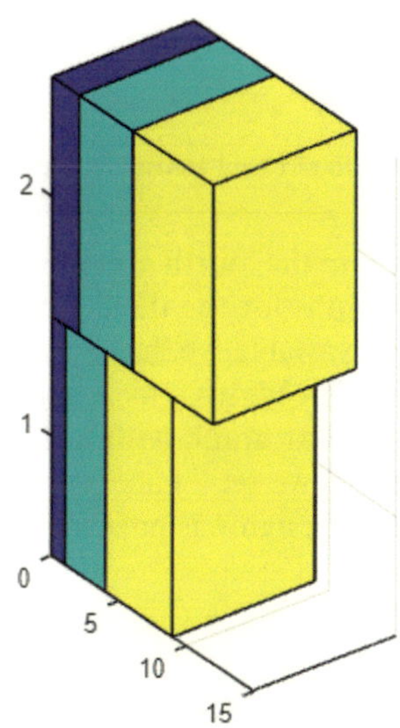

■ **Fig. 10.12** Stacked horizontal 3D bar chart

Here, after we enter 「stacked」 in the fourth argument of the **bar3h** function, the properties are stored in the variable 「b3h」. For the first group of 「b3h(1)」, we set 「FaceColor」 to 「red」, 「FaceAlpha」 to 0.5, and 「LineStyle」 to 「:」, creating a highly transparent red 3D horizontal bar graph with a dotted outline. For the second group of 「b3h(2)」, we set 「LineWidth」 to 2, resulting in a 3D horizontal bar graph with a thick stacked outline. See ◘ Fig. 10.13.

```
b3h = bar3h(Year,[Gold,Silver],1.2,"stacked");
    b3h(1).FaceColor = "red";
    b3h(1).FaceAlpha = 0.5;
    b3h(1).LineStyle = ":";
    b3h(2).LineWidth = 2;
```

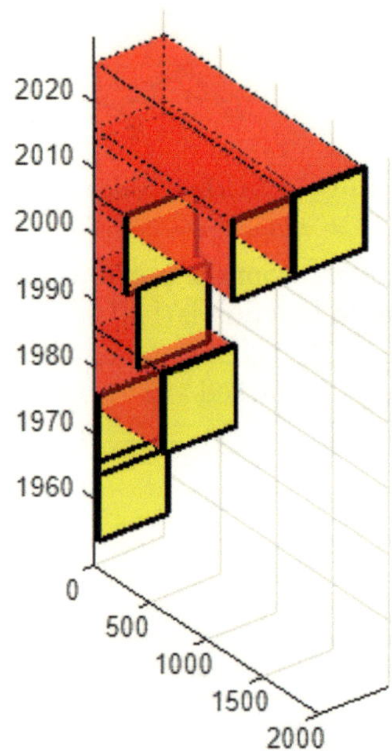

◘ **Fig. 10.13** Stacked 3D bar chart with red semi-transparent and yellow segments

Here, we specify 「grouped」 in the fourth argument of the **bar3h** function, resulting in the 3D horizontal bar graphs for the three groups, 「[Gold,Silver,Platinum]」. The properties are stored in the variable 「b3h」. For the first group of 「b3h(1)」, we set 「FaceColor」 to 「none」, 「LineWidth」 to 2, and 「LineStyle」 to 「-.」, creating a transparent 3D horizontal bar graph with a thick dash-dotted outline. Next, for the second group of 「b3h(2)」, we set 「FaceColor」 to 「red」, creating a red 3D horizontal bar graph. The resulting figure is presented in ◘ Fig. 10.14.

```
b3h = bar3h(Year,[Gold,Silver,Platinum],0.8,"grouped");
    b3h(1).FaceColor = "none";
    b3h(1).LineWidth = 2;
    b3h(1).LineStyle = "-.";
    b3h(2).FaceColor = "red";
```

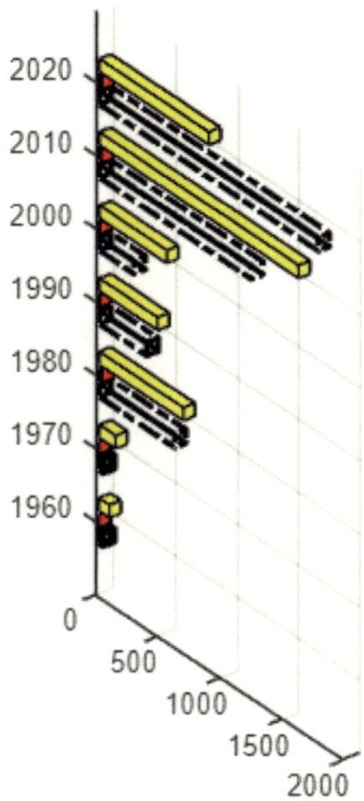

Fig. 10.14 Grouped horizontal 3D bar chart with one group of dash-dotted outlined transparent bars

Here, the properties of the **bar3h** function are stored in the variable 「b3h」. Next, using 「b3h(1).YData」, we retrieve the y-axis data of 「b3h(1)」 and store it in the variable 「ydata」. Then, by using 「b3h(1).CData=ydata」, we set the color data (CData) of the bar graph based on 「ydata」. Finally, by using 「b3h(1).FaceColor="interp"」, we change the color of the bar graph based on the YData values. See Fig. 10.15. The similar program using 「CData」 is applied for the **bar3** function. See also ▶ Figs. 8.19 and 8.20.

```
b3h = bar3h(Year,[Gold,Platinum],0.5);
    ydata = b3h(1).YData;
    b3h(1).CData = ydata;
    b3h(1).FaceColor = "interp";
```

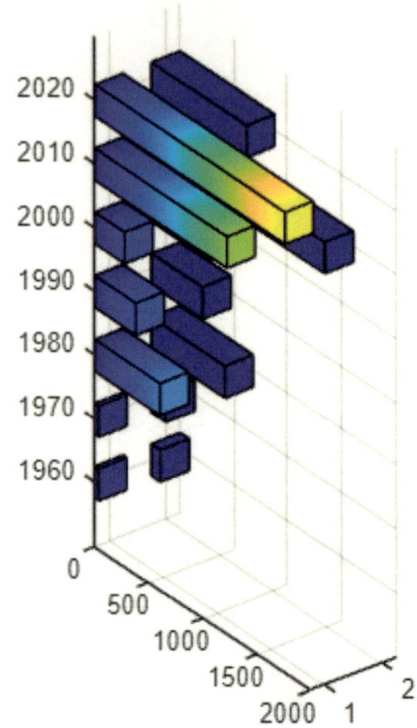

Fig. 10.15 Horizontal 3D bar chart with one group of bars having interpolated face colors

Here, as in Fig. 10.15, we use「CData」to change the color of the 3D horizontal bar graph. Specifically, we use a **for** loop to change the color of the two bar graphs. Compare Fig. 10.15 with Fig. 10.16.

```
b3h = bar3h(Year,[Gold,Platinum],0.5);
for i = 1:2
    ydata = b3h(i).YData;
        b3h(i).CData = ydata;
        b3h(i).FaceColor = "interp";
end
```

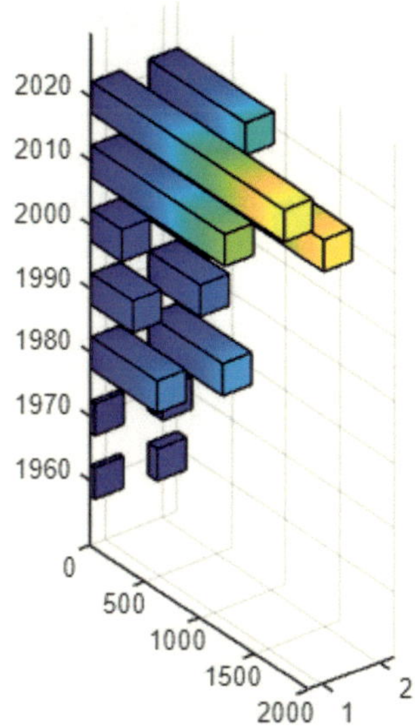

Fig. 10.16 Horizontal 3D bar chart with two groups having interpolated face colors by a for loop

Here, we set 「FaceAlpha」 to 0.2, 「Marker」 to 「o」, and 「LineStyle」 to 「:」, creating a highly transparent 3D horizontal bar graph with a dotted outline and circles at the corners of the bars. See Fig. 10.17.

```
b3h = bar3h(Year,Platinum,0.5);
    b3h.FaceAlpha = 0.2;
    b3h.Marker = "o";
    b3h.LineStyle = ":";
```

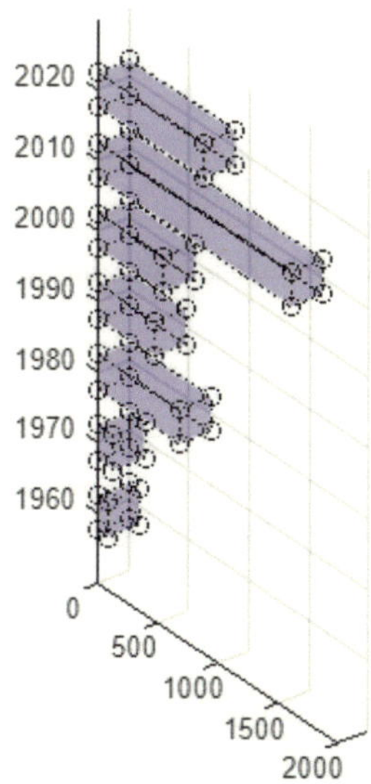

Fig. 10.17 Horizontal 3D bar chart with semi-transparent purple bars, circle markers, and dotted edges

10.5 Combining Optional Functions and the Bar3h Function

Here, the **xticklabels**, **yticks**, and **zticks** functions are used simultaneously. Unlike Fig. 10.4, the x-axis is labeled using ⌜xticklabels({"Silver","Gold","Platinum"})⌟. Compare Fig. 10.4 with Fig. 10.18. Additionally, the y-axis and z-axis ticks are adjusted using ⌜yticks(0:200:2000)⌟ and ⌜zticks(1960:30:2020)⌟, respectively. See Fig. 10.18.

```
bar3h(Year,[Silver,Gold,Platinum]);
xticklabels({"Silver","Gold","Platinum"});
yticks(0:200:2000);
zticks(1960:30:2020);
box on
```

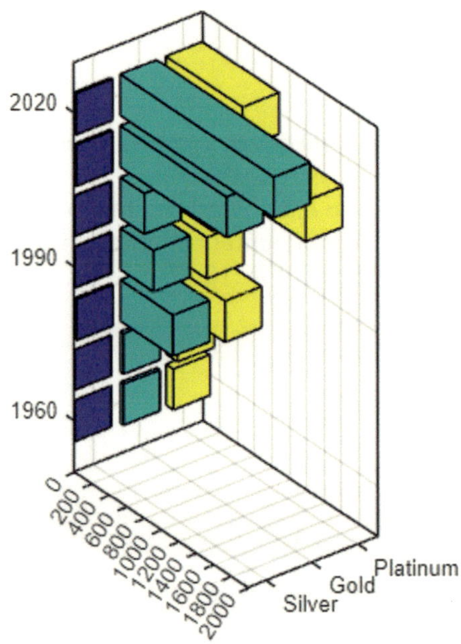

Fig. 10.18 Horizontal 3D bar chart with black frame, labeled X-ticks, and adjusted axes ticks

10

Here, using the **view** function, we adjust the viewing angle. Additionally, we do not group categories on the x-axis using ⌜xticklabels({})⌟. In contrast, the **legend** function is used to categorize the groups through the legend. Similar programs are shown in ► Figs. 8.26 and 8.27. See Fig. 10.19.

```
bar3h(Year,[Silver,Gold,Platinum],0.5);
xticklabels({});
yticks(0:200:2000);
zticks(1960:30:2020);
box on
view(120,-100);
legend(["Silver","Gold","Platinum"]);
```

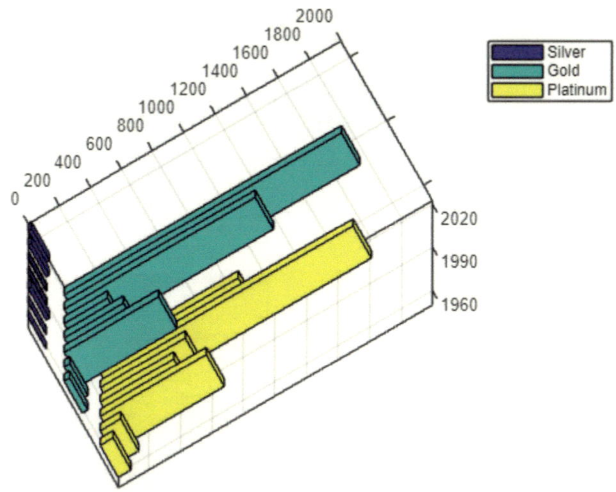

Fig. 10.19 Horizontal 3D bar chart from a different angle (cf. Fig. 10.18)

Here, the **annotation** function is used three times. In the first **annotation** function, 「ellipse」 is specified in the first argument. Specifically, the location is specified by the variable 「dim」, which encircles a specific bar graph with an ellipse. In the second **annotation** function, 「TextArrow」 is specified in the first argument, creating an arrow with the text 「High」. In the third **annotation** function, 「Arrow」 is specified, creating an arrow pointing to the bar. For more details on the **annotation** function, see ▶ Sect. 18.1. Finally, the **view** function is used to adjust the viewing angle. See Fig. 10.20.

```
bar3h(Year,[Silver,Gold,Platinum]);
xticklabels({"Silver","Gold","Platinum"});
dim = [0.55 0.15 0.2 0.25];
annotation("ellipse",dim,"Color","red");
str = "High";
x1 = [0.6 0.4];
y1 = [0.95 0.85];
annotation("TextArrow",x1,y1,"String",str);
x2 = [0.6 0.4];
y2 = [0.95 0.75];
an = annotation("Arrow",x2,y2);
    an.Color = "red";
view(120,10);
box on
```

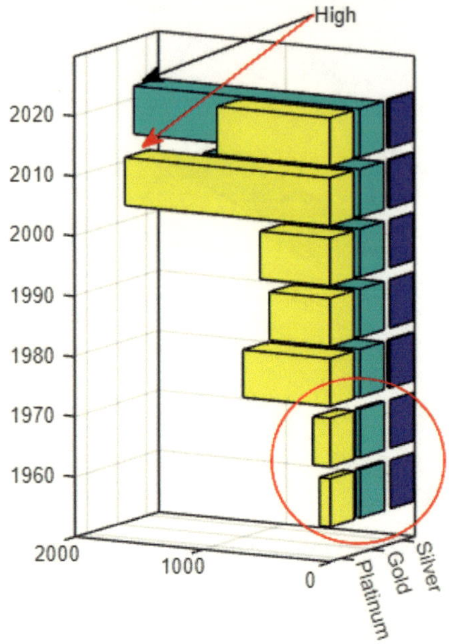

Fig. 10.20 3D bar chart from a different angle, with a text arrow, an arrow, and an ellipse

10

Here, using 「view(2)」, we convert the 3D horizontal bar graph created with the **bar3h** function into a 2D graph. In addition, we store the properties of the **legend** function in the variable 「lg」. We set 「Location」 to 「northoutside」 and 「NumColumns」 to 3, creating a legend with three columns positioned above the figure, outside the plot area. The resulting figure is shown in Fig. 10.21.

```
bar3h(Year,[Silver,Gold,Platinum]);
xticklabels({"Silver","Gold","Platinum"});
view(2);
lg = legend(["Silver","Gold","Platium"]);
    lg.Location = "northoutside";
    lg.NumColumns = 3;
```

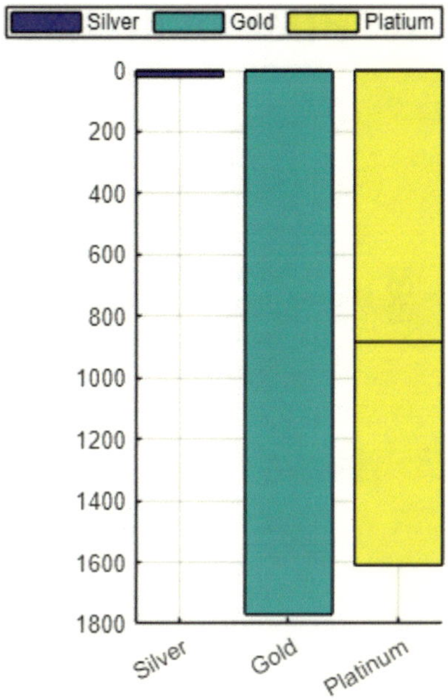

Fig. 10.21 Flat horizontal 3D bar chart from 2D top-down perspective

Here, using the **hold on** command, we overlay two 3D horizontal bar graphs created with the **bar3h** function. To make it easier to distinguish between the two bar graphs, we use 「FaceColor」 to specify different colors, and 「FaceAlpha」 to increase the transparency. The **title** and **zlabel** functions are additionally used. See Fig. 10.22.

```
figure
b1 = bar3h(Year,Gold,0.5);
    b1.FaceColor = "red";
    b1.FaceAlpha = 0.4;
    b1.LineWidth = 2;
hold on
b2 = bar3h(Year,Platinum,0.5);
    b2.FaceColor = "blue";
    b2.FaceAlpha = 0.2;
hold off
box on
ti = title("Price of metals");
    ti.Color = "blue";
    ti.FontSize = 20;
zl = zlabel("Year");
    zl.FontSize = 20;
```

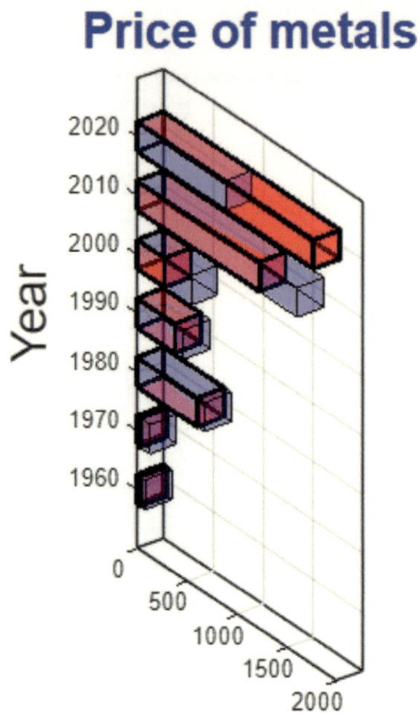

Fig. 10.22 3D bar chart with overlaid bars (not stacked)

10

The properties of the **bar3h** function are stored in the variable 「b3h1」. Next, the coordinates of the bar graph are obtained from 「XData」, 「YData」, and 「ZData」. For example, 「xdata1=b3h1(1).XData(:,2)」 retrieves the x-axis data of the second column from the 「Silver」 bar graph corresponding to 「b3h1(1)」 and stores it in the variable 「xdata1」. Similarly, the y-axis and z-axis data are stored in 「ydata1」 and 「zdata1」, respectively.

Then, using 「labels1=string(yp1)」, the data in 「yp1」 is converted to a string and stored in the variable 「labels1」. Finally, the first **text** function places text on the 3D horizontal bar graph created by 「b3h1(1)」. The same process is applied for 「b3h1(2)」. For more details on the **text** function, see ► Sect. 18.2. See Fig. 10.23.

```
figure
b3h1 = bar3h(Year,[Silver,Gold],0.5);
    b3h1(1).FaceAlpha = 0.2;
    xdata1 = b3h1(1).XData(:,2);
    ydata1 = b3h1(1).YData(:,2);
    zdata1 = b3h1(1).ZData(:,2);
xp1 = xdata1(2:6:end);
yp1 = ydata1(2:6:end);
zp1 = zdata1(2:6:end);
labels1 = string(yp1);
text(xp1,yp1,zp1,labels1);
    xdata2 = b3h1(2).XData(:,4);
    ydata2 = b3h1(2).YData(:,2);
    zdata2 = b3h1(2).ZData(:,2);
xp2 = xdata2(2:6:end);
yp2 = ydata2(2:6:end);
zp2 = zdata2(2:6:end);
labels2 = string(yp2);
text(xp2,yp2,zp2,labels2);
```

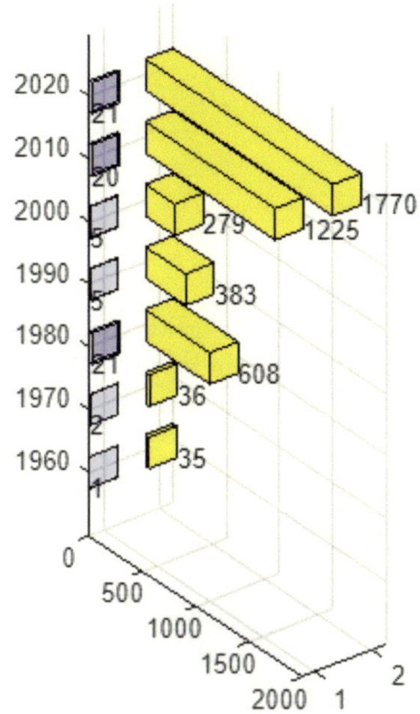

Fig. 10.23 3D bar chart with data labels

10.6 Questions

Question 1

Create a 3D horizontal bar graph as shown in Fig. 10.24, using 「Year」 as the first argument and 「[Silver, Gold]」 as the second argument. Then, create a frame for the boxes. Additionally, use the **view** function to change the viewing angle.

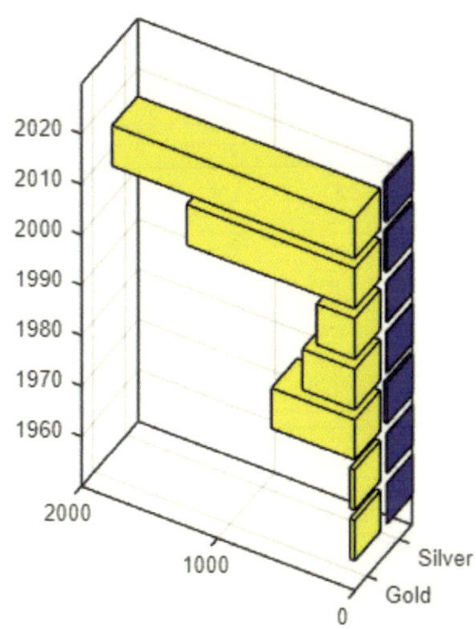

Fig. 10.24 3D bar chart (Question 1)

Answer 1

```
figure
bar3h(Year,[Silver,Gold]);
xticklabels({"Silver","Gold"});
box on
view(120,40);
```

Question 2

Create Fig. 10.25 using「Year」as the first argument and「[Gold, Platinum]」as the second argument. For the bars representing the「Gold」data, make the bars blue-filled and highly transparent, and give them a thick dotted outline.

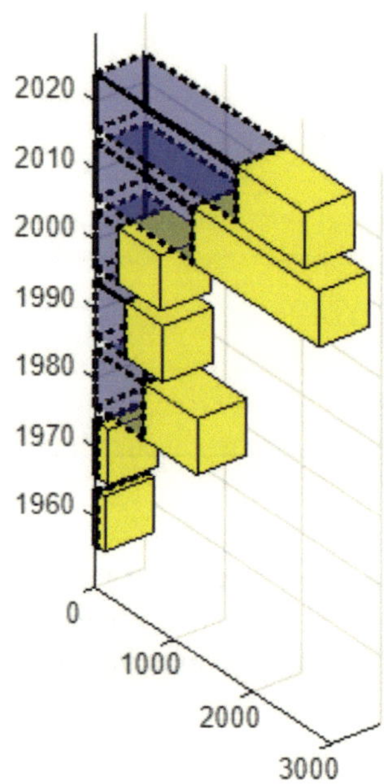

Fig. 10.25 3D bar chart (Question 2)

Answer 2

```
figure
Ex2 = bar3h(Year,[Gold,Platinum],"stacked");
    Ex2(1).FaceColor = "blue";
    Ex2(1).FaceAlpha = 0.2;
    Ex2(1).LineStyle = ":";
    Ex2(1).LineWidth = 2;
```

Question 3

Create Fig. 10.26 by using the variables 「Year」, 「Gold」, 「Silver」, and 「Platinum」. As shown in the figure below, restrict the bars to the three years: 2000, 2010, and 2020. Note that the variables 「Gold」, 「Silver」, and 「Platinum」 contain data for the years 1960, 1970, 1980, and 1990 in addition to 2000, 2010, and 2020.

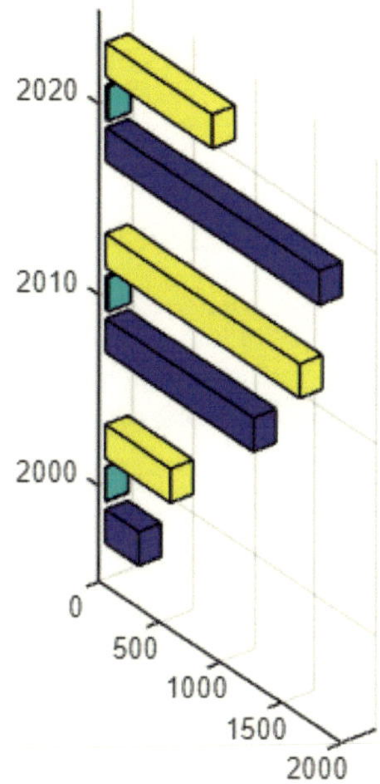

Fig. 10.26 3D bar chart (Question 3)

Answer 3

```
figure
Year2 = Year(5:7);
Gold2 = Gold(5:7);
Silver2 = Silver(5:7);
Platinum2 = Platinum(5:7);
bar3h(Year2,[Gold2,Silver2,Platinum2],"grouped");
```

Histogram Function

Contents

11.1 Details of the Data Used in this Chapter – 340

11.2 Excel Data of ▶ Section 11.1 – 342

11.3 Basic Syntax of the Histogram Function – 343

11.4 Setting and Controlling the Histogram Function – 352

11.5 Combining Optional Functions and the Histogram Function – 361

11.6 Questions – 371

Supplementary Information The online version contains supplementary material available at https://doi.org/10.1007/978-981-96-6213-5_11.

Y. Nakamoto, *A Comprehensive Guide to Creating Graphics with MATLAB Programming*, Classroom Companion: Economics, https://doi.org/10.1007/978-981-96-6213-5_11

Learning Objectives

This chapter creates histograms using the **histogram** function. By reading this chapter, you will learn how to use the **histogram** function and understand the similarities and differences in the created figures, syntax, and the type of data between the **histogram** function and the **bar** function (covered in ▶ Chap. 7). In particular, the beginning of ▶ Sect. 11.3 explains the key differences between histograms and bar graphs. Additionally, you will learn how to handle categorical data.

Overview

In this chapter, we will create a histogram by using the **histogram** function. A histogram is a type of graph that divides specific data into intervals and displays the number of occurrences or variations within each interval in a form similar to a bar graph. Since the **histogram** function and the **bar** function (▶ Chap. 7) both involve bars, readers unfamiliar with statistics may have difficulty understanding their differences. Therefore, we first explain these differences.

The main difference between a bar graph (▶ Chap. 7) and a histogram is the use of a vertical axis. In a bar graph, the heights of the bars are given by the elements of the vectors or matrices. For instance, the bar graphs in ▶ Chap. 7 show the percentage of the population aged 65 and over in Japan, Germany, the United States, Italy, and France from 2000 to 2050. As discussed in ▶ Sect. 7.1, the percentage for Japan in 2000 was 17.8%; this value, 17.8, corresponds directly to the height of the vertical bar. Conversely, a histogram treats the value of 17.8 as a single data point in a specific range. For example, suppose that a range is defined as the interval 15-20. If it contains a single data point, such as 17.8, a bar with a height of 1 is created. If there are two data points, such as 17.8 and 17.9, within the range, a bar with a height of 2 is created. As a result, by creating a histogram, the frequency distribution of the collected numerical data can be easily visualized. As an example of the data used in the **histogram** function, refer to the variables 「Mpop」 and 「Wpop」 generated in ▶ Sect. 11.1.

Additionally, it may be helpful to focus on the use of the horizontal axis between a bar graph and a histogram. In a bar graph, independent bars are displayed on each tick of the x-axis. Therefore, the horizontal axis is often given by different categories. For example, in the bar graphs from ▶ Chap. 7, each tick on the horizontal axis represents a country (see ▶ Figs. 7.38, 7.42, and 7.49). Alternatively, because the horizontal axis in a histogram represents a range of values, the histogram does not use the categories. In this chapter, we use data on the age distribution of the Australian population in 2019 to create a histogram. As the horizontal axis represents each age range, the created figures show the range divided into 10-year intervals on the x-axis.

11.1 Details of the Data Used in this Chapter

The variable 「Australia_total_population_2019」 represents the total population of Australia in 2019, approximately 25.37 million. The variable 「pop_100t」 is derived from the total population and is scaled down by a factor of 100,000. The variable 「Age_Exp」 denotes age ranges, such as 「0-9y.o.」 for ages 0-9 years. The variable 「Age」 assigns the numbers 1 through 8 to eight age categories, starting with 「0-9y.o.」 (corresponding to 「Age=1」) and ending with 「over 70」 (corresponding to 「Age=8」). The dataset 「dataMP」 shows the percentage of each age range in Australia's total male population. For example, a value of 6.5 in 「dataMP」 indicates that males in the 「0-9y.o.」 age range (Age=1) constitute 6.5% of the total male population. Conversely, 「dataWP」 represents the percentage of females in each age group in Australia. 「dataM」 is calculated by multiplying 「pop_100t」 with 「dataMP」 and dividing it by 100, yielding the male population in each age group. Similarly, 「dataW」 is the female population in each age group, calculated from 「dataWP」.

```
Australia_total_population_2019 = 25370000;

pop_100t = Australia_total_population_2019/100000;

Age_Exp = ["0-9y.o.","10-19y.o.","20-29y.o.","30-39y.o.",...
    "40-49y.o.","50-59y.o.","60-69y.o.","over 70"]';

Age = (1:8)';

dataMP = [6.5,6.2,7.4,7.2,6.4,6,5,5.1]';

dataWP = [6.1,5.9,7.1,7.3,6.6,6.2,5.3,6]';

dataM = pop_100t*dataMP/100';

dataW = pop_100t*dataWP/100';

pop = table(Age_Exp,Age,dataMP,dataWP,dataM,dataW);
```

In the following program, we explain the use of 「Mpop」 for males where 「Wpop」 denotes a variable for females with the same program content. Note that 「Mpop」 and 「Wpop」 are hypothetical data because we generate random numbers using the **randi** function.

First, we create an empty vector (or matrix) with 「Mpop=[ ]」. This empty vector will be filled with the ages of 126 males that we are about to generate. When we use a **for** loop, it is set up as 「for j=1:length(Age)」. Here, 「length(Age)=8」, which is the length of the variable 「Age」. In other words, 「j=1:length(Age)」 is the same as 「j=1:8」.

Let us consider the first iteration of the **for** loop. When 「j=1」, we can see that 「Mnum1=round(dataM(1))」 modifies the first number in 「dataM」, 16.4905, to an integer, which is then stored in the variable 「Mnum1」. This corresponds to the number of people aged 0 to 9. Since we use the **round** function, it rounds to the nearest integer, resulting in 「Mnum1=16」. Note that for 「j=1」, 「minAge=0」 and 「maxAge=9」, corresponding to ages 0 to 9. Finally, using the **randi** function, we generate random integers within the range of 「minAge=0」 and 「maxAge=9」 for the 16 people represented by 「Mnum1=16」. We then add these 16 people to the empty vector 「Mpop」 using 「Mpop=[Mpop; Mnum2']」. This completes the first iteration of the loop. For the second iteration of the loop, the newly generated ages of 10 to 19 years old are added to the previously created 16 people in 「Mpop=[Mpop; Mnum2']」.

```
Mpop = [];
Wpop = [];
for j = 1:length(Age)
    Mnum1 = round(dataM(j));
    Wnum1 = round(dataW(j));
    minAge = (j-1)*10;
    maxAge = j*9;
    Mnum2 = randi([minAge maxAge],1,Mnum1);
    Wnum2 = randi([minAge maxAge],1,Wnum1);
    Mpop = [Mpop;Mnum2'];
    Wpop = [Wpop;Wnum2'];
end
```

「Mpop」 is a column data set with 126 rows and 1 column, whereas 「Wpop」 has 128 rows and 1 column; therefore, the number of data points does not match. Hence, to adjust 「Mpop」 to 128 rows and 1 column, two data points are added to the 20-29 age group.

```
Mpop = [Mpop;[20;20]];
```

11.2 Excel Data of ▶ Section 11.1

▶ Section 11.2 involves importing the data used in ▶ Sects. 11.3, 11.4, 11.5, and 11.6 in Excel format. First, the file and sheet names are specified. For more details on how to import Excel data, refer to ▶ Chap. 2.

11

```
file = "alldata.xlsx";
sheet1_1 = "Chap11_Population_Australia";
data1_1 = readtable(file,"Sheet",sheet1_1);
```

The program below is the same as that used in the ▶ Sect. 11.1.

```
Australia_total_population_2019 = 25370000;
pop_100t = Australia_total_population_2019/100000;
Age_Exp = ["0-9y.o.","10-19y.o.","20-29y.o.","30-39y.o.","40-49y.o.",...
    "50-59y.o.","60-69y.o.","over 70"]';
Age = (1:8)';
dataMP = [6.5,6.2,7.4,7.2,6.4,6,5,5.1]';
dataWP = [6.1,5.9,7.1,7.3,6.6,6.2,5.3,6]';
dataM = pop_100t*dataMP/100';
dataW = pop_100t*dataWP/100';
pop = table(Age_Exp,Age,dataMP,dataWP,dataM,dataW);
```

The following program is the same as that in ▶ Sect. 11.1 and will create the simulated data.

```
Mpop = [];
Wpop = [];
for j = 1:length(Age)
    Mnum1 = round(dataM(j));
    Wnum1 = round(dataW(j));
    minAge = (j-1)*10;
    maxAge = j*9;
    Mnum2 = randi([minAge maxAge],1,Mnum1);
    Wnum2 = randi([minAge maxAge],1,Wnum1);
    Mpop = [Mpop;Mnum2'];
    Wpop = [Wpop;Wnum2'];
end
Mpop = [Mpop;[20;20]];
```

11.3 Basic Syntax of the Histogram Function

Before explaining the specific program for histograms, we explain the difference between a bar graph and a histogram. Here, we use the data from 「dataM」 and 「dataW」 to create a bar graph. 「dataM」 and 「dataW」 are row vectors with 8 columns. For example, the first element of 「dataM」, 16.4905, represents the population of Australian males aged 0 to 9 years, divided by 100,000. 「dataW」 represents the population of Australian females. See ◘ Fig. 11.1 for 「dataM」 and 「dataW」.

Workspace

Name ▲	Value
dataM	[16.4905,15.7294,18.7738,18.2664,16.2368,15.2220,12.6850,12.9387]
dataW	[15.4757,14.9683,18.0127,18.5201,16.7442,15.7294,13.4461,15.2220]

◘ **Fig. 11.1** Contents of Workspace

The following program is an application of bar graphs. 「tiledlayout(1,2)」 creates a 1-by-2 tiled layout. The **bar** function is used to plot the population distribution of Australian males in the first tile of the 1-by-2 tiled layout as a bar graph, and the population distribution of Australian females is plotted in the second tile. See ◘ Fig. 11.2.

```
t = tiledlayout(1,2);
title(t,"Bar: The number of population in Australia by age");
ax1 = nexttile;
bar(ax1,Age_Exp,dataM);
subtitle(ax1,"Males");
ylabel(ax1,"The number (0.1 Million)");
xlabel(ax1,"Age");
grid on
ax2 = nexttile;
bar(ax2,Age_Exp,dataW);
subtitle(ax2,"Females");
ylabel(ax2,"The number (0.1 Million)");
xlabel(ax2,"Age");
grid on
```

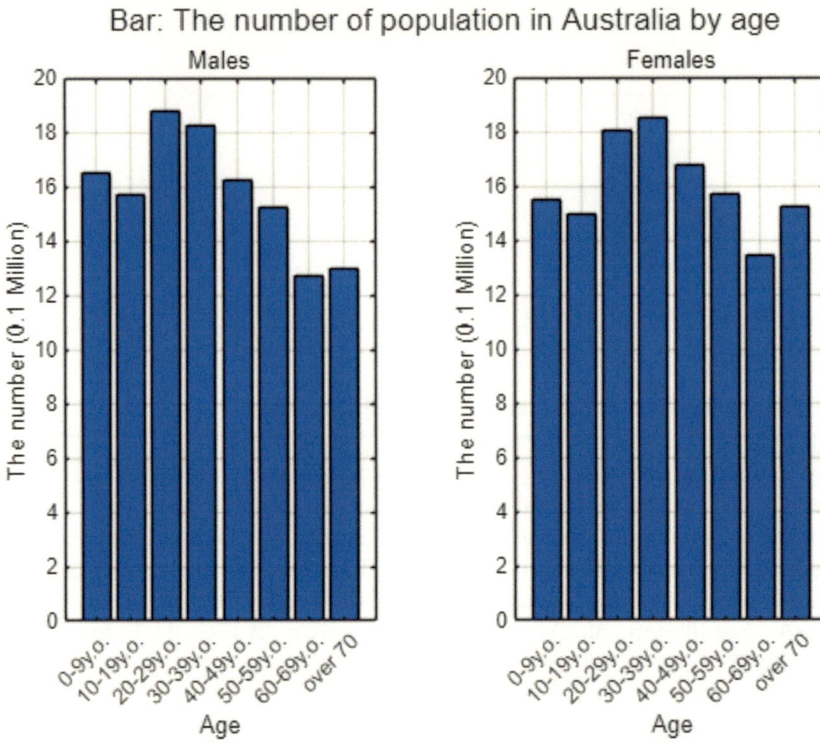

◘ **Fig. 11.2** Bar chart comparing age distribution by gender (2-column layout)

Attention

Error Program: Using the **histogram** function may result in an error because 「Age_Exp」 is a string vector representing categories, not numerical ranges. See ◘ Fig. 11.3 for the error message. Specifically, as 「Age_Exp」 is currently a 1-by-8 string row vector, we can use 「Age_Exp」 with the **bar** function but not with the **histogram** function, which requires numerical values to specify ranges.

```
%%% Error
% figure
% histogram(Age_Exp,dataM);
```

```
Error using histogram
Expected input number 1, x, to be one of these types:

double, single, uint8, uint16, uint32, uint64, int8, int16, int32, int64, logical, datetime, duration, categorical

Instead its type was string.

Error in histogram>parseinput (line 267)
        validateattributes(x,{'numeric','logical','datetime','duration','categorical'},...

Error in histogram (line 149)
[opts,passthrough,dispatchToCategorical] = parseinput(args,firstaxesinput);
```

◘ **Fig. 11.3** Error output of the histogram function: Use of string input

Since the **tiledlayout** function was used in the previous program that created ◘ Fig. 11.2, we modify the default layout using the **figure** command. Here, we use only 「dataM」 as the first argument of the **histogram** function where 「dataM」 is 1-by-8 data. Because the range of the horizontal axis is set automatically, the resulting figure displays three bars. See ◘ Fig. 11.4. Because the x-axis in the **histogram** function cannot be divided into categories but is divided into ranges of numbers, this is clearly different from ◘ Fig. 11.2.

Let us look again at 「dataM」 in ◘ Fig. 11.1. The data points around 13 on the x-axis consist of two data points from 「dataM」, the seventh and eighth data points (12.6850, 12.9387) in 「dataM」, which are counted as two. Therefore, the height of the bar graph is 2. See ◘ Fig. 11.4. Similarly, the data points around 16 on the x-axis consist of four data points from 「dataM」, the first, second, fifth, and sixth data points (16.4905, 15.7294, 16.2368, 15.2220) in 「dataM」, which are counted as four in ◘ Fig. 11.4. Finally, around 18.5 on the x-axis, there are two data points from 「dataM」, the third and fourth data points (18.7738, 18.2664) in 「dataM」, which are counted as two in ◘ Fig. 11.4.

```
figure
histogram(dataM);
```

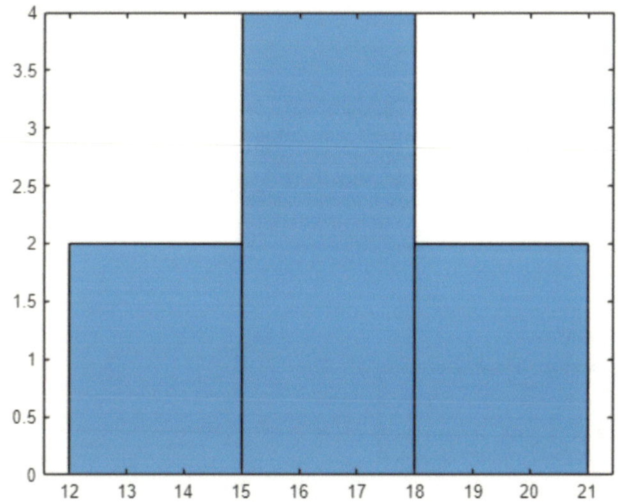

◘ **Fig. 11.4** Histogram using aggregate data (inappropriate input)

The syntax of the **histogram** function includes 「histogram(X,nbins)」, which allows you to divide the data. Because 「dataM」 is a 1-by-8 vector, 「histogram(data M,8)」 may divide the data into 8 bins. See ◘ Fig. 11.5 for the created figure using 「histogram(X,nbins)」. Compare it with ◘ Fig. 11.2 created by the **bar** function. We now conclude the explanation of the differences between histograms and bar graphs. For a more detailed comparison, it is recommended that you read ▶ Chap. 7 in conjunction with this chapter.

```
histogram(dataM,8);
ylim([0,5]);
```

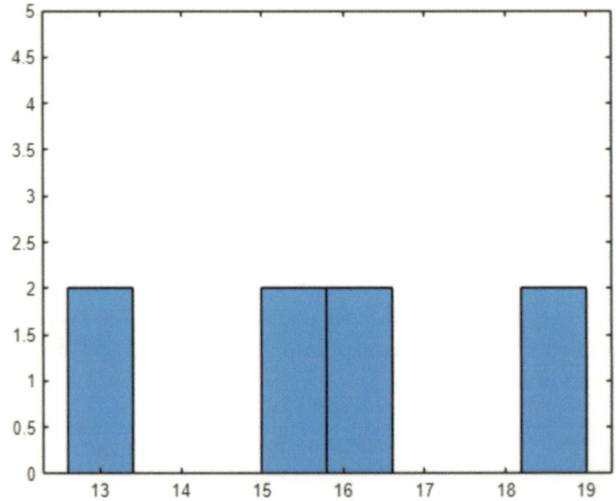

Fig. 11.5 Histogram using aggregate data (inappropriate input), with adjusted X-axis limits

Basic Syntax:

11

Now, we explain the three basic syntaxes of the **histogram** function in MATLAB, enclosed in the black rectangle of Figure 11.6. This can be reviewed by typing 「doc histogram」 in the Command Window. The first syntax inside the black rectangle is the basic form of the **histogram** function, 「histogram(X)」. Based on the basic syntax, the **histogram** function requires a single input, 「X」. The second syntax, 「histogram(X,nbins)」, allows us to specify the number of bins for the histogram, where 「nbins」 defines how many intervals the data will be divided into. The third syntax, 「histogram(X,edges)」, allows us to specify the edges of the bins where 「edges」 is a vector that defines the specific bin boundaries for the data in 「X」

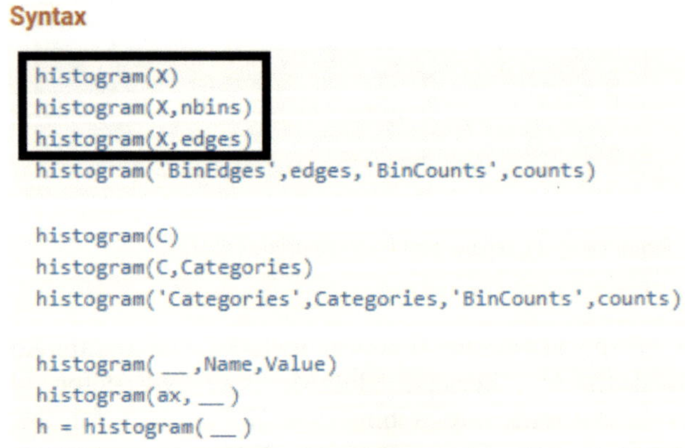

Fig. 11.6 The syntax of the histogram function

Using the **histogram** function, we confirm the distribution of 「Mpop」. The figure illustrates that the horizontal axis ranges from 0 to 80. For example, between 0 and 9, there is a bar with a height of 16. This means that within the 126-by-1 column vector of 「Mpop」, there are 16 occurrences of numbers within the range 0 to 9. See ◘ Fig. 11.7.

```
histogram(Mpop);
```

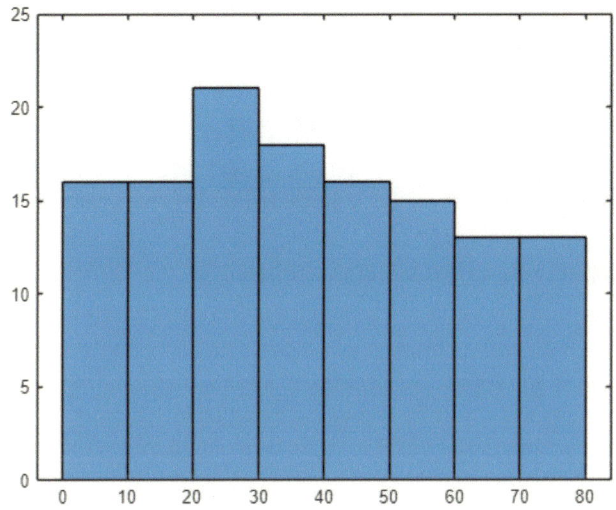

◘ **Fig. 11.7** Histogram using one variable

By combining 「Mpop」 and 「Wpop」, the data become 128 rows and 2 columns. However, without distinguishing between these two data sets, all of the data is used to create one histogram. See ◘ Fig. 11.8. If you want to represent 「Mpop」 and 「Wpop」 as separate histograms in the same figure, refer to ◘ Figs. 11.33, 11.34, and 11.35.

```
histogram([Mpop,Wpop]);
```

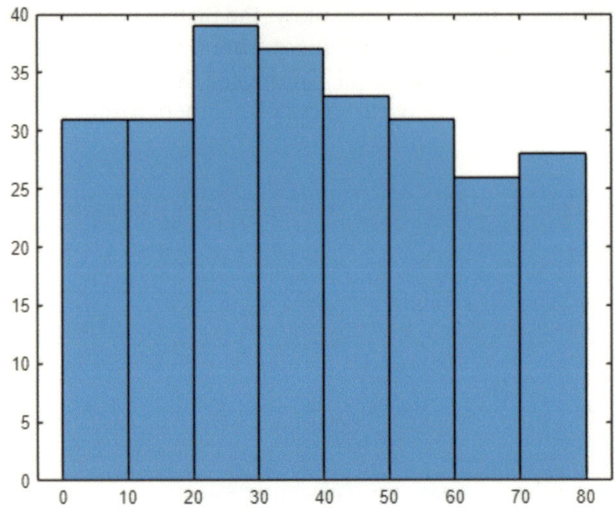

Fig. 11.8 Histogram using merged data of two variables

Attention
Error Program: If you try to concatenate two variables with different sizes in the row direction, an error will occur. In this case, the variable 「a1」 has 5 rows and 1 column, and the variable 「b1」 has 4 rows and 1 column. Since the number of rows differs between the two variables, we cannot concatenate them in the row direction. The error message is presented in Fig. 11.9.

```
%%% Error Program
% a1 = [1;2;3;4;5];
% b1 = [2;3;4;5];
% histogram([a1,b1]);
```

Fig. 11.9 Error output of the histogram function: Inconsistent vector lengths

In the previous error program in Fig. 11.9, we attempted to concatenate the column vectors of variables 「a1」 and 「b1」 in the row direction. However, they are now concatenated in the column direction. In other words, 「[a1; b1]」 becomes a 9-by-1 column vector. In this case, the histogram can be created. See Fig. 11.10.

```
a1 = [1;2;3;4;5];
b1 = [2;3;4;5];
histogram([a1;b1]);
```

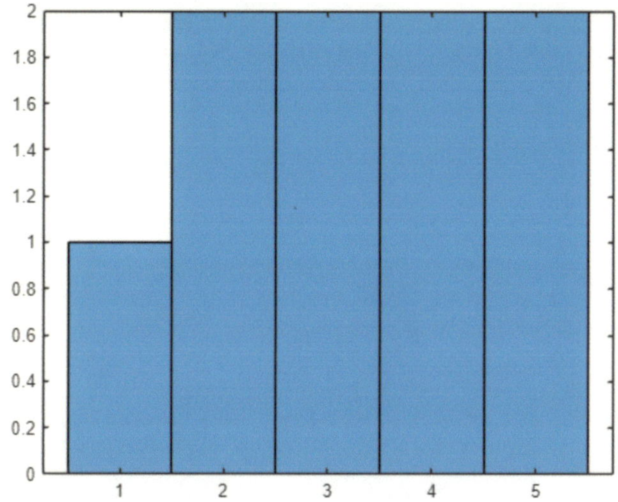

Fig. 11.10 Histogram using vertically concatenated column vectors (cf. Fig. 11.9)

Here, we create a histogram by combining the variables 「Mpop」 and 「Wpop」 along the column direction. See Fig. 11.11.

```
histogram([Mpop;Wpop]);
```

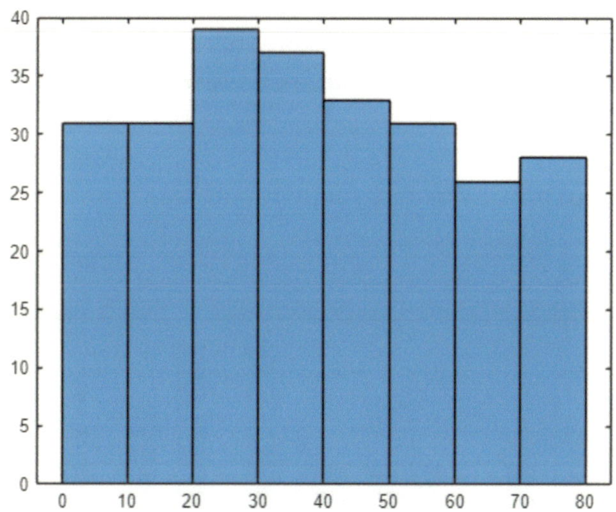

Fig. 11.11 Histogram using merged column vectors of two variables (cf. Fig. 11.8)

▶ Example of Table Variables

Here, using the **table** function with the variables 「Mpop」 and 「Wpop」, we create the table variable 「MWpop」. Then, we create a histogram of 「Mpop」. See Fig. 11.12.

```
MWpop = table(Mpop,Wpop);
histogram(MWpop.Mpop);
```

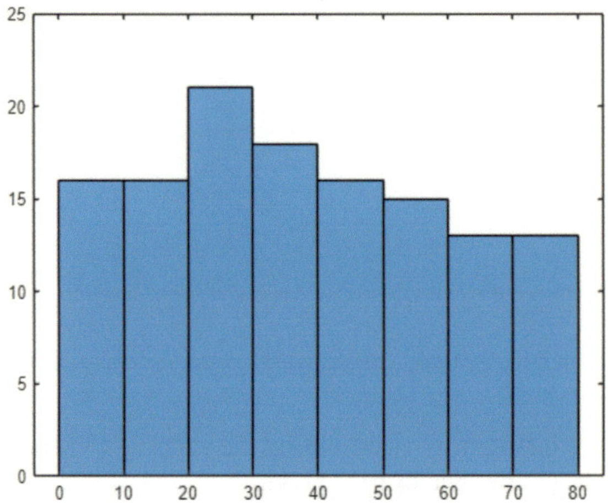

Fig. 11.12 Use of a table variable: Histogram

◀

Here, we specify the number of bins used in the histogram. More bins result in a histogram with a finer division of the data. Here, we set 「nbins」 to 20. The resulting figure is illustrated in Fig. 11.13.

```
nbins = 20;
histogram(Mpop,nbins);
```

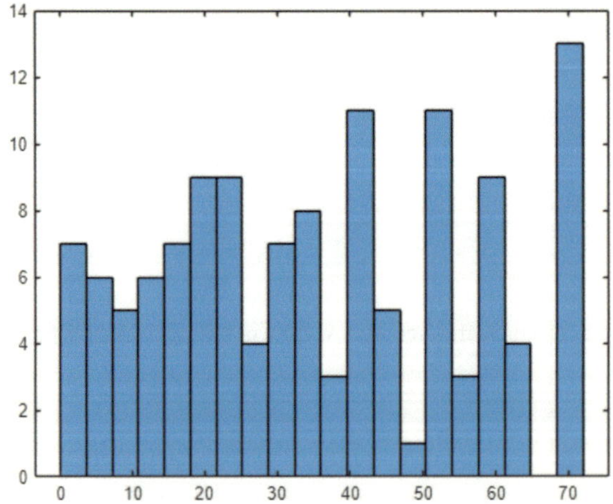

Fig. 11.13 Histogram using fine bin division

Here, we set the number of bins to 20 in the second argument. See ◘ Fig. 11.14 for the created figure.

```
histogram([Mpop;Wpop],20);
```

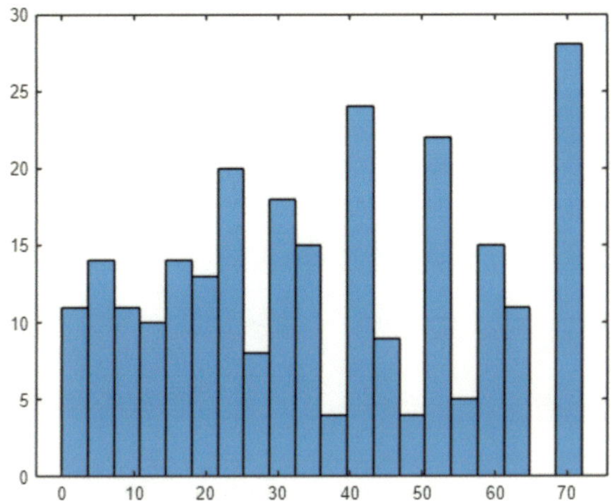

◘ **Fig. 11.14** Histogram using fine data division and merged column vectors

Here, the variable 「edges=[0 20:5:60 80]」 is created with specified bin edges. In other words, the range from 0 to 20 is treated as a single bin. Next, the range from 20 to 60 is divided into increments of 5. Finally, the range from 60 to 80 is treated as the last bin. See ◘ Fig. 11.15.

```
edges = [0 20:5:60 80];
histogram(Mpop,edges);
```

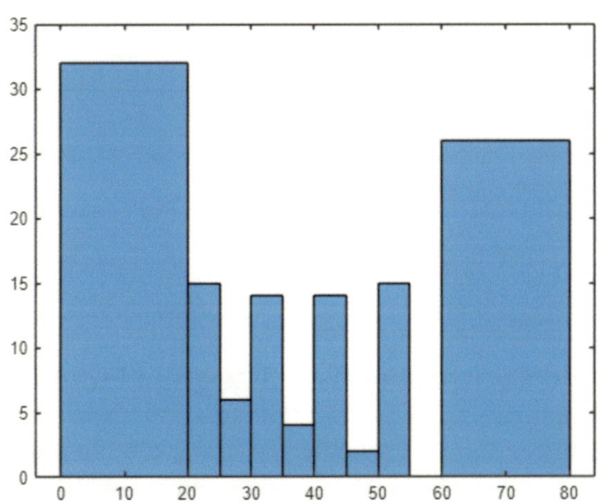

◘ **Fig. 11.15** Histogram with wide–narrow–wide bin intervals

Here, the variable 「edges」 includes a negative value. See Fig. 11.16.

```
edges = [-50 0:10:80 100];
histogram(Mpop,edges);
```

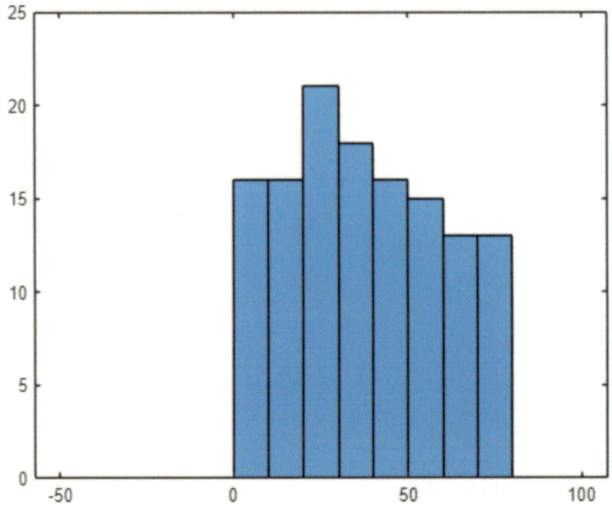

Fig. 11.16 Histogram with adjusted bin edges including a negative interval

11

Attention

Error Program: An error occurs because you cannot specify both the number of bins and the bin edges at the same time. For example, specifying 「edges」 as the third argument, as shown below, results in an error. The error message is presented in Fig. 11.17.

```
%%% Error program
%nbins = 50;
%edges = [0 20:5:60 80];
%histogram([Mpop,Wpop],nbins,edges);
```

```
Error using histogram>parseinput (line 219)
Incorrect number of input arguments. Each parameter name must be followed by a corresponding value.

Error in histogram (line 50)
[opts,passthrough,dispatchToCategorical] = parseinput(args,firstaxesinput);
```

Fig. 11.17 Error output of the histogram function: Simultaneous use of both number of bins and bin edges

11.4 Setting and Controlling the Histogram Function

Here, we create a histogram by explicitly specifying the bin edges and bin counts. 「BinEdges」 is set to values from 0 to 4. Additionally, 「BinCounts」 is set to 「[10 100 50 40]」, which represents the counts for each corresponding bin defined by the edges. See Fig. 11.18.

```
histogram("BinEdges",0:4,"BinCounts",[10 100 50 40]);
```

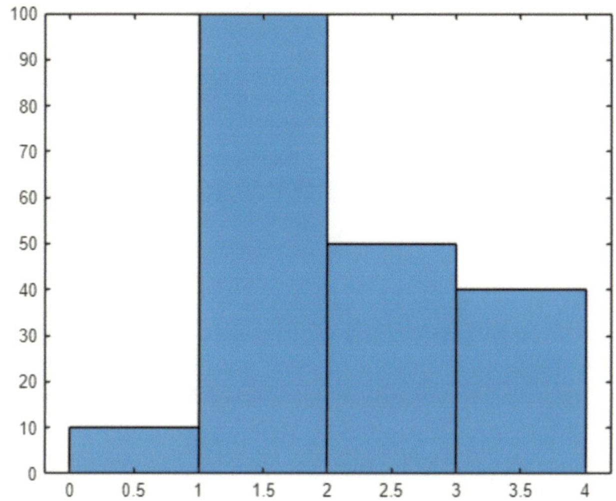

Fig. 11.18 Histogram with manually specified bin ranges and frequencies

Attention

Error Program: The following program will result in an error because the number of bin edges does not match the number of bin counts. In the command 「histogram(" BinEdges",0:10,"BinCounts", [10 100 50 40])」, there are 11 bin edges (from 0 to 10), but only 4 bin counts are provided. The error message is shown in Fig. 11.19.

```
%%% Error program
% histogram("BinEdges",0:10,"BinCounts",[10 100 50 40]);
```

```
Error using histogram>parseinput (line 376)
'BinCounts' is not supported if any of 'NumBins','BinWidth', 'BinMethod', or 'BinLimits' is sp

Error in histogram (line 50)
[opts,passthrough,dispatchToCategorical] = parseinput(args,firstaxesinput);
```

Fig. 11.19 Error output of the histogram function: Mismatch between number of bin edges and bin counts

Here, 「BinEdges」 is set to the values from -10 to 0 using 「-10:0」. Additionally, 「BinCounts」 is set to 「[1 2 3 4 5 6 7 8 9 10]」. This means that there are 10 bins, and the counts for each bin are given by the corresponding values in the 「BinCounts」 vector. See Fig. 11.20.

```
histogram("BinEdges",-10:0,"BinCounts",[1 2 3 4 5 6 7 8 9 10]);
```

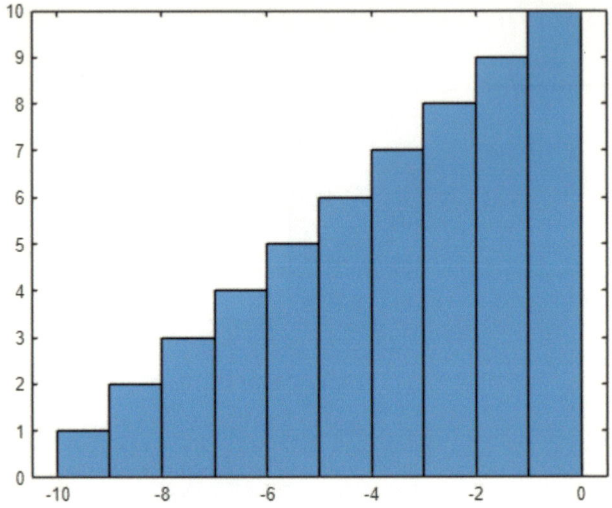

Fig. 11.20 Histogram with negative bin edges

Here, the properties of the **histogram** function are stored in the variable 「hist」. Then, 「FaceAlpha」 is set to 0.1 and 「FaceColor」 is set to 「red」, creating a semi-transparent red histogram. See Fig. 11.21.

11

```
hist = histogram("BinEdges",-10:0,"BinCounts",[1 2 3 4 5 6 7 8 9 10]);
    hist.FaceAlpha = 0.1;
    hist.FaceColor = "red";
```

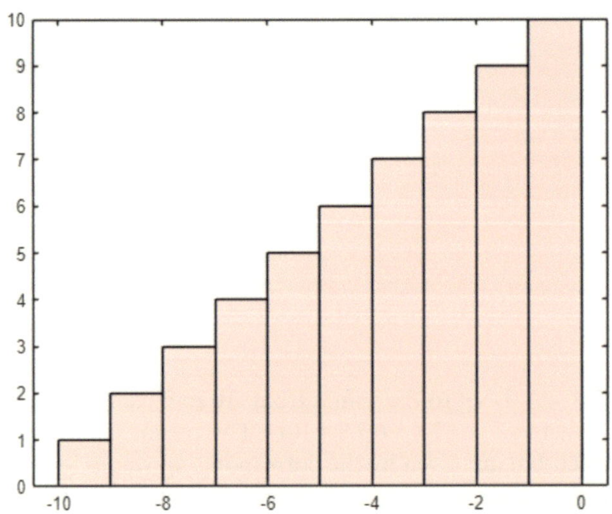

Fig. 11.21 Histogram with negative bin edges and semi-transparent red bins

Here, we specify the range of the x-axis. Specifically, we set 「BinLimits」 to 「[40,50]」, which means that the range is between 40 and 50. Note that when using 「BinLimits」, we must specify a two-element vector such as 「[40,50]」. See Fig. 11.22.

```
histogram(Mpop,"BinLimits",[40,50]);
```

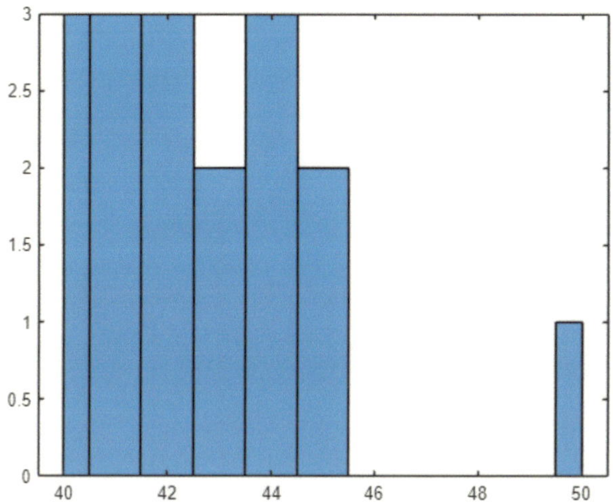

Fig. 11.22 Histogram using "BinLimits" property to restrict range

In the following program, we create a histogram using the data from the concatenated variables 「Mpop」 and 「Wpop」. The number of bins is set to 50. The x-axis range is set to 「[0, 100]」 using 「BinLimits」, meaning that only values between 0 and 100 are included in the histogram. See Fig. 11.23.

```
histogram([Mpop;Wpop],50,"BinLimits",[0,100]);
```

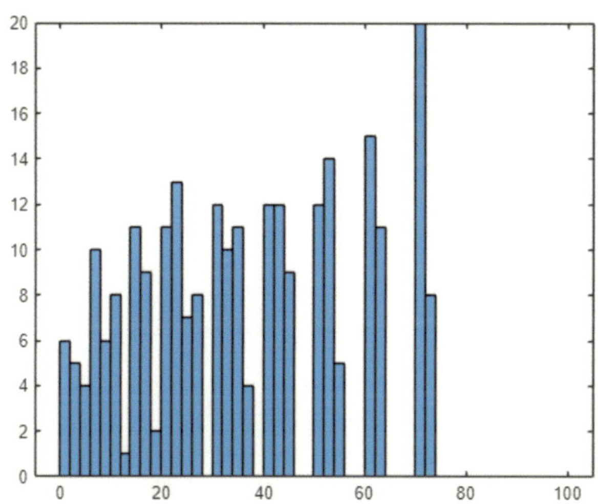

Fig. 11.23 Histogram using "BinLimits" property to expand range

Here, we set 「FaceColor」 to 「w (=white)」, which fills the bars with white. Additionally, we set 「EdgeColor」 to 「r (=red)」, making the outline color red. See ◘ Fig. 11.24 for the resulting figure.

```
histogram(Mpop,"FaceColor","w","EdgeColor","r");
```

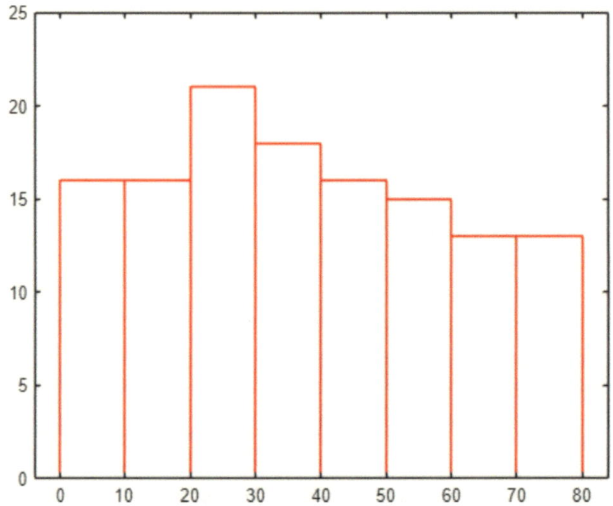

◘ **Fig. 11.24** Histogram with white bins and red edges

Here, the properties of the **histogram** function are stored in the variable 「hist」. Then, we set 「FaceColor」 to 「white」, 「EdgeColor」 to 「blue」, and 「LineWidth」 to 8, creating a white histogram with a thick blue outline. See ◘ Fig. 11.25 for the created figure.

```
hist = histogram(Mpop);
    hist.FaceColor = "white";
    hist.EdgeColor = "blue";
    hist.LineWidth = 8;
```

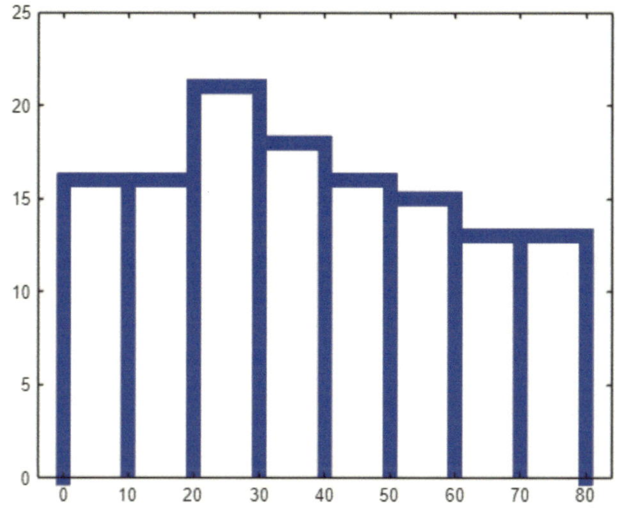

Fig. 11.25 Histogram with white bins and thick blue edges

In addition to the previous program that created Fig. 11.25, we set 「Bin Width」 to 15, making the width of the histogram bars wider. See Fig. 11.26.

```
hist = histogram(Mpop);
    hist.FaceColor = "white";
    hist.EdgeColor = "blue";
    hist.LineWidth = 8;
    hist.BinWidth = 15;
```

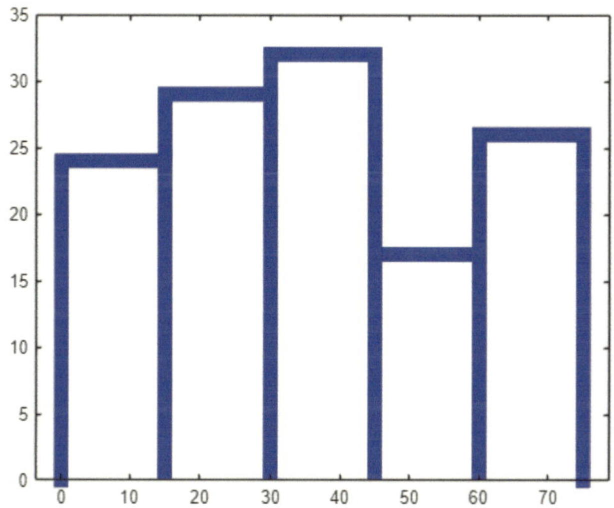

Fig. 11.26 Histogram with white large bins and thick blue edges

Here, we set 「DisplayStyle」 to 「stairs」, which creates only the outline of the histogram without filling the bars. See Fig. 11.27.

```
histogram(Mpop,"DisplayStyle","stairs");
```

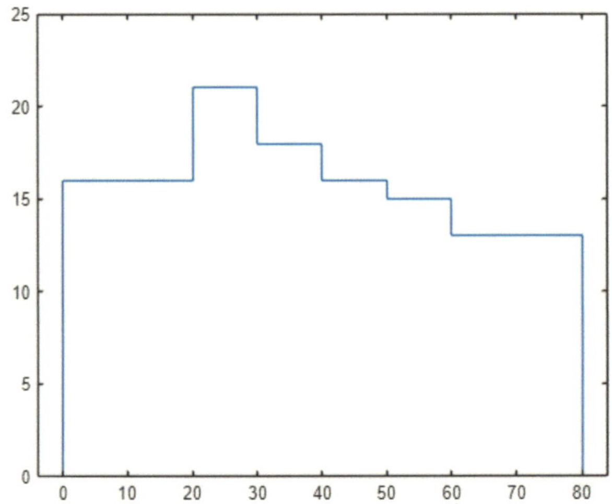

Fig. 11.27 Histogram with white bins and no internal edges

In the following program, we set 「DisplayStyle」 to 「stairs」, creating only the outline of the histogram without filling the bars. Additionally, we set 「LineStyle」 to 「:」 and 「LineWidth」 to 3, creating a wider, dotted outline. See Fig. 11.28.

```
hist = histogram(Mpop,"DisplayStyle","stairs");
    hist.LineStyle = ":";
    hist.LineWidth = 3;
```

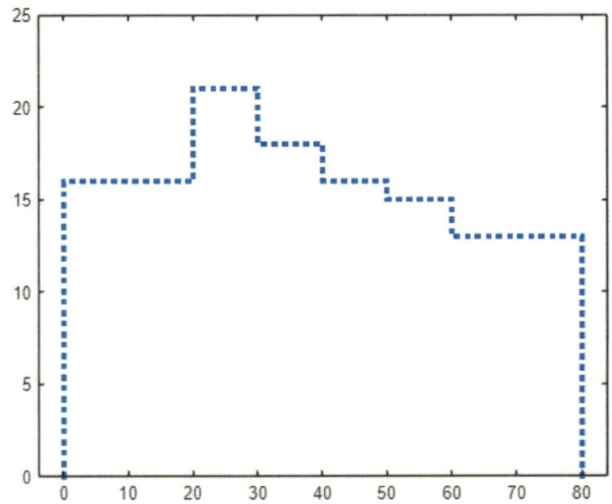

Fig. 11.28 Histogram with white bins, no internal edges, and dotted border

We specify the method for binning the data in the histogram. Here, we set 「Bin Method」 to 「integers」, which creates bins for each integer value in the data. See ◘ Fig. 11.29.

```
histogram(Mpop,"BinMethod","integers");
```

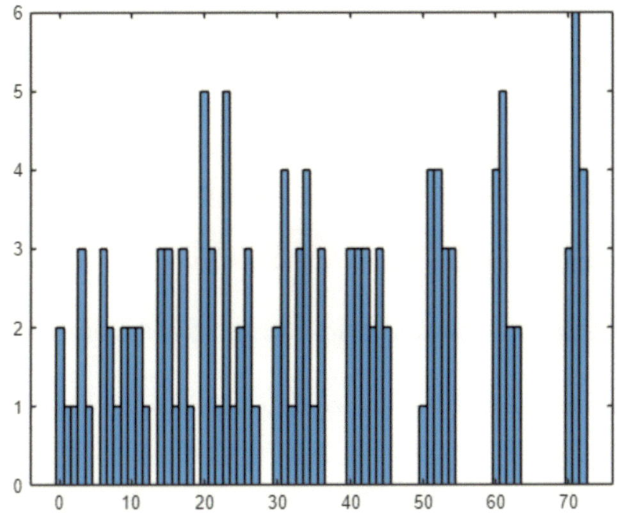

◘ **Fig. 11.29** Histogram with integer bins

Here, we specify the method for binning the data in the histogram. Specifically, we set 「BinMethod」 to 「scott」, which uses Scott's rule to automatically determine an appropriate bin width based on the size of the data. Additionally, we set 「FaceColor」 to 「red」, 「EdgeColor」 to 「blue」, and 「LineWidth」 to 3, creating a red histogram with a thick blue outline. See ◘ Fig. 11.30.

```
hist = histogram(Mpop,"BinMethod","scott");
    hist.FaceColor = "red";
    hist.EdgeColor = "blue";
    hist.LineWidth = 3;
```

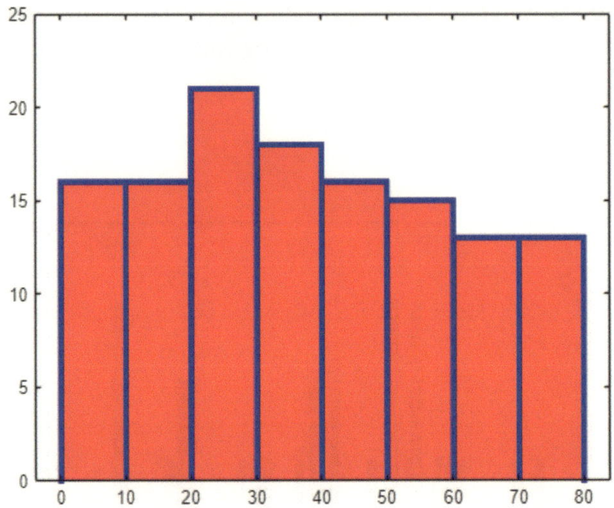

Fig. 11.30 Histogram using "BinMethod" property set to "scott" for automatic bin width

Here, 「Orientation」 is set to 「horizontal」, which creates a horizontal bar chart. See Fig. 11.31.

```
histogram(Mpop,"Orientation","horizontal");
```


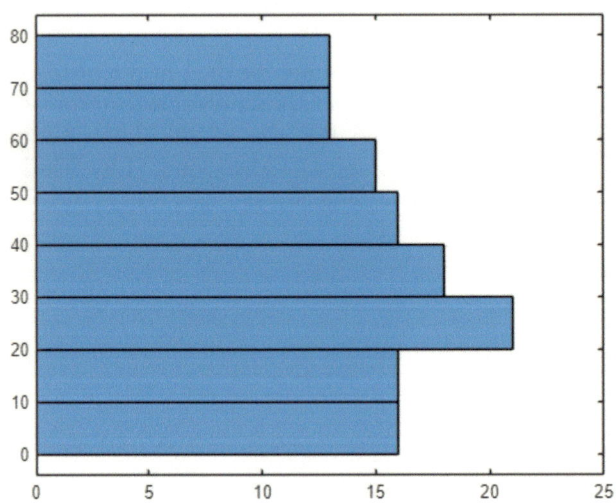

Fig. 11.31 Horizontal histogram

By setting ⌜Orientation⌟ to ⌜horizontal⌟, we create a horizontal histogram using the **histogram** function. The properties of the **histogram** function are stored in the variable ⌜hist⌟. After that, we set ⌜FaceColor⌟ to ⌜red⌟, ⌜EdgeColor⌟ to ⌜blue⌟, and ⌜LineWidth⌟ to 2, creating a red histogram with a thick blue outline. See ◘ Fig. 11.32.

```
nbins = 5;
hist = histogram(Mpop,nbins,"Orientation","horizontal");
    hist.FaceColor = "red";
    hist.EdgeColor = "blue";
    hist.LineWidth = 2;
```

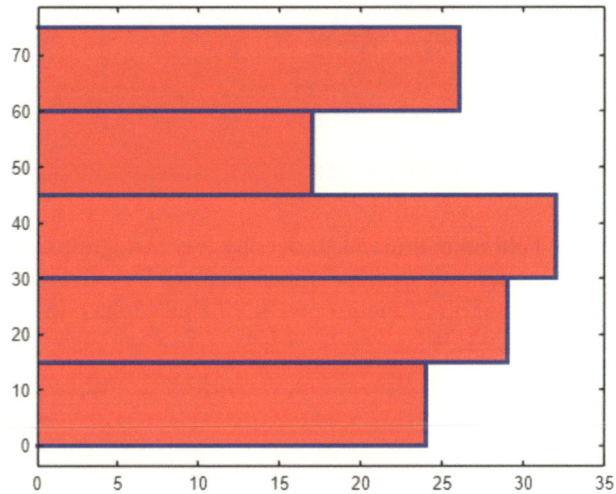

◘ **Fig. 11.32** Horizontal histogram with red bins and blue edges

11.5 Combining Optional Functions and the Histogram Function

Here, we use the **hold on** command to overlay another histogram (⌜histogram(Wpop) ⌟) on the existing histogram created by ⌜histogram(Mpop)⌟. This enables us to observe two sets of bars on the same axis. We also use the **legend** function to create a legend that shows the correspondence between colors and histograms. See ◘ Fig. 11.33 for the created figure.

```
histogram(Mpop);
hold on
histogram(Wpop);
hold off
legend("Men","Women","Location","northeast");
```

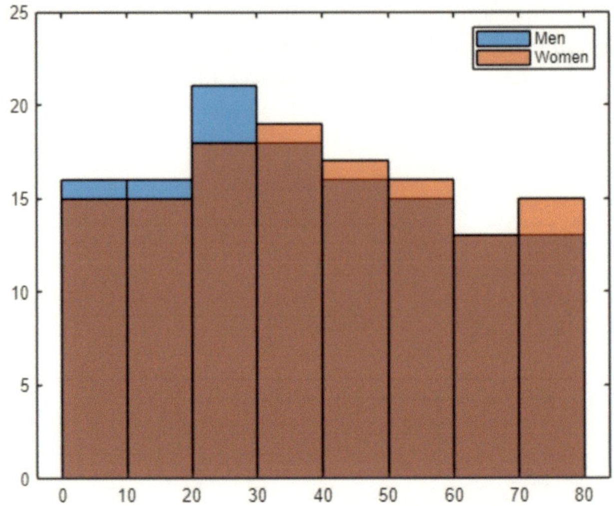

Fig. 11.33 Overlaid histograms with a legend

Here, we use the **hold on** command to overlay two histograms. Additionally, for the first histogram function, we specify 「LineStyle」 as 「--」 to create a dashed outline. For the second histogram function, we set 「FaceAlpha」 to 0.2 to adjust the transparency. See Fig. 11.34.


```
histogram(Mpop,"LineStyle","--");
hold on
histogram(Wpop,"FaceAlpha",0.2);
hold off
legend("Men","Women","Location","northeast");
```

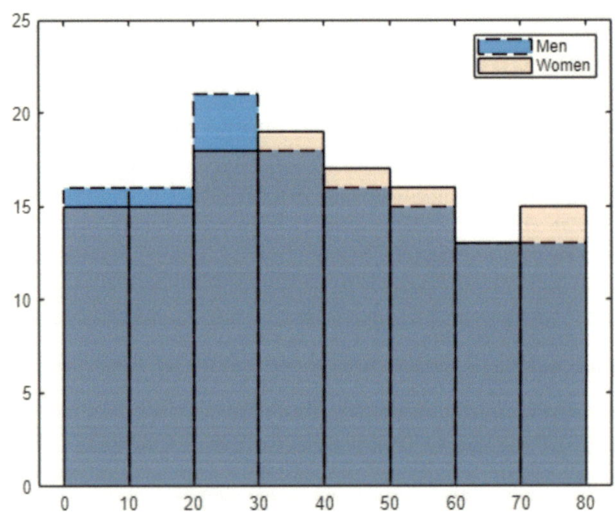

Fig. 11.34 Overlaid histograms with dashed edges and a legend

In the following program, we first create a histogram for 「Mpop」 with 15 bins and store its properties in the variable 「h1」. Then, we use the **hold on** command to overlay another histogram for 「Wpop」 with 15 bins. Additionally, we set 「Face Alpha」 to 0.2 and 「FaceColor」 to 「green」, creating a green histogram with high transparency.

By using the **legend** function, a legend is added with the labels 「Men」 and 「Women」, where 「Location」 is set to 「Southwest」. Finally, we activate the fine grid with the **grid minor** command. See Fig. 11.35.

```
h1 = histogram(Mpop,15);
hold on
histogram(Wpop,15,"FaceAlpha",0.2);
hold off
    h1.FaceColor = "green";
legend("Men","Women","Location","Southwest");
grid minor
```

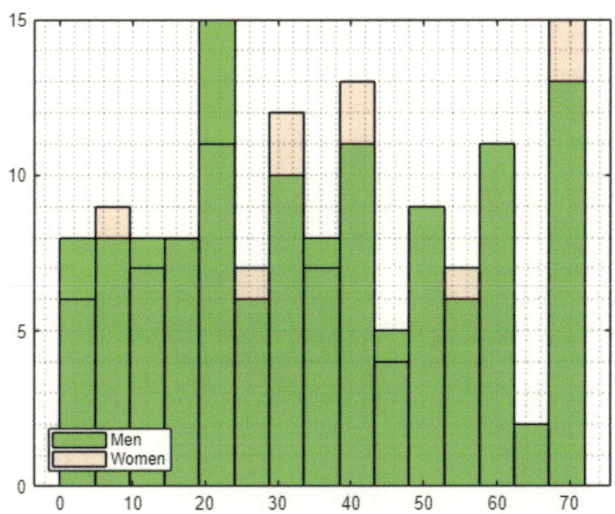

Fig. 11.35 Overlaid histograms with fine gridlines and a legend (bottom-left position)

Here, we specify 「tiledlayout(1,3)」. Then, we create three histograms side by side. Specifically, in the second and third **histogram** functions, we set 「Normalization」 to 「percentage」 and 「probability」, respectively. As a result, although the same 「Mpop」 data is used in the **histogram** functions, the vertical axis scales differ for each histogram. See Fig. 11.36.

```
ti = tiledlayout(1,3);
title(ti,"Type of normalization");
nexttile;
histogram(Mpop);
ylabel("count");
grid on
nexttile;
histogram(Mpop,"Normalization","percentage");
ylabel("percentage");
grid on
nexttile;
histogram(Mpop,"Normalization","probability");
ylabel("probability");
grid on
```

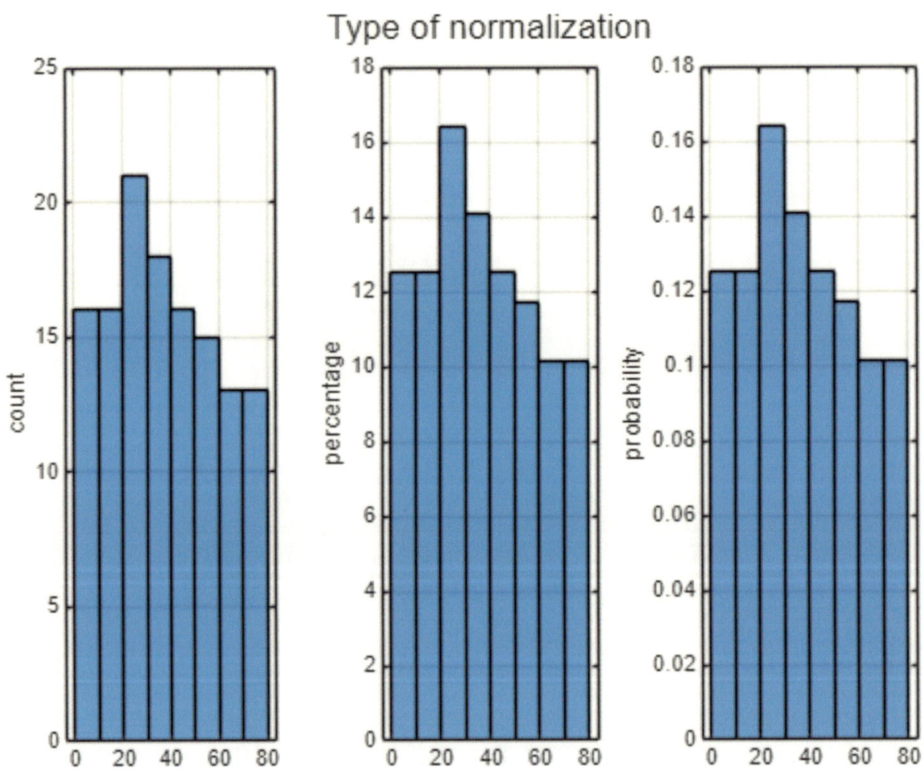

Fig. 11.36 Histogram (3-row layout) with different settings of "Normalization" properties

11

Here, we use the **hold on** command to overlay the figures generated by the **histogram** and **plot** functions. Specifically, the **histcounts** function is used to return the histogram bin counts and bin edges for the data in 「Mpop」, storing the counts in the variable 「N」 and the edges in the variable 「edges」. Then, a new variable 「x」 is created by adding 5 to each value in 「edges」. Finally, the **plot** function is used to create a line plot with circles at each data point, plotting 「x」 on the x-axis and 「N」 on the y-axis. The resulting figure is shown in Fig. 11.37.

```
figure
histogram(Mpop,"FaceAlpha",0.2);
hold on
[N,edges] = histcounts(Mpop);
x = edges + 5;
pl = plot(x(1:end-1),N,"-o");
    pl.MarkerSize = 10;
    pl.MarkerFaceColor = "red";
hold off
```

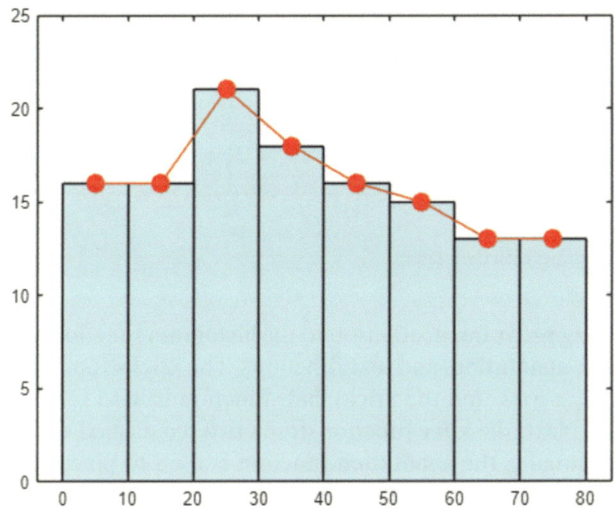

Fig. 11.37 Histogram overlaid with a red line connecting the data points

As in the previous program that created Fig. 11.37, the **histcounts** function returns the histogram bin counts and bin edges for the data in ⌜Mpop⌟, storing the counts in ⌜N⌟ and the edges in ⌜edges⌟. Additionally, ⌜bin_centers⌟ is calculated by taking the edges from the first to the second-to-last element (⌜edges(1:end-1)⌟) and adding half the width of the bin (⌜diff(edges)/2⌟). This gives the center points of each bin.

Finally, the **text** function and a **for** loop are used to place the values of ⌜N(i) ⌟ one by one, from ⌜i=1⌟ to ⌜length(N) (= 8)⌟. For more details of the **text** function, see ▶ Sect. 18.2. The resulting figure is illustrated in Fig. 11.38.

```
histogram(Mpop,"FaceAlpha",0.2);
[N,edges] = histcounts(Mpop);
bin_centers = edges(1:end-1)+diff(edges)/2;
hold on
for i = 1:length(N)
    tx = text(bin_centers(i),N(i),num2str(N(i)));
        tx.HorizontalAlignment = "center";
        tx.VerticalAlignment = "bottom";
        tx.FontSize = 15;
end
hold off
```

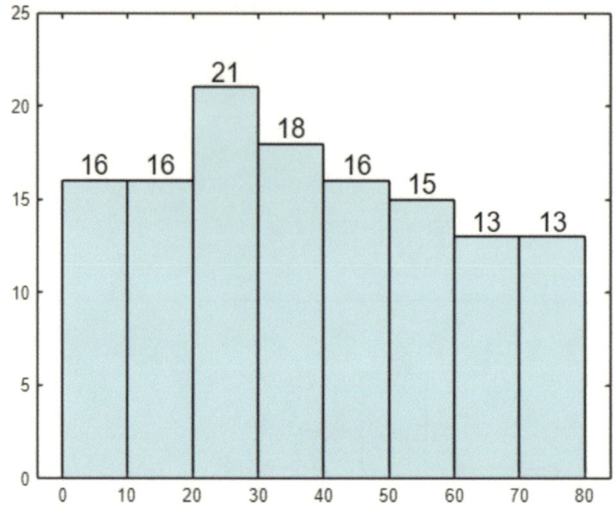

Fig. 11.38 Histogram with data labels

In the following program, in addition to the **histogram** function, we use the **xticks**, **xticklabels**, **xline, annotation**, and **text** functions. The **xticks** function is used to place tick marks on the x-axis, and the **xticklabels** function is used to place text labels on each of the ticks. Next, the **xline** function draws two red dashed vertical lines at x=40 and x=60. Additionally, the **annotation** function is used to place a double arrow on the figure by specifying 「doublearrow」. Finally, the **text** function places the text 「41-60 years old」 on the figure. For more details on the **annotation**, **text**, and **xline** functions, see ▶ Sects. 18.1, 18.2, and 18.3. Figure 11.39 shows the created figure.

11

```
hist = histogram(Mpop);
    hist.FaceColor = "red";
    hist.FaceAlpha = 0.2;
xticks([5 15 25 35 45 55 65 75]);
xticklabels({"0-10","11-20","21-30","31-40","41-50",...
    "51-60","61-70","71-80"});
xlabel("AGE");
hold on
xline(40,"--r","LineWidth",2);
xline(60,"--r","LineWidth",2);
xx = [0.52 0.69];
yy = [0.85 0.85];
annotation("doublearrow",xx,yy);
text(41,21,"41-60 years old");
hold off
```

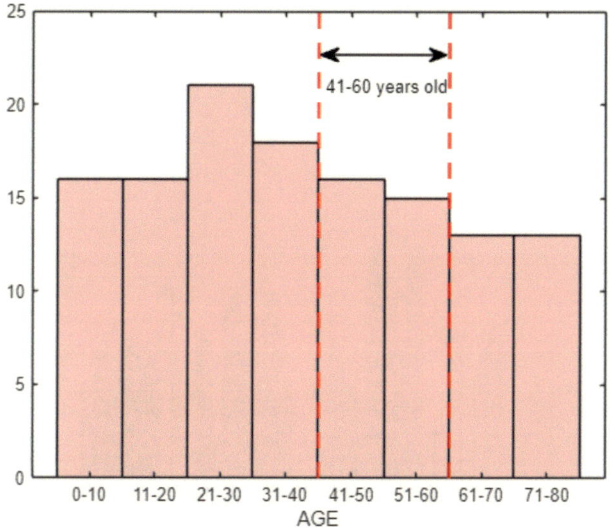

Fig. 11.39 Histogram with two vertical dashed lines and a double arrow

Here, we specify the bin edges as 20 to 50 for the first histogram using ⌜h1.BinEdges=20:50⌟. Alternatively, for the second histogram, we specify the bin edges as 10 to 40 using ⌜h2.BinEdges=10:40⌟. This means that the range from 20 to 40 is the overlapping area between the first and second histograms. We use the **xline** function to draw boundary lines at x=20 and x=40. Additionally, we use different line styles, such as dash-dotted and dashed lines. The text annotations are positioned to the left or within the boundary lines to indicate the ranges. For more details on the **xline** function, see ▶ Sect. 18.3. See Fig. 11.40 for the created figure.

```
h1 = histogram(Mpop);
hold on
h2 = histogram(Wpop,"FaceAlpha",0.1);
hold off
    h1.BinEdges = 20:50;
    h2.BinEdges = 10:40;
xline(20,"--r",{"The overlapping","area"},"LineWidth",2);
x1 = xline(40,"-.b");
    x1.LabelHorizontalAlignment = "center";
ylim([0,10]);
```

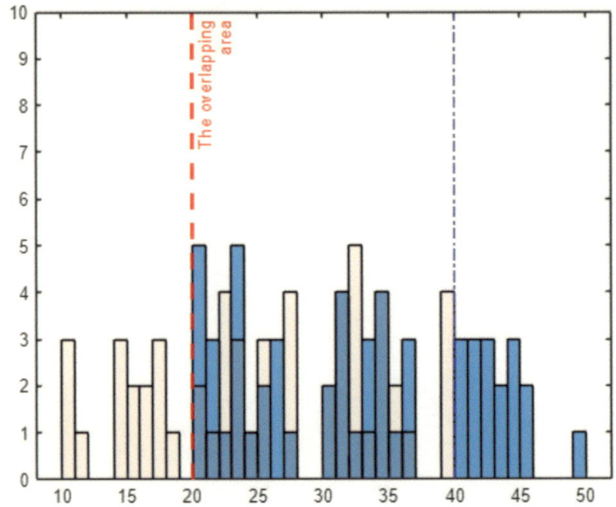

Fig. 11.40 Overlaid histograms with vertical dashed and dash-dotted lines

We create a figure similar to Fig. 11.39 using the **yline** and **annotation** functions. The detailed explanation can be found in the previous programs that created Figs. 11.39 and 11.40. Here, we set 「Orientation」 to 「horizontal」, creating a horizontal histogram. Additionally, we use the **hold on** command to overlay two histograms. See Fig. 11.41.

11

```
h1 = histogram(Mpop,"Orientation","Horizontal");
hold on
h2 = histogram(Wpop,"FaceAlpha",0.1,"Orientation","Horizontal");
    h1.BinEdges = 20:50;
    h2.BinEdges = 10:40;
yline(20,":r","LineWidth",4);
yline(40,"-b");
xlim([0,10]);
xx = [0.75 0.75];
yy = [0.35 0.7];
anta = annotation("doublearrow",xx,yy);
    anta.LineWidth = 5;
grid minor
hold off
```

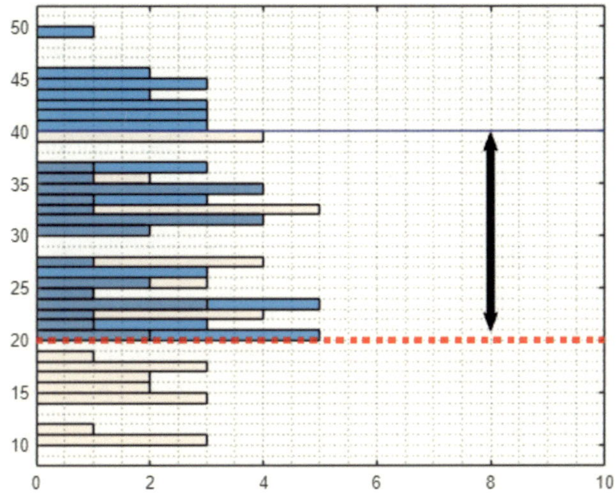

Fig. 11.41 Overlaid horizontal histograms with two horizontal lines and a double arrow

Here is a program that uses the **histogram** function to handle categorical data. Categorical data refers to data that is represented by categories, such as young and old, rather than numerical values. In the following program, the **discretize** function is used to categorize the data in 「Mpop」 into two groups. Concretely, the two categories are labeled as 「young」 and 「old」. Consequently, the data in 「Mpop」 is divided into these two categories. The created figure is shown in Fig. 11.42.

```
C = discretize(Mpop,2,"categorical",["young","old"]);
histogram(C);
```

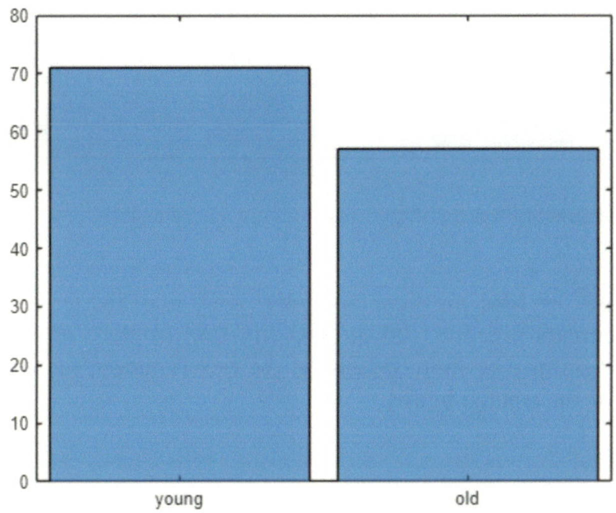

Fig. 11.42 Histogram with two categorical age groups

Here, using the **discretize** function, we create three groups: 「young」, 「middle-aged」, and 「old」. Then, we create a bar chart using the **histogram** function. Additionally, we use the **countcats** function to count the number of data points in each of the three groups, which is stored in the variable 「N」. Finally, we use the **text** function to place the numbers above the bars in the chart. For more details on the **text** function, see ▶ Sect. 18.2. See ◘ Fig. 11.43 for the resulting figure.

```
C = discretize(Mpop,3,"categorical",["young","middle-aged","old"]);
histogram(C);
hold on
N = countcats(C);
for i = 1:length(N)
    tx = text(i,N(i),num2str(N(i)));
        tx.HorizontalAlignment = "center";
        tx.VerticalAlignment = "bottom";
        tx.FontSize = 15;
end
title("Categorical histogram: Australian male");
hold off
```

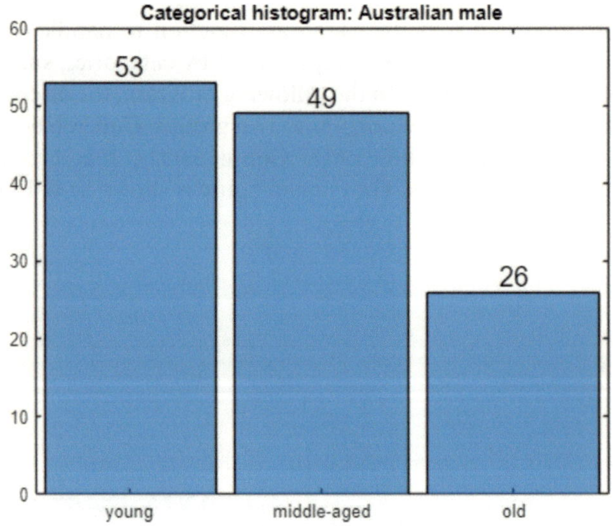

◘ **Fig. 11.43** Histogram with three categorical age groups and data labels

Here, we use the **hold on** command to overlay two histograms. Additionally, we use the **countcats** function and the **text** function to place numbers above each bar in the histograms. For more details on the **text** function, see ▶ Sect. 18.2. See ◘ Fig. 11.44 for the created figure.

```
C1  = discretize(Mpop,2,"categorical",["young","old"]);
hist1 = histogram(C1);
    hist1.FaceAlpha = 0.1;
hold on
C2  = discretize(Wpop,2,"categorical",["young","old"]);
hist2 = histogram(C2);
    hist2.LineStyle = "--";
    hist2.FaceAlpha = 0.2;
N1  =  countcats(C1);
N2  =  countcats(C2);
for i = 1:length(N1)
    tx1 = text(i, N1(i), num2str(N1(i)));
        tx1.HorizontalAlignment = "center";
        tx1.VerticalAlignment = "bottom";
    tx2 = text(i, N2(i), num2str(N2(i)),"Color","red");
        tx2.HorizontalAlignment = "center";
        tx2.VerticalAlignment = "bottom";
end
title("Categorical histogram: Australian male and female");
legend("Male","Female");
hold off
```

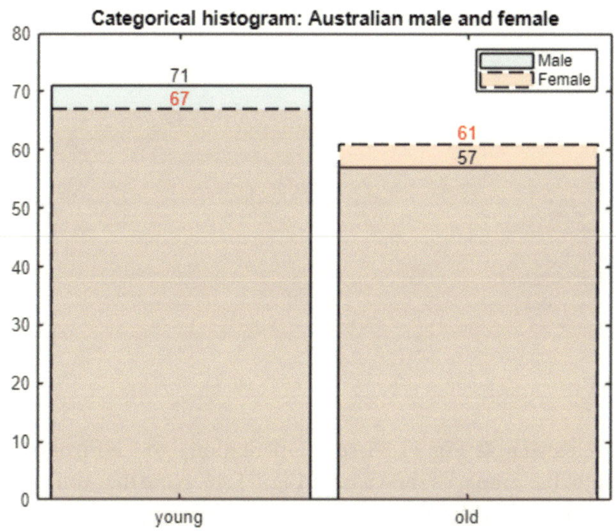

Fig. 11.44 Overlaid histograms with two categorical age groups and data labels

11.6 Questions

Question 1

Create Fig. 11.45 using the **hold on** command. Specifically, after creating the histogram for the variable ⌜Mpop⌟, overlay the histogram for the variable ⌜Wpop⌟. Store the properties of the **histogram** function created for ⌜Wpop⌟ in the variable ⌜Ex1⌟. Then, set the outline to a thick dashed line. Additionally, increase the transparency of the bars and fill them with blue.

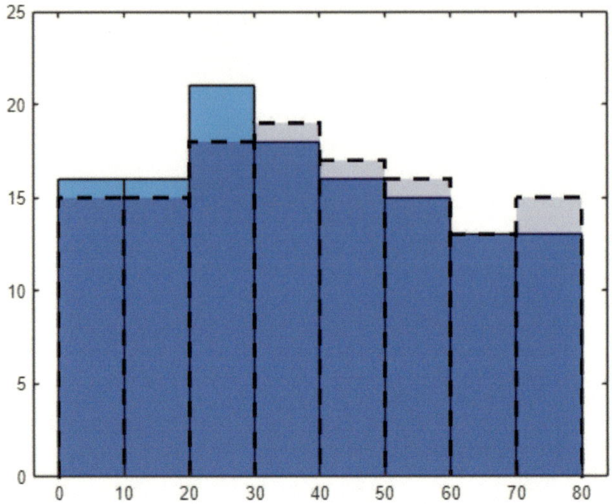

Fig. 11.45 Histogram (Question 1)

Answer 1

```
figure
histogram(Mpop);
hold on
Ex1 = histogram(Wpop);
    Ex1.LineStyle = "--";
    Ex1.LineWidth = 2;
    Ex1.FaceColor = "blue";
    Ex1.FaceAlpha = 0.2;
hold off
```

11

Question 2

For the figure created in Fig. 11.38, set 「Orientation」 to 「horizontal」 and change it to a horizontal bar graph, as shown in Fig. 11.46. Note that you need to place the corresponding values to the right of each bar in the bar graph.

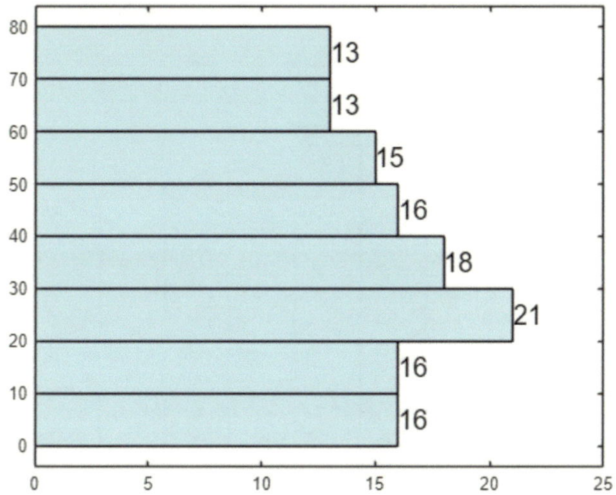

Fig. 11.46 Histogram (Question 2)

Answer 2

```
figure
histogram(Mpop,"Orientation","horizontal","FaceAlpha",0.2);
[N,edges] = histcounts(Mpop);
bin_centers = edges(1:end-1)+diff(edges)/2;
hold on
for i = 1:length(N)
    tx = text(N(i),bin_centers(i),num2str(N(i)));
    tx.HorizontalAlignment = "left";
    tx.VerticalAlignment = "middle";
    tx.FontSize = 15;
end
hold off
```

Question 3

Using the variables 「Mpop」 and 「Wpop」, create Fig. 11.47 which overlays two histograms. Then, as shown in the figure below, use the **plot** function to overlay the lines connecting the heights of each bar. Finally, set 「MarkerSize」 to 10, and keep the color of the lines as the default.

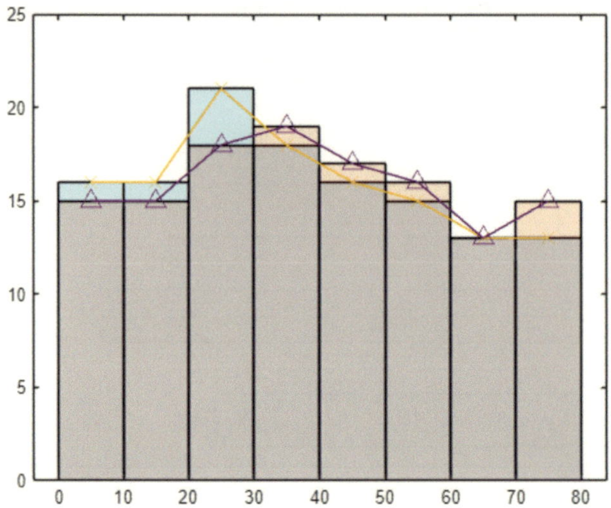

Fig. 11.47 Histogram (Question 3)

Answer 3

```
figure
histogram(Mpop,"FaceAlpha",0.2);
hold on
histogram(Wpop,"FaceAlpha",0.2);
[NM,edgesM] = histcounts(Mpop);
[NW,edgesW] = histcounts(Wpop);
x = edges + 5;
plM = plot(x(1:end-1),NM,"-x");
    plM.MarkerSize = 10;
plW = plot(x(1:end-1),NW,"-^");
    plW.MarkerSize = 10;
hold off
```

Histogram2 Function

Contents

12.1 Details of the Data Used in this Chapter – 376

12.2 Excel Data of ▶ Section 12.1 – 377

12.3 Basic Syntax of the Histogram2 Function – 378

12.4 Setting and Controlling the Histogram2 Function – 384

12.5 Combining Optional Functions and the Histogram2 Function – 389

12.6 Questions – 403

Supplementary Information The online version contains supplementary material available at https://doi.org/10.1007/978-981-96-6213-5_12.

Y. Nakamoto, *A Comprehensive Guide to Creating Graphics with MATLAB Programming*, Classroom Companion: Economics, https://doi.org/10.1007/978-981-96-6213-5_12

Learning Objectives

This chapter creates 3D histograms using the **histogram2** function. By reading this chapter, you will learn how to use the **histogram2** function. In particular, ▶ Sect. 12.5 teaches you how to create 2D tiles that effectively enhance data understanding. Moreover, you will understand the similarities and differences in the created figures, syntax, and the type of data between the **histogram2** function and the **histogram** function (covered in ▶ Chap. 11).

Overview

In this chapter, we use the **histogram2** function to create a 3D histogram. The difference between the **histogram2** function and the **histogram** function (▶ Chap. 11) is that the **histogram2** function shows the distribution based on two variables at the same time, while the **histogram** function uses one variable. Specifically, the **histogram2** function uses both the horizontal and vertical axes to specify the ranges of the two variables, and the height of the bars corresponds to the frequency or density. Consequently, the **histogram2** function may potentially reveal complex relationships that a one-dimensional histogram cannot. This is the advantage of the **histogram2** function. Alternatively, a disadvantage is that the increase in dimensions may make the figures more difficult to understand visually than a one-dimensional histogram. This disadvantage is the same as for 3D plots. In such cases, it may be helpful to use the optional function. Refer to ▶ Sect. 12.5 for details.

12

12.1 Details of the Data Used in this Chapter

This chapter uses 2010 and 2020 data on GDP per capita (expressed in current US dollars) and population for African countries. Note that countries for which either 2010 or 2020 data were missing are excluded. The variable 「country」 represents the name of the countries, 「gdp_2010」 represents the GDP per capita for 2010, 「gdp_2020」 represents the GDP per capita for 2020, 「pop_2010」 is the population for 2010, and 「pop_2020」 is the population for 2020. Note that 「pop_2010」 and 「pop_2020」 are rounded to the nearest thousand, and that 「gdp_2010」 and 「gdp_2020」 are rounded to the nearest tenth and displayed as integers.

```
country = ["Algeria","Angola","Benin","Botswana","Burkina Faso",...
    "Burundi","Cameroon","Cape Verde","Chad","Comoros","Congo",...
    "Djibouti","Egypt","Eswatini","Ethiopia","Gabon","Gambia",...
    "Ghana","Guinea","Guinea-Bissau","Kenya","Lesotho","Liberia",...
    "Libya","Madagascar","Malawi", "Mali","Mauritania","Mauritius",...
    "Morocco","Mozambique","Namibia","Niger","Nigeria","Rwanda",...
    "Senegal","Seychelles","Sierra Leone","South Africa","Sudan",...
    "Tanzania","Togo","Tunisia","Uganda","Zambia","Zimbabwe"]';

gdp_2010 = [4481,3587,1010,6042,628,232,1353,3164,895,...
    1323,3598,1837,2923,4264,341,8933,861,1757,631,644,...
    1176,1222,505,12478,493,623,689,1716,8108,3135,472,...
    5411,476,2328,612,1273,10805,402,8130,1637,732,723,...
    4343,809,1469,976]';

pop_2010 = [35856000,23364000,9446000,2092000,...
    16117000,9127000,19878000,492000,11895000,656000,...
    4438000,919000,87252000,1100000,89238000,1711000,...
    1937000,25575000,10271000,1567000,41518000,2023000,...
    4020000,6492000,21731000,14718000,1529000,3419000,...
    1283000,32465000,23074000,2099000,16648000,160953000,...
    10309000,12530000,92000,6437000,51785000,33740000,...
    45111000,6572000,10895000,32342000,13792000,12840000]';

gdp_2020 = [3322,1709,1240,5863,834,260,1539,3268,655,1357,2453,...
    3224,3802,3535,969,7276,748,2275,1009,839,2068,1035,600,7063,...
    477,568,831,2076,9005,3375,454,4226,569,2083,803,1465,14199,...
    509,5672,777,1130,877,3574,920,957,1771]';

 pop_2020 = [43452000,33428000,12643000,2546000,...
     21523000,12220000,26491000,556000,16645000,...
    806000,5702000,1090000,107465000,1181000,...
    117191000,2293000,2574000,32180000,13205000,...
    2016000,51986000,2254000,5088000,6654000,...
    28225000,19377000,21224000,4499000,1298000,...
    36689000,31178000,2489000,24334000,208327000,...
    13146000,16436000,106000,8234000,58802000,...
    44440000,61705000,8443000,12162000,44405000,...
    18928000,15670000]';

africa_gdp_pop = table(country,gdp_2010,gdp_2020,pop_2010,pop_2020);
```

12.2 Excel Data of ▶ Section 12.1

Here, we extract the GDP per capita and population data for African countries used in ▶ Sects. 12.3, 12.4, 12.5, and 12.6 from the Excel data. We use the **renamevars** function to change the variable names in the table variable from 「GDP_2010」 and 「GDP_2020」 to 「gdp_2010」 and 「gdp_2020」, respectively. Additionally, the data type of the 「country」 column in the table is converted to a string. Furthermore, using 「africa_gdp_pop=rmmissing(africa_gdp_pop)」, rows with missing values are removed from the table variable 「africa_gdp_pop」. Finally, each column of the table variable 「africa_gdp_pop」 is extracted.

```
file = "alldata.xlsx";
sheet = "Chap12_Africa_GDP_pop";
africa_gdp_pop = readtable(file,"Sheet",sheet);
africa_gdp_pop = renamevars(africa_gdp_pop,...
    ["GDP_2010","GDP_2020"],["gdp_2010","gdp_2020"]);
africa_gdp_pop.country = string(africa_gdp_pop.country);
africa_gdp_pop = rmmissing(africa_gdp_pop);
gdp_2010 = africa_gdp_pop.gdp_2010;
gdp_2020 = africa_gdp_pop.gdp_2020;
pop_2010 = africa_gdp_pop.pop_2010;
pop_2020 = africa_gdp_pop.pop_2020;
```

12.3 Basic Syntax of the Histogram2 Function

Basic Syntax:

Here, we explain the three basic syntaxes of the **histogram2** function. See the syntaxes enclosed in the black box of Figure 12.1. First, 「histogram2(X,Y)」 is the syntax for creating a histogram based on two variables, 「X」 and 「Y」. The x-axis corresponds to the data of 「X」, and the y-axis corresponds to the data of 「Y」.
Next, 「histogram2(X,Y,nbins)」 uses the third argument, 「nbins」, thereby specifying the number of bins for both the x-axis and y-axis.
Lastly, 「histogram2(X,Y,Xedges,Yedges)」 allows us to specify the bin edges for the x-axis and y-axis, enabling us to finely control the range of the bins.

Syntax

```
histogram2(X,Y)
histogram2(X,Y,nbins)
histogram2(X,Y,Xedges,Yedges)
histogram2('XBinEdges',Xedges,'YBinEdges',Yedges,'BinCounts',counts)

histogram2(___,Name,Value)
histogram2(ax,___)
h = histogram2(___)
```


Fig. 12.1 The syntax of the histogram2 function

Based on the 「histogram2(X,Y)」 syntax, a 3D histogram is created with 「gdp_2020」 as 「X」 and 「pop_2020」 as 「Y」. The ticks on the x-axis are automatically generated based on the data from 「gdp_2020」, and the ticks on the y-axis are also automatically generated based on the data from 「pop_2020」. See Fig. 12.2.

```
histogram2(gdp_2020,pop_2020);
```

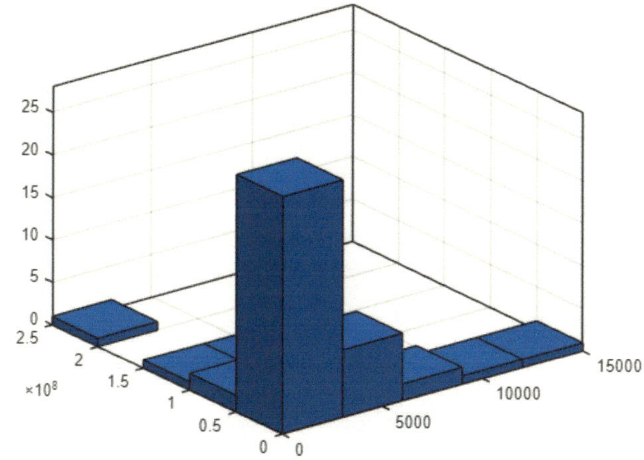

Fig. 12.2 3D histogram from two variables

Attention

Error Program: Here, based on the syntax 「histogram2(X,Y)」, suppose that we want to use both 「gdp_2020」 and 「gdp_2010」 for 「X」 and both 「pop_2020」 and 「pop_2010」 for 「Y」. The following program uses 「gdp_2020」 as the first argument, 「gdp_2010」 as the second argument, 「pop_2020」 as the third argument, and 「pop_2010」 as the fourth argument. This will result in an error because it does not follow the syntax of 「histogram2(X,Y)」. See Fig. 12.3 for the specific error message. Additionally, compare the following program with the one shown in Fig. 12.4.

```
%%% Error program
% histogram2(gdp_2020,gdp_2010,pop_2020,pop_2010);
```

```
Error using histogram2
Expected input number 3, xedges, to be monotonically increasing.

Error in histogram2>parseinput (line 158)
            validateattributes(in,{'numeric','logical'},{'vector', ...

Error in histogram2 (line 39)
[opts,args,vectorinput] = parseinput(args, firstaxesinput);
```

Fig. 12.3 Error output of the histogram2 function: Incorrect use of four variables

In contrast to the program that generated the error in Fig. 12.3, here we set the first argument to 「[gdp_2020;gdp_2010]」, and the second argument to 「[pop_2020;pop_2010]」, creating a 92-by-1 vector, respectively. The result is shown in Fig. 12.4.

```
histogram2([gdp_2020;gdp_2010],[pop_2020;pop_2010]);
```

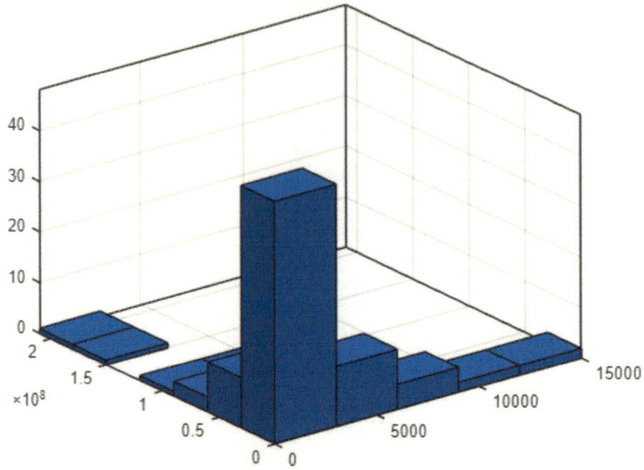

Fig. 12.4 3D histogram using merged column vectors

By setting the first argument to 「[gdp_2020,gdp_2010]」, we horizontally concatenate the column vectors of 「gdp_2020」 and 「gdp_2010」, creating a 46-by-2 matrix. Similarly, we set the second argument to 「[pop_2020,pop_2010]」, horizontally concatenating the column vectors. The resulting figure (Fig. 12.5) is the same as Fig. 12.4. Compare the previous program that created Fig. 12.4 with the following program.

```
histogram2([gdp_2020,gdp_2010],[pop_2020,pop_2010]);
```

12

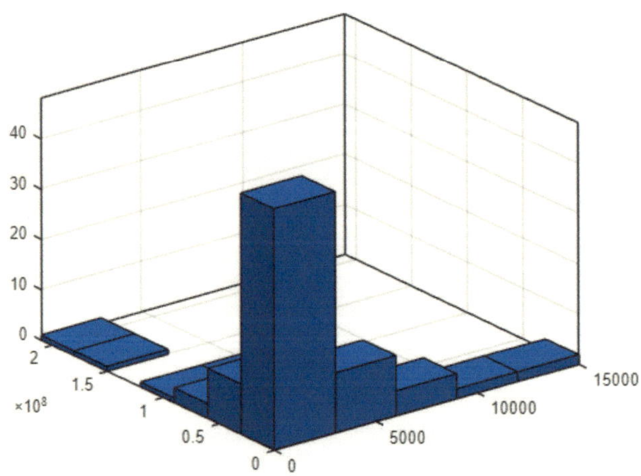

Fig. 12.5 3D histogram using merged 2-column matrix

Here, by specifying 20 as the third argument, we set the number of bins to 20. As a consequence, 20 bins can be placed along both the x-axis and y-axis. See Fig. 12.6.

```
histogram2(gdp_2020,pop_2020,20);
```

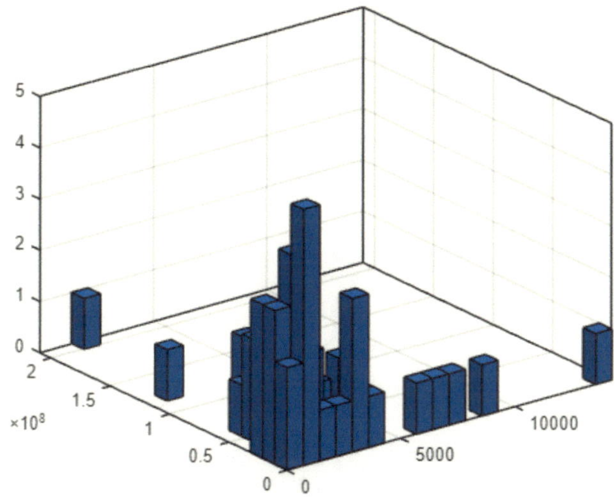

Fig. 12.6 3D histogram using fine bin division

Here, as in Fig. 12.4, the first and second arguments in the **histogram2** function use 92-by-1 column vectors, respectively. Additionally, we change the number of bins by specifying 20 as the third argument. The resulting figure is presented in Fig. 12.7.

```
histogram2([gdp_2020;gdp_2010],[pop_2020;pop_2010],20);
```

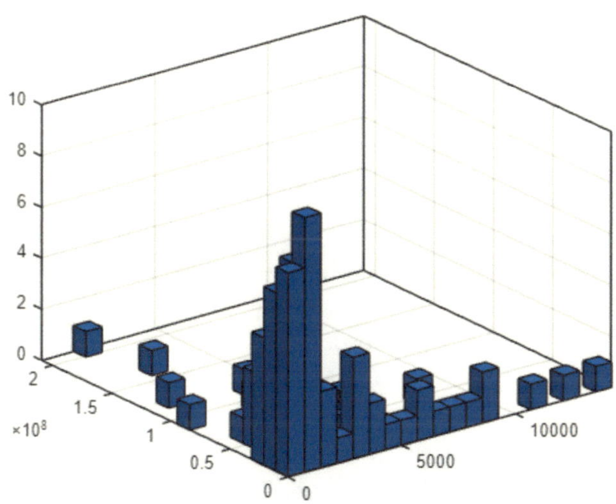

Fig. 12.7 3D histogram using fine bin division and merged column vectors

For example, if a 3D figure is difficult to interpret, you can change the order of the arguments. Here, we swap the first and second arguments of the **histogram2** function used in Fig. 12.7. See Fig. 12.8.

```
histogram2([pop_2020;pop_2010],[gdp_2020;gdp_2010],20);
```

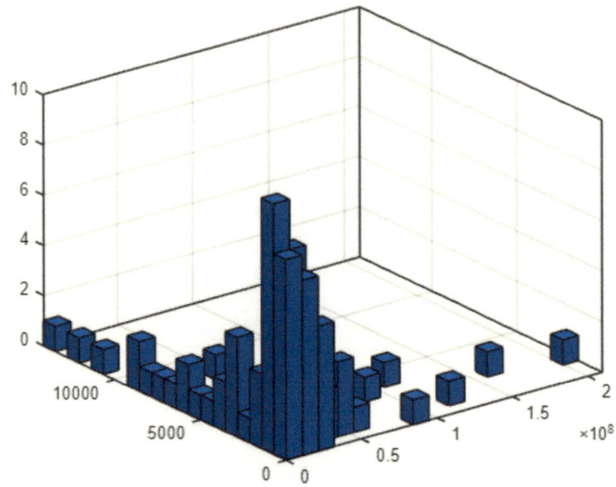

Fig. 12.8 3D histogram using fine bin division and merged column vectors (X- and Y-data reversed, cf. Fig. 12.7)

Here, specifying 「20 5」 in the third argument of the **histogram2** function assigns different numbers of bins to the x-axis and y-axis. Specifically, the number of bins on the x-axis is set to 20, and the number of bins on the y-axis is set to 5. See Fig. 12.9 for the created figure.

12

```
histogram2([pop_2020;pop_2010],[gdp_2020;gdp_2010],[20 5]);
```

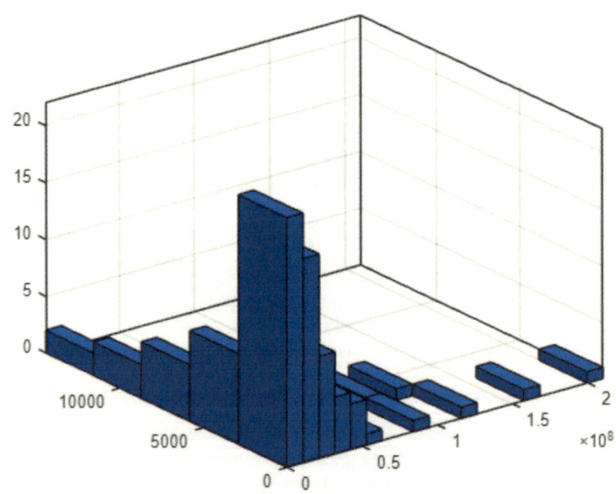

Fig. 12.9 3D histogram with different bin numbers in X- and Y-axes

Here, by specifying 「Xedges」 in the third argument of the **histogram2** function and 「Yedges」 in its fourth argument, we adjust the bin edges for the x-axis and y-axis.

Specifically, 「Xedges=[-20000 0:2000:20000 Inf]」 means that the range from -20000 to 0 is treated as a single bin. It also divides the next range, from 0 to 20000, into increments of 2000. Finally, the range from 20000 to Inf (= Infinity) is treated as the last bin. In other words, the bins are structured as follows: Bin 1: -20000 to 0; Bin 2: 0 to 2000; Bin 3: 2000 to 4000; Bin 4: 4000 to 6000...Bin 12: 20000 to Inf. Similarly, we can consider 「Yedges=[-2000000 0:100000:2000000 Inf]」. See ◘ Fig. 12.10.

```
Xedges = [-20000 0:2000:20000 Inf];
Yedges = [-2000000 0:100000:2000000 Inf];
histogram2([gdp_2020;gdp_2010],[pop_2020;pop_2010],Xedges,Yedges);
```

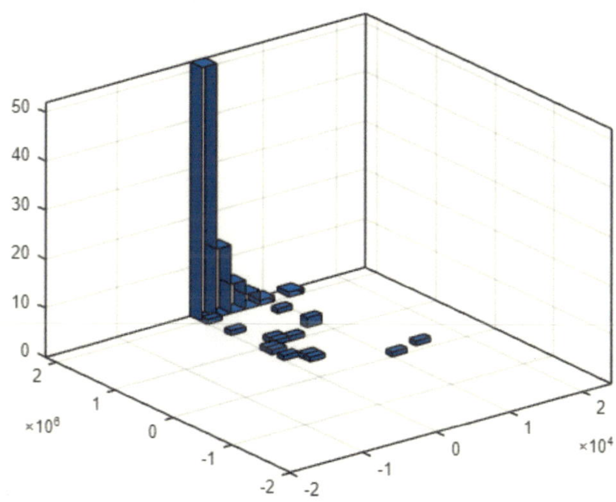

◘ **Fig. 12.10** 3D histogram with different bin edges in X- and Y-axes

Attention

Error Program: Here, note that we specify only 「Xedges」. That is, we do not specify any 「Yedges」. This results in an error. See ◘ Fig. 12.11 for details of the error.

```
%%% Error program
% Xedges = [-20000 0:2000:20000 Inf];
% histogram2([gdp_2020;gdp_2010],[pop_2020;pop_2010],Xedges);
```

```
Error using histogram2
Expected input number 3, m, to be integer-valued.

Error in histogram2>parseinput (line 151)
            validateattributes(in,{'numeric','logical'},{'integer', 'positive', ...

Error in histogram2 (line 39)
[opts,args,vectorinput] = parseinput(args, firstaxesinput);
```

Fig. 12.11 Error output of the histogram2 function: Incomplete bin edge specification

12.4 Setting and Controlling the Histogram2 Function

Here, we create a 3D histogram using the specified bin edges and bin counts. First, 「C=[1 1 1 1; 2 2 2 2; 3 4 5 6; 10 20 30 40]」 creates a 4-by-4 matrix, which represents the count of data in the 16 bins used in the **histogram2** function. In the **histogram2** function, we set 「XBinEdges」 from -2 to 2 and 「YBinEdges」 from -2 to 2, adjusting the bin edges for both the x-axis and y-axis. Additionally, by setting 「BinCounts」 to 「C」, the heights of the histogram bins are determined based on the values in the variable 「C」. See Fig. 12.12.

```
C = [1 1 1 1; 2 2 2 2; 3 4 5 6; 10 20 30 40];
histogram2("XBinEdges",-2:2,"YBinEdges",-2:2,"BinCounts",C);
```


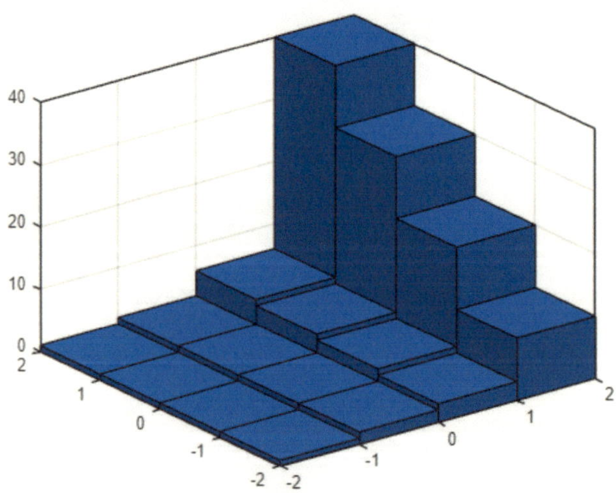

Fig. 12.12 3D histogram with manually specified bin ranges and frequencies

Here, the properties of the **histogram2** function are stored in the variable 「hist2」. Then, by setting 「BinWidth」 to 「[1 1]」, we adjust the bin widths of the created histogram to 1 for both axes. As a result, each bin becomes narrower. See Fig. 12.13.

```
hist2 = histogram2(gdp_2020,pop_2020);
    hist2.BinWidth = [1 1];
```

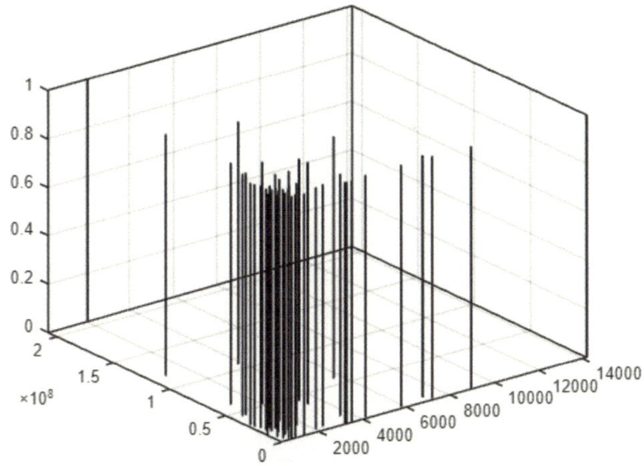

Fig. 12.13 3D histogram with narrow bins

Here, using 「BinWidth」, we set different bin widths for the x-axis and y-axis. First, the value of 2 times 10 to the power of 7 (「2*10^7」) is assigned to the variable 「bwmax」. This value is later used as the bin width for the y-axis. Next, by using 「hist2.BinWidth=[2000 bwmax]」, we set the bin width for the x-axis to 2000 and the bin width for the y-axis to 「bwmax (= 20000000)」. The resulting figure is shown in Fig. 12.14.

```
hist2 = histogram2(gdp_2020,pop_2020);
bwmax = 2*10^7;
    hist2.BinWidth = [2000 bwmax];
```

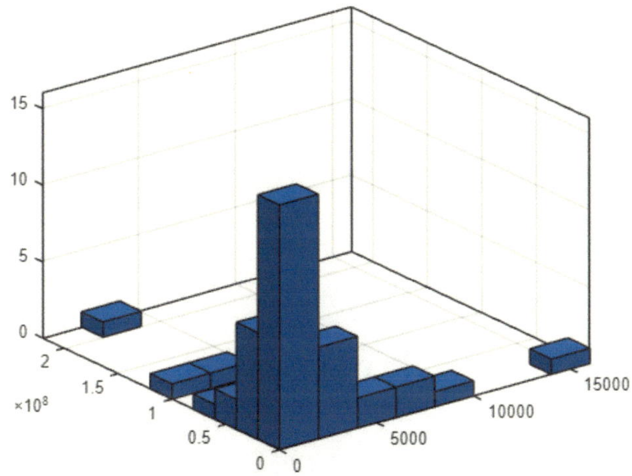

Fig. 12.14 3D histogram with different bin widths in X- and Y-axes

Here, the properties of the **histogram2** function are stored in the variable 「hist2」. Then, by setting 「XBinLimits」 to 「[-10000 30000]」, we modify the boundaries of the x-axis. See ◘ Fig. 12.15.

```
hist2 = histogram2(gdp_2020,pop_2020);
    hist2.XBinLimits = [-10000 30000];
```

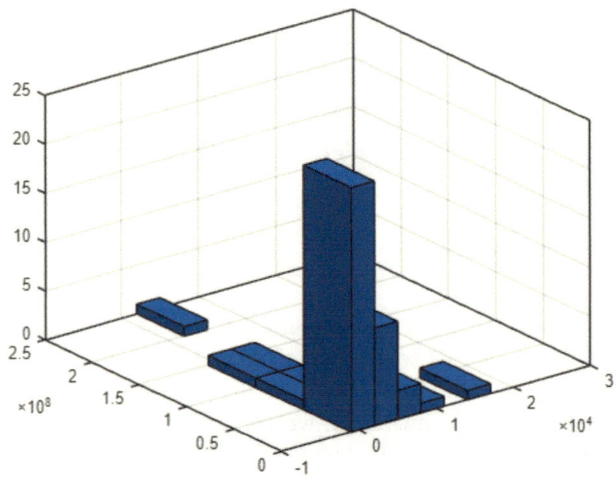

◘ **Fig. 12.15** 3D histogram with extended bin limits in X-axis

12

Here, by setting 「ShowEmptyBins」 to 「on」, bins are created even in empty areas, with the bin height set to zero. The default setting for 「ShowEmptyBins」 is 「off」. Compare ◘ Fig. 12.2 with ◘ Fig. 12.16.

```
histogram2(gdp_2020,pop_2020,"ShowEmptyBins","on");
```

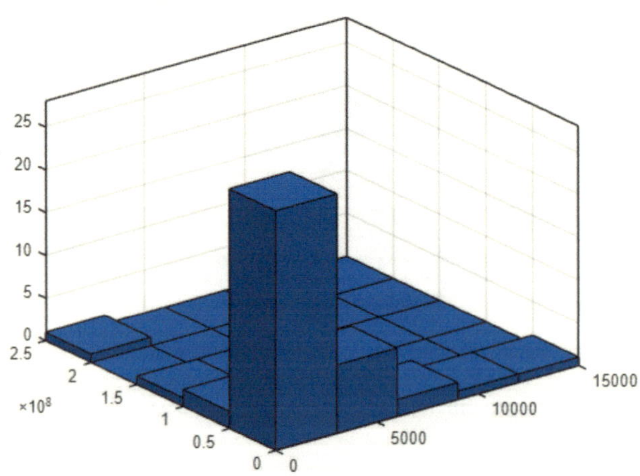

◘ **Fig. 12.16** 3D histogram with visible empty bins

As in the previous program that created ◘ Fig. 12.16, we set 「ShowEmptyBins」 to 「on」. Additionally, here, we set 「FaceLighting」 to 「none」, removing the lighting effect from the histogram. By default, the lighting effect on the histogram is set to 「lit」. Compare ◘ Fig. 12.16 with ◘ Fig. 12.17.

```
hist2 = histogram2(gdp_2020,pop_2020,"ShowEmptyBins","on");
    hist2.FaceLighting = "none";
```

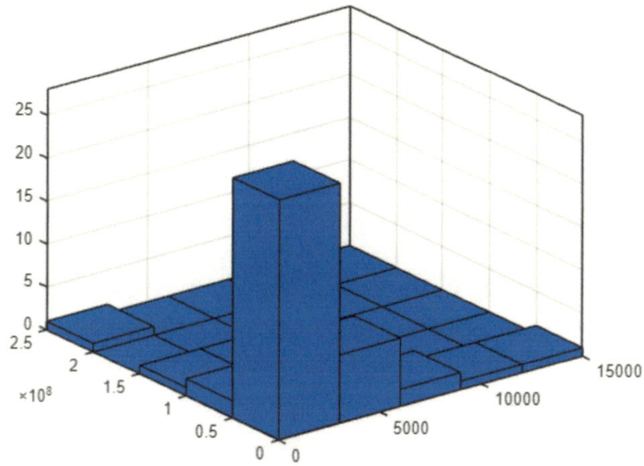

◘ **Fig. 12.17** 3D histogram with no face lighting and visible empty bins

Here, the properties of the **histogram2** function are stored in the variable 「hist2」. Then, we set 「LineStyle」 to 「:」, 「LineWidth」 to 3, 「FaceColor」 to 「red」, and 「FaceAlpha」 to 0.5, creating a semi-transparent red histogram with a thick dotted outline. See ◘ Fig. 12.18.

```
hist2 = histogram2(gdp_2020,pop_2020,"ShowEmptyBins","on");
    hist2.LineStyle = ":";
    hist2.LineWidth = 3;
    hist2.FaceColor = "red";
    hist2.FaceAlpha = 0.5;
```

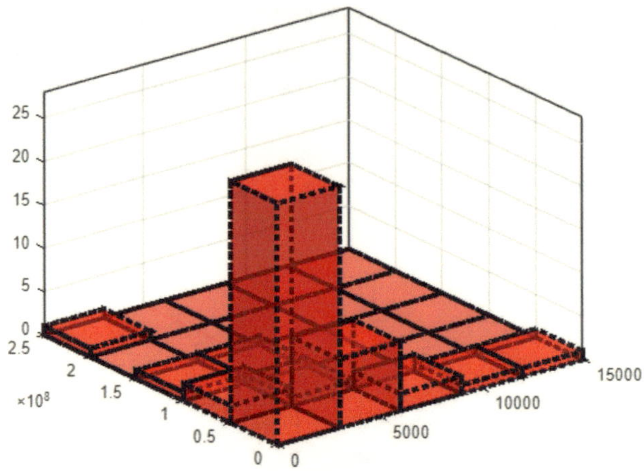

Fig. 12.18 3D histogram with semi-transparent red bins and dotted black edges

Here, we set 「LineStyle」 to 「:」, 「LineWidth」 to 3, and 「FaceAlpha」 to 0.5, creating a histogram with a thick dotted outline and high transparency. Additionally, setting 「FaceColor」 to 「flat」 changes the color of the histogram based on frequency. See Fig. 12.19.

```
hist2 = histogram2(gdp_2020,pop_2020,"ShowEmptyBins","on");
    hist2.LineStyle = ":";
    hist2.LineWidth = 3;
    hist2.FaceAlpha = 0.5;
    hist2.FaceColor = "flat";
```

12

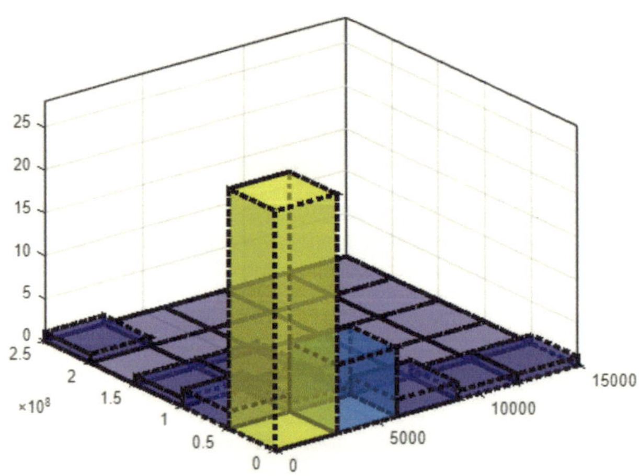

Fig. 12.19 3D histogram with colors based on frequency

Here, the properties of the **histogram2** function are stored in the variable 「h」. The resulting figure is presented in ◘ Fig. 12.20. Additionally, by using 「Values」, we store the values (frequency or counts) contained in each bin of the created histogram into the variable 「z」. In ◘ Fig. 12.21, we can check the number of observations contained in each bin.

```
h = histogram2(gdp_2020,pop_2020);
```

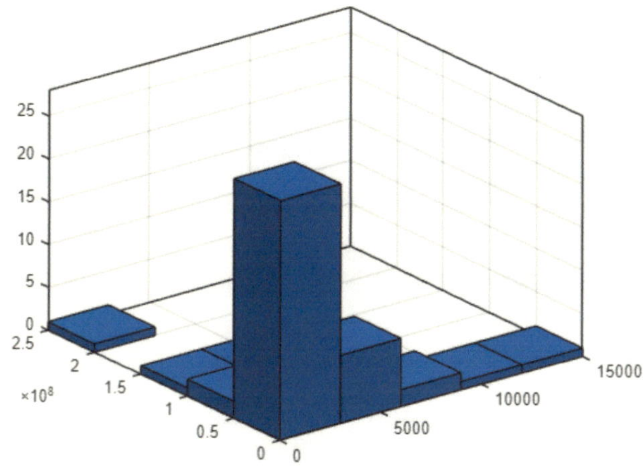

◘ **Fig. 12.20** 3D histogram and the creation of the variable "h"

```
z = h.Values
```

```
z = 5×5
    28     2     1     0     1
     8     1     1     0     0
     2     0     0     0     0
     1     0     0     0     0
     1     0     0     0     0
```

◘ **Fig. 12.21** Contents of the bin counts using the variable "h"

12.5 Combining Optional Functions and the Histogram2 Function

Here, we set 「DisplayStyle」 to 「tile」. This allows the histogram to be displayed as a flat 2D tile, rather than a 3D histogram. Additionally, by adding the **colorbar** command, we visually represent the frequency or count of each bin using colors. See ◘ Fig. 12.22.

```
histogram2(gdp_2020,pop_2020,"Displaystyle","tile");
colorbar;
```

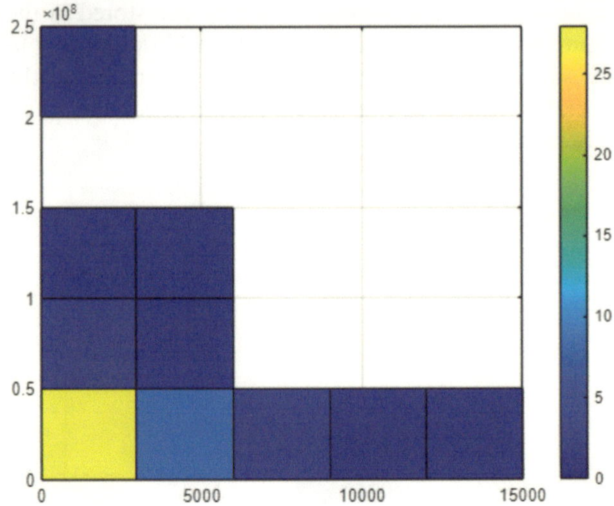

Fig. 12.22 Tiled 3D histogram with a colorbar

As in the previous program that created Fig. 12.22, we set 「DisplayStyle」 to 「tile」, which displays the histogram as a flat 2D tile. Additionally, the **colorbar** command is used. Unlike the previous program that generated Fig. 12.22, we set the third argument of the **histogram2** function to 20, which adjusts the number of bins. Compare Fig. 12.22 with Fig. 12.23.

```
histogram2(gdp_2020,pop_2020,20,"Displaystyle","tile");
colorbar;
```

12

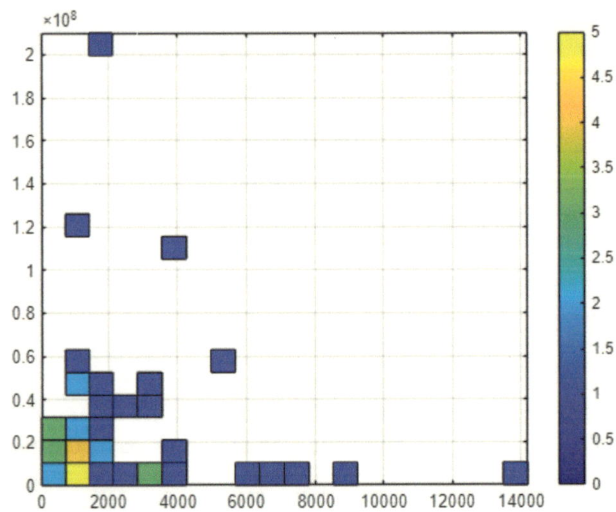

Fig. 12.23 Tiled 3D histogram with small tiles and a colorbar

Here, by setting the third argument of the **histogram2** function to 「[50 4]」, we adjust the number of bins for the x-axis and y-axis, respectively. Compare ◘ Fig. 12.23 with ◘ Fig. 12.24.

```
histogram2(gdp_2020,pop_2020,[50 4],"Displaystyle","tile");
colorbar;
```

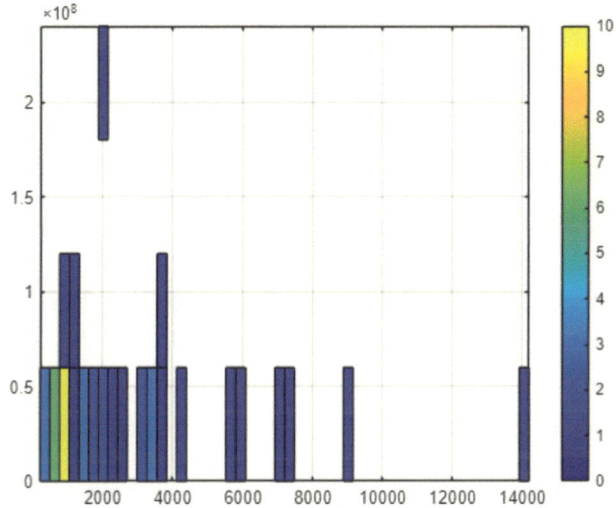

◘ **Fig. 12.24** Tiled 3D histogram with elongated tiles and a colorbar

Here, the **xlabel**, **ylabel**, and **zlabel** functions are used simultaneously. Specifically, for the y-axis label, we set 「FontSize」 to 20 and 「Color」 to 「blue」, creating a large, blue-colored label (「Population 2020」). For the z-axis label, we set 「Rotation」 to 60 and 「Color」 to 「red」, which rotates the z-axis label slightly and displays it in red. See ◘ Fig. 12.25.

```
histogram2(gdp_2020,pop_2020);
xlabel("GDP per capita 2020");
yl = ylabel("Population 2020");
     yl.FontSize  =  20;
     yl.Color = "blue";
zl  = zlabel({"The number of","countries"});
     zl.Rotation = 60;
     zl.Color = "red";
```

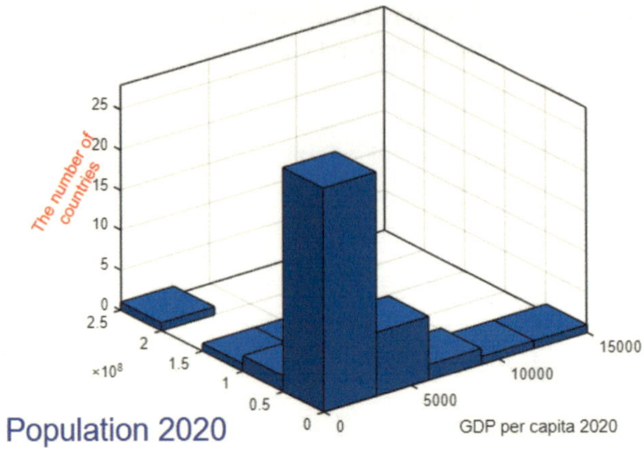

Fig. 12.25 3D histogram with various axes labels

Here, by setting 「hist2.FaceColor」 to 「flat」, we change the color of the higher frequency bars. Additionally, by using the **view** function, we adjust the viewing angle. See Fig. 12.26.

```
hist2 = histogram2(gdp_2020,pop_2020);
    hist2.FaceColor = "flat";
view(70,50);
title("Bivariate histogram from a rotated perspective");
yl = ylabel("Population 2020");
    yl.FontSize = 20;
    yl.Color = "blue";
```

12

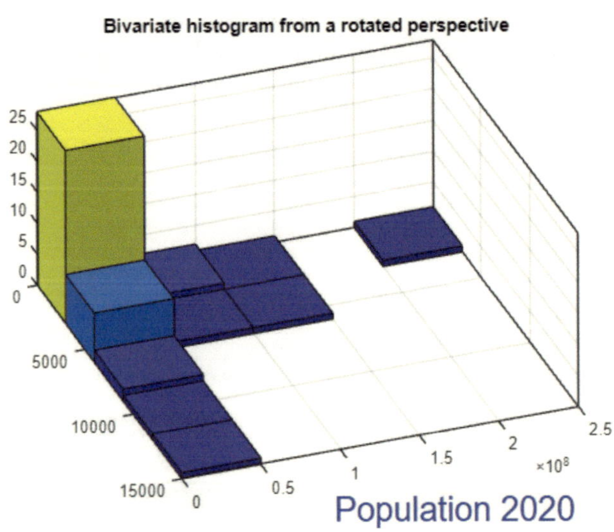

Fig. 12.26 3D histogram from a different angle, with colors based on frequency

Here, by using the **morebins** command, we increase the number of bins in the histogram. Specifically, after storing the properties of the **histogram2** function in the variable 「h」, we set 「morebins(h)」. Additionally, we adjust the viewing angle using the **view** function. See ◘ Fig. 12.27.

```
h = histogram2([gdp_2020;gdp_2010],[pop_2020;pop_2010]);
view(30,70);
morebins(h);
```

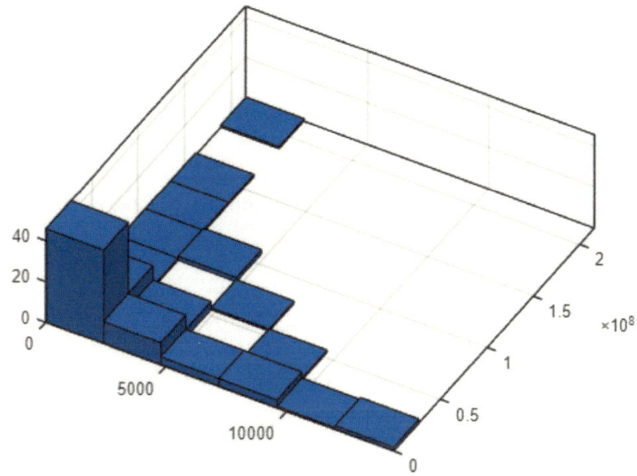

◘ **Fig. 12.27** 3D histogram from a different angle, with adjusted number of bins

In contrast to the previous program that created ◘ Fig. 12.27, here, we set 「morebins(h,"x")」. This increases the number of bins along the x-axis of the histogram stored in the variable 「h」. Note that as the viewing angle changes due to the **view** function, the position of the x-axis also changes. Compare ◘ Fig. 12.27 with ◘ Fig. 12.28.

```
h = histogram2([gdp_2020;gdp_2010],[pop_2020;pop_2010]);
view(30,70);
morebins(h,"x");
```

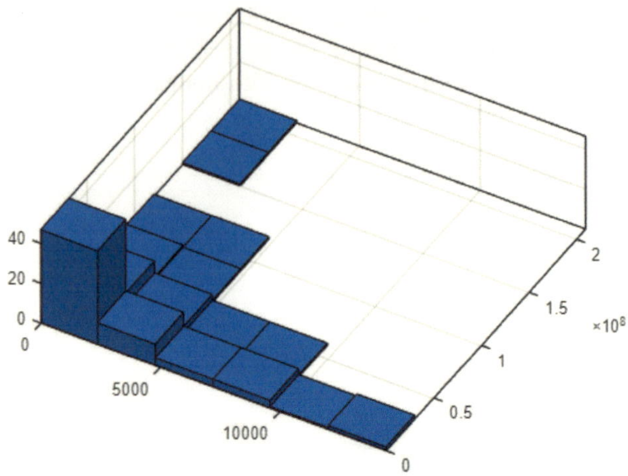

Fig. 12.28 3D histogram from a different angle, with adjusted number of X-axis bins

Here, we use the **morebins** and **fewerbins** commands simultaneously. Specifically, by using 「morebins(h,"y")」, we increase the number of bins in the y-axis direction of the histogram stored in the variable 「h」. Instead, by using 「fewerbins(h,"x")」, we decrease the number of bins in the x-axis direction. See Fig. 12.29.

```
h = histogram2([gdp_2020;gdp_2010],[pop_2020;pop_2010]);
view(30,70);
morebins(h,"y");
fewerbins(h,"x");
```

12

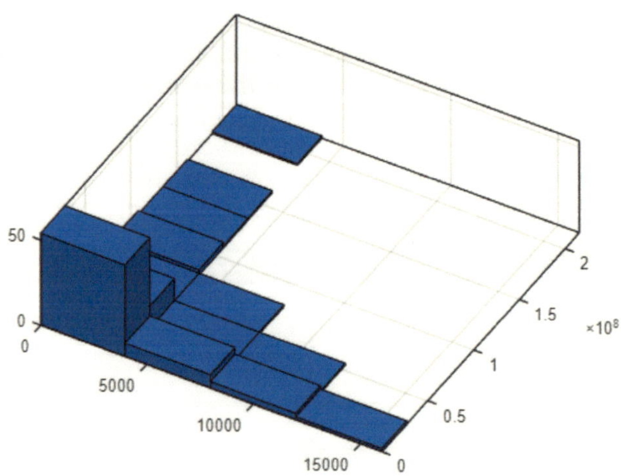

Fig. 12.29 3D histogram with more Y-axis bins and fewer X-axis bins

First, we set 「DisplayStyle」 to 「tile」, changing the 3D view created by the **histogram2** function to a 2D angle. Additionally, with 「colorbar("southoutside")」, we adjust the position of the colorbar. Furthermore, with 「cl.Label.String = "The number of countries"」, we set the label of the colorbar to 「The number of countries」. See ◘ Fig. 12.30.

```
histogram2(gdp_2020,pop_2020,"Displaystyle","tile");
cl = colorbar("southoutside");
    cl.Label.String = "The number of countries";
    cl.FontSize = 15;
```

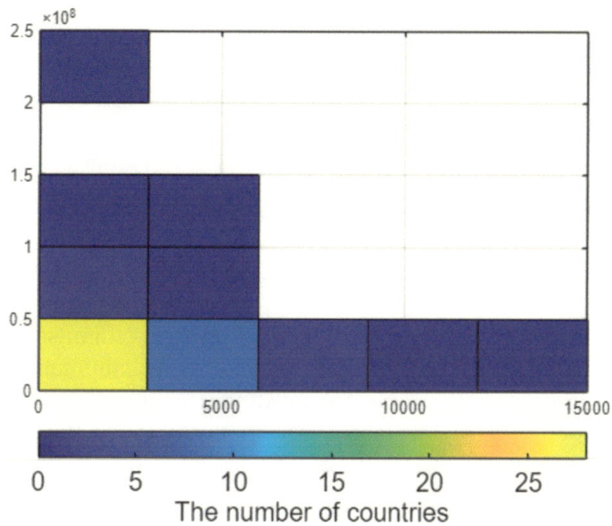

◘ **Fig. 12.30** Tiled 3D histogram with a labeled colorbar (outside-bottom-center position)

Here, as in the previous program that created ◘ Fig. 12.30, we use the **colorbar** command. Specifically, after storing the properties of the colorbar in the variable 「cl」, we set the label to 「Reverse color scale: The number of countries」 by using 「cl.Label.String = "Reverse color scale: The number of countries"」. Additionally, we set 「FontSize」 to 15 and 「Color」 to 「red」, creating a red colorbar with a large text label. Finally, we set 「Direction」 to 「reverse」, reversing the display of the colorbar. Compare the created colorbar in ◘ Fig. 12.30 with the one in ◘ Fig. 12.31.

```
histogram2(gdp_2020,pop_2020,"Displaystyle","tile");
cl = colorbar("southoutside");
    cl.Label.String = "Reverse color scale: The number of countries";
    cl.FontSize = 15;
    cl.Color = "red";
    cl.Direction = "reverse";
```

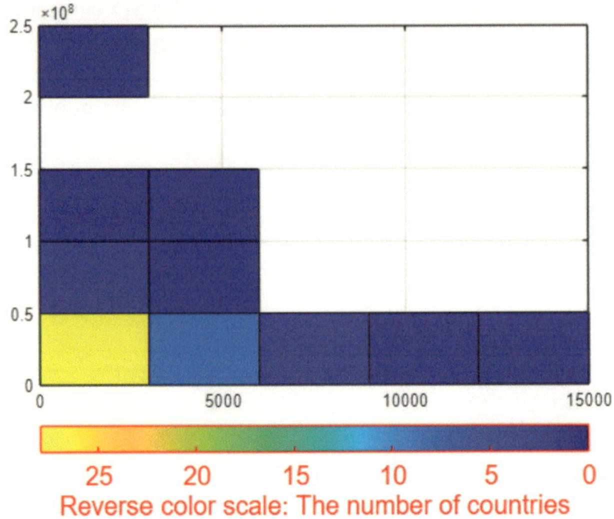

Fig. 12.31 Tiled 3D histogram with a red-labeled, reversed color scale (outside-bottom-center position)

Here, by using 「colorbar("westoutside")」, we set the position of the colorbar to the left outside of the figure. Next, 「c.Ticks=[c.Limits(1),c.Limits(2)]」 sets the ticks of the colorbar. Specifically, 「c.Limits(1)」 represents the minimum value of the colorbar, and 「c.Limits(2)」 represents the maximum value of the colorbar. Finally, using 「c.TickLabels」, we set the labels corresponding to the ticks of the colorbar. The label for the minimum value in the colorbar is given by 「A small number」, and the label for the maximum value is given by 「A large number」. See **Fig. 12.32**.

12

```
histogram2(gdp_2020,pop_2020,"Displaystyle","tile");
c = colorbar("westoutside");
    c.Ticks = [c.Limits(1), c.Limits(2)];
    c.TickLabels = {"A small number","A large number"};
```

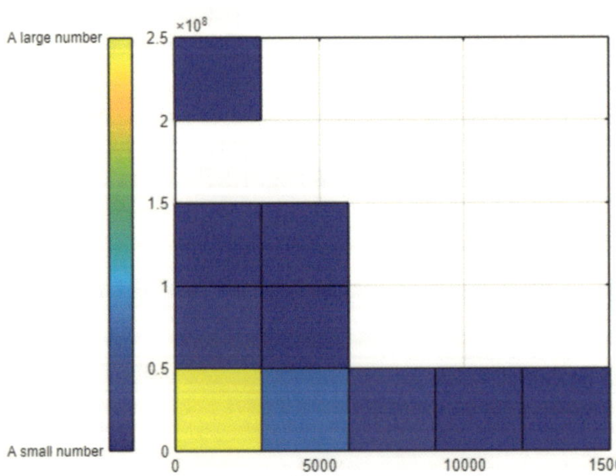

Fig. 12.32 Tiled 3D histogram with a tick-labeled colorbar (outside-left position)

For details on how to use the **colorbar** command, refer to the previous program that created ◘ Fig. 12.32. Here, by specifying 「TextArrow」 as the first argument to the **annotation** function, we create an arrow with text. Note that 「xx」 and 「yy」 specify the start and end points of the arrow. Concretely, the start coordinate is (x,y)=(0.6,0.5), and the end coordinate is (x,y)=(0.3,0.2). For more details on the **annotation** function, see ► Sect. 18.1. The resulting figure is presented in ◘ Fig. 12.33.

```
histogram2(gdp_2020,pop_2020,"Displaystyle","tile");
c = colorbar("westoutside");
    c.Ticks = [c.Limits(1),mean(c.Limits),c.Limits(2)];
    c.TickLabels = {"A small num","A mean num","A large num"};
xx = [0.6 0.3];
yy = [0.5 0.2];
annotation("TextArrow",xx,yy,"String","28 countries");
```

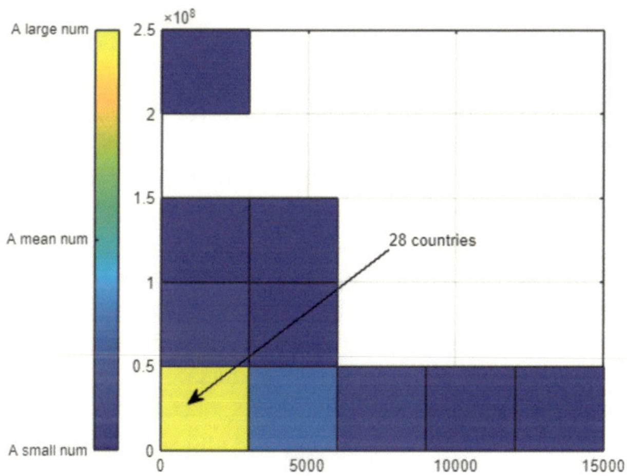

◘ **Fig. 12.33** Tiled 3D histogram with a tick-labeled colorbar and a text arrow

Here, by using the **hold on** command, we overlay two figures created by the **histogram2** function. Specifically, by setting 「DisplayStyle」 to 「tile」, we switch to a 2D view and perform the overlay. Additionally, we use the **legend** function to create a legend. See ◘ Fig. 12.34.

```
h1 = histogram2(gdp_2020,pop_2020,"Displaystyle","tile");
    h1.FaceAlpha = 0.1;
    h1.EdgeColor = "red";
    h1.LineWidth = 3;
hold on
h2 = histogram2(gdp_2010,pop_2010,"Displaystyle","tile");
    h2.FaceAlpha = 0.4;
    h2.LineStyle = ":";
    h2.LineWidth = 3;
hold off
legend("2020","2010");
```

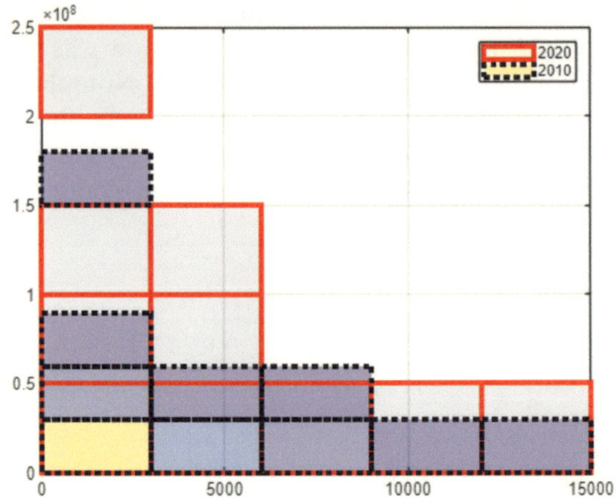

■ **Fig. 12.34** Overlaid tiled 3D histograms for 2010 and 2020

In the previous program that created ■ Fig. 12.34, when overlaying the two figures, the figures were formed tiles of different sizes. Here, we address this issue. Specifically, we retrieve the minimum and maximum values from two data sets and set the bin edges for the x-axis and y-axis based on them. That is, we retrieve the minimum and maximum values of each of 「gdp_2010」 and 「gdp_2020」, then store the minimum and maximum values among them in the variables 「x_min」 and 「x_max」. Similarly, we retrieve the minimum and maximum values of each of 「pop_2010」 and 「pop_2020」, then store the minimum and maximum values among them in the variables 「y_min」 and 「y_max」.

12

After that, using 「x_edges=linspace(x_min,x_max,6)」, we divide the range between 「x_min」 and 「x_max」 into 6 equal parts and store them in the variable 「x_edges」, which is used as bin edges for the x-axis. Similarly, 「y_edges」 is used as bin edges for the y-axis. Finally, in the **histogram2** function, we set 「XBinEdges」 to 「x_edges」 and 「YBinEdges」 to 「y_edges」, ensuring that the tile sizes of the two histogram figures are the same. Compare ■ Fig. 12.34 with ■ Fig. 12.35.

```
x_min  = min([min(gdp_2010),min(gdp_2020)]);
x_max  = max([max(gdp_2010),max(gdp_2020)]);
y_min  = min([min(pop_2010),min(pop_2020)]);
y_max  = max([max(pop_2010),max(pop_2020)]);
x_edges = linspace(x_min,x_max,6);
y_edges = linspace(y_min,y_max,6);
h1 = histogram2(gdp_2020,pop_2020,"XBinEdges",x_edges,...
    "YBinEdges",y_edges,"Displaystyle","tile");
    h1.FaceAlpha = 0.1;
    h1.EdgeColor = "red";
    h1.LineWidth = 3;
hold on
h2 = histogram2(gdp_2010,pop_2010,"XBinEdges",x_edges,...
    "YBinEdges",y_edges,"Displaystyle","tile");
    h2.FaceAlpha = 0.4;
    h2.LineStyle = ":";
    h2.LineWidth = 3;
hold off
legend("2020","2010");
```

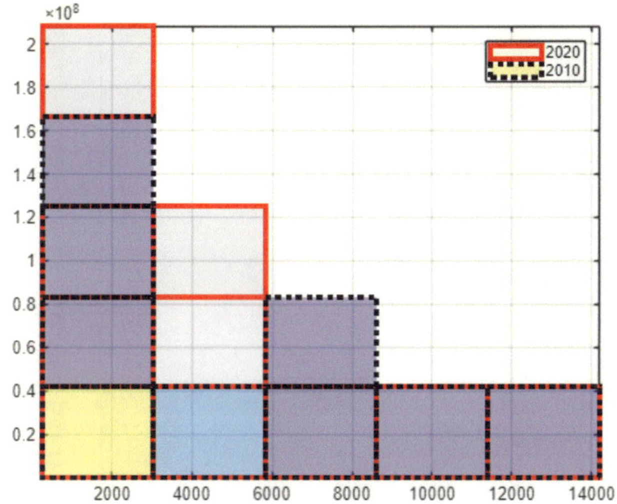

Fig. 12.35 Overlaid tiled 3D histograms with adjusted tile sizes

Here, by setting ｢DisplayStyle｣ to ｢tile｣, we create 2D tiles, calculate the center coordinates of the tiles in a 2D histogram, and display the frequency values in each tile. Specifically, using ｢counts=h.Values｣, we retrieve the frequency values. See Fig. 12.21 for the details of ｢Values｣. Next, with ｢x_edges=h.XBinEdges｣, the bin edge data for the x-axis of the histogram is stored in ｢x_edges｣. ｢y_edges=h.YBinEdges｣ corresponds to the data for the y-axis. Then, using ｢x_centers=(x_edges(1:end-1)+x_edges(2:end))/2｣, we calculate the midpoints between adjacent bin edges in ｢x_edges｣ and store the center coordinates of each tile on the x-axis in ｢x_centers｣.

Finally, in the **for** loop, we use the **text** function to place the frequency values obtained in the variable ｢counts｣ at the coordinates (X,Y)=(x_centers(i),y_centers(j)). For more details of the **text** function, see ▶ Sect. 18.2. The resulting figure is shown in Fig. 12.36.

```
h = histogram2(gdp_2020,pop_2020,"DisplayStyle","tile");
colorbar;
counts = h.Values;
x_edges = h.XBinEdges;
y_edges = h.YBinEdges;
x_centers = (x_edges(1:end-1)+x_edges(2:end))/2;
y_centers = (y_edges(1:end-1)+y_edges(2:end))/2;
for i = 1:length(x_centers)
    for j = 1:length(y_centers)
        if counts(i,j) > 0
            tx = text(x_centers(i),y_centers(j),num2str(counts(i,j)),...
            "HorizontalAlignment","center","VerticalAlignment","middle");
            tx.Color = "red";
            tx.FontSize = 20;
        end
    end
end
```

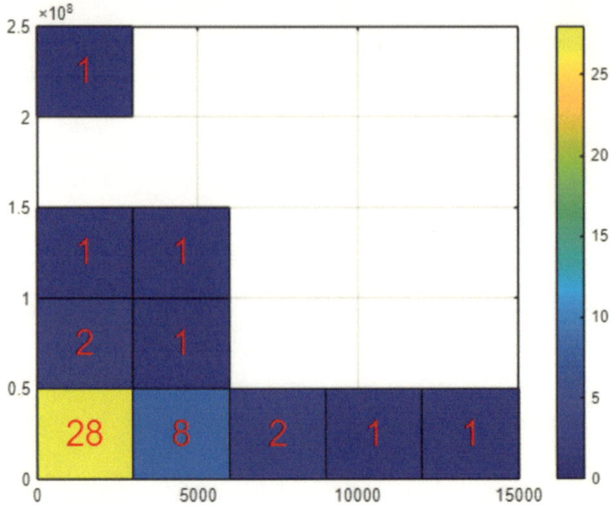

Fig. 12.36 Tiled 3D histogram with bin counts as text labels in red

The following program is the same as Fig. 12.36, except for one point. The difference is that 「hist2.Normalization="percentage"」 is used to represent the histogram as percentages. For example, the yellow tile shows 「60.8696」, which means 60.8696%. Note that in Fig. 12.36, the same tile shows the number 「28」. Compare Fig. 12.36 with Fig. 12.37.

12

```
hist2 = histogram2(gdp_2020,pop_2020,"DisplayStyle","tile");
hist2.Normalization = "percentage";
colorbar;
counts = hist2.Values;
x_edges = hist2.XBinEdges;
y_edges = hist2.YBinEdges;
x_centers = (x_edges(1:end-1)+x_edges(2:end))/2;
y_centers = (y_edges(1:end-1)+y_edges(2:end))/2;
for i = 1:length(x_centers)
    for j = 1:length(y_centers)
        if counts(i,j) > 0
            tx = text(x_centers(i),y_centers(j),num2str(counts(i,j)),...
                "HorizontalAlignment","center",...
                "VerticalAlignment","middle");
            tx.Color = "red";
            tx.FontSize = 15;
        end
    end
end
```

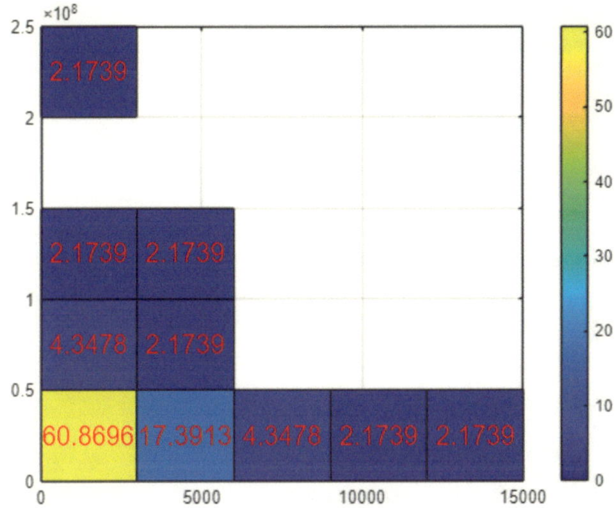

Fig. 12.37 Tiled 3D histogram with percentages as text labels in red

Here, we use a **for** loop to display the frequencies contained in two different histograms. The basic program is almost the same as the one that created Fig. 12.36. The differences are as follows. First, as confirmed in 「x_edges=linspace(x_min,x_max,10)」 and 「y_edges=linspace(y_min,y_max,10)」, we divide the range into 10 equal parts. Second, within the **for** loop, we use the **text** function twice. In particular, with 「count1=h1.Values(i,j)」, we retrieve the frequency of the data contained in the tile of the variable 「h1」, which is stored in the variable 「count1」. Then, to display 「count1」, we use the first **text** function. To avoid overlapping text, we set 「HorizontalAlignment」 to 「left」. Next, by using 「count2=h2.Values(i,j)」, we make use of the second **text** function. Here we set 「HorizontalAlignment」 to 「right」. See Fig. 12.38.

```
x_min  =  min([min(gdp_2010),min(gdp_2020)]);
x_max  =  max([max(gdp_2010),max(gdp_2020)]);
y_min  =  min([min(pop_2010),min(pop_2020)]);
y_max  =  max([max(pop_2010),max(pop_2020)]);
x_edges = linspace(x_min,x_max,10);
y_edges = linspace(y_min,y_max,10);
h1 = histogram2(gdp_2020,pop_2020,"XBinEdges",x_edges,...
    "YBinEdges",y_edges,"Displaystyle","tile");
    h1.FaceAlpha = 0.3;
    h1.EdgeColor = "red";
    h1.LineWidth = 3;
hold on
h2 = histogram2(gdp_2010,pop_2010,"XBinEdges",x_edges,...
    "YBinEdges",y_edges,"Displaystyle","tile");
    h2.FaceAlpha = 0.1;
    h2.LineStyle = ":";
    h2.LineWidth = 3;
    h2.EdgeColor = "black";
for i = 1:length(x_edges)-1
    for j = 1:length(y_edges)-1
        x_center  =  (x_edges(i)+x_edges(i+1))/2;
        y_center  =  (y_edges(j)+y_edges(j+1))/2;
        count1 = h1.Values(i,j);
        if count1 > 0
            text(x_center,y_center,num2str(count1),...
                "Color","red","FontSize",14,"HorizontalAlignment","left");
        end
        count2 = h2.Values(i,j);
        if count2 > 0
            text(x_center,y_center,num2str(count2), ...
                "Color","blue","FontSize",14,"HorizontalAlignment","right");
        end
    end
end
legend("GDP 2020","GDP 2010");
grid on
hold off
```


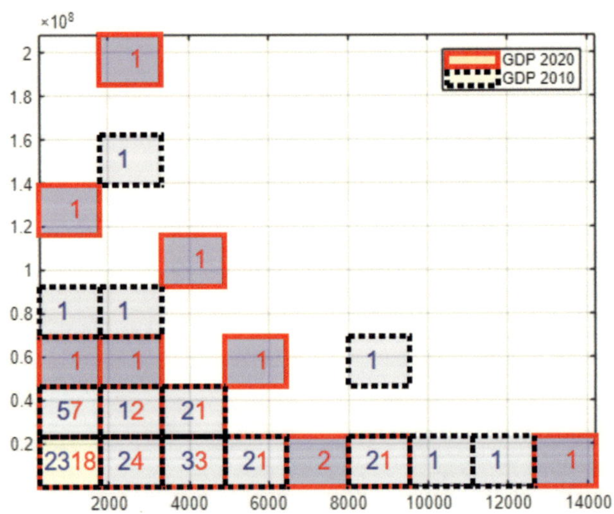

Fig. 12.38 Overlaid tiled 3D histograms with bin counts as text labels

12.6 Questions

Question 1

Create Fig. 12.39 using the **histogram2** function with 「gdp_2020」 as the first argument and 「pop_2020」 as the second argument. Concretely, store the properties of the **histogram2** function in the variable 「Ex1」. Next, create a thick dotted outline. Additionally, set 「FaceColor」 to 「flat」. Finally, use the **xlabel** and **ylabel** functions to add labels (GDP for the x-axis and Population for the y-axis).

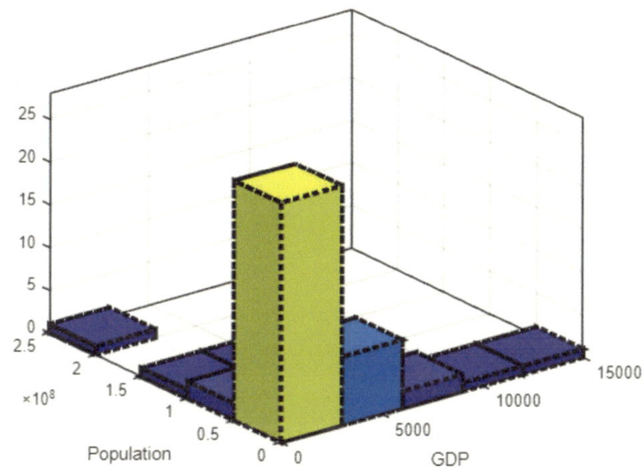

Fig. 12.39 3D histogram (Question 1)

Answer 1

```
Ex1 = histogram2(gdp_2020,pop_2020);
    Ex1.LineStyle = ":";
    Ex1.LineWidth = 3;
    Ex1.FaceColor = "flat";
xlabel("GDP");
ylabel("Population");
```

Question 2

Create Fig. 12.40 using the **histogram2** function with 「gdp_2020」 as the first argument and 「pop_2020」 as the second argument. Display the 3D histogram as a 2D tile. Additionally, create a colorbar outside the top of the graph and set its label to 「The number of countries」.

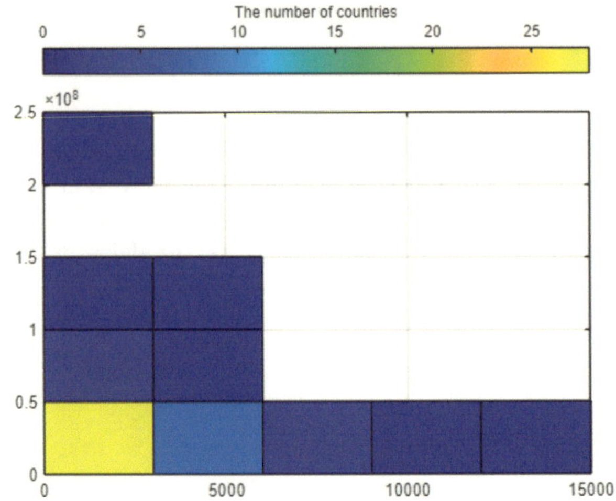

Fig. 12.40 3D histogram (Question 2)

Answer 2

```
figure
histogram2(gdp_2020,pop_2020,"Displaystyle","tile");
clb = colorbar("northoutside");
    clb.Label.String = "The number of countries";
```

Question 3

12

Create Fig. 12.41 using the **histogram2** function with 「gdp_2020」 as the first argument and 「pop_2020」 as the second argument. After displaying the histogram in tiled form, create a colorbar. In the colorbar, include the tick marks for the upper limit, lower limit, and mean, as well as the midpoint between the upper limit and mean, and the midpoint between the lower limit and mean.

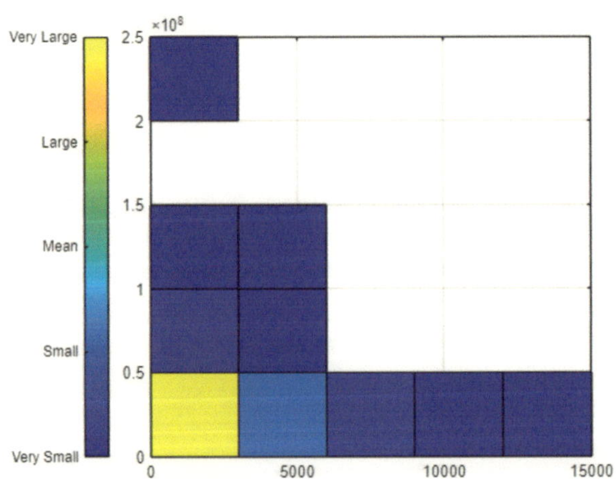

Fig. 12.41 3D histogram (Question 3)

Answer 3

```
histogram2(gdp_2020,pop_2020,"Displaystyle","tile");
Ex3 = colorbar("westoutside");
    Ex3.Ticks = [Ex3.Limits(1),median([Ex3.Limits(1),mean(Ex3.Limits)]),...
        mean(Ex3.Limits),median([mean(Ex3.Limits),Ex3.Limits(2)]),...
        Ex3.Limits(2)];
    Ex3.TickLabels = {"Very Small","Small","Mean","Large","Very Large"};
```

Boxchart Function

Contents

13.1 Details of the Data Used in this Chapter – 408

13.2 Excel Data of ► Section 13.1 – 409

13.3 Basic Syntax of the Boxchart Function – 409

13.4 Setting and Controlling the Boxchart Function – 419

13.5 Combining Optional Functions and the Boxchart Function – 426

13.6 Questions – 435

Supplementary Information The online version contains supplementary material available at https://doi.org/10.1007/978-981-96-6213-5_13.

Y. Nakamoto, *A Comprehensive Guide to Creating Graphics with MATLAB Programming*, Classroom Companion: Economics, https://doi.org/10.1007/978-981-96-6213-5_13

Learning Objectives

This chapter creates box plots using the **boxchart** function. By reading this chapter, you will learn how to use the **boxchart** function and gain a deeper interest in statistical terms such as the median, interquartile range, minimum, maximum, and outliers. You will understand a different way of representing data distribution compared to histograms (covered in ▶ Chap. 11). Moreover, you will learn about the notch feature of the **boxchart** function. This may increase your interest in the median. However, note that this book does not cover more advanced statistical topics.

Overview

In this chapter, we create a box plot using the **boxchart** function. The box plot visually represents the distribution of the data, including the median, interquartile range, minimum, maximum, and outliers. If you install the Statistics and Machine Learning Toolbox, you can create a box plot using the **boxplot** function instead of the **boxchart** function. It is important to note that while many features are common between the two functions, the **boxchart** function does not include all of the features of the more specialized **boxplot** function. According to the documentation for the **boxplot** function, it differs in the following four points:

- The **boxplot** function supports the use of categorical rulers on the group axis.
- The **boxplot** function provides legend options.
- The **boxplot** function supports the use of the **hold on** command.
- The **boxplot** function includes an improved visual design for notches.

13.1 Details of the Data Used in this Chapter

13

The following data represent the GDP growth rates for the G20 countries (excluding the European Union) and their GDP growth rates for the year 2021.

```
countries = ["Argentina","Australia","Brazil","Canada","China",...
    "Germany","France","United Kingdom","Indonesia","India","Italy",...
    "Japan","Korea, Rep.","Mexico","Russian Federation","Saudi Arabia",...
    "Turkiye","United States","South Africa"]';

area1 = ["other_than_euro","other_than_euro","other_than_euro",...
    "other_than_euro","other_than_euro","euro","euro","other_than_euro",...
    "other_than_euro","other_than_euro","euro","other_than_euro",...
    "other_than_euro","other_than_euro","other_than_euro","other_than_euro",...
    "other_than_euro","other_than_euro","other_than_euro"]';

area2 = ["non_eu_asia","non_eu_asia","non_eu_asia","non_eu_asia","Asia",...
    "Europe","Europe","Europe","Asia", "Asia","Europe","Asia","Asia",...
    "non_eu_asia","Europe","Asia","Asia","non_eu_asia","Asia"]';

gdp_growth_2021 = [-1.55,3.02,2.91,1.07,5.20,-0.30,0.70,0.10,5.05,7.58,...
    0.92,1.92,1.36,3.23,3.60,-0.75,4.52,2.54,0.60]';
```

Based on the data in 「gdp_growth_2021」, the following data include synthetic modifications. Some numbers differ from 「gdp_growth_2021」.

```
gdp_growth_nan = [-1.55,3.02,2.91,1.07,5.20,-0.30,0.70,0.10,25.00,7.58,...
    0.92,1.92,1.36,3.23,3.60,-0.75,20.00,2.54,22.00]';

g19_gdp = table(countries,area1,area2,gdp_growth_2021,gdp_growth_nan);
```

13.2 Excel Data of ▶ Section 13.1

▶ Section 13.2 involves importing the data used in ▶ Sects. 13.3, 13.4, 13.5, and 13.6, formatted as an Excel file. We identify the file name (「alldata.xlsx」) and the sheet name (「Chap13_gdp」). For more information about importing data from Excel, see ▶ Chap. 2.

```
file = "alldata.xlsx";
sheet1 = "Chap13_gdp";
g19_gdp = readtable(file,"Sheet",sheet1);
```

We extract the data from the table variable 「g19_gdp」 and store it in the corresponding variables.

```
g19_gdp.Country = string(g19_gdp.Country);
countries = g19_gdp.Country;
area1 = g19_gdp.area1;
area2 = g19_gdp.area2;
gdp_growth_2021 = g19_gdp.gdp_growth_2021;
gdp_growth_nan = g19_gdp.gdp_growth_nan;
```

13.3 Basic Syntax of the Boxchart Function

Basic Syntax:

The **boxchart** function creates a box plot figure that visualizes the statistical distribution of data. Various syntax variations can be viewed by typing 「doc boxchart」 in the Command Window. Figure 13.1 shows the syntaxes for the **boxchart** function.

The first syntax inside the black box is the basic syntax of the **boxchart** function, 「boxchart(ydata)」 . This command creates an ungrouped box plot for a single data set, where 「ydata」 represents the values you want to visualize using the **boxchart** function.

In the second syntax, 「boxchart(xgroupdata,ydata)」 , we can use two types of data. Here, 「xgroupdata」 represents the grouping variable, while 「ydata」 represents the data values. This syntax allows you to create grouped box plots where the data is categorized based on 「xgroupdata」

Fig. 13.1 The syntax of the boxchart function

Here, we create a box plot of the data for the variable「gdp_growth_2021」, which is a 1-by-19 vector. The box plot provides a summary of the distribution of a data set. Figure 13.2 explains the key elements: Box, median, whiskers, and outliers. The box represents the interquartile range between the 25th percentile (the lower edge of the box) and the 75th percentile (the upper edge of the box). Next, the horizontal line inside the box represents the median (50th percentile) of the data, which is located around y=2. The whiskers extend from the box to the minimum and maximum values, which are not considered outliers. There are no outliers in Fig. 13.2. Outliers, if present, would typically be marked as single points beyond the ends of the whiskers.

```
boxchart(gdp_growth_2021);
```

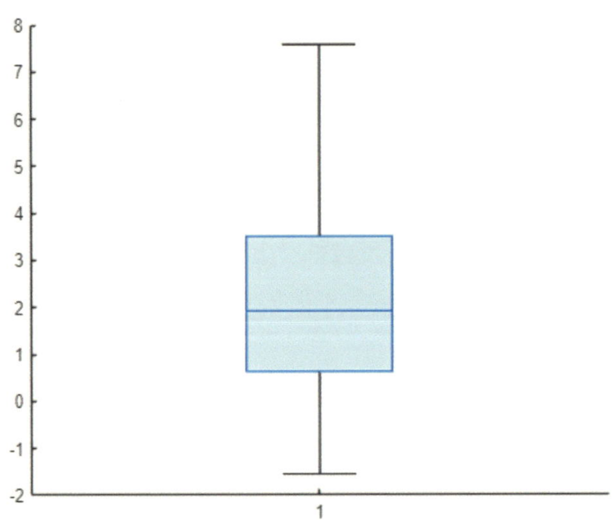

Fig. 13.2 Box plot of a single variable

Here, a 19-by-2 matrix 「gdp_col」, which consists of 「gdp_growth_2021」 and 「gdp_growth_nan」, is used as the first argument of the **boxchart** function. As a result, two types of box plots are created. Notably, in the box plot created using the hypothetical data from 「gdp_growth_nan」, three circular points appear as outliers. See ◘ Fig. 13.3.

```
gdp_col = [gdp_growth_2021,gdp_growth_nan];
boxchart(gdp_col);
```

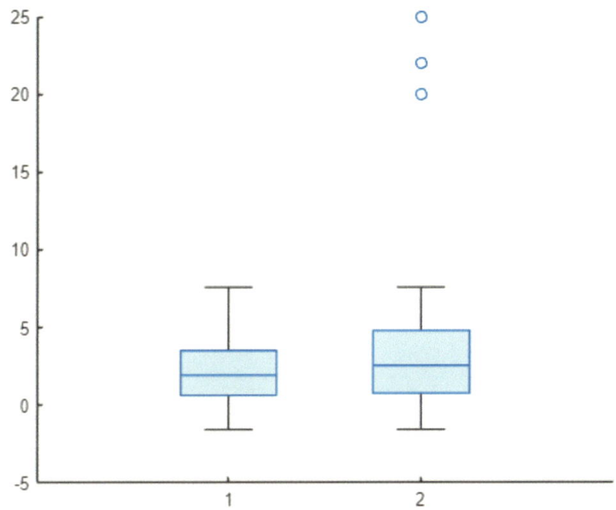

◘ **Fig. 13.3** Box plots of two variables

Attention

Error Program: This error occurs because the data used for 「ydata」 of the basic syntax 「boxchart(ydata)」 is of type string, which is not the expected type in the **boxchart** function. The **boxchart** function requires numeric data (such as double, single, uint8, int32, etc.) for 「ydata」. However, it currently uses string data, which causes the error presented in ◘ Fig. 13.4.

```
%%% Error program
% boxchart(area1);
```

```
Error using boxchart
Expected ydata to be one of these types:

double, single, uint8, uint16, uint32, uint64, int8, int16, int32, int64

Instead its type was string.

Error in boxchart (line 100)
validateattributes(ydata,{'numeric'},{'2d','real'},mfilename,'ydata');
```

◘ **Fig. 13.4** Error output of the boxchart function: Incorrect use of data

Attention

Error Program: The following error occurs because the data used for 「xgroupdata」 in the basic syntax 「boxchart(xgroupdata,ydata)」 is of type string, which is not the type expected by the **boxchart** function. The **boxchart** function requires numeric or categorical data (such as double, single, uint8, int32, int64, categorical, etc.) for 「xgroupdata」. However, it currently uses string data. See Fig. 13.5 for the error message. In the program shown in Fig. 13.6, the **categorical** function is used to convert the data type of 「xgroupdata」 to an appropriate type to resolve the error.

```
%%% Error program
% boxchart(area1,gdp_growth_2021);
```

```
Error using boxchart
Expected xgroupdata to be one of these types:

double, single, uint8, uint16, uint32, uint64, int8, int16, int32, int64, categorical

Instead its type was string.

Error in boxchart (line 101)
validateattributes(xgroupdata,{'numeric','categorical'},{'vector','real'},mfilename,'xgroupdata');
```

Fig. 13.5 Error output of the boxchart function: Incorrect use of data (two arguments)

The **categorical** function is used. Since 「area1」 consists of two groups: 「euro」 and 「other_than_euro」, two box plots are created as shown in Fig. 13.6. Labels are automatically added to the x-axis. When you look at the labels on the x-axis, some of the text in 「other_than_euro」 appears smaller. This is because there is an underscore in the variable name. The correction for this is explained in the program that creates Fig. 13.29. See Fig. 13.6 for the created figure.


```
group1 = categorical(area1);
boxchart(group1,gdp_growth_2021);
```

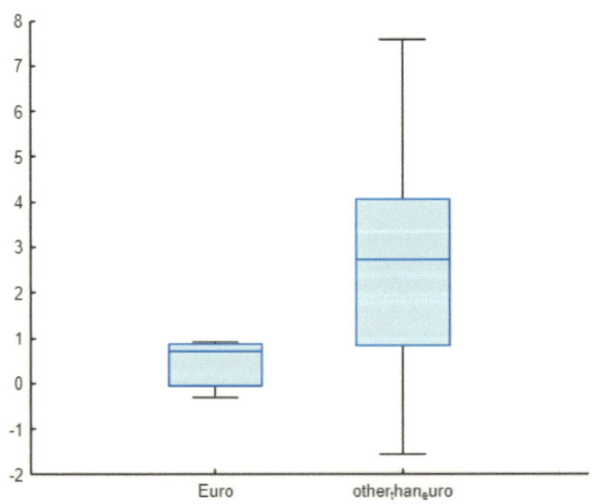

Fig. 13.6 Box plots with two categorical groups

Here, the variable 「countries」 is used. When using the **categorical** function, each country name is categorized. Since there is only one data point for each country, only a horizontal line (median) is provided, and neither the box nor the whiskers are created. See ◘ Fig. 13.7 for the created figure.

```
countries = categorical(countries);
boxchart(countries,gdp_growth_2021);
```

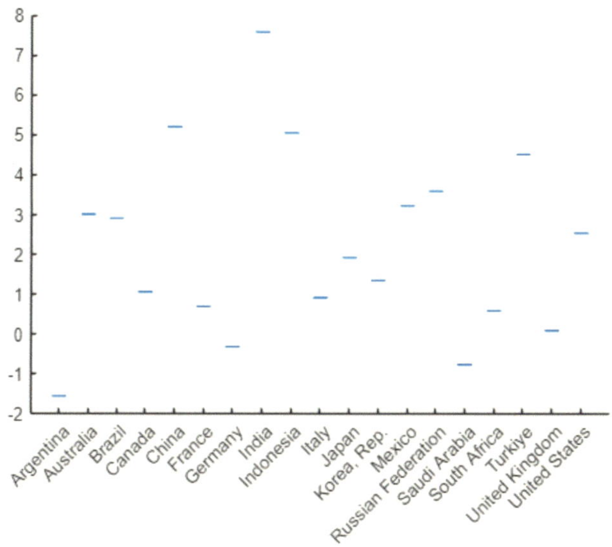

◘ **Fig. 13.7** Box plots with only median lines

The variable 「area2」 is used. Since there are three groups: 「Asia」, 「Europe」, and 「non_eu_asia」, three box plots are created. When looking at the labels on the x-axis, some of the text in 「non_eu_asia」 appears smaller. This is because there is an underscore in the variable name. The underscore problem is solved in the program that generates ◘ Fig. 13.29. The resulting figure is presented in ◘ Fig. 13.8.

```
group2 = categorical(area2);
boxchart(group2,gdp_growth_2021);
```

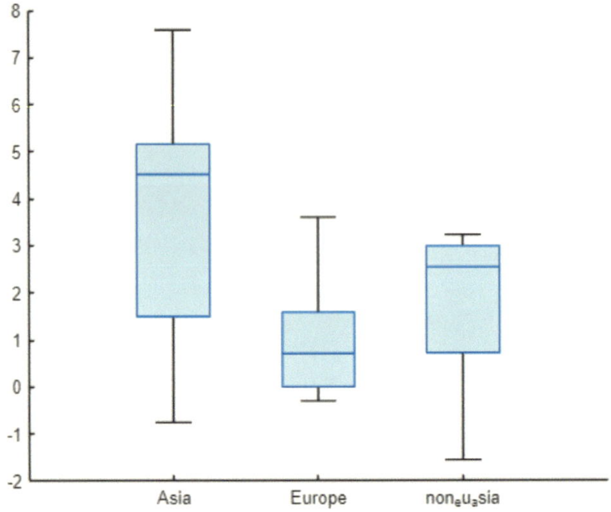

Fig. 13.8 Box plots with multiple categorical groups

Attention

Error Program: This error occurs because 「xgroupdata」 used in the **boxchart** function is not a vector. Specifically, here 「[group1,group2]」 in the first argument of the **boxchart** function becomes a 19-by-2 matrix of categorical data, which is not a vector. See Fig. 13.9 for the error message.

```
%%% Error program
% group1 = categorical(area1);
% group2 = categorical(area2);
% boxchart([group1,group2],gdp_growth_2021);
```

13

```
Error using boxchart
Expected xgroupdata to be a vector.

Error in boxchart (line 101)
validateattributes(xgroupdata,{'numeric','categorical'},{'vector','real'},mfilename,'xgroupdata');
```

Fig. 13.9 Error output of the boxchart function: Invalid matrix input

Attention

Error Program: Here, the error occurs because of the mismatch in the sizes of 「xgroupdata」 and 「ydata」 in the **boxchart** function. The error message is shown in Fig. 13.10.

```
%%% Error program
% group1 = categorical(area1);
% group2 = categorical(area2);
% group_col = [group1;group2];
% boxchart(group_col,gdp_growth_2021);
```

```
Error using boxchart (line 105)
xgroupdata must be the same size as ydata.
```

Fig. 13.10 Error output of the boxchart function: Inconsistent input lengths

As in the previous program in Fig. 13.10, a column vector is created for 「group_col」. Then, by creating a column vector that uses 「gdp_growth_2021」 twice, as seen in 「gdp_col=[gdp_growth_2021;gdp_growth_2021]」, the lengths of the column vectors 「group_col」 and 「gdp_growth_2021」 become equal, creating five box plots. See Fig. 13.11.

```
group1 = categorical(area1);
group2 = categorical(area2);
group_col = [group1;group2];
gdp_col = [gdp_growth_2021;gdp_growth_2021];
boxchart(group_col,gdp_col);
```

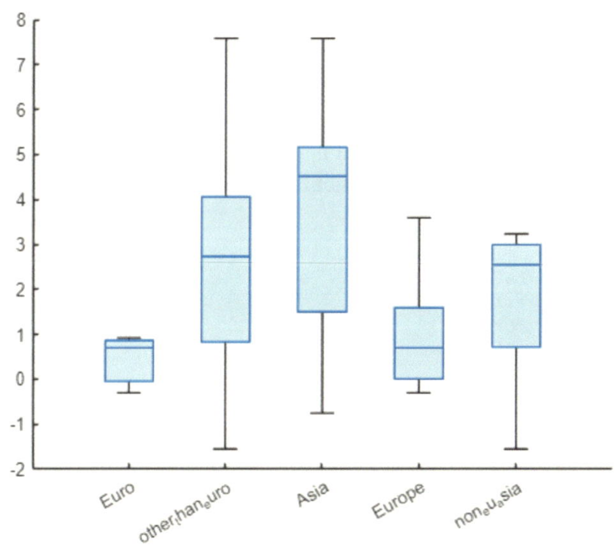

Fig. 13.11 Box plots with multiple categorical groups from combined variables

Suppose that you want to extract the data for 「Asia」 from 「group2」. As shown in Fig. 13.12, within 「group2」, 「Asia」 appears after 「non_eu_asia」, so that it is regarded as the second category.

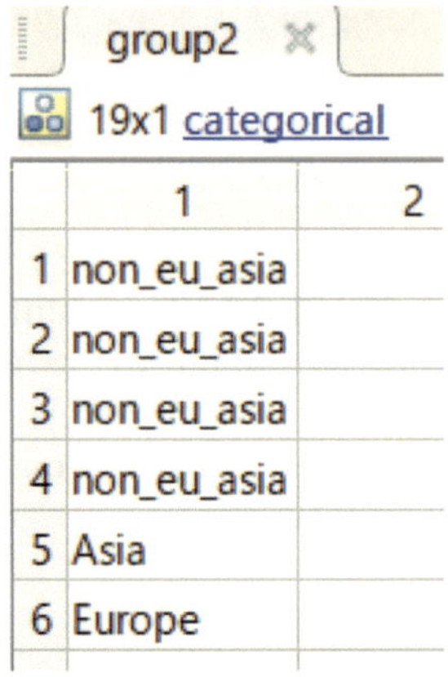

Fig. 13.12 Contents of a categorical variable

Thus, suppose that you want to extract the data for 「Asia」 by using 「group2(2)」. This is incorrect. In this case, 「group2(2)」 takes only one data, which is the second data of the column vector 「group2」. That is 「group2(2)=non_eu_asia」. Therefore, 「group_col=[group1;group2(2)]」 becomes a 20-by-1 column vector consisting of 19 rows and 1 column from 「group1」 and 1 row and 1 column from 「group2(2)」. The resulting figure is presented in Fig. 13.13. To extract the data for 「Asia」, you need to use a logical index. For instructions on how to use it, see the program in Fig. 13.15.

```
group1 = categorical(area1);
group2 = categorical(area2);
group_col = [group1;group2(2)];
gdp_col = [gdp_growth_2021;gdp_growth_2021(2)];
boxchart(group_col,gdp_col);
```

13

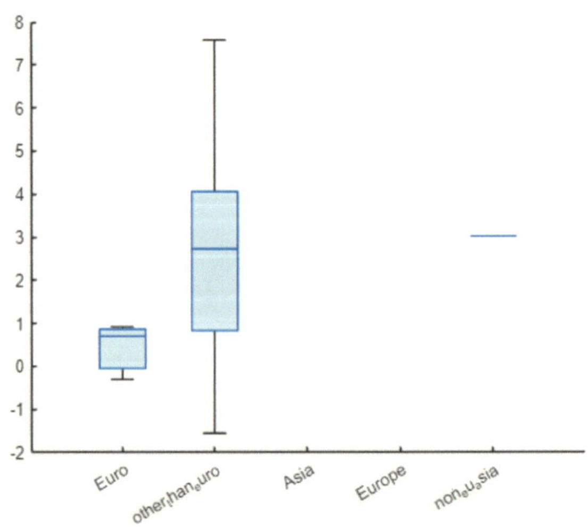

Fig. 13.13 Box plots due to incorrect data extraction

By using a logical index, you can extract only the data for 「Asia」 as follows. In 「asia_idx=group2=="Asia"」, 「Asia」 in 「group2」 are set to 1 (=TRUE), and all others are set to 0 (=FALSE). See ◘ Fig. 13.14. Note that 「asia_idx」 is a 19-by-1 logical array, containing true or false values.

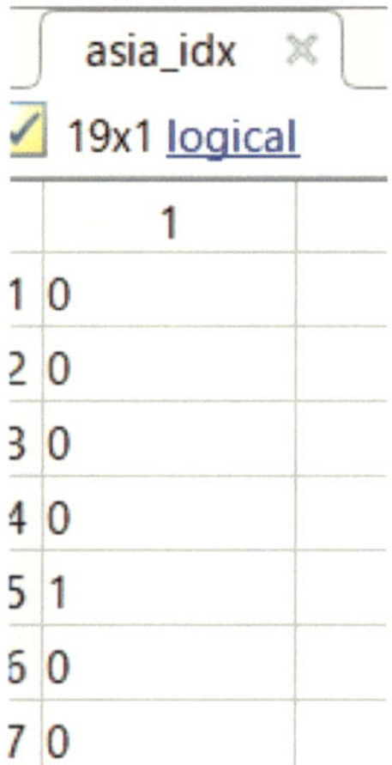

◘ **Fig. 13.14** Contents of a logical index

With 「group2_asia=group2(asia_idx)」 and 「gdp_growth_asia=gdp_growth_2021(asia_idx)」, only the data where 「asia_idx」 equals 1 (=TRUE) is extracted. In other words, the variables 「group2_asia」 and 「gdp_growth_asia」 contain only the data for 「Asia」. The **boxchart** function creates three box plots for 「Euro」, 「other_than_euro」, and 「Asia」. The created figure is shown in ◘ Fig. 13.15.

```
group1 = categorical(area1);
group2 = categorical(area2);
asia_idx = group2 == "Asia";
group2_asia = group2(asia_idx);
gdp_growth_asia = gdp_growth_2021(asia_idx);
group_col = [group1;group2_asia];
gdp_col = [gdp_growth_2021;gdp_growth_asia];
boxchart(group_col,gdp_col);
```

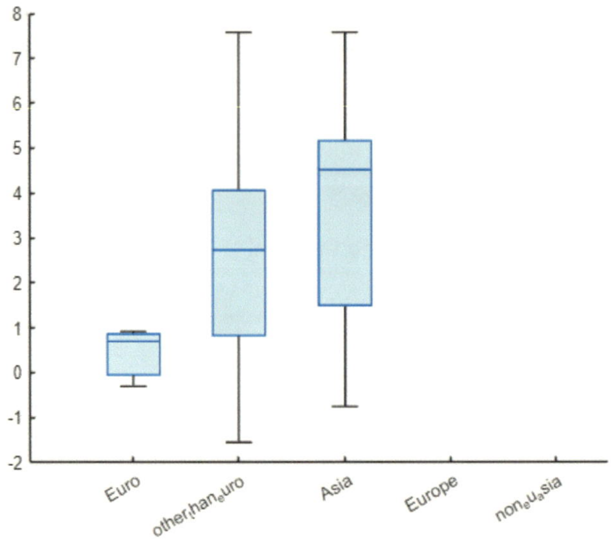

Fig. 13.15 Box plots with extra categories on the X-axis

▶ **Example of Table Variables**

Here, we create a box plot using the table variable「g19_gdp」. See Fig. 13.16.

```
group2 = categorical(g19_gdp.area2);
boxchart(group2,g19_gdp.gdp_growth_2021);
```

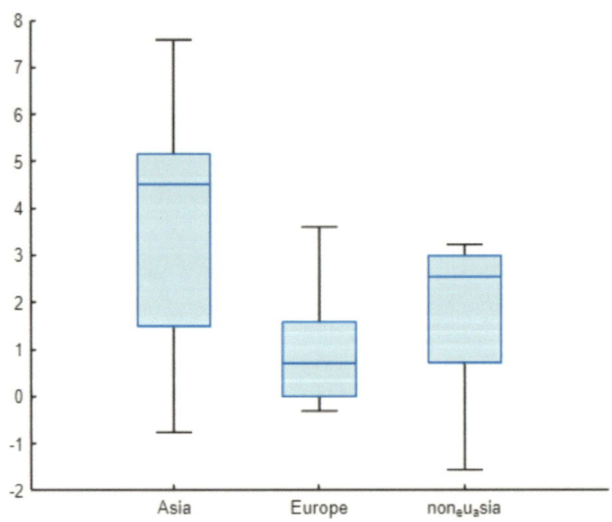

Fig. 13.16 Use of a table variable: Box plots

◀

13.4 Setting and Controlling the Boxchart Function

The properties created by the **boxchart** function are stored in the variable 「bxc」. When setting 「BoxFaceColor」 to 「red」, we change the color of the box to red. See Fig. 13.17.

```
group1 = categorical(area1);
bxc = boxchart(group1,gdp_growth_2021);
    bxc.BoxFaceColor = "red";
```

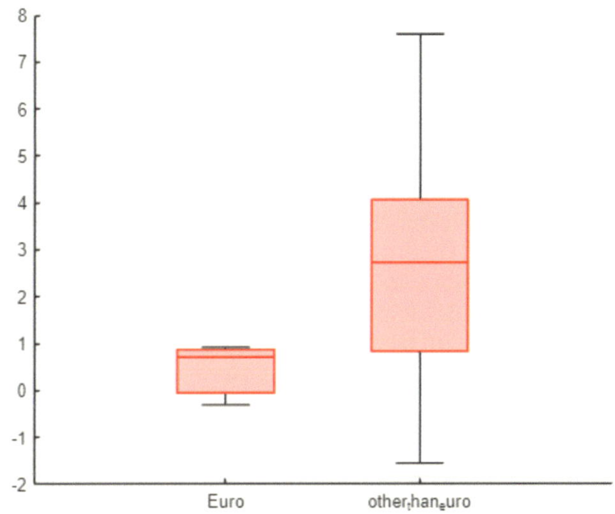

Fig. 13.17 Box plots with red boxes

Here, we set 「BoxFaceColor」 to 「cyan」, 「WhiskerLineColor」 to 「blue」, 「WhiskerLineStyle」 to 「--」, 「LineWidth」 to 2, and 「BoxMedianLineColor」 to 「black」. This creates a box plot where the box is cyan, the median line is black, the whiskers are blue with a dashed line, and all lines are wider. See Fig. 13.18.

```
group1 = categorical(area1);
bxc = boxchart(group1,gdp_growth_2021);
    bxc.BoxFaceColor = "cyan";
    bxc.WhiskerLineColor = "blue";
    bxc.WhiskerLineStyle = "--";
    bxc.LineWidth = 2;
    bxc.BoxMedianLineColor = "black";
```

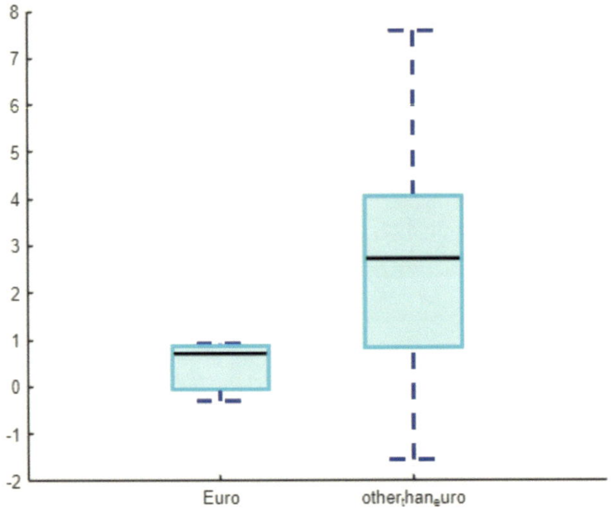

Fig. 13.18 Box plots with cyan boxes, black median lines, and dashed blue whiskers

Here, we set 「BoxFaceColor」 to 「red」, 「BoxFaceAlpha」 to 0.8, 「BoxMedian LineColor」 to 「black」, and 「WhiskerLineStyle」 to 「:」. This creates a box plot where the box color is red with less transparency, the median line is black, and the whiskers are a black dotted line. See Fig. 13.19.

```
group1 = categorical(area1);
bxc = boxchart(group1,gdp_growth_2021);
    bxc.BoxFaceColor = "red";
    bxc.BoxFaceAlpha = 0.8;
    bxc.BoxMedianLineColor = "black";
    bxc.WhiskerLineStyle = ":";
```

13

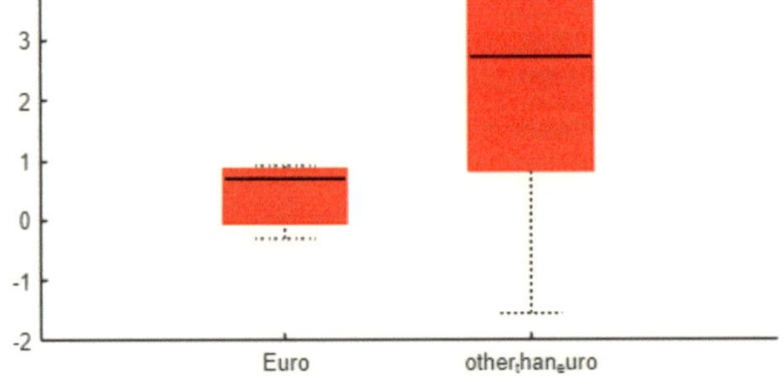

Fig. 13.19 Box plots with red boxes, black median lines, and dotted whiskers

Attention

Error Program: The following program will result in an error. Because the **boxchart** function creates only one box plot, accessing individual box plots using an index like 「bxc(2)」 will cause an error. For details about the error, refer to Fig. 13.20. If you want to do something like this, you need to use 「GroupByColor」 as displayed in Fig. 13.21.

```
%%% Error program
% group1 = categorical(area1);
% bxc = boxchart(group1,gdp_growth_2021);
% bxc(1).BoxFaceColor = "red";
% bxc(2).BoxFaceColor = "green";
```

```
Unrecognized property 'BoxFaceColor' for class 'matlab.graphics.GraphicsPlaceholder'.
```

Fig. 13.20 Error output of the boxchart function: Unsupported use of different colors for each box

Here, 「GroupByColor」 is used to color the box plots by group. See Fig. 13.21. Note that the box plots are not created directly above the x-axis ticks. The reason for this is likely to allow multiple box plots to be placed side by side above a single x-axis tick. For instance, see Fig. 13.23, which shows multiple box plots on a single x-axis tick. Additionally, the abovementioned issue in Fig. 13.21 is resolved in the program that will create Fig. 13.31.

```
group1 = categorical(area1);
boxchart(group1,gdp_growth_2021,"GroupByColor",group1);
```

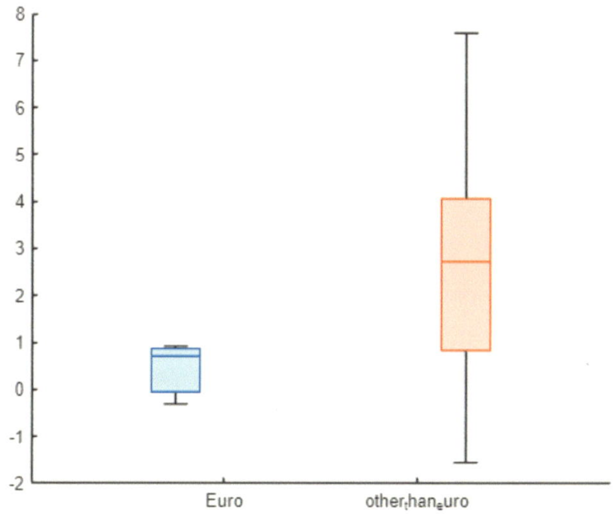

Fig. 13.21 Two-tick single box plots using "GroupByColor" property

Here, the colors for five box plots are changed using 「GroupByColor」. Note that the box plots are not created directly above the x-axis ticks. This issue is resolved in the program that will create ◘ Fig. 13.31. See ◘ Fig. 13.22 for the created figure.

```
group1 = categorical(area1);
group2 = categorical(area2);
group_col = [group1;group2];
gdp_col = [gdp_growth_2021;gdp_growth_2021];
boxchart(group_col,gdp_col,"GroupByColor",group_col);
```

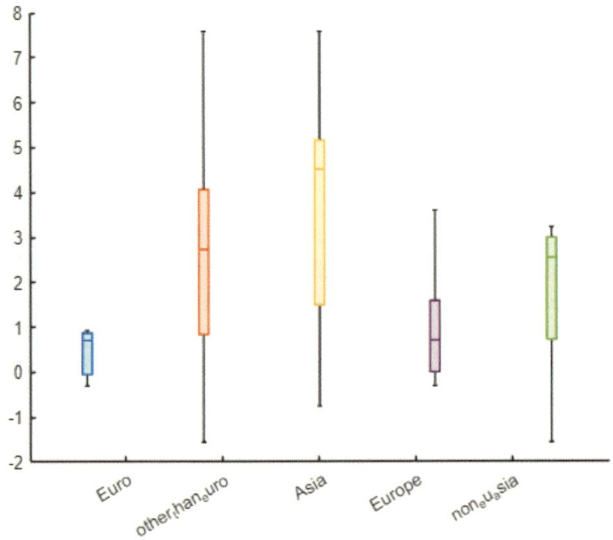

◘ **Fig. 13.22** Multiple-tick single box plots using "GroupByColor" property

Here, five box plots are arranged on a single x-axis tick. The colors are then changed using 「GroupByColor」. The resulting figure is presented in ◘ Fig. 13.23.

```
group1 = categorical(area1);
group2 = categorical(area2);
group_col = [group1;group2];
gdp_col = [gdp_growth_2021;gdp_growth_2021];
boxchart(gdp_col,"GroupByColor",group_col);
```

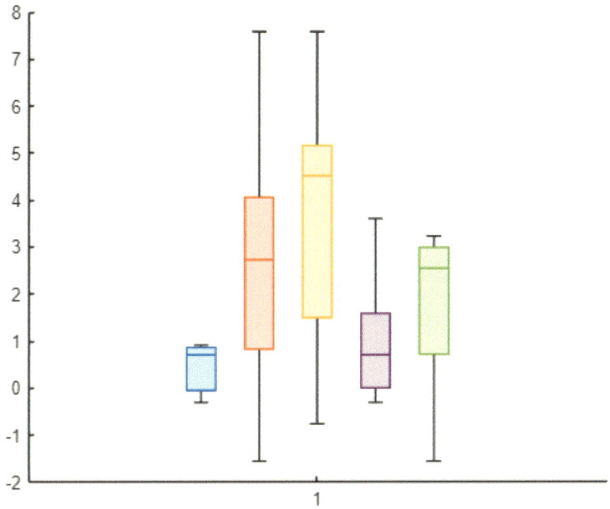

Fig. 13.23 Single-tick grouped box plots using "GroupByColor" property

Here, we set 「Notch」 to 「on」, creating notched box plots. Each box contains a tapered, shaded region called a notch. For the details about the notch, refer to the MATLAB documentation. The resulting figure is presented in Fig. 13.24.

```
group2 = categorical(area2);
boxchart(group2,gdp_growth_2021,"Notch","on");
```

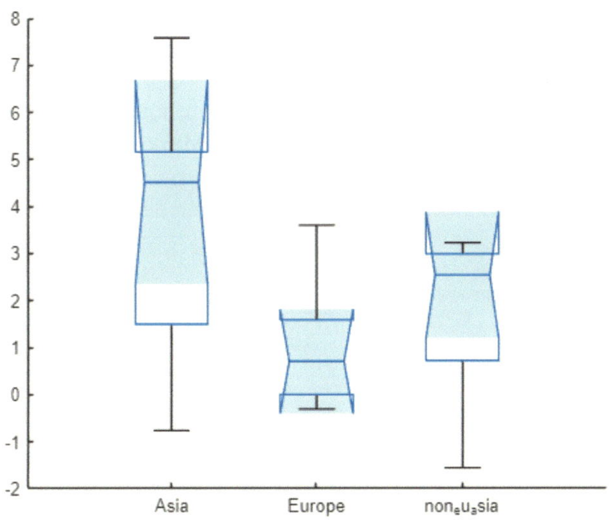

Fig. 13.24 Notched box plots

Here, the properties of the notched box plot are stored in 「bxc」. Then, 「WhiskerLineStyle」, 「LineWidth」, and 「BoxMedianLineColor」 are modified. Compare ◘ Fig. 13.24 with ◘ Fig. 13.25.

```
group2 = categorical(area2);
bxc = boxchart(group2,gdp_growth_2021,"Notch","on");
    bxc.WhiskerLineStyle = "--";
    bxc.LineWidth = 2;
    bxc.BoxMedianLineColor = "red";
```

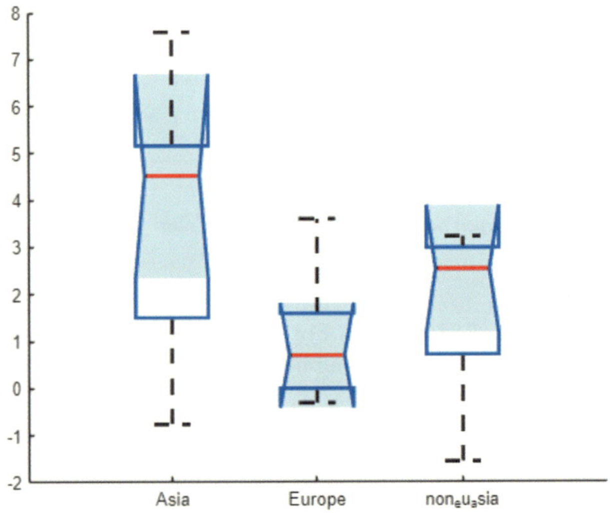

◘ **Fig. 13.25** Notched box plots with red median lines and black whiskers

13

The following program uses data containing outliers. There are three outliers, represented by circle markers. Specifically, we set 「MarkerSize」 to 15 and 「Marker Color」 to 「blue」, making the outlier markers large and their color blue. See ◘ Fig. 13.26.

```
group1 = categorical(area1);
bxc = boxchart(group1,gdp_growth_nan);
    bxc.MarkerSize = 15;
    bxc.MarkerColor = "blue";
```

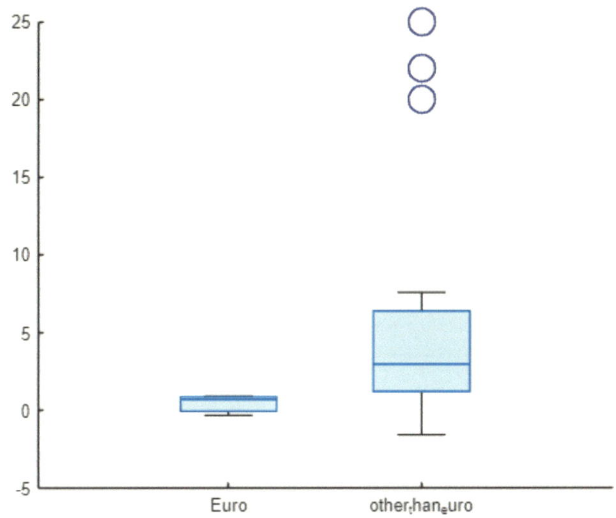

Fig. 13.26 Box plots with large blue outliers

Here, we set 「MarkerStyle」 to 「x」 and 「JitterOutliers」 to 「on」, creating outliers with cross markers and enabling the jitter effect for outliers. See **Fig. 13.27**.

```
group1 = categorical(area1);
bxc = boxchart(group1,gdp_growth_nan);
    bxc.MarkerStyle = "x";
    bxc.JitterOutliers = "on";
```

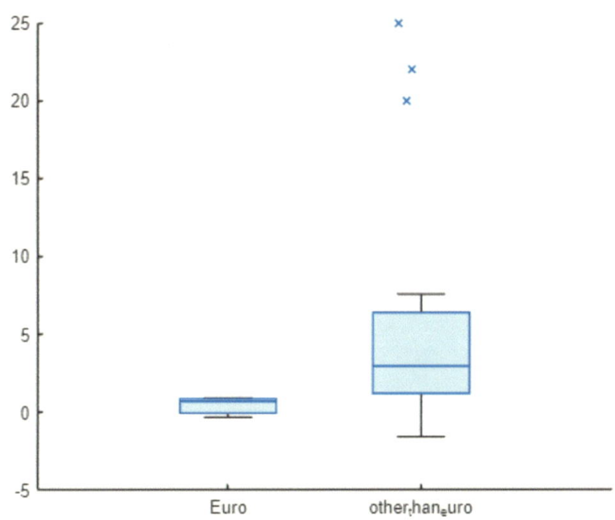

Fig. 13.27 Box plots with jittered-cross outliers

Here, 「Orientation」 is set to 「horizontal」, which creates a horizontal box plot. See Fig. 13.28.

```
group1 = categorical(area1);
boxchart(group1,gdp_growth_2021,"Orientation","horizontal");
```

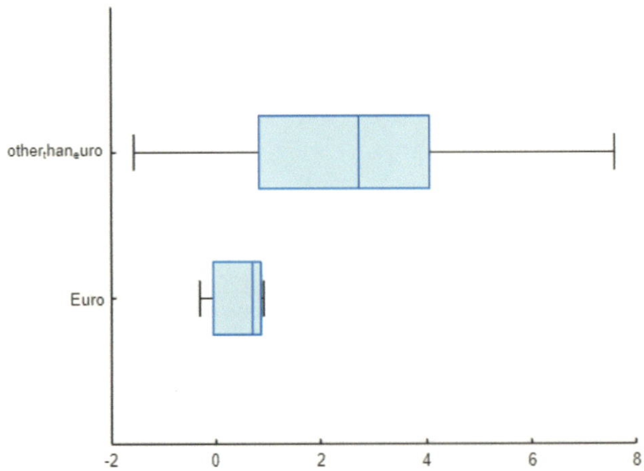

Fig. 13.28 Horizontal box plots

13.5 Combining Optional Functions and the Boxchart Function

Here, the issue of smaller text on the x-axis in Figs. 13.6 and 13.8 is resolved. Specifically, the **xticklabels** function is used. See Fig. 13.29. Additionally, note that if you use 「boxchart(gdp_growth_2021)」 instead of 「boxchart(group1,gdp_growth_2021)」, only one box plot will be created.

```
group1 = categorical(area1);
boxchart(group1,gdp_growth_2021);
xticklabels({"Euro","Other than Euro"});
```

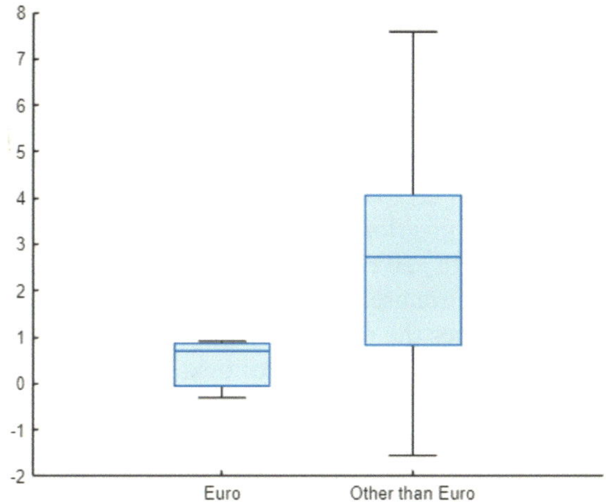

Fig. 13.29 Box plots with manually created X-axis labels

The **xticklabels** and **xlabel** functions are used to place text on the x-axis. Additionally, the **legend** function is used to create a legend. Specifically, check that the colors changed by 「GroupByColor」 are reflected in the legend. See Fig. 13.30.

```
group2 = categorical(area2);
boxchart(gdp_growth_2021,"GroupByColor",group2);
xticklabels({"xticklabels:region"});
xlabel("xlabel:region");
ylabel("GDP Growth 2023");
title("GDP Growth for different regions");
legend("Others","Asia","Europe");
grid on
```

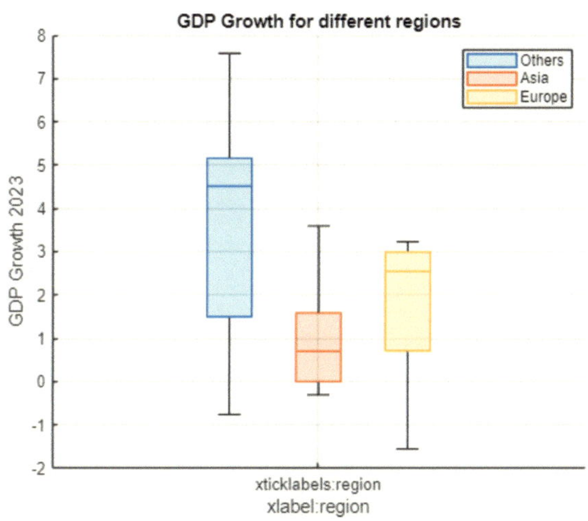

Fig. 13.30 Single-tick grouped box plots with a two-line X-axis label and a legend

Here, we address the issue in ◘ Figs. 13.21 and ◘ 13.22, where trying to create a box plot on an x-axis tick using「GroupByColor」did not result in the box plot above the x-axis tick. See the position of the box plots in ◘ Figs. 13.21 and ◘ 13.22. We now use the **hold on** command to overlay the box plots created by the **boxchart** function. Note that because the **hold on** command causes the box plots to be drawn at the same location, we need to specify the positions. Specifically, we set「x1=1」and「x2=3」to define the positions on the x-axis, and then use「x_position1=x1*ones(size(gdp_growth_2021))」and「x_position2=x2*ones(size(gdp_growth_nan))」. These values are used in the first argument of the **boxchart** function for each plot, which determines the positions of each box.

Finally, the **xticks** function is used to place ticks at positions x=1 and x=3, and the **xticklabels** function is used to change the tick labels to the text strings「gdp 2023」and「hypothetical data」. See ◘ Fig. 13.31.

```
x1 = 1;
x2 = 3;
x_position1 = x1*ones(size(gdp_growth_2021));
x_position2 = x2*ones(size(gdp_growth_nan));
bxc1 = boxchart(x_position1, gdp_growth_2021);
    bxc1.BoxFaceColor = "red";
hold on
bxc2 = boxchart(x_position2,gdp_growth_nan);
    bxc2.BoxFaceColor = "blue";
xticks([1,3]);
xticklabels({"gdp 2023","hypothetical data"});
hold off
```

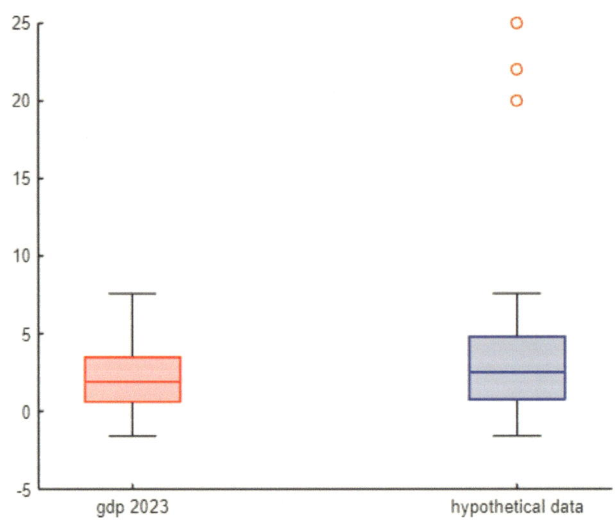

◘ **Fig. 13.31** Box plots with a large space between boxes

13

In ◘ Fig. 13.31, box plots were created at custom x-axis positions. Additionally, the **hold on** command was used to create the plots, but the **hold on** command is not used here. First, 「x_positions=[ones(size(gdp_growth_2021));3*ones(size(gdp_growth_2021))]」 is used to specify the x-axis positions. The created variable 「x_positions」 becomes a 38-by-1 column vector, where the first 19 elements are set to 1, and the remaining 19 elements are set to 3. 「x_positions」 is used as the first argument in the **boxchart** function, and the 38-by-1 column vector 「gdp_col」 is used as the second argument. Then, the **xticks** and **xticklabels** functions are used to place text labels at x=1 and x=3 on the axis. See ◘ Fig. 13.32 for the created figure.

```
group2 = categorical(area2);
group_col = [group2;group2];
gdp_col = [gdp_growth_2021;gdp_growth_nan];
x_positions = [ones(size(gdp_growth_2021));3*ones(size(gdp_growth_2021))];
boxchart(x_positions,gdp_col,"GroupByColor",group_col);
xticks([1 3]);
xticklabels({"gdp 2023","Hypothetical data"});
legend(["Asia","Europe","Others"]);
```

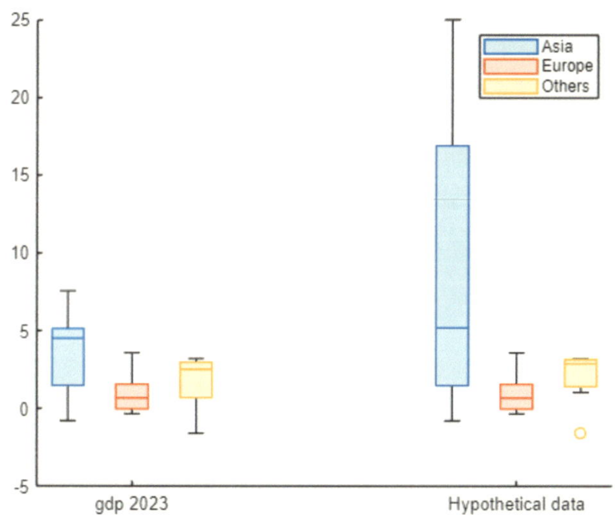

◘ **Fig. 13.32** Two-tick grouped box plots with a large space between groups

Here, text is placed at the specified locations of the box plots. Specifically, after the properties of the **boxchart** function are stored in the variable 「bxc」, we retrieve the categories and corresponding values using 「XData」 and 「YData」. Then, the list of groups is obtained using 「uniqueGroups=unique(xData)」. In this case, 「uniqueGroups」 becomes a 2-by-1 categorical array containing 「Euro」 and 「other_than_euro」.

Inside the **for** loop, the text is displayed repeatedly. When 「i=1」, 「groupIdx=xData==uniqueGroups(1)」 checks if the data corresponds to whether 「uniqueGroups(1)」 is 「Euro」 or not. The elements in the vector 「xData」 corresponding to 「Euro」 are set to 「groupIdx=1 (=TRUE)」, and the other elements are set to 「groupIdx=0 (=FALSE)」. In 「maxVal=max(yData(groupIdx))」, the maximum value of 「yData」 where 「groupIdx=1 (=TRUE)」 is stored in the variable 「maxVal」. Additionally, in 「countVal=sum(groupIdx)」, the number of elements where 「groupIdx=1 (=TRUE)」 is summed.

Finally, at the position (x,y) where 「x=uniqueGroups(1)」 and 「y=maxVal」, the **text** function displays 「The number: 3 (=num2str(countVal))」. For more details on the **text** function, see ▶ Sect. 18.2. The resulting figure is presented in ◘ Fig. 13.33.

```
group1 = categorical(area1);
bxc = boxchart(group1,gdp_growth_2021);
    xData = bxc.XData;
    yData = bxc.YData;
uniqueGroups = unique(xData);
hold on
for i = 1:length(uniqueGroups)
    groupIdx = xData == uniqueGroups(i);
    maxVal = max(yData(groupIdx));
    countVal = sum(groupIdx);
    tx = text(uniqueGroups(i),maxVal,strcat("The number: ",num2str(countVal)),...
        "HorizontalAlignment","center","VerticalAlignment","bottom");
        tx.Color = "red";
end
hold off
title("Boxchart with the number of countries");
xticklabels({"Euro","Other than Euro"});
```

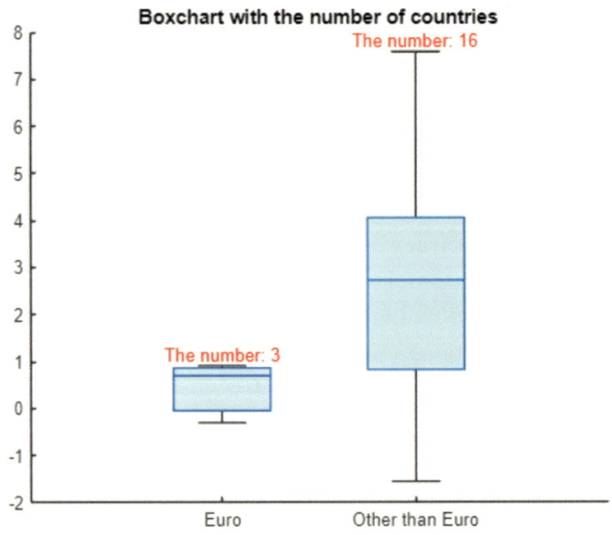

◘ **Fig. 13.33** Box plots with text labels

Here, after retrieving the maximum, minimum, and average values, the **text** function is used to insert these values into the box plot. Specifically, the maximum, minimum, and average values are calculated using 「max(gdp_growth_2021)」, 「min(g dp_growth_2021)」, and 「mean(gdp_growth_2021)」. These values are su-bsequen tly stored in the variables 「max_val」, 「min_val」, and 「mean_val」, respectively.

The **text** function is then used to place these values at specified locations on the box plot. By specifying 「max_val+0.5」, 「min_val-0.5」, and 「x_position+0.5」 as the first and second arguments of each **text** function, we adjust the position of the text. For more details on the **text** function, see ▸ Sect. 18.2. See ◘ Fig. 13.34 for the created figure.

```
boxchart(gdp_growth_2021);
max_val = max(gdp_growth_2021);
min_val = min(gdp_growth_2021);
mean_val = mean(gdp_growth_2021);
x_position = 1;
text(x_position,max_val+0.5,sprintf("MAX: %.1f",max_val),...
    "HorizontalAlignment","center","FontSize",12);
text(x_position,min_val-0.5,sprintf("MIN: %.1f",min_val),...
    "HorizontalAlignment","center","FontSize",12);
text(x_position+0.5,mean_val,sprintf("MEAN: %.1f",mean_val),...
    "HorizontalAlignment","center","VerticalAlignment","bottom",...
    "FontSize",12);
title("Boxchart with Max, Min, and Mean values");
xticklabels({"G19 countries"});
ylim([-4 9]);
grid minor
```

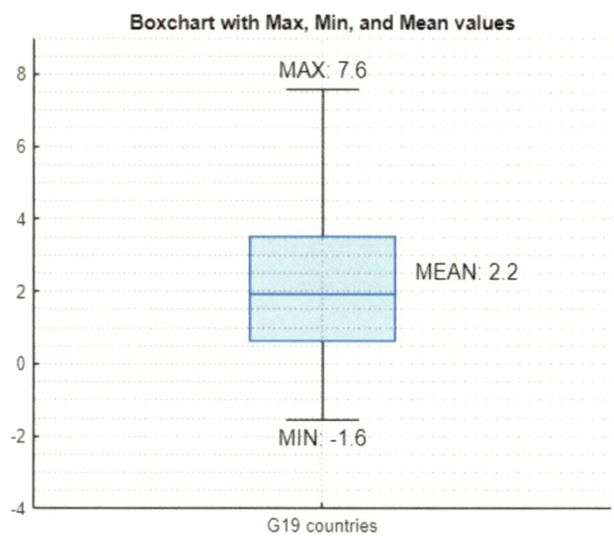

◘ **Fig. 13.34** Box plots with text labels (maximum, mean, and minimum values)

Here, the **hold on** command overlays the figures generated by the **boxchart** and **plot** functions. Specifically, the **groupsummary** function is used to calculate the mean value for each group, and these values are stored in the variable 「meanpoint」. Note that 「meanpoint」 is a 2-by-1 double array containing the mean values for the 「Euro」 and 「other_than_euro」 groups in 「group1」.

Then, by using 「plot(meanpoint,":kx")」, these two means are connected by a dotted black line with cross markers. Additionally, using 「meanpoint(1)」 and 「meanpoint(2)」, the **text** function places text at the specified positions. For more details on the **text** function, see ▶ Sect. 18.2. ◘ Fig. 13.35 shows the created figure.

```
figure
group1 = categorical(area1);
meanpoint = groupsummary(gdp_growth_2021,group1,"mean");
boxchart(group1,gdp_growth_2021);
xticklabels({"Euro","Other than Euro"});
hold on
pl = plot(meanpoint,":kx");
    pl.MarkerSize = 8;
x_position_euro = 1;
x_position_other = 2;
text(x_position_euro-0.55,meanpoint(1),sprintf("MEAN: %.1f",meanpoint(1)),...
    "HorizontalAlignment","center","VerticalAlignment","bottom",...
    "FontSize",12);
text(x_position_other+0.55,meanpoint(2),sprintf("MEAN: %.1f",meanpoint(2)),...
    "HorizontalAlignment","center","VerticalAlignment","bottom",...
    "FontSize",12);
hold off
legend(["gdp 2023","mean gdp"]);
```

13

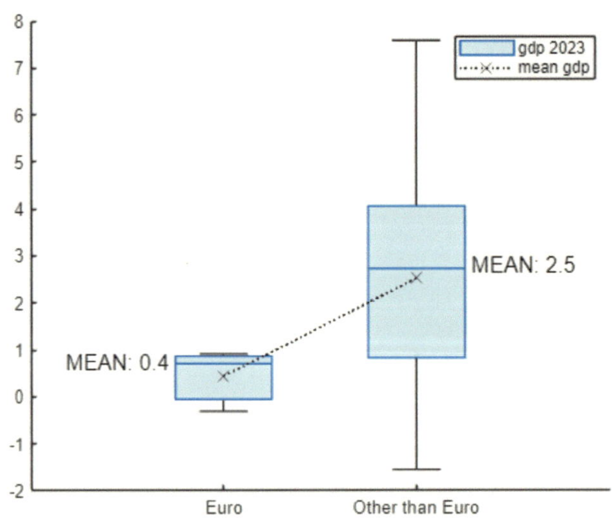

◘ **Fig. 13.35** Box plots with two boxes connected by a dotted line and mean value labels

Here, similar to the program used to create ◘ Fig. 13.35, the **hold on** command overlays the box plots created by the **boxchart** function and the line created by the **plot** function. Unlike ◘ Fig. 13.35, we use the **plot** function three times to create the average, maximum, and minimum lines. For example, in 「groupsummary(gdp_col,group_col,"mean")」, since the variable 「group_col」 is a 38-by-1 vector with five groups, the average values for each group are calculated based on the corresponding elements in 「gdp_col」. See ◘ Fig. 13.36 for the created figure.

```
group1 = categorical(area1);
group2 = categorical(area2);
group_col = [group1;group2];
gdp_col = [gdp_growth_2021;gdp_growth_2021];
meanpoint = groupsummary(gdp_col,group_col,"mean");
maxpoint = groupsummary(gdp_col,group_col,"max");
minpoint = groupsummary(gdp_col,group_col,"min");
boxchart(group_col,gdp_col);
hold on
plmean = plot(meanpoint,":rx");
    plmean.LineWidth = 2;
plmax = plot(maxpoint,":kx");
    plmax.LineWidth = 2;
plmin = plot(minpoint,":bx");
    plmin.LineWidth = 2;
hold off
legend(["gdp 2023","mean","max","min"]);
```

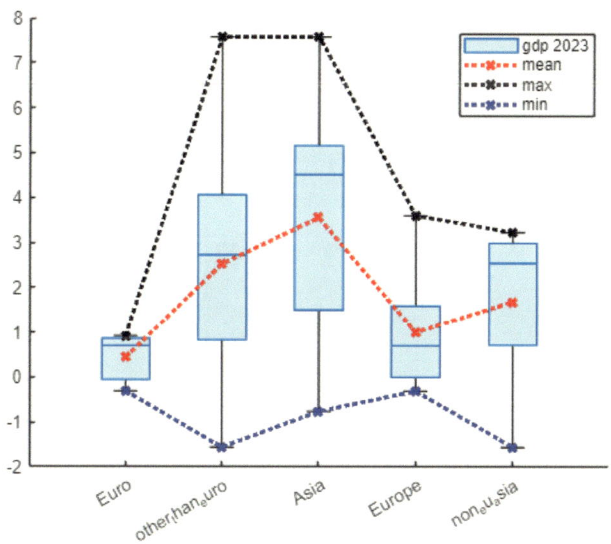

◘ **Fig. 13.36** Box plots with lines connecting maximum, mean, and minimum values

Here, 「tiledlayout(1,2)」 is used to create a 1-by-2 tiled layout. Then, the **boxchart** function is used to create a box plot for each space. See ◘ Fig. 13.37.

```
ti = tiledlayout(1,2);
title(ti,"tiledlayout function: Boxchart");
ax1 = nexttile;
group1 = categorical(area1);
boxchart(group1,gdp_growth_2021,"GroupByColor",group1);
xlabel(ax1,"GDP 2023");
xticklabels({"Euro","Other"});
ylim([-5 25]);
legend;
grid minor
ax2 = nexttile;
group1 = categorical(area1);
boxchart(group1,gdp_growth_nan,"GroupByColor",group1);
xlabel(ax2,"Hypothetical GDP");
xticklabels({"Euro","Other"});
legend;
grid minor
```

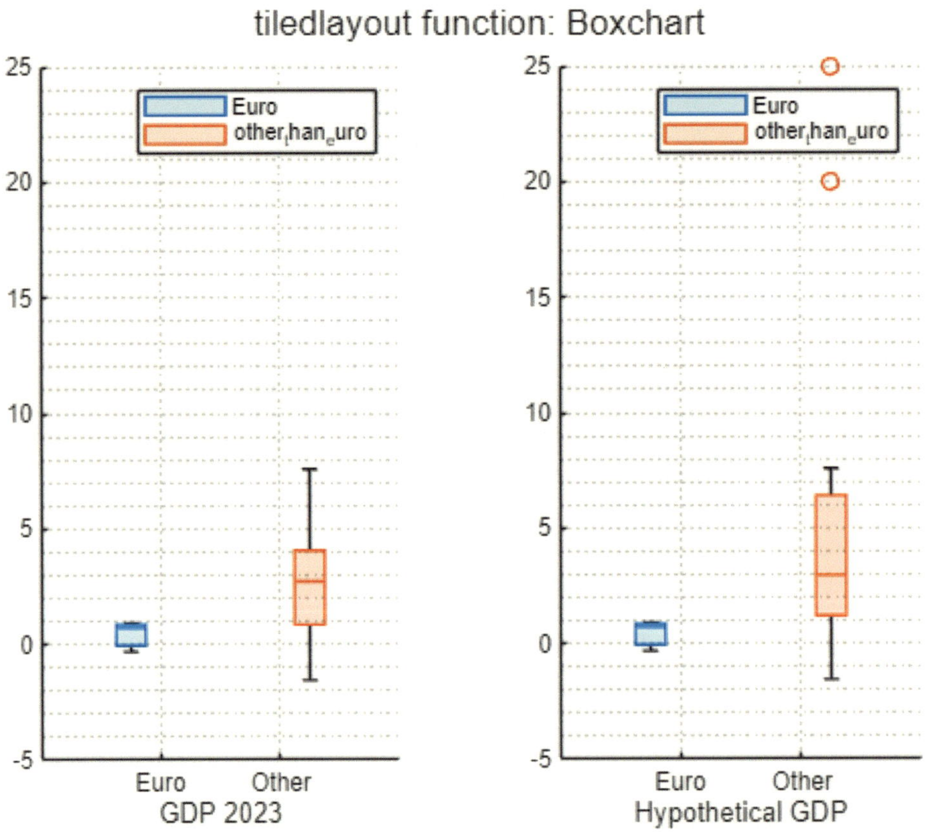

◘ **Fig. 13.37** Box plots (2-column layout)

13.6 Questions

Question 1

Create ■ Fig. 13.38 using the **boxchart** function with 「categorical(area1)」 as the first argument and 「gdp_growth_nan」 as the second argument. Concretely, use red upward-pointing triangles for the outliers. Additionally, using the **xticklabels** function label the ticks on the x-axis of the box plots with 「Euro」 and 「Other than Euro」. Finally, use the **grid minor** command.

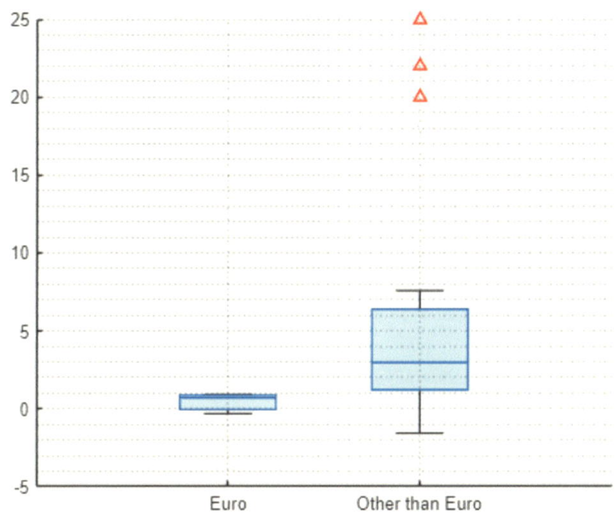

■ **Fig. 13.38** Box plot (Question 1)

Answer 1

```
figure
Ex1 = boxchart(categorical(area1),gdp_growth_nan);
    Ex1.MarkerStyle = "^";
    Ex1.MarkerColor = "red";
xticklabels({"Euro","Other than Euro"});
grid minor
```

Question 2

Create ■ Fig. 13.39 using the **boxchart** function with 「categorical(area2)」 as the first argument and 「gdp_growth_2021」 as the second argument. Set 「Notch」 to 「on」 to add notches to the box plot. Additionally, set 「BoxFaceColor」 to 「blue」, 「BoxFaceAlpha」 to 0.5, 「LineWidth」 to 4, and 「BoxEdgeColor」 to 「green」.

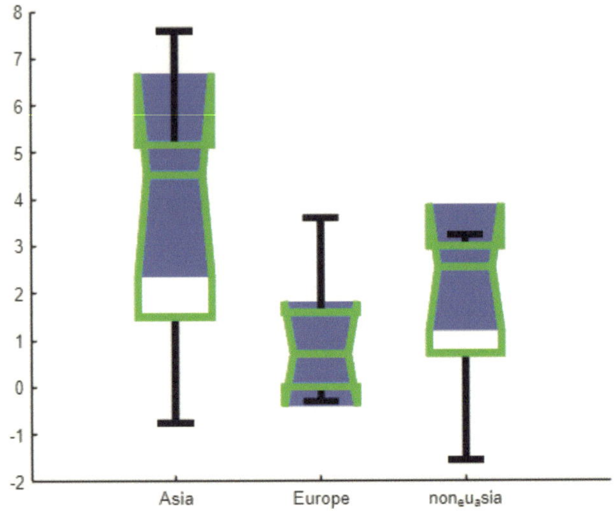

Fig. 13.39 Box plot (Question 2)

Answer 2

```
figure
Ex2 = boxchart(categorical(area2),gdp_growth_2021,"Notch","on");
    Ex2.BoxFaceColor = "blue";
    Ex2.BoxFaceAlpha = 0.5;
    Ex2.LineWidth = 4;
    Ex2.BoxEdgeColor = "green";
```

Question 3

13

Obtain the maximum and minimum values from each of the box plots, and create Fig. 13.40 by connecting the maximum value of the first box plot and the minimum value of the second box plot with a dashed line. Specifically, use the **hold on** command to overlay the line connecting the two points with the **plot** function on the box plots created by the **boxchart** function.

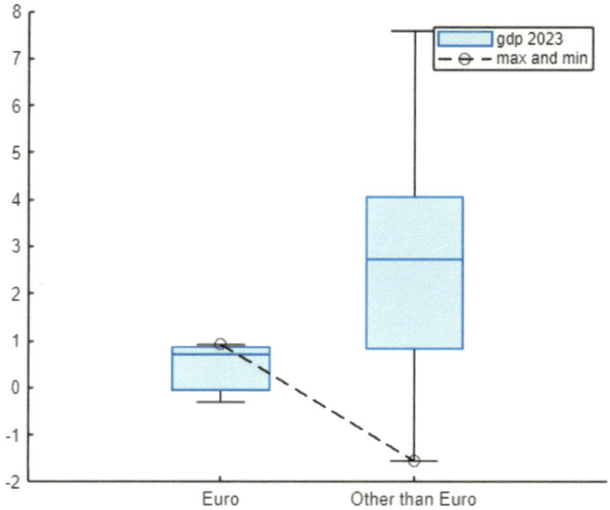

Fig. 13.40 Box plot (Question 3)

Answer 3

```
figure
group1 = categorical(area1);
maxpoints = groupsummary(gdp_growth_2021,group1,"max");
minpoints = groupsummary(gdp_growth_2021,group1,"min");
max_min_points = [maxpoints(1),minpoints(2)];
boxchart(categorical(area1),gdp_growth_2021);
xticklabels({"Euro","Other than Euro"});
hold on
plot(max_min_points,"--ko");
hold off
legend(["gdp 2023","max and min"]);
```

Piechart and Donutchart Functions

Contents

14.1 Details of the Data Used in this Chapter – 440

14.2 Excel Data of ► Section 14.1 – 440

14.3 Basic Syntaxes of the Piechart and Donutchart Functions – 441

14.4 Setting and Controlling the Piechart and Donutchart Functions – 446

14.5 Combining Optional Functions and the Piechart and Donutchart Functions – 453

14.6 Questions – 462

Supplementary Information The online version contains supplementary material available at https://doi.org/10.1007/978-981-96-6213-5_14.

Y. Nakamoto, *A Comprehensive Guide to Creating Graphics with MATLAB Programming*, Classroom Companion: Economics, https://doi.org/10.1007/978-981-96-6213-5_14

Learning Objectives

This chapter creates two kinds of pie charts using the **piechart** and **donutchart** functions. By reading this chapter, you will learn how to use the **piechart** and **donutchart** functions. Additionally, in ▶ Sect. 14.5, you will gain insights and tips into how to utilize the central hole in a donut chart effectively.

Overview

MATLAB provides two functions for creating pie charts: the **pie** and **piechart** functions. The **piechart** function is recommended over the **pie** function. Therefore, in this chapter, we will use the **piechart** function to create pie charts. The **piechart** function visualizes the relative proportions of data. Specifically, the **piechart** function represents each category's or group's proportion as a slice of the pie, making it easier to understand each category's or group's percentage. In contrast, the **donutchart** function introduced in this chapter resembles a standard pie chart, but with a hole in the center. Both charts are valuable for comparing categories and groups and understanding their relative proportions.

14.1 Details of the Data Used in this Chapter

The data below shows the production and supply shares of primary energy. The regions are Asia, Africa, Europe, North America, South America, and Oceania. The variable 「Production」 represents the share of each region's production in the world total. For example, the first element of the 「Production」 variable, 45.9, means that Asia's share is 45.9%. Note that global primary energy production in 2019 was 14,719 million tons. The variable 「Supply」 represents the share of each region's energy supply in the world total. Note that the global primary energy supply in 2019 was 14,519 million tons.

```
Area = ["Asia","Africa","Europe","North America","South America","Oceania"]';

Production = [45.9,8.1,17.4,20.8,4.7,3.1]';

Supply = [52,5.9,17.7,19.2,4.2,1]';

data = table(Area,Production,Supply);
```

14.2 Excel Data of ▶ Section 14.1

Using the **readtable** function, we read the table-formatted data and store it in the variable 「data」.

```
file = "alldata.xlsx";
sheet = "Chap14_Energy_Production_Supply";
data = readtable(file,"Sheet",sheet);
```

The table variable 「data」 will be a matrix with 3 columns and 6 rows. The variables 「Area」, 「Production」, and 「Supply」 are extracted from the table variable 「data」, respectively.

```
Area = data.Area;
Production = data.Production;
Supply = data.Supply;
```

14.3 Basic Syntaxes of the Piechart and Donutchart Functions

Basic Syntax:

We explain the basic syntax of the **piechart** function, 「piechart(data)」, enclosed in a black box in Figure 14.1, and the basic syntax of the **donutchart** function, 「donutchart(data)」, enclosed in a black box in Figure 14.2. These are used to create pie and donut charts. By entering 「doc piechart」 and 「doc donutchart」 in the Command Window, we can see the syntaxes in Figure 14.1 and Figure 14.2. Each element of 「data」 corresponds to a slice of the pie or donut chart and determines its size. Specifically, the **donutchart** function creates a variation of the pie chart with a blank center.

Syntax

```
piechart(data)
piechart(data,names)
piechart(categoricaldata)

piechart(tbl,datavar)
piechart(tbl,datavar,namesvar)
piechart(tbl,categoricalvar)

piechart(parent,___)
piechart(___,Name=Value)
p = piechart(___)
```

Fig. 14.1 The syntax of the piechart function

Syntax

```
donutchart(data)
donutchart(data,names)
donutchart(categoricaldata)

donutchart(tbl,datavar)
donutchart(tbl,datavar,namesvar)
donutchart(tbl,categoricalvar)

donutchart(parent,___)
donutchart(___,Name=Value)
d = donutchart(___)
```

Fig. 14.2 The syntax of the donutchart function

Here, we create a pie chart using the vector 「Production」 in the **piechart** function. Colors are automatically assigned, and units are given in percentages. See ◘ Fig. 14.3.

```
piechart(Production);
```

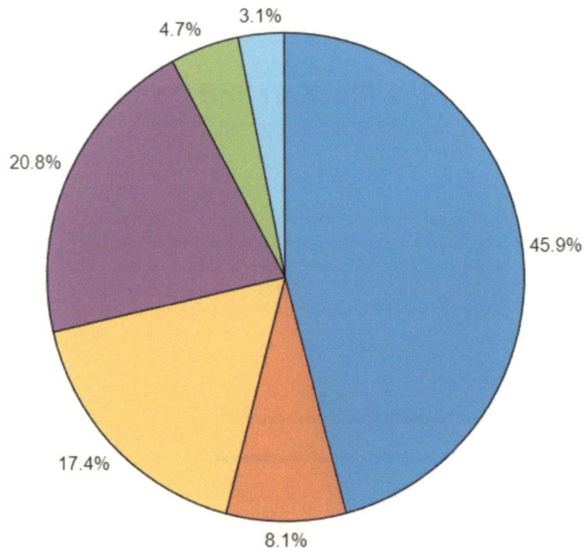

◘ **Fig. 14.3** Pie chart showing proportions within a single variable

The variable 「Production」 sums up to 100, as shown in ◘ Fig. 14.4.

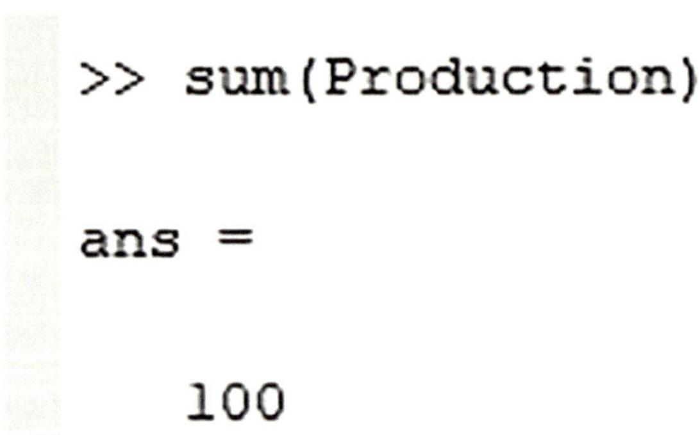

◘ **Fig. 14.4** The sum of the "Production" variable in Command Window

Here, we use the **piechart** function with a vector 「[100,200,300]」 in the first argument. The sum of its elements is 600, which is greater than 100. Upon execution, each slice is adjusted by the values of 16.7% for 100/600, 33.3% for 200/600, and 50% for 300/600, thus determining the area of each slice of the pie. See ◘ Fig. 14.5.

```
piechart([100,200,300]);
```

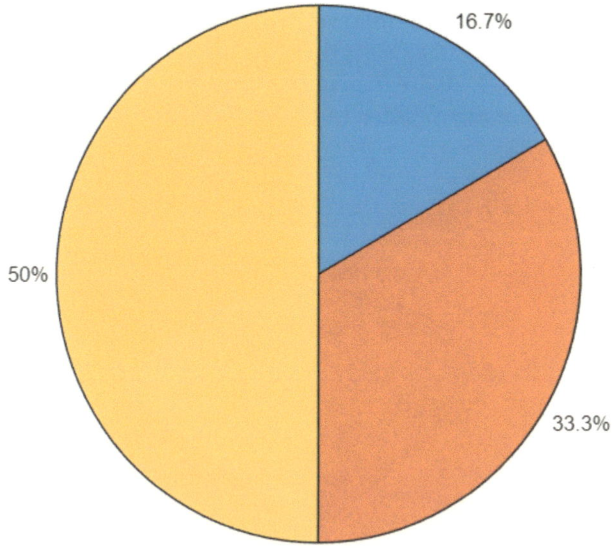

Fig. 14.5 Pie chart showing proportions for three categories

Here, unlike the previous program that created Fig. 14.5, we place the numbers in a different order. As a result, we can change the order of the slices of the previous pie chart. Compare Fig. 14.5 with Fig. 14.6.

```
piechart([300,100,200]);
```

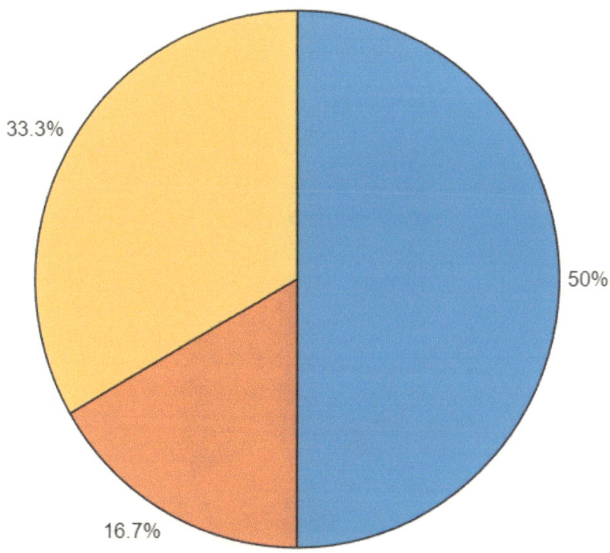

Fig. 14.6 Pie chart showing proportions for three categories (cf. Fig. 14.5)

Here, each slice is calculated as follows: 「15% (≒ 8.1/54.1)」, 「32.2% (≒ 17.4/54.1)」, 「38.4% (≒ 20.8/54.1)」, 「8.69% (≒ 4.7/54.1)」, and 「5.73% (≒ 3.1/54.1)」 where 54.1 is the sum of 8.1, 17.4, 20.8, 4.7, and 3.1. Additionally, note that the percentages written next to each slice are rounded to the third decimal place. See ◘ Fig. 14.7 for the created figure.

```
piechart([8.1,17.4,20.8,4.7,3.1]);
```

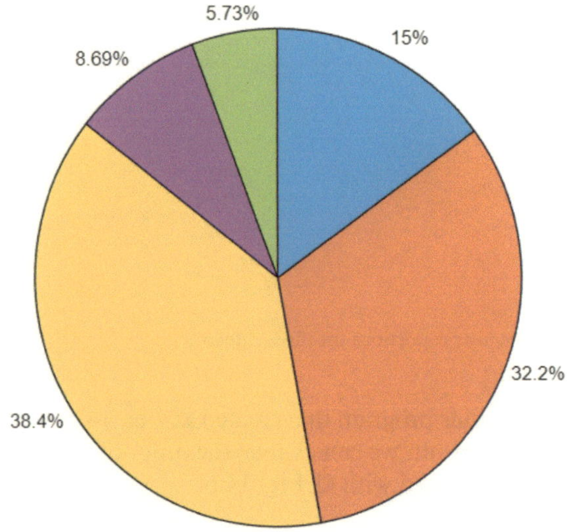

◘ **Fig. 14.7** Pie chart showing proportions for multiple categories

Here, the sum of each element is less than 1. In this case, a partial pie chart is created. See ◘ Fig. 14.8.

```
piechart([0.3,0.1]);
```

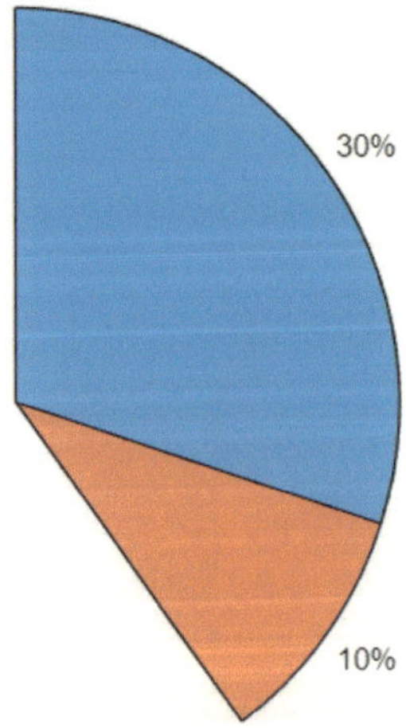

◘ **Fig. 14.8** Partial pie chart showing proportions for two categories

Here, we use the **donutchart** function to create a donut chart for 「Production」. Each slice is labeled with the percentage. While the **piechart** function is more commonly used, the **donutchart** function can be used when you want to create a chart with a blank center. The resulting figure is presented in ◻ Fig. 14.9.

```
donutchart(Production);
```

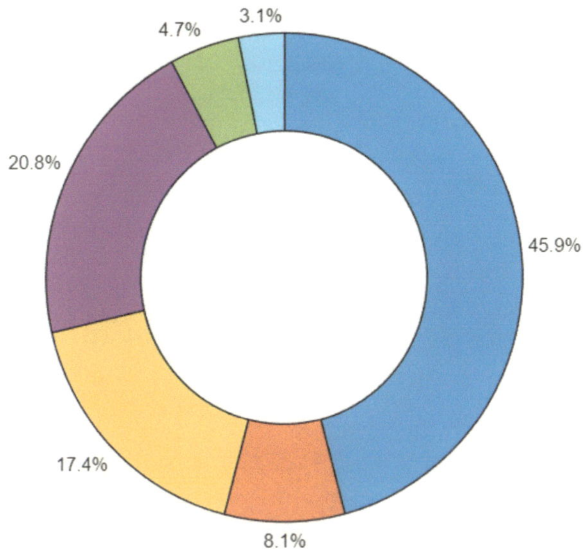

◻ **Fig. 14.9** Donut chart showing proportions within a single variable

As in ◻ Fig. 14.8, we use the data where the sum of each element is less than 1 in the first argument of the **donutchart** function. As in ◻ Fig. 14.8, a partial donut chart is created. See ◻ Fig. 14.10.

```
donutchart([0.3,0.1]);
```

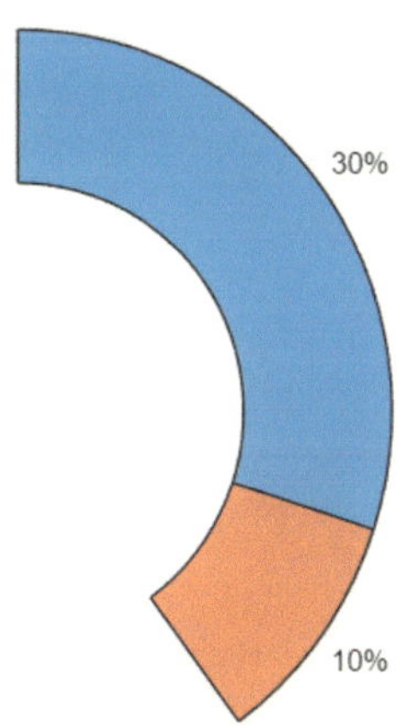

◻ **Fig. 14.10** Partial donut chart showing proportions for two categories

14.4 Setting and Controlling the Piechart and Donutchart Functions

Here, we use the variable 「Area」 to label the areas in the second argument of the **piechart** function. See Fig. 14.11 for the created figure.

```
piechart(Production,Area);
```

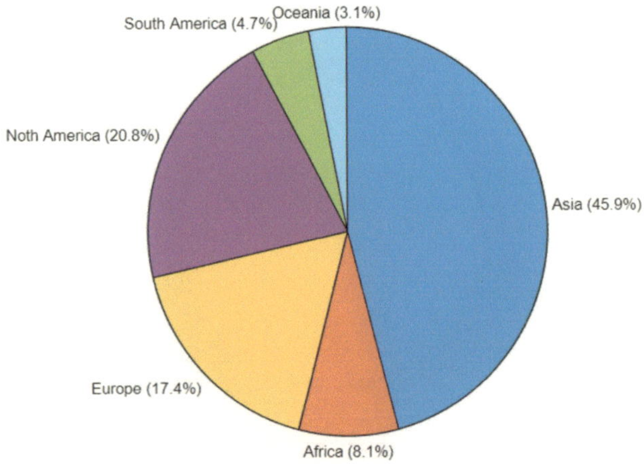

Fig. 14.11 Pie chart showing proportions for multiple categories with labels

Unlike the previous program that created Fig. 14.11, here we use the **donutchart** function. See Fig. 14.12.

```
donutchart(Production,Area);
```


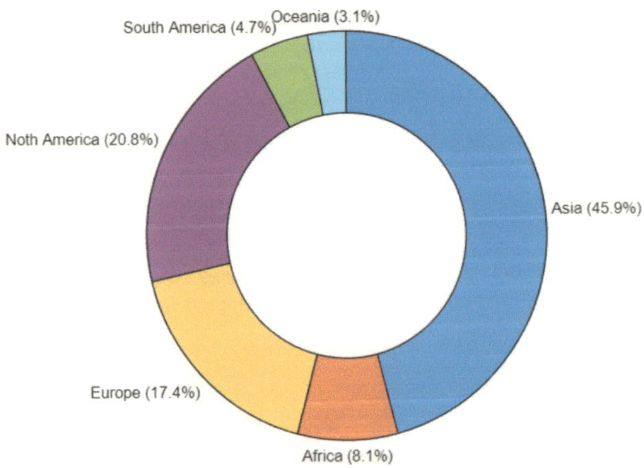

Fig. 14.12 Donut chart showing proportions for multiple categories with labels

We store the properties of the **piechart** function in a variable 「p」. We set 「EdgeColor」 to 「red」, making the outline red. See ◘ Fig. 14.13.

```
p = piechart(Production,Area);
    p.EdgeColor = "red";
```

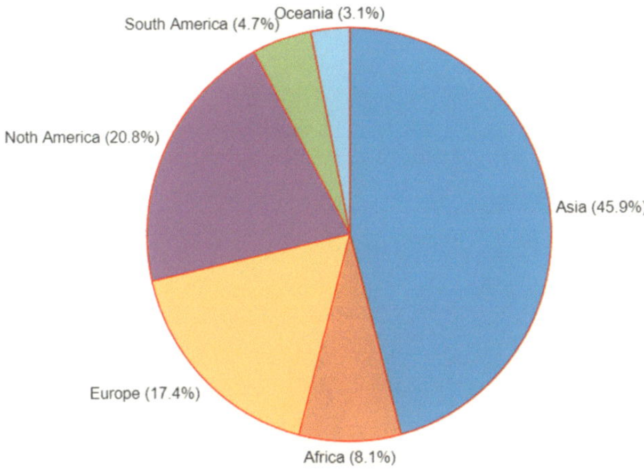

◘ **Fig. 14.13** Pie chart with red edges

By setting 「LineWidth」 to 15, we make the outline of a pie chart thicker. Note that the default is 0.5. See ◘ Fig. 14.14.

```
piechart(Production,Area,"LineWidth",15);
```

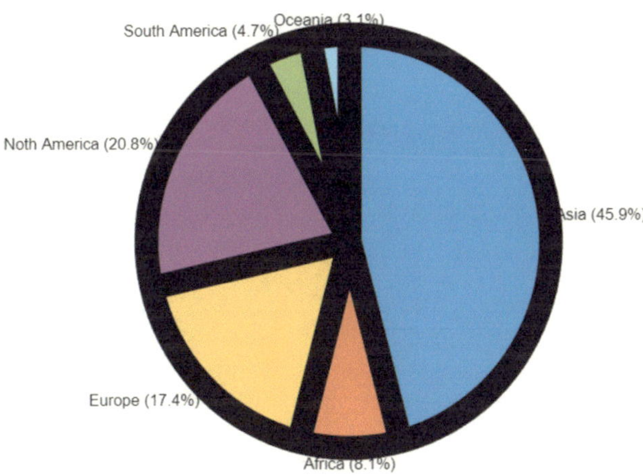

◘ **Fig. 14.14** Pie chart with thick black edges

We store the properties of the **piechart** function in a variable「p」. We set「Edge Color」 to「g (=green)」, making the outline green. We also set「LineWidth」 to 10, making the outline lines thicker. See ◘ Fig. 14.15.

```
p = piechart(Production,Area);
    p.EdgeColor = "g";
    p.LineWidth = 10;
```

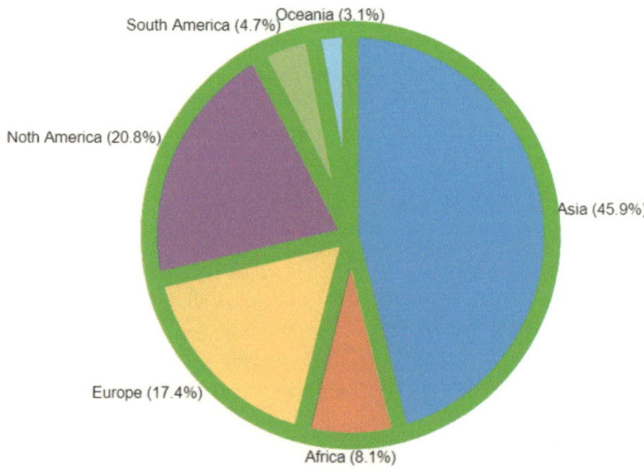

◘ **Fig. 14.15** Pie chart with thick green edges

We set「FontColor」 to「m (=magenta)」 and「FontSize」 to 15, creating large text in magenta. However, as the text size is large, it overlaps. See ◘ Fig. 14.16.

```
p1 = piechart(Production,Area);
    p1.FontColor = "m";
    p1.FontSize = 15;
```


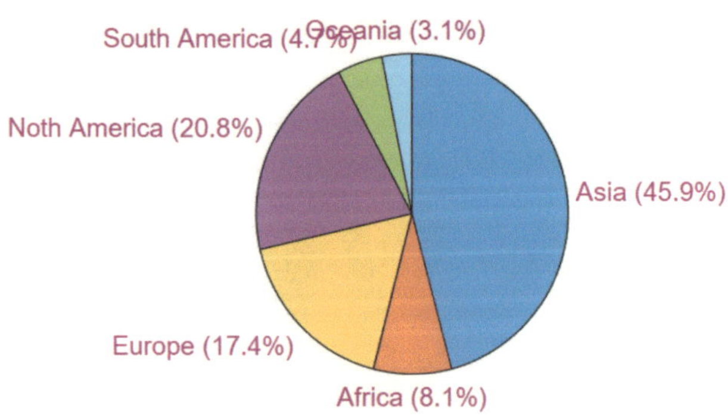

◘ **Fig. 14.16** Pie chart with large magenta labels

Here, after storing the properties of the pie chart in the variable 「pic」, we set 「FaceAlpha」 to 0.2 and 「Position」 to 「[0.2 0.1 0.1 0.5]」, creating a highly transparent pie chart with adjusted position and size. See Fig. 14.17.

```
pic = piechart(Production,Area);
    pic.FaceAlpha = 0.2;
    pic.Position = [0.2 0.1 0.1 0.5];
```

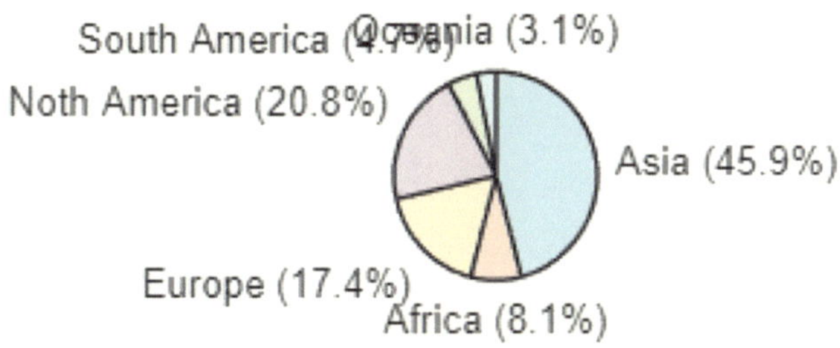

Fig. 14.17 Small semi-transparent pie chart at specified position

The property 「pic.Position」 in the previous program that created Fig. 14.17 is retained as is. Therefore, by using the **figure** command, we reset the position and size to their default values. Additionally, in the **piechart** function, we set 「Direction」 to 「counterclockwise」, and as a result, the slices of the pie chart appear in counterclockwise order. See Fig. 14.18.

```
figure
piechart(Production,Direction="counterclockwise");
```

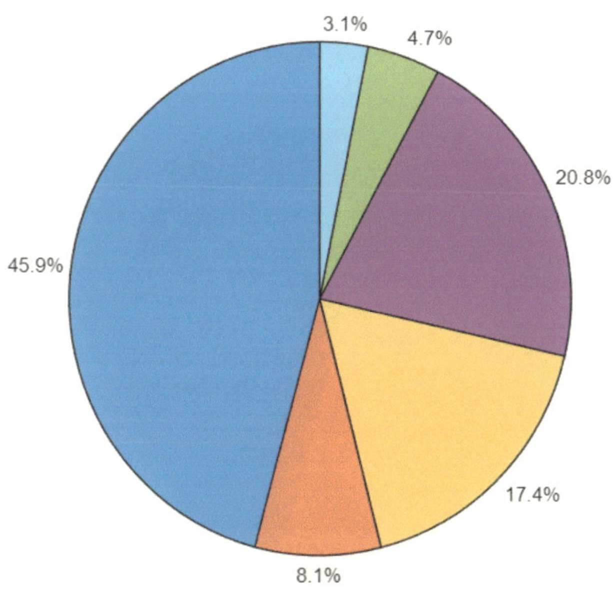

Fig. 14.18 Pie chart in counterclockwise slice order

Using the second argument of the **piechart** function, we separate and display one or more slices within the pie chart. We set 「ExplodedWedges=3」, so that the third slice is separated from the center. See Fig. 14.19.

```
piechart(Production,ExplodedWedges=3);
```

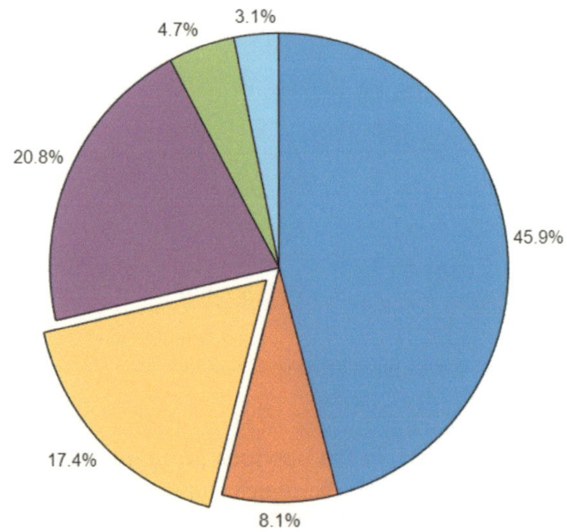

Fig. 14.19 Pie chart with one separated slice

Here, we separate the first, third, and fifth slices. See Fig. 14.20 for the resulting figure.

```
piechart(Production,ExplodedWedges=[1 3 5]);
```

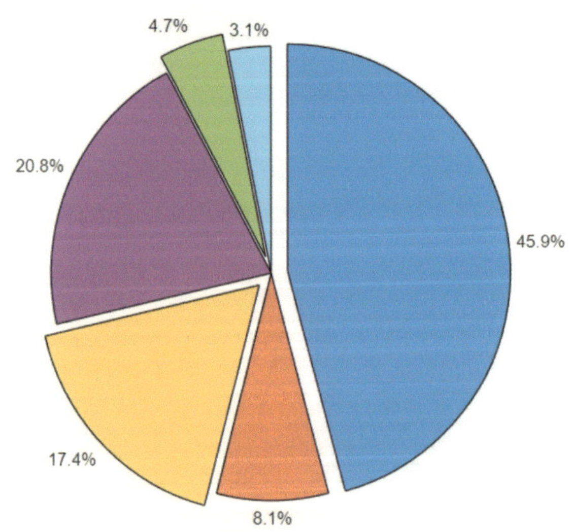

Fig. 14.20 Pie chart with multiple separated slices

Here, using the **piechart** function, we set 「StartAngle」 to 45 to shift the slices 45 degrees clockwise. Additionally, we separate all slices by using 「ExplodedWedges=[1 2 3 4 5 6]」. See ◘ Fig. 14.21.

```
piechart(Production,ExplodedWedges=[1 2 3 4 5 6],StartAngle=45);
```

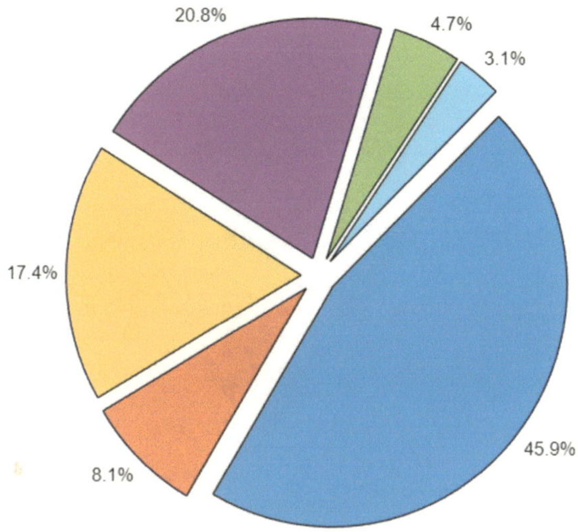

◘ **Fig. 14.21** Pie chart with rotated and multiple separated slices

Here, we use 「ExplodedWedges=[1 3 5]」 to separate the first, third, and fifth slices. Additionally, we specify 「Direction」 as 「counterclockwise」 to rotate them in a counterclockwise direction. See ◘ Fig. 14.22.

```
donutchart(Production,ExplodedWedges=[1 3 5],Direction="counterclockwise");
```

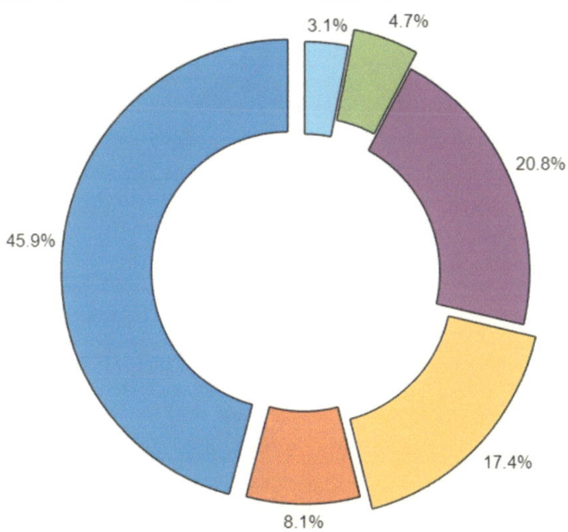

◘ **Fig. 14.22** Donut chart with separated slices in counterclockwise order

Here, we create the same chart as before. Additionally, setting 「Labels=[ ]」, we hide the labels for 30% and 10%. Compare ◘ Fig. 14.8 with ◘ Fig. 14.23.

```
piechart([0.3,0.1],Labels=[]);
```

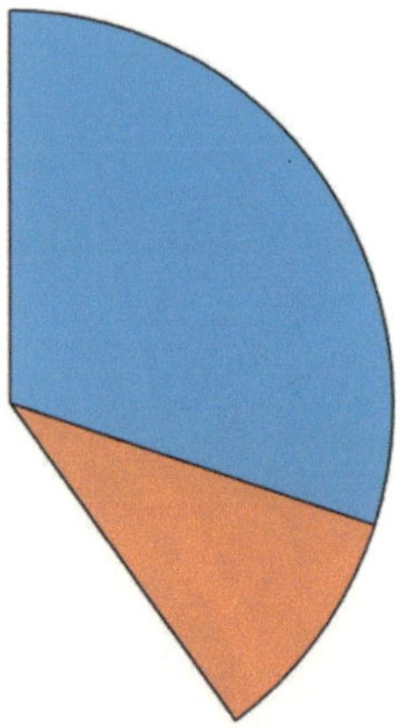

◘ **Fig. 14.23** Partial pie chart with no labels

Here, using 「Labels=["A:0.3","B:0.1"]」, we display the text specified on each slice. See ◘ Fig. 14.24 for the resulting figure. Moreover, compare this figure with ◘ Figs. 14.8 and 14.23.

```
piechart([0.3,0.1],Labels=["A:0.3","B:0.1"]);
```

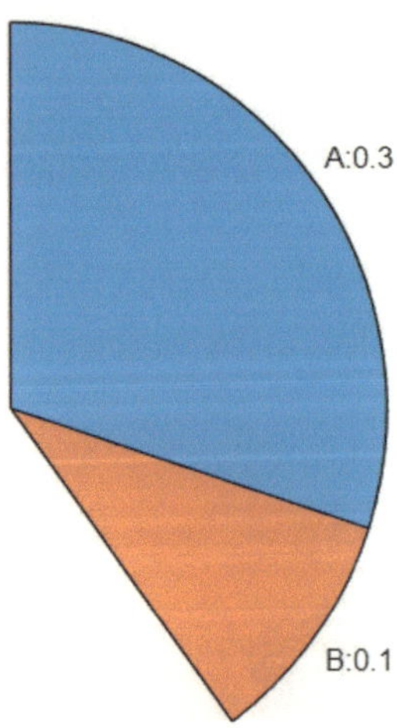

◘ **Fig. 14.24** Partial pie chart with manually specified labels

Here, we store the properties of the **piechart** function in the variable 「pic」. We then set 「FontColor」 to 「m (=magenta)」 and 「FontSize」 to 15, creating a magenta-colored text with a larger font size. Additionally, we display each label as two lines of text by using 「Labels」. Specifically, we display the text name of the variable 「Area」 on the first line and the ratio 「Production」 on the second line, where we convert the variable 「Production」 to a string type. We use 「\newline」 to add a line break. See Fig. 14.25.

```
pic = piechart(Production,Area);
    pic.FontColor = "m";
    pic.FontSize = 15;
    pic.Labels = Area+"¥newline¥newline" +"("+string(Production)+"%)";
```

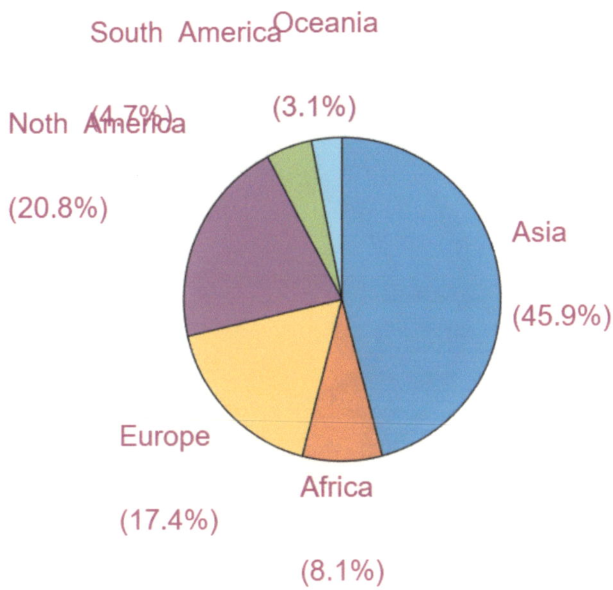

Fig. 14.25 Pie chart with space between specified categories and percentage labels

14.5 Combining Optional Functions and the Piechart and Donutchart Functions

Attention

First of all, although we often use the **hold on** command to overlay plots in MATLAB, as of the 2024b version, the **hold on** command is not supported for the **piechart** and **donutchart** functions, making it impossible to overlay multiple pie charts. The error message is shown in Fig. 14.26. The **hold on** command is commonly used to retain the current plot and add new plots to it, but in this case, you would need to find alternative methods to achieve similar effects.

```
Error using hold (line 61)
Using hold with donutchart is not supported.
```

Fig. 14.26 Error output of the donutchart function: Unsupported use of the "hold on" command

Here, we use the **title** function. Additionally, in the **donutchart** function, we set 「StartAngle」 to -90 and 「LineWidth」 to 10, rotating the chart and giving it a thicker outline. See **Fig. 14.27**.

```
d = donutchart(Production,Area,StartAngle=-90);
    d.LineWidth = 10;
title("StartAngle=-90");
```

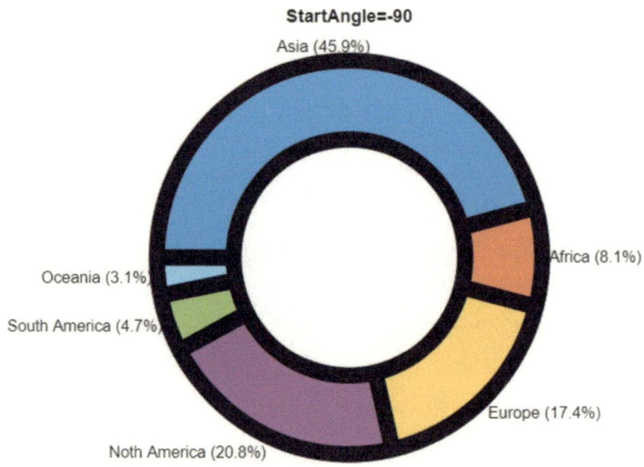

Fig. 14.27 Donut chart with thick black edges and rotated slices

This program creates a pie chart based on the variable 「Production」, with the slices separated according to 「ExplodedWedges=[1 2 3 4 5 6]」, meaning that all six slices are pulled away from the center. The color order for the created chart is set to meadow using 「colororder meadow」. See **Fig. 14.28**.

```
piechart(Production,ExplodedWedges=[1 2 3 4 5 6]);
colororder meadow;
title("color: meadow");
```

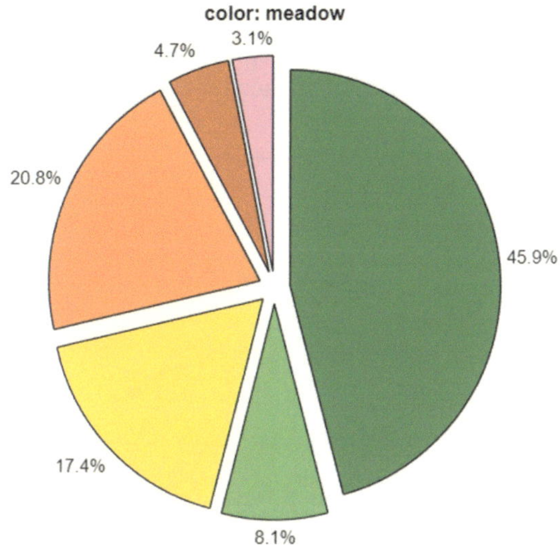

Fig. 14.28 Pie chart with separated slices and predefined color order ("meadow")

The use of 「ExplodedWedges=[1 2 3 4 5 6]」 is the same as in the previous program that created Fig. 14.28. Additionally, the 「newcolors」 vector defines a custom set of colors: 「r (=red)」, 「k (=black)」, 「b (=blue)」, 「m (=magenta)」, 「y (=yellow)」, and 「w (=white)」. The 「colororder」 function is then used to apply this custom color set to the pie chart using 「colororder(newcolors)」, so that each slice is displayed with the specified colors. See Fig. 14.29 for the created figure.

```
piechart(Production,ExplodedWedges=[1 2 3 4 5 6]);
newcolors = ["r" "k" "b" "m" "y" "w"];
colororder(newcolors);
title("color:self-made");
```

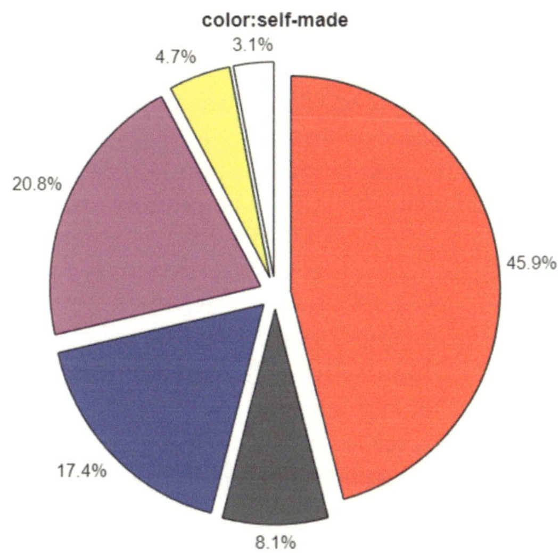

Fig. 14.29 Pie chart with separated slices and user-specified color order

Here, the **colororder** function includes color palettes such as 「meadow」 in ◘ Fig. 14.28, as well as 「gem」, 「glow」, 「sail」, 「reef」, 「dye」, and 「earth」. Here, 「reef」 is specified. See ◘ Fig. 14.30 for the created figure.

```
piechart(Production,ExplodedWedges=[1 2 3 4 5 6]);
colororder reef;
title("color: reef");
```

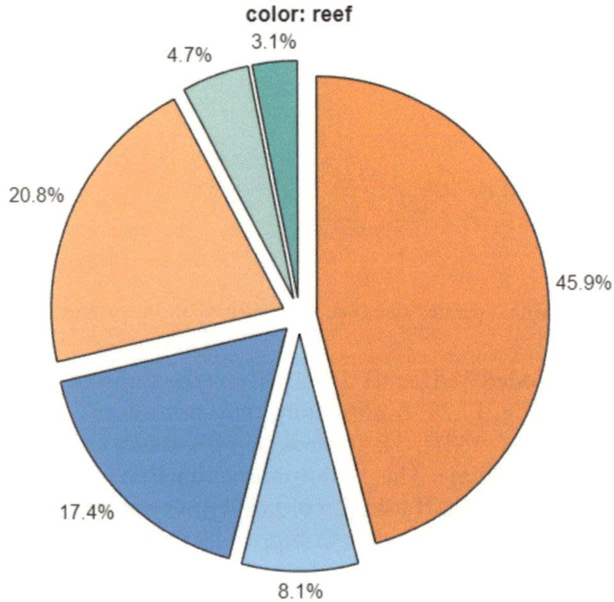

◘ **Fig. 14.30:** Pie chart with separated slices and predefined color order ("reef")

14

Here, we use the **annotation** function three times. The use of the **annotation** function is essentially the same in the first and second instances, where 「textarrow」 creates the arrow with a text. The variables 「xa」, 「ya」, 「xaf」, and 「yaf」 specify the start and end points of these text arrows.

In the third instance of the **annotation** function, we place a text string in the center hole created by the **donutchart** function. The position and size of the text box are specified by the position vector 「[0.4,0.45,0.1,0.1]」, where the first two values 「0.4,0.45」 specify the coordinates, and the last two values 「0.1,0.1」 determine the width and height of the text box. The text displayed in the box consists of two lines: 「Total Production」 on the first line and 「10000000」 on the second line. The box is adjusted to fit the text by setting 「FitBoxToText」 to 「on」. We set 「FontSize」 to 14 and 「FontWeight」 to 「bold」, creating bold text with a larger font size. Additionally, 「EdgeColor」 is set to 「none」, which removes the border around the text box. For more details on the **annotation** function, see ► Sect. 18.1. The resulting figure is shown in ◘ Fig. 14.31.

Attention

Error Program: Finally, note that in MATLAB version 2024b, the **text** function cannot be used with the **donutchart** function. The error message is presented in Fig. 14.32.

```
figure
dtc = donutchart(Production,Labels=[]);
    dtc.FaceAlpha = 0.2;
xa = [0.8 0.75];
ya = [0.9 0.8];
annotation("textarrow",xa,ya,"String","Asia (45.90%)");
xaf = [0.65 0.55];
yaf = [0.05 0.1];
annotation("textarrow",xaf,yaf,"String","Africa (8.1%)");
annotation("textbox",[0.4,0.45,0.1,0.1],"String",...
    {"Total Production","10000000"},"FitBoxToText","on",...
    "FontSize",14,"FontWeight","bold","EdgeColor","none");
title("Primary energy production");
```

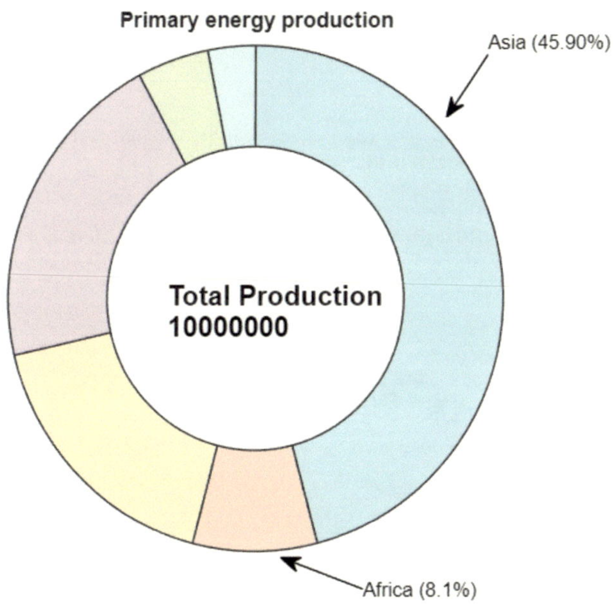

Fig. 14.31 Donut chart with a textbox placed in the center hole and text arrows

```
%%% Error program
% text(0,0,"Total Production");
```

```
Error using text
Text cannot be a child of DonutChart.
```

Fig. 14.32 Error output of the donutchart function: Unsupported use of the text function

Here, we use the **annotation** function four times. The first **annotation** function places a text string in the center of the circle created by the **donutchart** function, which is the same as the previous program that created ◘ Fig. 14.31.

The other three instances create a text box by using 「textbox」 in the first argument of the **annotation** function. Specifically, by using 「annotation("textbox", [0.45,0.1,0.1,0.1],"String", "Africa (8.1%)", "FitBoxToText","on", "FontSize",8,"FontWeight","bold","EdgeColor","none")」, the text displayed by 「Africa (8.1%)」 is generated in a slice of the donut chart, positioned at the location, size, and width specified by 「[0.45,0.1,0.1,0.1]」 on the figure. The box is adjusted to fit the text using 「FitBoxToText」 set to 「on」. Moreover, we set 「FontSize」 to 8 and 「Font Weight」 to 「bold」, creating bold text with a smaller font size. Additionally, we set 「EdgeColor」 to 「none」 to remove the border around the text box. The rest annotations are used similarly. For more details on the **annotation** function, see ► Sect. 18.1. The resulting figure is displayed in ◘ Fig. 14.33.

```
donutchart(Production,Labels=[]);
annotation("textbox",[0.45,0.45,0.1,0.1],"String",...
    {"Total Production","10000000"},"FitBoxToText","on",...
    "HorizontalAlignment","center","FontSize",14,...
    "FontWeight","bold","EdgeColor","none");
annotation("textbox",[0.45,0.1,0.1,0.1],"String",...
    "Africa (8.1%)","FitBoxToText","on",...
    "FontSize",8,"FontWeight","bold","EdgeColor","none");
annotation("textbox",[0.66,0.6,0.1,0.1],"String",...
    "Asia (45.9%)","FitBoxToText","on",...
    "FontSize",8,"FontWeight","bold","EdgeColor","none");
ant = annotation("textbox",[0.25,0.25,0.1,0.1],"String",...
    "Europe (17.4%)","FitBoxToText","on",...
    "FontSize",8,"FontWeight","bold","EdgeColor","none");
    ant.Color = "red";
```

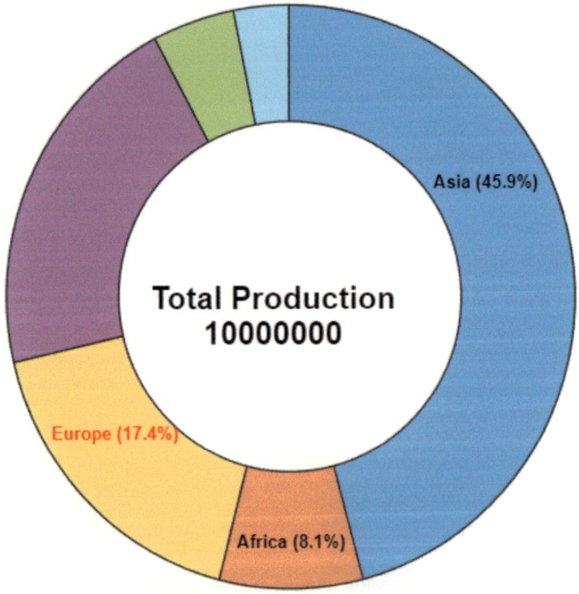

◘ **Fig. 14.33** Donut chart with a textbox placed in the center hole and three textboxes on the donut

As in ◘ Figs. 14.31 and 14.33, we place text in the center of the circle created by the **donutchart** function. Here, we set the background to yellow. To do so, we use the **annotation** function twice. First, we use 「rectangle」 as the first argument of the **annotation** function to create a yellow box. Then, we use the **annotation** function again to output the text using 「textbox」. One important point to note is the order in which these are used. If you use 「textbox」 first and then 「rectangle」, the text will be overwritten. For more details of the **annotation** function, see ▶ Sect. 18.1. The resulting figure is presented in ◘ Fig. 14.34.

```
donutchart(Production);
dim = [0.35,0.43,0.3,0.12];
annotation("rectangle",dim,"FaceColor","yellow");
annotation("textbox",[0.45,0.45,0.1,0.1],"String",...
    {"Total Production","10000000"},"FitBoxToText","on",...
    "HorizontalAlignment","center","FontSize",14,...
    "FontWeight","bold","EdgeColor","none");
```

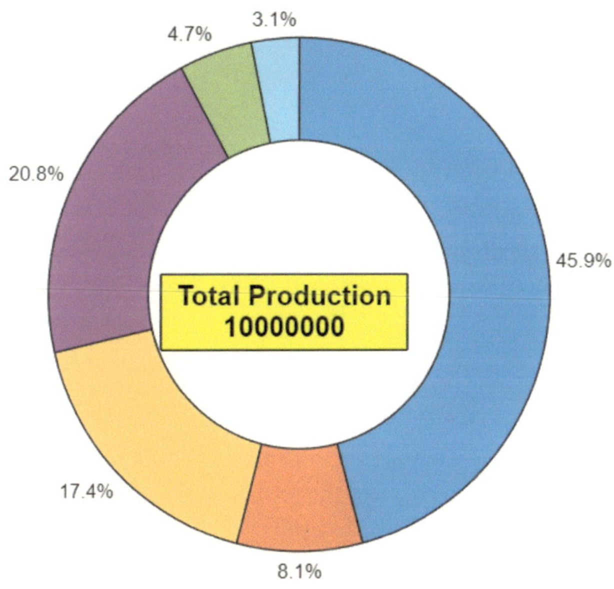

◘ **Fig. 14.34** Donut chart with a yellow textbox placed in the center hole

Here, using the **piechart** function, we create a figure similar to ◘ Fig. 14.34. First, we use the **annotation** function with 「rectangle」 and then we use 「textbox」. As noted in ◘ Fig. 14.34, the order in which these are used is important. That is, if you use 「textbox」 first and then 「rectangle」, the text will be overwritten and will not appear. For more details on the **annotation** function, see ▶ Sect. 18.1. The resulting figure is presented in ◘ Fig. 14.35.

```
piechart(Production,Area);
dim = [0.3,0.45,0.4,0.1];
ant = annotation("rectangle",dim,"FaceColor","blue");
annotation("textbox", [0.3,0.45,0.1,0.1],"String",...
    "Total Production","FitBoxToText","on",...
    "FontSize",20,"FontWeight","bold","EdgeColor","none");
```

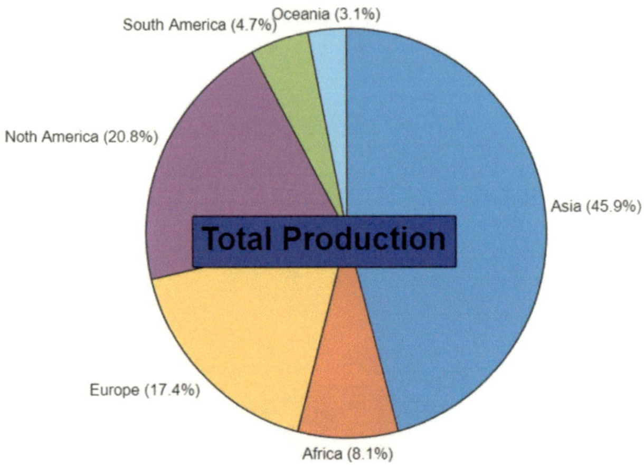

Fig. 14.35 Pie chart with a blue textbox placed at the center

Here, we use the **tiledlayout** function to arrange two pie charts side by side. We also specify「ExplodedWedges=2」to separate the second slice from the center. The resulting figure is shown in Fig. 14.36.

```
figure
a = tiledlayout(1,2);
title(a,"piechart","FontWeight","bold");
nexttile;
piechart(Production,ExplodedWedges=2);
title("Primary energy production");
nexttile;
piechart(Supply,ExplodedWedges=2);
title("Primary energy supply");
```

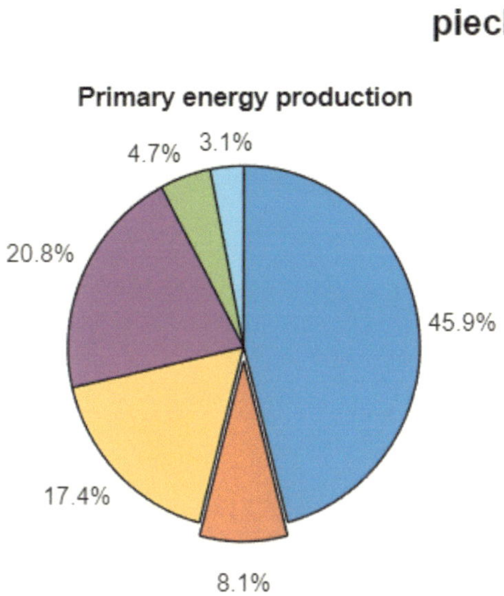

Fig. 14.36 Pie charts (2-column layout)

「InnerRadius」 is 0.6 by default and takes a number between 0 and 1. Specifying 0 will create a pie chart without a hole, and specifying 1 will create a thin ring with a large hole in the center. Here, in the 1-by-1 space, 「InnerRadius」 is set to 0.2, and in the 1-by-2 space, it is set to 0.9. See Fig. 14.37.

```
figure
tl = tiledlayout(1,2);
title(tl,"InnerRadius");
nexttile;
donutchart(Production,"InnerRadius",0.2);
title("InnerRadius=0.2");
nexttile;
donutchart(Production,"InnerRadius",0.9);
title("InnerRadius=0.9");
```

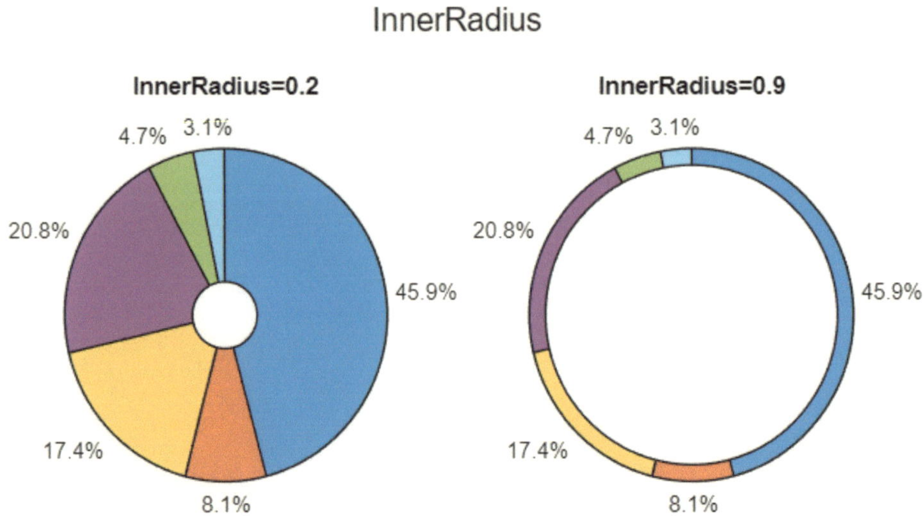

Fig. 14.37 Donut charts with large and small holes (2-column layout)

14.6 Questions

Question 1

Create Fig. 14.38 using the **piechart** function with 「Production」 as the first argument and 「Area」 as the second argument. Set a red outline with high transparency and slightly thicker lines. Additionally, set the font size of the labels to 10, and set the chart title to 「Piechart」.

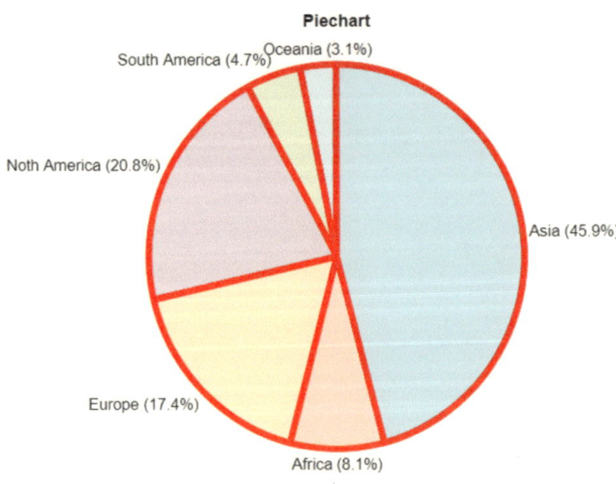

Fig. 14.38 Pie chart (Question 1)

14

Answer 1

```
figure
Ex1 = piechart(Production,Area);
    Ex1.EdgeColor = "red";
    Ex1.LineWidth = 4;
    Ex1.FaceAlpha = 0.2;
    Ex1.FontSize = 10;
title("Piechart");
```

Question 2

Create Fig. 14.39 using the **donutchart** function with 「Production」 as the first argument. In detail, use 「InnerRadius」 to set the inner radius of the donut chart to 0.1, making the hole smaller. Additionally, use the variable 「Area」 to create labels for each section of the pie chart. The labels should display the region on the first line and the percentage of the total on the second line.

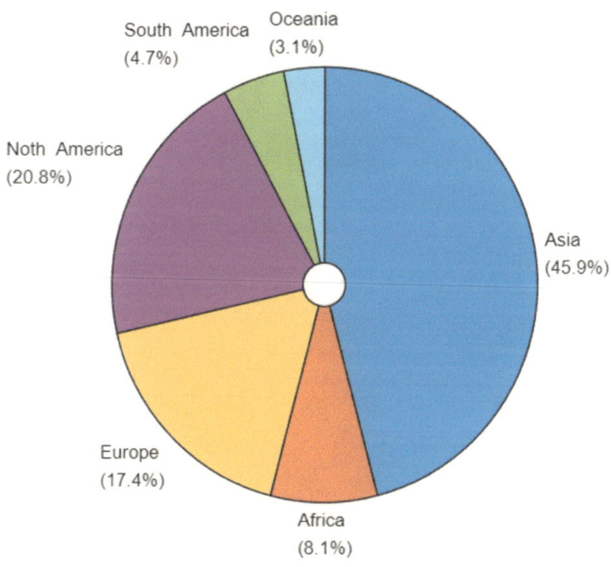

Fig. 14.39 Donut chart (Question 2)

Answer 2

```
figure
Ex2 = donutchart(Production,"InnerRadius",0.1);
    Ex2.Labels = Area+"¥newline" +"("+string(Production)+"%)";
```

Question 3

Create Fig. 14.40 using the **piechart** function with 「Production」 as the first argument.

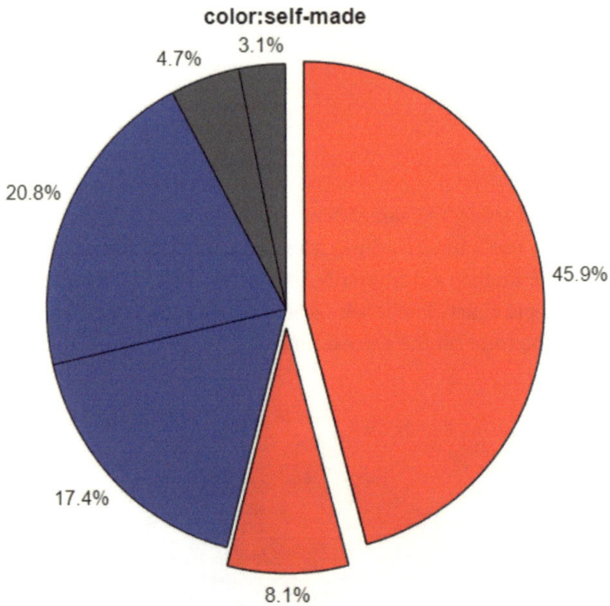

Fig. 14.40 Pie chart (Question 3)

Answer 3

```
figure
piechart(Production,ExplodedWedges=[1 2]);
newcolors = ["red" "red" "blue" "blue" "black" "black"];
colororder(newcolors);
title("color:self-made");
```

Plotmatrix Function

Contents

15.1 Details of the Data Used in this Chapter – 466

15.2 Excel Data of ► Section 15.1 – 467

15.3 Basic Syntax of the Plotmatrix Function – 469

15.4 Setting and Controlling the Plotmatrix Function – 477

15.5 Combining Optional Functions and the Plotmatrix Function – 483

15.6 Questions – 493

Supplementary Information The online version contains supplementary material available at https://doi.org/10.1007/978-981-96-6213-5_15.

Y. Nakamoto, *A Comprehensive Guide to Creating Graphics with MATLAB Programming*, Classroom Companion: Economics, https://doi.org/10.1007/978-981-96-6213-5_15

Learning Objectives

This chapter creates a matrix figure, which consists of scatter plots (covered in ▶ Chap. 3) and histograms (covered in ▶ Chap. 11) that help visualize relationships between multiple variables. By reading this chapter, you will learn how to use the **plotmatrix** function. In ▶ Sect. 15.5, you will learn about the importance of displaying the correlation coefficient in figures to handle the complexity of multiple variables.

Overview

The **plotmatrix** function visualizes pairwise relationships between multiple variables, making it especially useful for exploring correlations in a multivariate data set. One of its key advantages is that it not only generates scatter plots for each pair of variables but also displays histograms on the diagonal for each individual variable. See ▶ Chaps. 3 and 11 for explanations of the **scatter** and **histogram** functions.

15.1 Details of the Data Used in this Chapter

In this chapter, we use large amounts of data. The data include GDP per capita, fertility rate, infant mortality, and population. GDP per capita, derived by dividing a country's total GDP by its population, indicates the average productivity and income level per person in the country. Next, fertility rate, infant mortality, and population are important indicators of health, demographics, and socioeconomic conditions, which are crucial for a country's development and well-being. Specifically, infant mortality indicates the proportion of children under a certain age who die.

In this section, we focus on two countries, Belgium and Singapore, and use GDP per capita, fertility rate, child mortality, and population data from 1980 to 2021. Specifically, we use 「BEL_Fertility」 and 「SGP_Fertility」 to represent the fertility rates for Belgium and Singapore, and 「BEL_Mortality」 and 「SGP_Mortality」 to represent child mortality rates for both countries, all rounded to three decimal places. Next, 「BEL_Pop」 and 「SGP_Pop」 represent the populations of Belgium and Singapore, rounded to the nearest ten thousand. Finally, 「BEL_GDP」 and 「SGP_GDP」 represent the GDP per capita of Belgium and Singapore, rounded to the nearest hundred.

15

In ▶ Sect. 15.2, the data refer to various 188 countries worldwide in Excel, spanning different regions and continents. Additionally, ▶ Sect. 15.2 uses the original unrounded data in the Excel file 「alldata.xlsx」.

```
BEL_Fertility = [1.68,1.66,1.62,1.57,1.54,1.51,1.54,1.54,1.57,...
    1.58,1.62,1.66,1.65,1.61,1.56,1.56,1.59,1.60,1.60,1.62,1.66,...
    1.67,1.65,1.67,1.72,1.76,1.80,1.82,1.85,1.84,1.86,1.82,1.80,...
    1.76,1.74,1.70,1.68,1.65,1.62,1.59,1.56,1.58]';

BEL_Mortality = [1.47,1.37,1.31,1.25,1.2,1.14,1.12,1.12,1.07,...
    1.00,0.97,0.97,0.95,0.92,0.87,0.74,0.64,0.66,0.64,0.60,0.60,...
    0.58,0.55,0.52,0.47,0.48,0.50,0.50,0.46,0.44,0.44,0.42,0.46,...
    0.43,0.40,0.39,0.38,0.42,0.44,0.42,0.37,0.40]';

BEL_Pop = [9830000,9840000,9860000,9870000,9870000,9880000,...
    9890000,9900000,9920000,9940000,9960000,9990000,10020000,...
    10040000,10070000,10100000,10120000,10150000,10190000,...
    10220000,10260000,10310000,10360000,10400000,10460000,...
    10520000,10580000,10650000,10730000,10800000,10880000,...
    10960000,11030000,11100000,11180000,11250000,11320000,...
    11380000,11450000,11510000,11560000,11610000]';

BEL_GDP = [12500,10400,9100,8600,8200,8600,11900,14700,16000,16100,...
    20100,20600,22800,21700,23600,28500,27500,24900,25400,25300,23100,...
    23100,25000,30700,35500,36900,38800,44500,48500,44900,44400,47600,...
    44800,46800,47900,41100,42100,44300,47700,46800,45500,51500]';

SGP_Fertility = [1.81,1.71,1.67,1.55,1.55,1.59,1.37,1.50,1.73,1.56,...
    1.80,1.52,1.50,1.51,1.49,1.65,1.45,1.58,1.30,1.27,1.56,1.23,1.36,...
    1.28,1.26,1.26,1.28,1.29,1.28,1.22,1.16,1.20,1.28,1.20,1.25,1.23,...
    1.00,0.95,0.95,0.96,1.00,1.02]';

SGP_Mortality = [1.48,1.42,1.32,1.23,1.15,1.09,1.03,0.98,0.92,...
    0.84,0.77,0.69,0.63,0.57,0.53,0.50,0.48,0.46,0.44,0.42,...
    0.39,0.36,0.34,0.32,0.30,0.29,0.29,0.29,0.28,0.28,0.28,...
    0.28,0.28,0.28,0.27,0.27,0.27,0.26,0.25,0.24,0.22,0.21]';

SGP_Pop = [2401000,2452000,2507000,2565000,2624000,2686000,2750000,...
    2815000,2881000,2946000,3022000,3121000,3228000,3334000,3439000,...
    3544000,3652000,3761000,3866000,3966000,4054000,4121000,4177000,...
    4226000,4270000,4345000,4487000,4663000,4838000,5011000,5164000,...
    5281000,5381000,5478000,5571000,5650000,5712000,5764000,5815000,...
    5866000,5910000,5941000]';

SGP_GDP = [5000,5700,6000,6700,7200,6800,6900,7800,9300,10700,12800,...
    14500,16100,18300,21600,24900,26200,26400,21800,21800,23900,21700,...
    22200,23700,27600,30000,33800,39400,40000,38900,47200,53900,55500,...
    57000,57600,55600,56900,61200,66800,66100,61300,77700]';

tbl_Belgium = table(BEL_GDP,BEL_Fertility,BEL_Mortality,BEL_Pop);

tbl_Singapore= table(SGP_GDP,SGP_Fertility,SGP_Mortality,SGP_Pop);
```

15.2 Excel Data of ► Section 15.1

We use the **readtable** function to read the table formatted data in MATLAB. The data is stored in the variable「tbl_Chap15」, which is a matrix with 7896 rows and 7 columns.

```
tbl_Chap15 = readtable("alldata.xlsx","Sheet","Chap15_gdp_etal");
```

```
Warning: Column headers from the file were modified to make them valid MATLAB identifiers before creating
variable names for the table. The original column headers are saved in the VariableDescriptions property.
Set 'VariableNamingRule' to 'preserve' to use the original column headers as table variable names.
```

「tbl_Chap15」 contains a large amount of data for 188 countries. Due to the large size of the data set, it is difficult to easily understand the relationships using the **plotmatrix** function. To solve this issue, we first extract data for two countries: Belgium and Singapore. Using the **strcmp** function, we generate a column with the string 「BEL」, which returns a logical array where 1 indicates a match and 0 indicates no match. Therefore, the variable 「Belgium」 is of type logical. Next, we select rows where 「Belgium」 is 1 using 「tbl_Chap15(Belgium,:)」, storing the selected rows in the variable 「tbl_Belgium」. Additionally, 「tbl_Belgium.Entity=[ ]」 creates an empty array for the 「Entity」 column in the table variable 「tbl_Belgium」, indicating that this column has no data. Similar steps are applied to 「tbl_Belgium.Code=[ ]」 and 「tbl_Belgium.Year=[ ]」. Finally, the variable 「tbl_Belgium」 is a table with 42 rows and 4 columns, containing only data for Belgium.

```
Belgium = strcmp(tbl_Chap15.Code,"BEL");
tbl_Belgium = tbl_Chap15(Belgium,:);
tbl_Belgium.Entity = [];
tbl_Belgium.Code = [];
tbl_Belgium.Year = [];
tbl_Belgium = renamevars(tbl_Belgium, ...
    ["GDPPerCapita","FertilityRate","ChildMortalityRate","Population"],...
    ["BEL_GDP","BEL_Fertility","BEL_Mortality","BEL_Pop"]);
```

The four variables 「BEL_Fertility」, 「BEL_Mortality」, 「BEL_Pop」, and 「BEL_GDP」 are the vectors that contain the columns extracted from the table variable 「tbl_Belgium」. While 「tbl_Belgium」 is a table array, these variables are individual column vectors with 42 rows and 1 column.

```
BEL_Fertility = tbl_Belgium.BEL_Fertility;
BEL_Mortality = tbl_Belgium.BEL_Mortality;
BEL_Pop = tbl_Belgium.BEL_Pop;
BEL_GDP = tbl_Belgium.BEL_GDP;
```


We perform similar programming for Singapore as we did for Belgium.

```
Singapore = strcmp(tbl_Chap15.Code,"SGP");
tbl_Singapore = tbl_Chap15(Singapore,:);
tbl_Singapore.Entity = [];
tbl_Singapore.Code = [];
tbl_Singapore.Year = [];
tbl_Singapore = renamevars(tbl_Singapore,...
    ["GDPPerCapita","FertilityRate","ChildMortalityRate","Population"],...
    ["SGP_GDP","SGP_Fertility","SGP_Mortality","SGP_Pop"]);
SGP_Fertility = tbl_Singapore.SGP_Fertility;
SGP_Mortality = tbl_Singapore.SGP_Mortality;
SGP_Pop = tbl_Singapore.SGP_Pop;
SGP_GDP = tbl_Singapore.SGP_GDP;
```

15.3 Basic Syntax of the Plotmatrix Function

Basic Syntax:

The two syntaxes enclosed in a black box are the basic uses of the **plotmatrix** function. See Figure 15.1. For more information on the various syntaxes of the **plotmatrix** function, type 「doc plotmatrix」 in the Command Window. First, 「plotmatrix(X,Y)」 is used to create a matrix of scatter plots between two data sets, 「X」 and 「Y」. The scatter plots show the relationships between the corresponding variables in 「X」 and 「Y」. Next, 「plotmatrix(X)」 is used to create a matrix of scatter plots for a single data set, 「X」. The diagonal of the matrix shows histograms for each variable in 「X」, and the off-diagonal plots show scatter plots of the variables against each other.

Syntax

```
plotmatrix(X,Y)
plotmatrix(X)
plotmatrix(___,LineSpec)
plotmatrix(ax,___)

[S,AX,BigAx,H,HAx] = plotmatrix(___)
```

Fig. 15.1 The syntax of the plotmatrix function

We now examine the relationship between the fertility rate and infant mortality rate in Belgium using the **plotmatrix** function. The syntax 「plotmatrix(X,Y)」 creates a scatter plot of column 「X」 against column 「Y」. In the first and second arguments of the **plotmatrix** function, we use the variables 「BEL_Fertility」 and 「BEL_mortality」, creating a scatter plot figure. See Fig. 15.2.

```
plotmatrix(BEL_Fertility,BEL_Mortality);
```

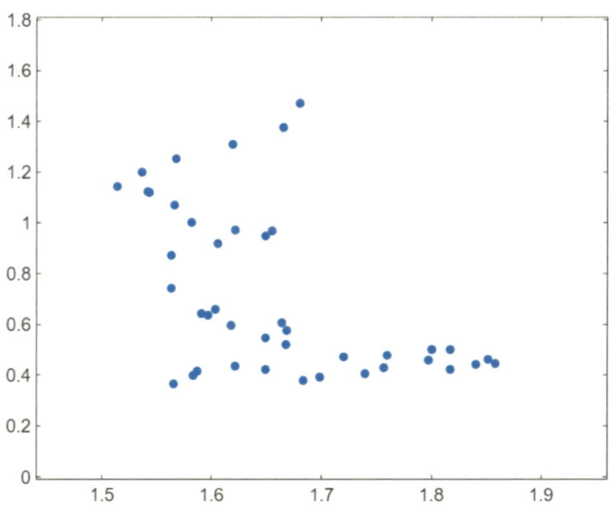

Fig. 15.2 Scatter plot of two variables using the plotmatrix function

▶ Example of Table Variables

Here, we use two different table variables: 「tbl_Belgium」 and 「tbl_Singapore」. See ◘ Fig. 15.3.

```
plotmatrix(tbl_Belgium.BEL_Fertility,tbl_Singapore.SGP_Fertility);
```

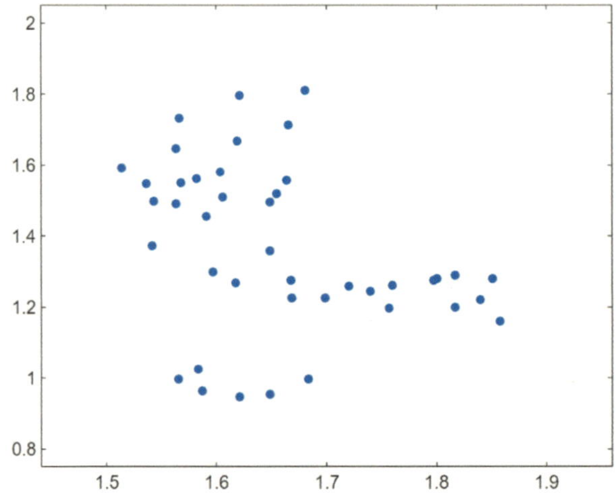

◘ **Fig. 15.3** Use of table variables: Scatter plot using the plotmatrix function

◀

Here, we use the syntax 「plotmatrix(X)」 with only one argument. In this case, a histogram is created for the data in 「X」. See ◘ Fig. 15.4.

```
plotmatrix(BEL_Fertility);
```

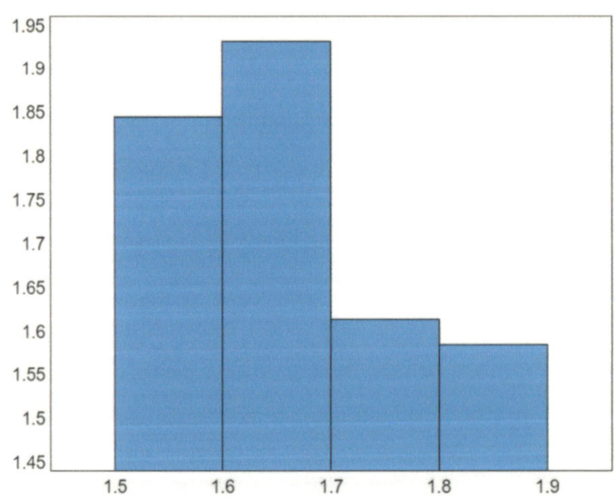

◘ **Fig. 15.4** Histogram using the plotmatrix function

Here, we use ⌜[BEL_Fertility, BEL_Mortality]⌟ as the first argument of the **plotmatrix** function. Consequently, two plots are created. The plot on the left shows the relationship between ⌜BEL_Fertility⌟ and ⌜BEL_Pop⌟, while the plot on the right shows the relationship between ⌜BEL_Mortality⌟ and ⌜BEL_Pop⌟. The resulting figure is shown in Fig. 15.5.

```
plotmatrix([BEL_Fertility,BEL_Mortality],BEL_Pop);
```

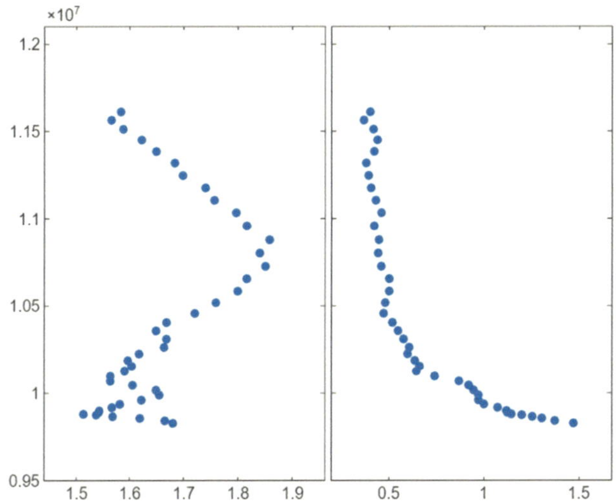

Fig. 15.5 Scatter plots using the plotmatrix function (2-column layout)

Attention

Error Program: Unlike the previous program, we do not use the square brackets of ⌜[]⌟. In this case, using ⌜plotmatrix(X,Y,Z)⌟ with three variables will result in an error. See Fig. 15.6 for the error message.

```
%%% Error program
% plotmatrix(BEL_Fertility,BEL_Mortality,BEL_Pop);
```

```
Error using plotmatrix
Invalid marker specification. Type 'help plot'.
```

Fig. 15.6 Error output of the plotmatrix function: Use of three variables

Here, we use ⌜[BEL_Mortality,BEL_Fertility]⌟ as the second argument of the **plotmatrix** function. As a result, the top plot shows the relationship between ⌜BEL_GDP⌟ and ⌜BEL_Mortality⌟, while the bottom plot shows the relationship between ⌜BEL_GDP⌟ and ⌜BEL_Fertility⌟. See Fig. 15.7. Additionally, compare this figure with Fig. 15.5.

```
plotmatrix(BEL_GDP,[BEL_Mortality,BEL_Fertility]);
```

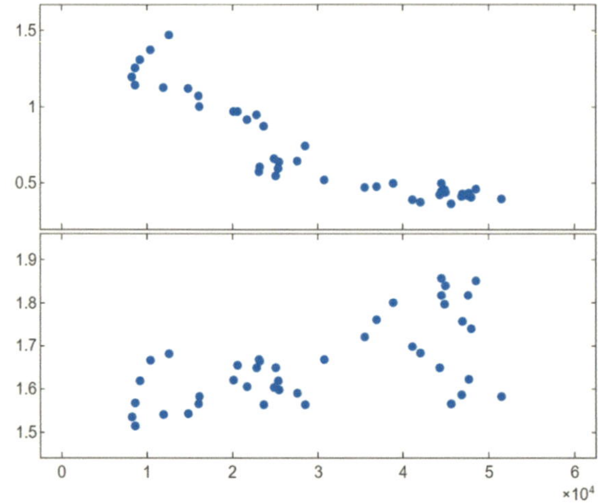

Fig. 15.7 Scatter plots using the plotmatrix function (2-row layout)

Here, we use 「[SGP_Mortality,SGP_Pop]」 as the first argument of the **plotmatrix** function and 「[SGP_GDP,SGP_Fertility]」 as the second argument. As a consequence, four plots are created. The top left plot shows the relationship between 「SGP_Mortality」 and 「SGP_GDP」, the top right plot shows the relationship between 「SGP_Pop」 and 「SGP_GDP」, the bottom left plot shows the relationship between 「SGP_Mortality」 and 「SGP_Fertility」, and the bottom right plot shows the relationship between 「SGP_Pop」 and 「SGP_Fertility」. See Fig. 15.8 for the created figure.

```
plotmatrix([SGP_Mortality,SGP_Pop],[SGP_GDP,SGP_Fertility]);
```

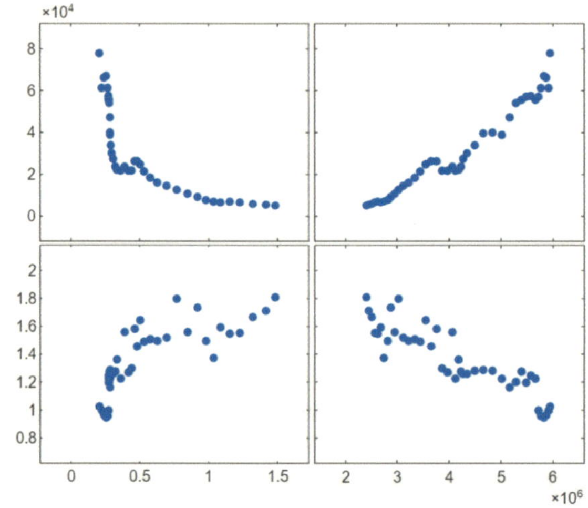

Fig. 15.8 Scatter plots using the plotmatrix function (2-row, 2-column layout)

Here, we use 「[SGP_Mortality,SGP_Fertility,SGP_GDP]」 as both the first and second arguments of the **plotmatrix** function. Consequently, a 3-by-3 matrix of scatter plots is created. The diagonal of the matrix shows a 45-degree linear plot for each variable (「SGP_Mortality」, 「SGP_Fertility」, and 「SGP_GDP」). The off-diagonal plots show the relationships between each pair of variables. Specifically, the top row shows the relationship between 「SGP_Mortality」 and the other two variables (「SGP_Fertility」 and 「SGP_GDP」). The middle row shows the relationship between 「SGP_Fertility」 and the other two variables. Finally, the bottom row shows the relationship between 「SGP_GDP」 and the other two variables. See ◘ Fig. 15.9.

```
plotmatrix([SGP_Mortality,SGP_Fertility,SGP_GDP],...
    [SGP_Mortality,SGP_Fertility,SGP_GDP]);
```

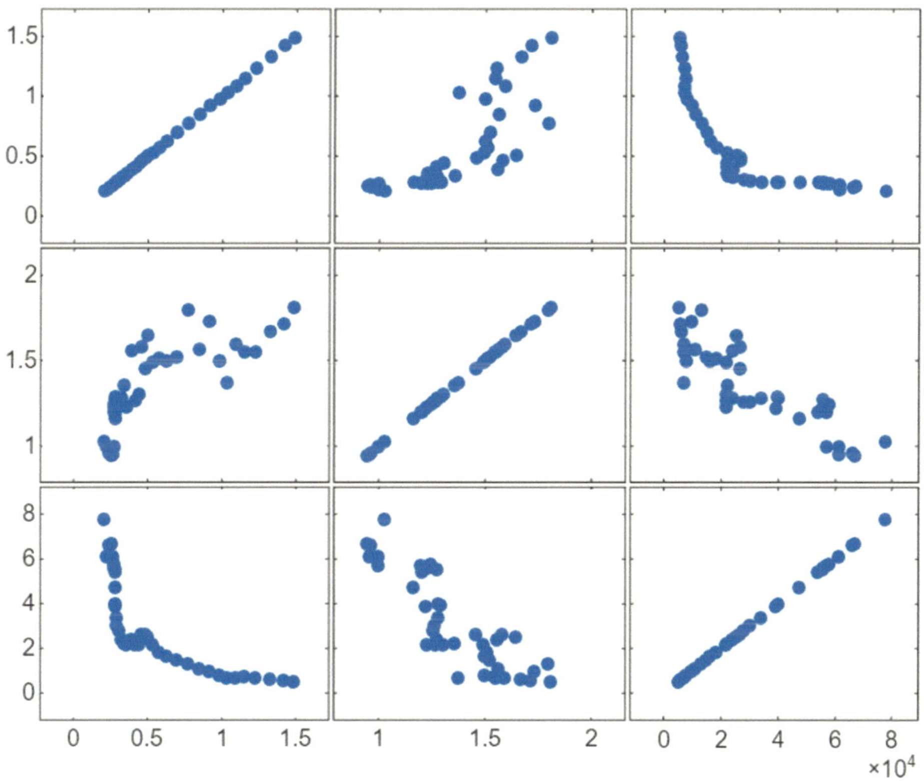

◘ **Fig. 15.9** Scatter plots using the plotmatrix function (3-row, 3-column layout)

Here, we define two matrices, 「SGP」 and 「BEL」, where 「SGP」 contains the variables 「[SGP_Mortality;SGP_Fertility;SGP_GDP]」 and 「BEL」 contains the variables 「[BEL_Mortality;BEL_Fertility;BEL_GDP]」. The **plotmatrix** function is used with 「SGP」 as the first argument and 「BEL」 as the second argument. Since both 「SGP」 and 「BEL」 are 126-by-1 vectors, we can create a single plot figure.

Note, however, that because each vector, 「SGP」 and 「BEL」, consists of three variables, it is difficult to understand the relationships among infant mortality, fertility, and GDP between Belgium and Singapore. See Fig. 15.10 for the resulting figure.

```
SGP  = [SGP_Mortality;SGP_Fertility;SGP_GDP];
BEL  = [BEL_Mortality;BEL_Fertility;BEL_GDP];
plotmatrix(SGP,BEL);
```

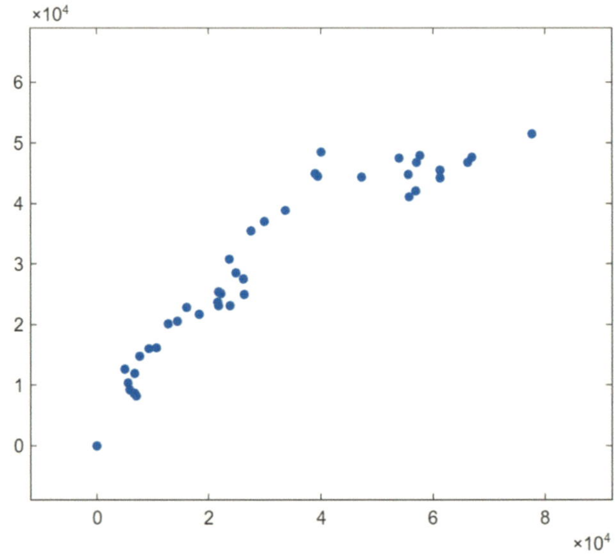

Fig. 15.10 Scatter plot using two merged column vectors

Here, we use the syntax 「plotmatrix(X)」. Specifically, we use 「X」 with the three variables 「BEL_Pop」, 「BEL_Fertility」, and 「BEL_Mortality」. In this case, the diagonal of the matrix displays histograms. In the first row and first column, the histogram for the first variable 「BEL_Pop」 is shown. In the second row and second column, the histogram for the variable 「BEL_Fertility」 is displayed. In the third row and third column, the histogram for the variable 「BEL_Mortality」 is displayed.

Additionally, the off-diagonal plots show the relationship between each pair of variables. For instance, in the first row and second column, a scatter plot of 「BEL_Pop」 against 「BEL_Fertility」 is shown. In the first row and third column, a scatter plot of 「BEL_Pop」 against 「BEL_Mortality」 is displayed. See Fig. 15.11.

```
plotmatrix([BEL_Pop,BEL_Fertility,BEL_Mortality]);
```

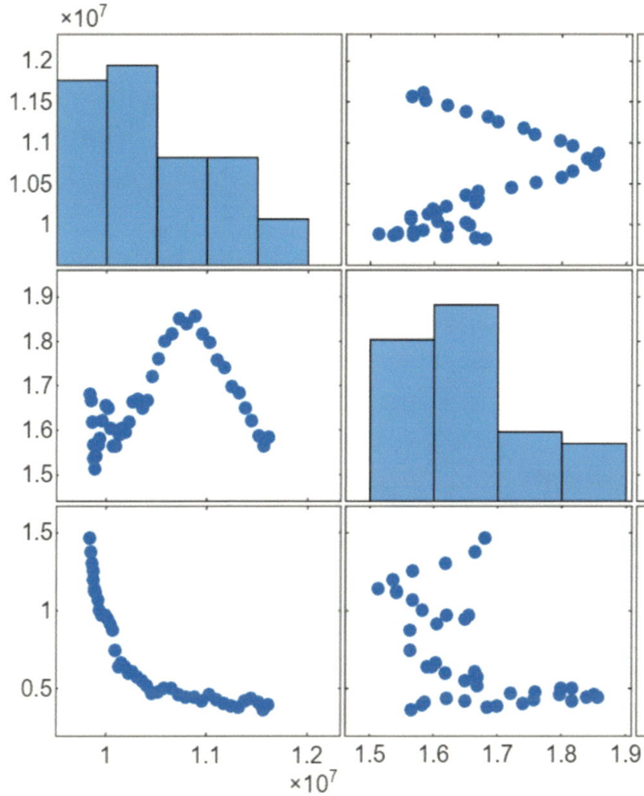

Fig. 15.11 Histograms and scatter plots using the plotmatrix function (3-row, 3-column layout)

Here, using the basic syntax 「plotmatrix(X,Y)」, we use the **plotmatrix** function with the variables 「SGP_Mortality」, 「SGP_Fertility」, 「SGP_GDP」, and 「SGP_Pop」 for both the first and second arguments. This creates a 4-by-4 matrix of scatter plots. See Fig. 15.12.

```
plotmatrix([SGP_Mortality,SGP_Fertility,SGP_GDP,SGP_Pop],...
    [SGP_Mortality,SGP_Fertility,SGP_GDP,SGP_Pop]);
```

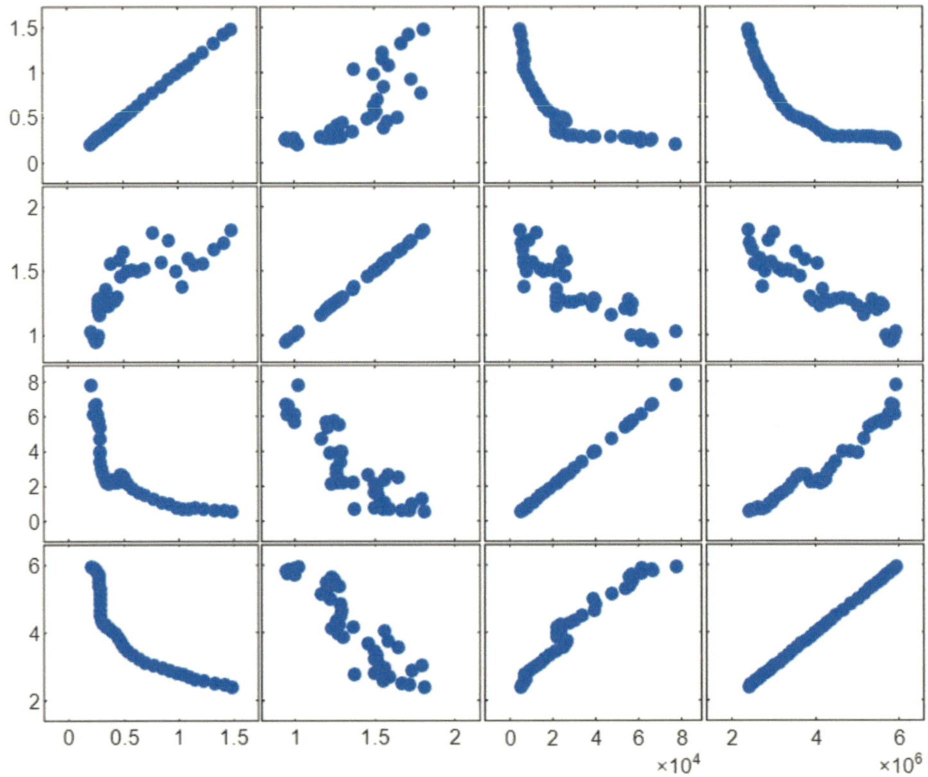

Fig. 15.12 Scatter plots using the plotmatrix function (4-row, 4-column layout)

Unlike the previous program that created Fig. 15.12, here we use the basic syntax of 「plotmatrix(X)」 where we use the four variables 「SGP_Mortality」, 「SGP_Fertility」, 「SGP_GDP」, and 「SGP_Pop」 as in Fig. 15.12. Then, the diagonal of the matrix shows histograms for each of the above variables. Compare Fig. 15.12 with Fig. 15.13.

15

```
plotmatrix([SGP_Mortality,SGP_Fertility,SGP_GDP,SGP_Pop]);
```

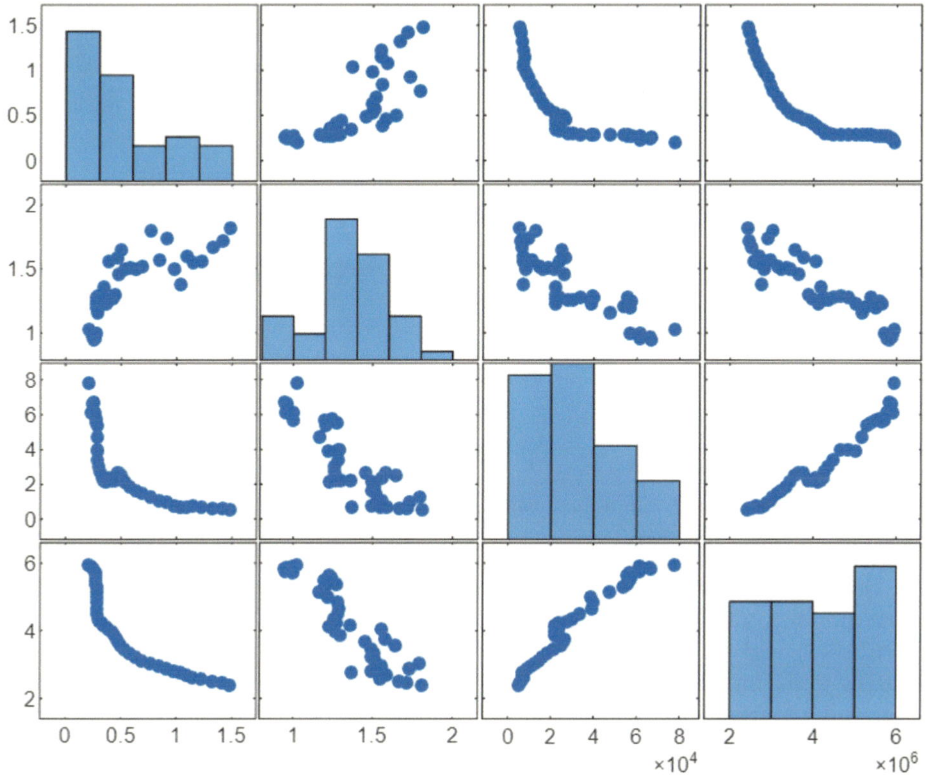

Fig. 15.13 Histograms and scatter plots using the plotmatrix function (4-row, 4-column layout)

15.4 Setting and Controlling the Plotmatrix Function

Here, we use the **plotmatrix** function with three variables: 「BEL_Fertility」, 「BEL_Mortality」, and 「BEL_GDP」. In the syntax 「plotmatrix(X,Y)」, we use 「X」 with the variables 「BEL_Fertility」 and 「BEL_Mortality」, such as 「X=[BEL_Fertility,BEL_Mortality]」. Then, 「plx(2)」 corresponds to the scatter plot figure of 「BEL_Mortality」 and 「BEL_GDP」 in the right figure. We set 「Color」 to 「red」 and 「Marker」 to 「x」, resulting in the red cross markers in the right figure. See Fig. 15.14.

```
plx = plotmatrix([BEL_Fertility,BEL_Mortality],BEL_GDP);
    plx(2).Color = "red";
    plx(2).Marker = "x";
```

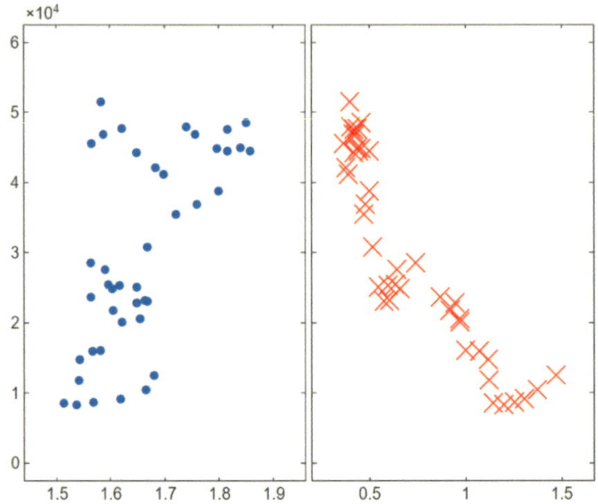

Fig. 15.14 Scatter plots with default markers on the left and red crosses on the right (2-column layout)

Here, we store the properties of the **plotmatrix** function in 「plx2」. 「plx2(1)」 corresponds to the upper figure (the scatter plot between 「BEL_Pop」 and 「BEL_Fertility」), and 「plx2(2)」 corresponds to the lower figure (the scatter plot between 「BEL_Pop」 and 「BEL_Mortality」). In the upper figure, we set 「MarkerSize」 to 5 and 「Marker」 to 「^」, resulting in upward-pointing triangles with a slightly different point size from the default. For the lower figure, we set 「Marker」 to 「pentagram」 and 「Color」 to 「green」, resulting in green pentagrams. See **Fig. 15.15** for the created figure.

```
plx2 = plotmatrix(BEL_Pop,[BEL_Fertility,BEL_Mortality]);
    plx2(1).MarkerSize = 5;
    plx2(1).Marker = "^";
    plx2(2).Marker = "pentagram";
    plx2(2).Color = "green";
```

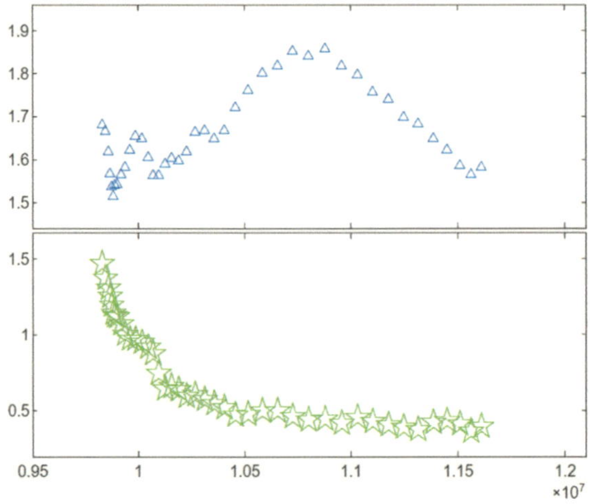

Fig. 15.15 Scatter plots with upward triangles on the top and green stars on the bottom (2-row layout)

Here, we set the markers of the scatter plot to asterisks 「*」 and the color to black (「k (=black)」). See Fig. 15.16.

```
plotmatrix([BEL_Pop,BEL_Fertility,BEL_Mortality],"*k");
```

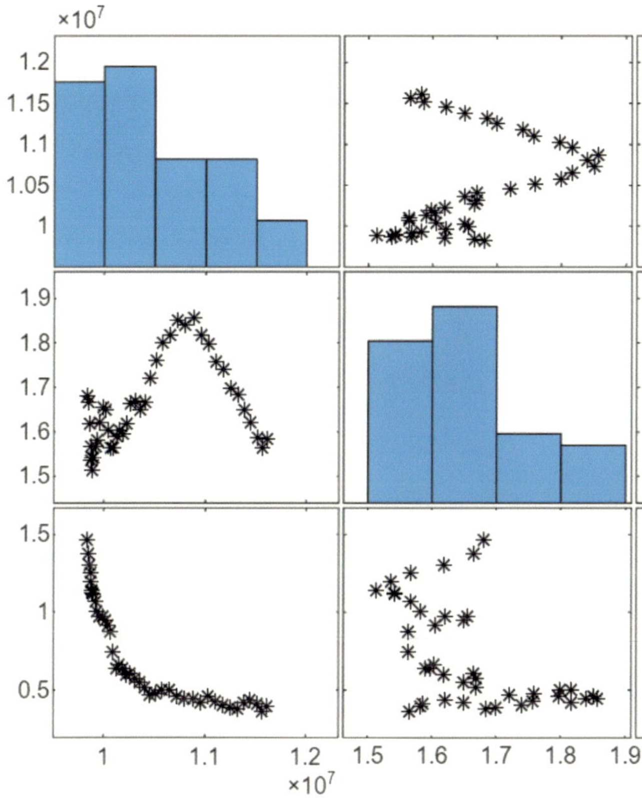

Fig. 15.16 Histograms and scatter plots with black asterisks (3-row, 3-column layout)

Here, we specify 「-k」, resulting in a solid black line without any marker type. See Fig. 15.17.

```
plotmatrix([BEL_Pop,BEL_Fertility,BEL_Mortality],"-k");
```

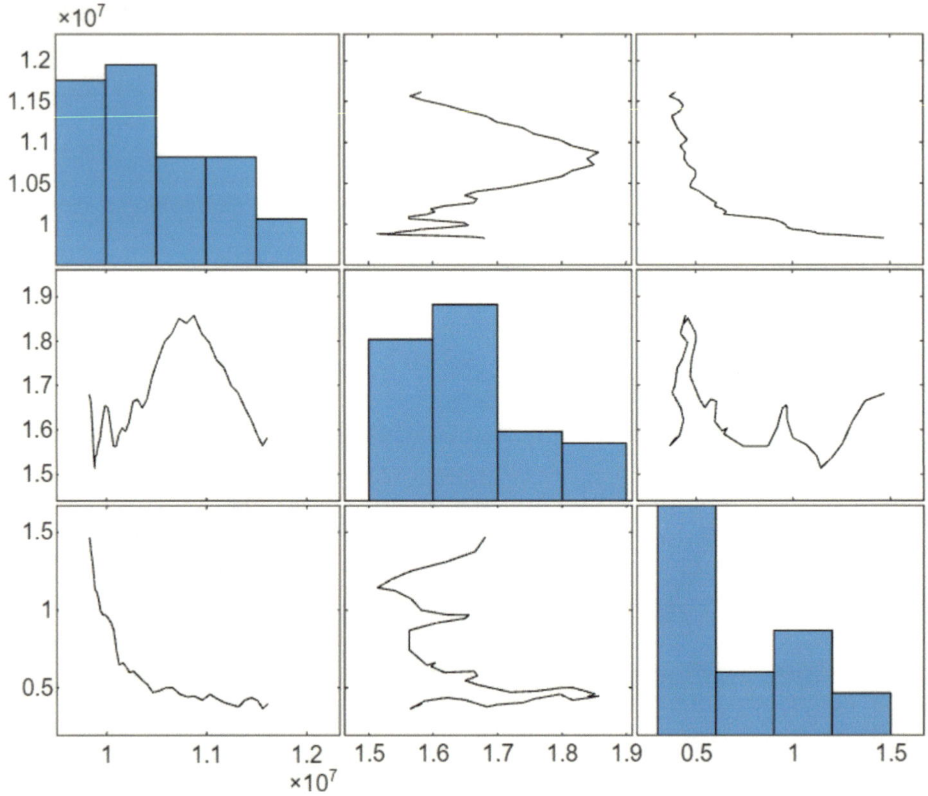

Fig. 15.17 Histograms and black line plots (3-row, 3-column layout)

The syntax for the **plotmatrix** function includes 「[S,AX,BigAx,H,HAx]=plotmatrix(___)」, which allows you to modify the scatter plot and histogram objects. For more details on this syntax, enter 「doc plotmatrix」. 「S」 is the variable that stores the scatter plot objects, 「AX」 is the variable that stores the axes handles for each subplot, 「BigAx」 is the variable that stores the handle for the main background axes, 「H」 is the variable that stores the histogram objects, and 「HAx」 is the variable that stores the axes handles for the histograms.

15

Here, we use the variable 「S」 to change the properties of the scatter plots. For instance, the command 「S(2).Color="red"」 changes the color of the second plot (in the second row and first column) to red, while 「S(6).Color="green"」 changes the color of the sixth plot (in the third row and second column) to green. See Fig. 15.18 for the created figure.

```
S = plotmatrix([BEL_Pop,BEL_Fertility,BEL_Mortality]);
    S(2).Color = "red";
    S(6).Color = "green";
```

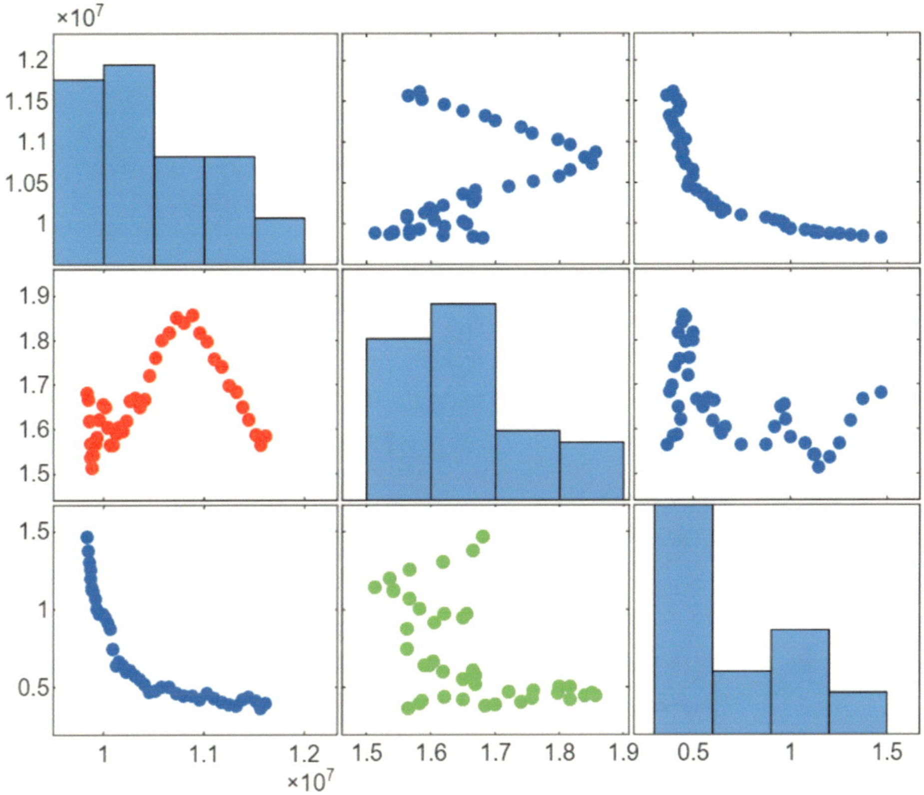

Fig. 15.18 Histograms and multi-colored scatter plots (3-row, 3-column layout)

Here, we use the syntax ⌜[S,AX,BigAx,H,HAx]=plotmatrix(___)⌟. Using ⌜H⌟, we modify the properties of the histograms. Concretely, ⌜H(1).FaceAlpha=0.2⌟ sets ⌜FaceAlpha⌟ of the first histogram (in the first row and first column) to 0.2. Additionally, ⌜H(2).EdgeAlpha=0⌟ sets ⌜EdgeAlpha⌟ of the second histogram (in the second row and second column) to 0, making the edges transparent. Finally, ⌜H(3).DisplayStyle="stairs"⌟ displays only the outline of the third histogram (in the third row and third column) without filling the interior.

Next, using the variable ⌜S⌟, we modify the plot objects. In detail, we set ⌜S(6). Color⌟ to ⌜red⌟ and ⌜S(7). Marker⌟ to ⌜x⌟, making the sixth plot (in the third row and second column) red, and creating the seventh plot (in the first row and third column) with cross markers. See Fig. 15.19 for the created figure.

```
[S,~,~,H] = plotmatrix([BEL_Pop,BEL_Fertility,BEL_Mortality]);
    H(1).FaceAlpha = 0.2;
    H(2).EdgeAlpha = 0;
    H(3).DisplayStyle = "stairs";
    S(6).Color = "red";
    S(7).Marker = "x";
```

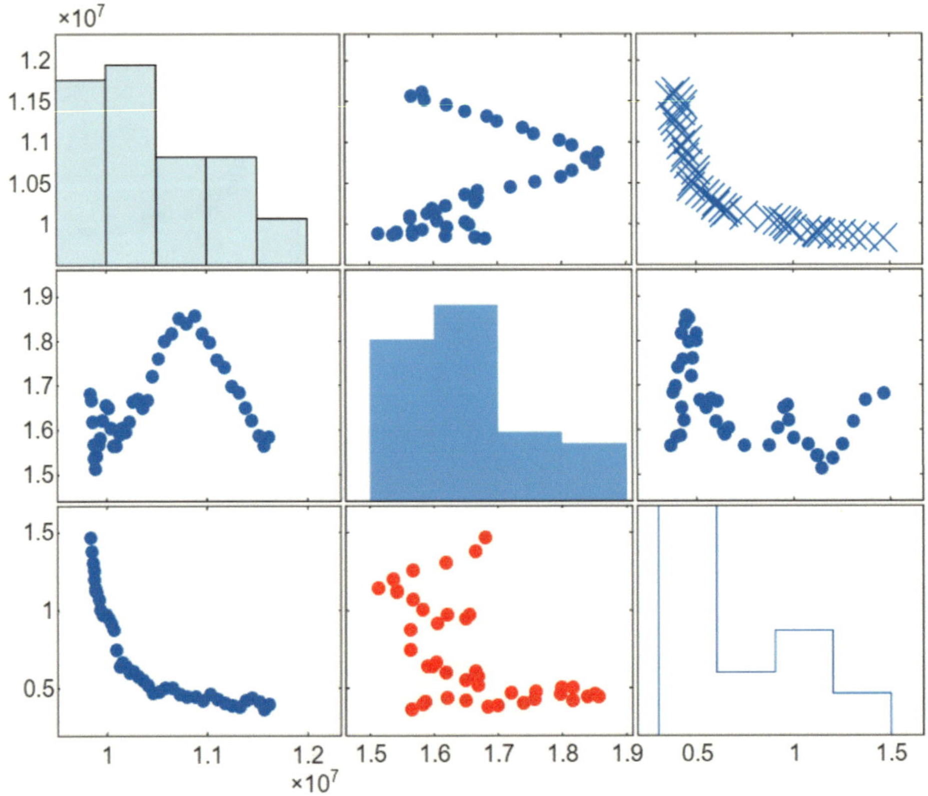

Fig. 15.19 Various histograms and scatter plots (3-row, 3-column layout)

The variable 「H」 contains the properties of the objects corresponding to the histograms. In the **for** loop, for each of the three histograms, 「H(i).FaceColor="m"」 sets the face color of the histograms to 「m (=magenta)」, 「H(i).FaceAlpha=0.2」 sets the transparency of the histograms to 0.2, and 「H(i).LineWidth=4」 increases the line width of the histogram borders to 4. This loop changes the appearance of all three histograms. See Fig. 15.20.


```
[~,~,~,H] = plotmatrix([BEL_Pop,BEL_Fertility,BEL_Mortality]);
for i = 1:3
    H(i).FaceColor = "m";
    H(i).FaceAlpha = 0.2;
    H(i).LineWidth = 4;
end
```

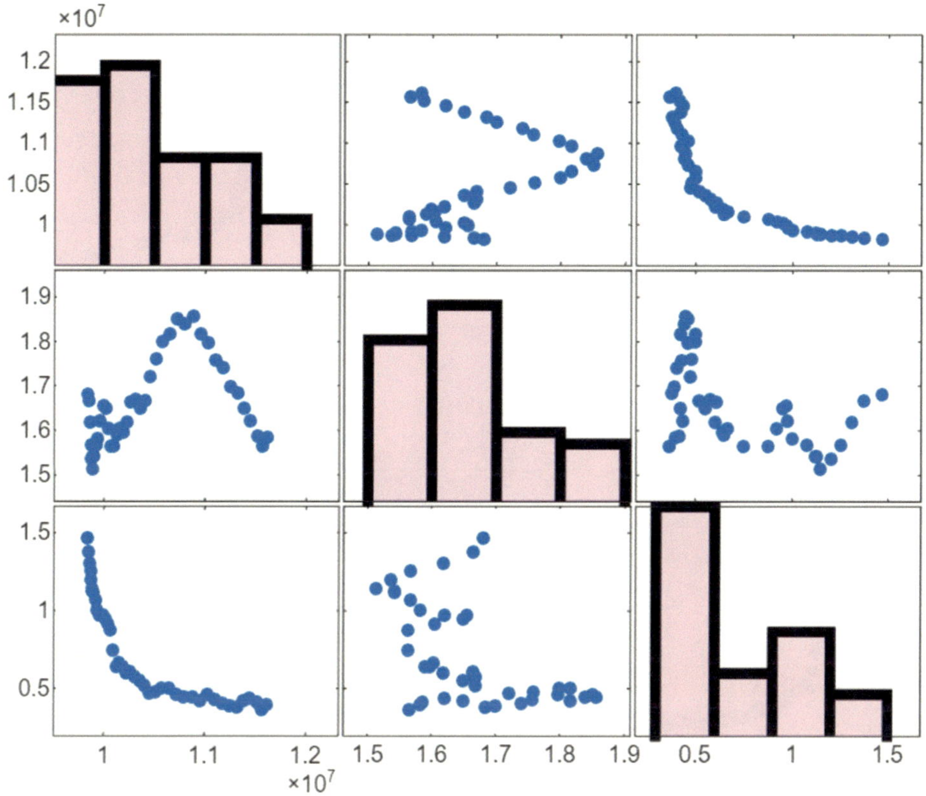

Fig. 15.20 Pink histograms with thick edges and scatter plots (3-row, 3-column layout)

15.5 Combining Optional Functions and the Plotmatrix Function

Here, we set「Color」to「k (=black)」for the fourth plot in the first row and second column. We also set the marker to「+」for the eighth plot in the second row and third column. In the **title** function, we set「FontSize」to 20 and「Color」to「red」, creating a large red title. See Fig. 15.21.

```
BEL_mat = plotmatrix([BEL_Pop,BEL_Fertility,BEL_Mortality]);
    BEL_mat(4).Color = "k";
    BEL_mat(8).Marker = "+";
t = title("Belguim");
    t.Color = "red";
    t.FontSize = 20;
```

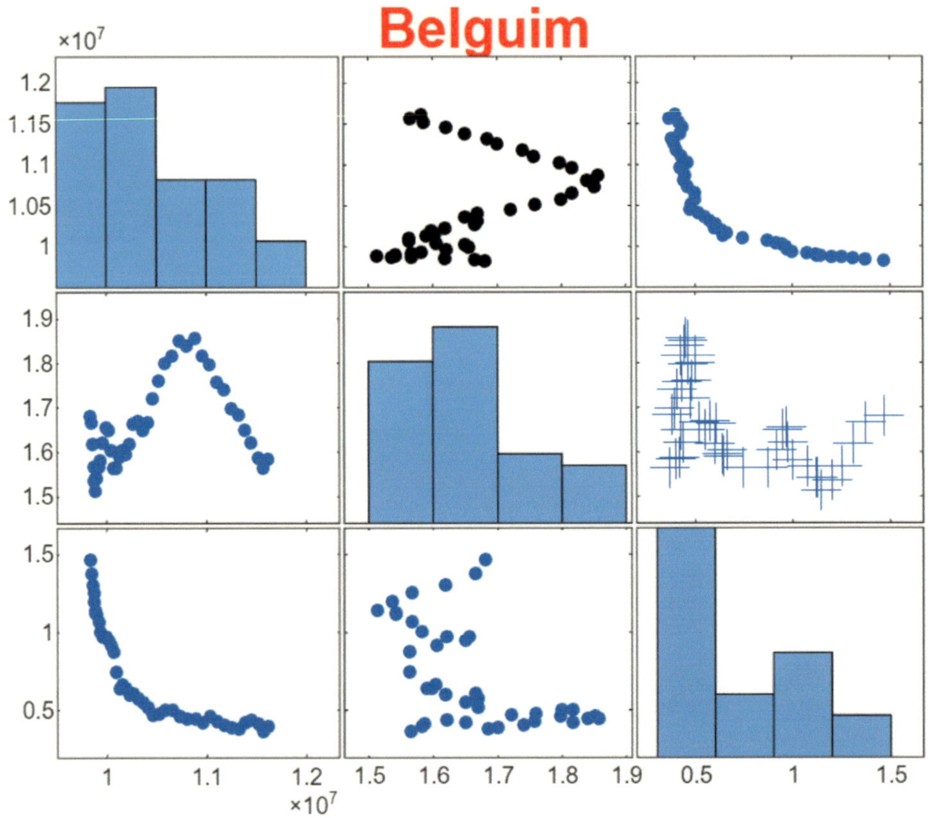

Fig. 15.21 Histograms and scatter plots with a large red title (3-row, 3-column layout)

In the following program, we use 「AX」 and 「BigAx」, from the syntax 「[S,AX,BigAx,H,HAx]=plotmatrix(___)」. Since there are four small axes, we use a **for** loop to iterate from 「i=1」 to 「i=4」, setting 「labels{i}」 for each axis. As a result, each axis is labeled 「Fertility」, 「Mortality」, 「Population」, and 「GDP」. Additionally, using 「BigAx」, we label the entire large horizontal axis as 「x-axis」, where we set 「FontSize」 to 25 and 「Color」 to 「green」. See Fig. 15.22.

15

```
labels = {"Fertility" "Mortality" "Population" "GDP"};
data = [SGP_Fertility SGP_Mortality SGP_Pop SGP_GDP];
[~,AX,BigAx] = plotmatrix(data);
for i = 1:4
  xlabel(AX(4,i),labels{i});
  ylabel(AX(i,1),labels{i});
end
xl = xlabel(BigAx,"x-axis");
    xl.FontSize = 25;
    xl.Color = [0.5 1 0];
title("Singapore");
```

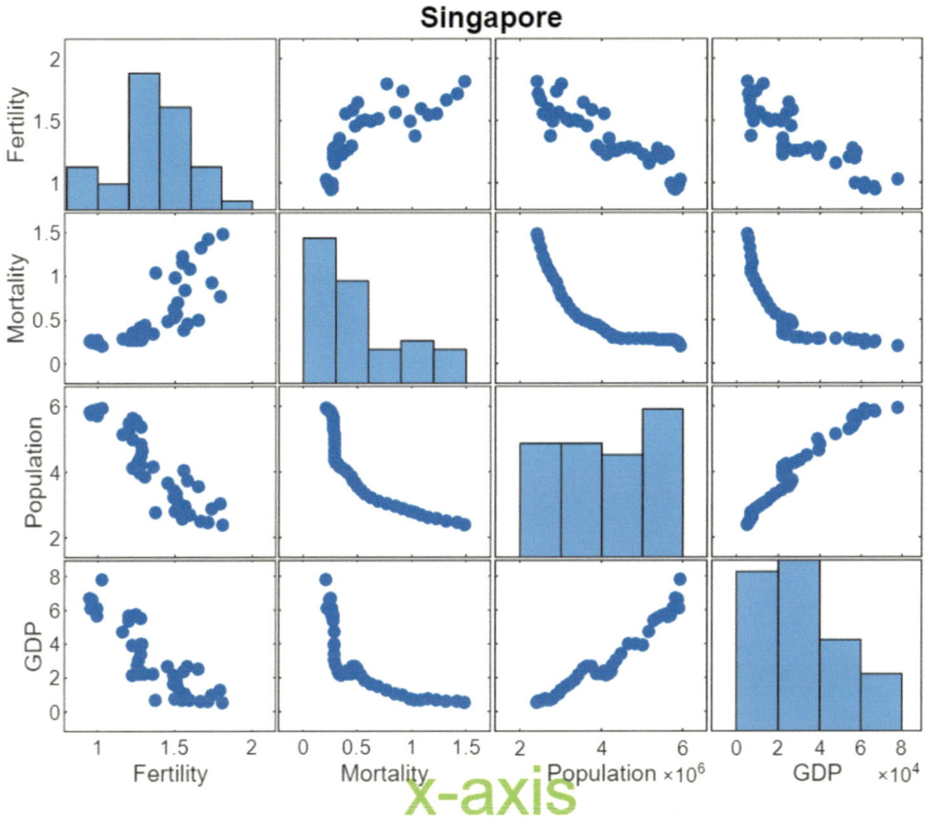

Fig. 15.22 Histograms and scatter plots with subplot labels (4-row, 4-column layout)

Here, we start by creating a figure with a 1-by-2 tiled layout using 「tiledlayout (1,2)」. After the first use of 「nexttile」, we generate a matrix of scatter plots and histograms for Belgium's population, fertility, and mortality using the **plotmatrix** function. Additionally, we modify the second and third histograms (「H1(2)」 and 「H1(3)」) by setting 「Orientation」 to 「horizontal」. Moreover, we change the edge color of the third histogram to green.

Next, we use 「nexttile」 again. Then, we create the scatter plot matrix for Singapore's population, fertility, and mortality. We customize the histograms by setting 「FaceColor」 of 「H2(1)」, 「H2(2)」, and 「H2(3)」 to 「r (=red)」, 「g (= green)」, and 「k (=black)」, respectively. For the scatter plots, we modify 「Marker」 in several plots: 「S2(2)」 is set to an upward-pointing triangle 「^」; 「S2(3)」 is set to a left-pointing triangle 「<」; 「S2(4)」 is set to a circle 「o」; 「S2(6)」 is set to an asterisk 「*」; 「S2(7)」 is set to a dot 「.」; and 「S2(8)」 is set to a plus sign 「+」. Finally, for specific plots (from 「S2(2)」 to 「S2(4)」 and 「S2(6)」 to 「S2(8)」), we set 「MarkerSize」 to 3. See Fig. 15.23.

```
figure
tiledlayout(1,2);
t1 = nexttile;
[S1,AX1,BigAx1,H1,HAx1] = plotmatrix([BEL_Pop,BEL_Fertility,BEL_Mortality]);
    H1(2).Orientation = "horizontal";
    H1(3).Orientation = "horizontal";
    H1(3).EdgeColor = "green";
title(t1,"Belguim");
t2 = nexttile;
[S2,AX2,BigAx2,H2,HAx2] = plotmatrix([SGP_Pop,SGP_Fertility,SGP_Mortality]);
    H2(1).FaceColor = "r";
    H2(2).FaceColor = "g";
    H2(3).FaceColor = "k";
    S2(2).Marker = "^";
    S2(3).Marker = "<";
    S2(4).Marker = "o";
    S2(6).Marker = "*";
    S2(7).Marker = ".";
    S2(8).Marker = "+";
for i = [2:4,6:8]
    S2(i).MarkerSize = 3;
end
title(t2,"Singapore");
```

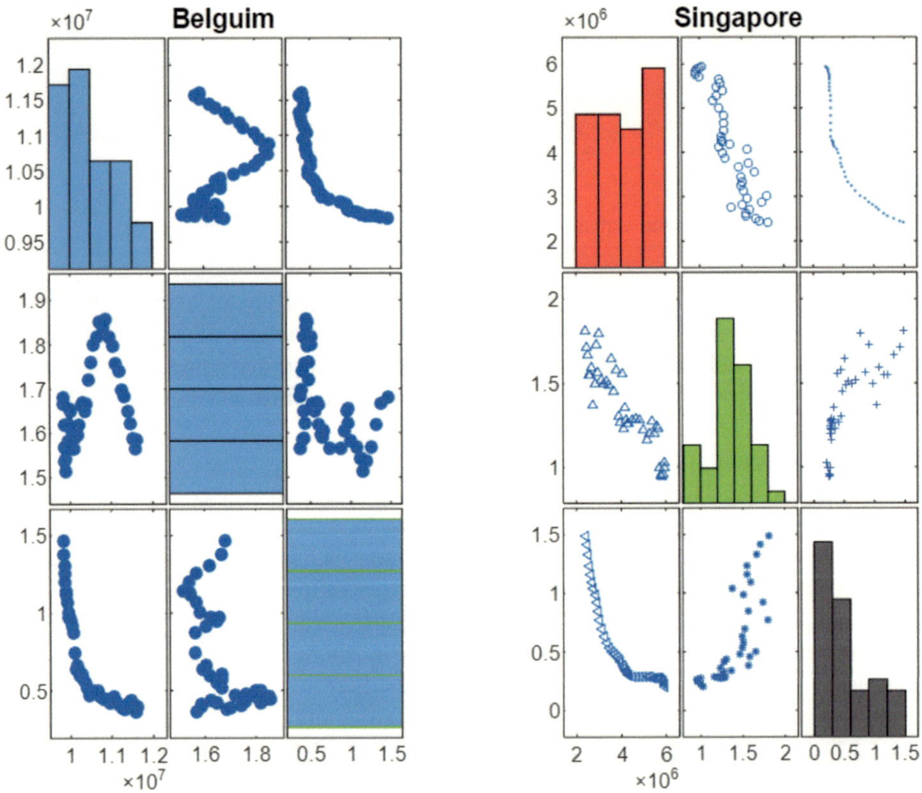

Fig. 15.23 Two-column display of histograms and scatter plots (3-row, 3-column layout)

Here, using the **corr** function, we measure the correlation coefficients between the variables ⌜BEL_GDP⌟, ⌜BEL_Pop⌟, and ⌜BEL_Fertility⌟. The correlation coefficient between ⌜BEL_Pop⌟ and ⌜BEL_GDP⌟ is extracted from the variable ⌜correlation_matrix⌟ and stored in the variable ⌜corr_value_1⌟. The value, 0.92, is then displayed on the plot using the **annotation** function, which creates a text box at the specified location, size, and width by ⌜[0.5 0.75 0.05 0.05]⌟. For more details on the **annotation** function, see ► Sect. 18.1. See ◘ Fig. 15.24 for the created figure.

```
figure
BEL_Data = [BEL_GDP,BEL_Pop,BEL_Fertility];
correlation_matrix = corr(BEL_Data);
plotmatrix(BEL_Data);
corr_value_1 = correlation_matrix(2,1);
annotation("textbox",[0.5 0.75 0.05 0.05],"String",...
    num2str(corr_value_1,"%.2f"),"FitBoxToText","on",...
    "FontSize",24,"FontWeight","bold","EdgeColor","none");
```

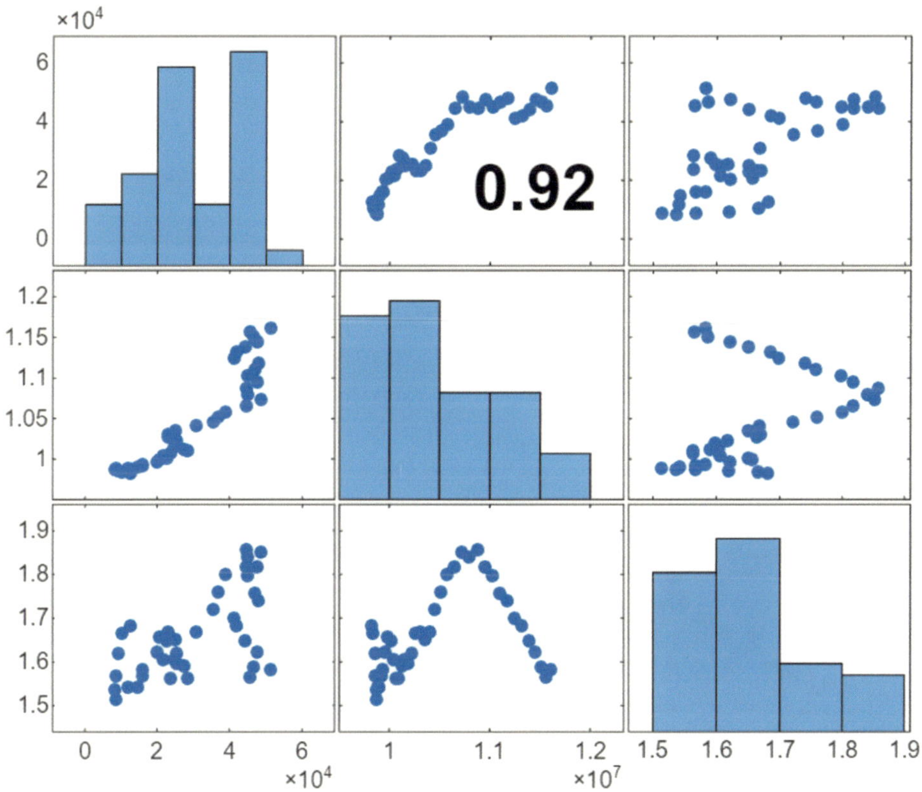

◘ **Fig. 15.24** Histograms and scatter plots with a bold black numeric label (3-row, 3-column layout)

As shown in ◘ Fig. 15.24, we place the correlation coefficients within the plots created by the **plotmatrix** function. For a detailed explanation of how to use the **corr** and **annotation** functions, see the previous program that created ◘ Fig. 15.24.

Now, we explain the **for** loop. The loop runs 9 times with ⌜for i=1:3⌟ and ⌜for j=1:3⌟. Note that the program within the ⌜if⌟ statement is executed only when the values of i and j are not equal, that is, 「if i~=j」. In other words, the program is executed for all off-diagonal plots. For these cases, the correlation coefficient value is stored in 「corr_value」 and placed in the specified location using the **annotation** function. For more details on the **annotation** function, see ▶ Sect. 18.1.

On the other hand, for the diagonal plots where 「i=j」 (i.e., i=j=1, i=j=2, i=j=3), note that the program inside 「if i~=j」 is not executed. The resulting figure is presented in ◘ Fig. 15.25.

```
BEL_Data = [BEL_GDP,BEL_Pop,BEL_Fertility];
correlation_matrix = corr(BEL_Data);
[h,ax] = plotmatrix(BEL_Data);
for i = 1:3
    for j = 1:3
        if i ~= j
            corr_value = correlation_matrix(i,j);
            ax_pos = get(ax(i,j),"Position");
            xPos = ax_pos(1);
            yPos = ax_pos(2);
            annotation("textbox",[xPos yPos 0.05 0.05],"String",...
                num2str(corr_value,"%.2f"),"FitBoxToText","on",...
                "FontSize",10,"FontWeight","bold","EdgeColor","red");
        end
    end
end
title("Correlation coefficients");
```

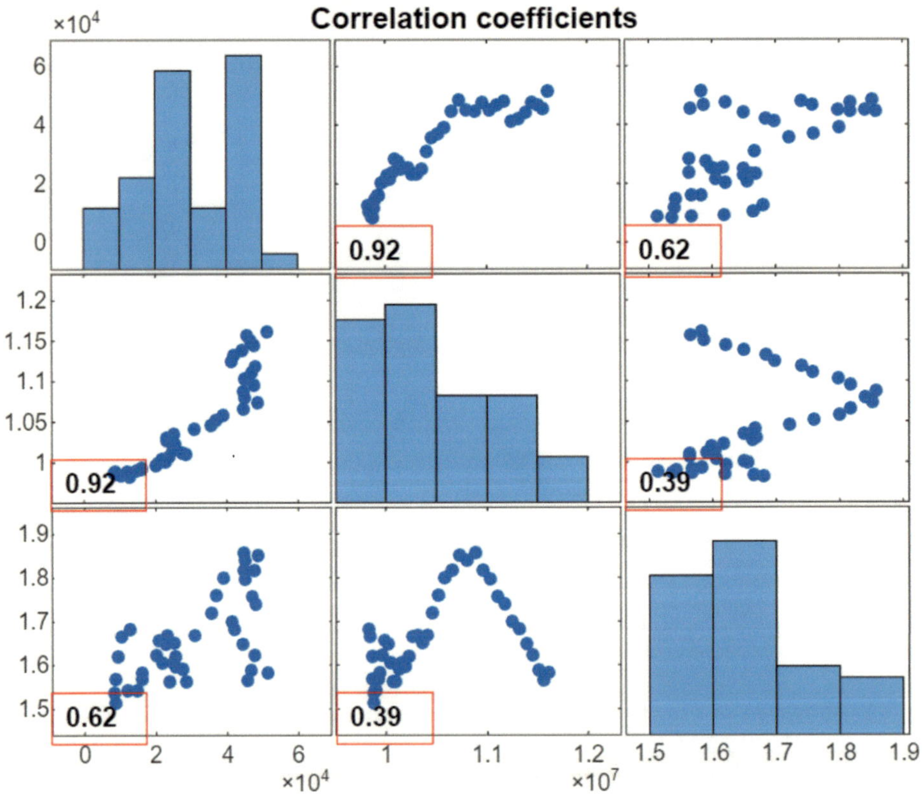

◘ **Fig. 15.25** Histograms and scatter plots with red-boxed labels (3-row, 3-column layout)

In the baseline of the programs of ◘ Figs. 15.24 and 15.25, we add three new points. In the first point, using a **for** loop, we assign labels to the x-axis of the fourth row and the y-axis of the first column for each subplot in the variable 「AX」.

In the second point, we use the **delete** function to remove the scatter plots in the top right corner of the figure. Specifically, the outer loop variable 「plot_row」 iterates from 1 to 4, while the inner loop variable 「plot_col」 starts at the current value of 「plot_row」 and iterates up to 4. Specifically, at each iteration of the inner loop, the **delete** function removes the axes object at the position (「plot_row」, 「plot_col」) in the variable 「AX」. It is essential to use the **delete** function after creating axis labels with the **xlabel** and **ylabel** functions. If the order between the first and second points is reversed, the target axis for the labels will no longer exist, resulting in an error ("Invalid target").

In the third point, using 「AX」, we apply the **legend** function to add a legend to the figure. See ◘ Fig. 15.26.

```
figure
labels = {"Fertility" "Mortality" "Population" "GDP"};
data = [SGP_Fertility SGP_Mortality SGP_Pop SGP_GDP];
[~,AX] = plotmatrix(data);
for i = 1:4
  xlabel(AX(4,i),labels{i});
  ylabel(AX(i,1),labels{i});
end
for plot_row = 1:4
    for plot_col = plot_row:4
        delete(AX(plot_row,plot_col));
    end
end
legend(AX(4,1),"GDP  Fertility");
legend(AX(4,2),"GDP  Mortality");
legend(AX(4,3),"GDP Population");
```

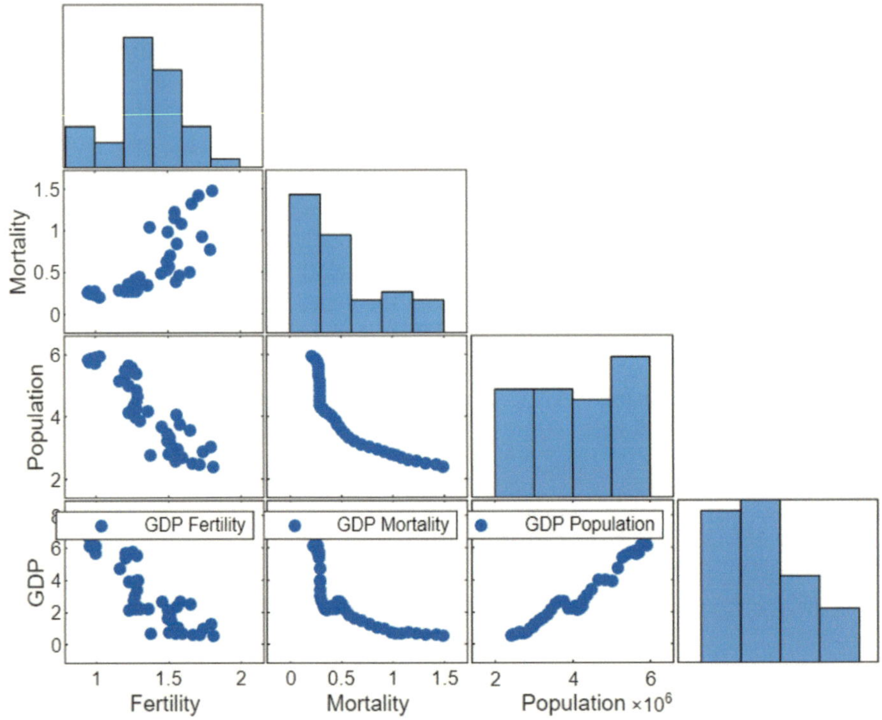

Fig. 15.26 Histograms and scatter plots with an empty upper triangle and legends (3-row, 3-column layout)

Here, using the **xlabel**, **ylabel**, and **delete** functions, we create a plotmatrix figure similar to the one in the program for Fig. 15.26. As mentioned in the previous program that created Fig. 15.26, we use the **delete** function after the **xlabel** and **ylabel** functions. In addition, we use the **set** function to outline specified boxes in red or blue. With 「set(AX(3,1),"XColor","red","YColor","red","LineWidth",4)」, we outline the box in the third row and first column with a thick red line. Similarly, 「set(AX(4,3), "XColor","blue","YColor","blue","LineWidth", 4)」 outlines the box in the fourth row and third column with a thick blue line. The resulting figure is presented in Fig. 15.27.

```
figure
labels = {"Fertility","Mortality","Population","GDP"};
data = [BEL_Fertility,BEL_Mortality,BEL_Pop,BEL_GDP];
[~,AX] = plotmatrix(data);
for i = 1:4
  xlabel(AX(4,i),labels{i});
  ylabel(AX(i,1),labels{i});
end
for plot_row = 1:4
    for plot_col = plot_row:4
        delete(AX(plot_row,plot_col));
    end
end
set(AX(3,1),"XColor","red","YColor","red","LineWidth",4);
set(AX(4,3),"XColor","blue","YColor","blue","LineWidth",4);
```

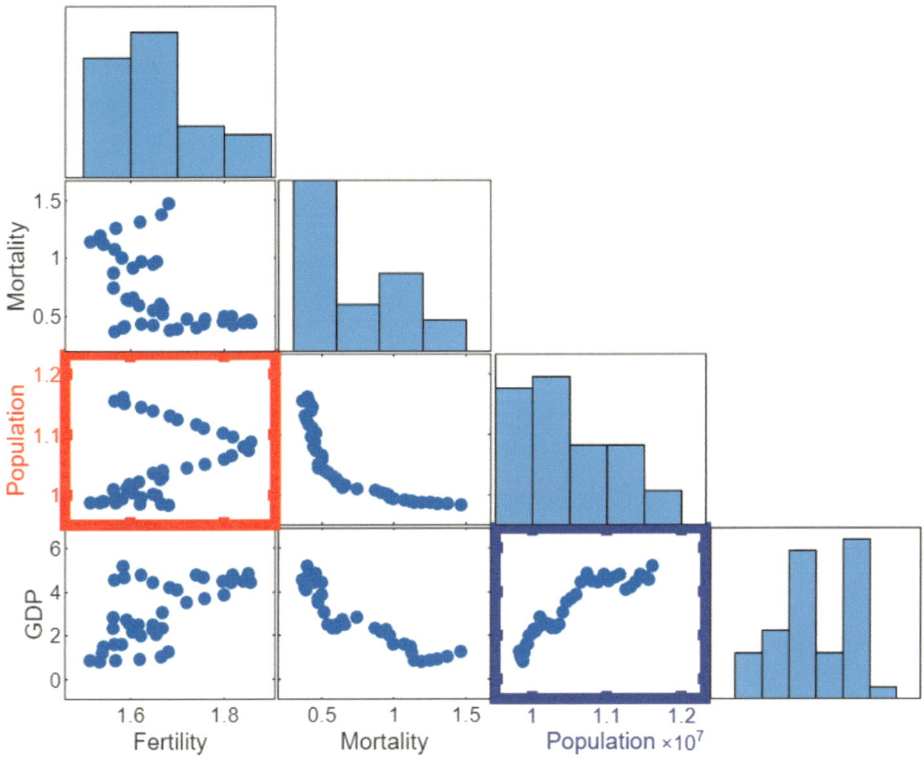

Fig. 15.27 Histograms and scatter plots with two subplots outlined in red and blue (4-row, 4-column layout)

Here, similar to Figs. 15.26 and 15.27, we use the **delete** function to create a plotmatrix figure. Additionally, we calculate the correlation coefficients and place them in the specified locations using the **annotation** function. The correlation coefficients for the off-diagonal elements (where the value of 「row」 is greater than that of 「col」) are calculated and displayed in the appropriate positions using the **annotation** function. For more details on the **annotation** function, see ► Sect. 18.1. The coordinates for placing the correlation coefficients are determined by extracting the position of the corresponding axes from 「AX」 with 「ax_pos=get(AX(col,row), "Position")」. The 「xPos」 and 「yPos」 values are computed based on the size and position of each plot, and the correlation coefficient is displayed in the center of each plot using the **annotation** function. For details on how the correlation coefficients are handled, refer to the explanations for Figs. 15.24 and 15.25.

One important point to note is that the **delete** function should be used last. If you delete certain parts first, you will not be able to retrieve the coordinates of the deleted spaces using 「ax_pos=get(AX(col,row),"Position")」. See Fig. 15.28.

```
figure
labels = {"Fertility","Mortality","Population","GDP"};
data = [SGP_Fertility,SGP_Mortality,SGP_Pop,SGP_GDP];
correlation_matrix = corr(data);
[S,AX,BigAx] = plotmatrix(data);
for i = 1:4
  xlabel(AX(4,i),labels{i});
  ylabel(AX(i,1),labels{i});
end
for row = 2:4
    for col = 1:row-1
        corr_value = correlation_matrix(row,col);
        ax_pos = get(AX(col,row),"Position");
        xPos = ax_pos(1)+ax_pos(3)/2;
        yPos = ax_pos(2)+ax_pos(4)/2 ;
        annotation("textbox",[xPos yPos 0.05 0.05],...
            "String",num2str(corr_value,"%.2f"),...
            "FitBoxToText","on","FontSize",20,...
            "FontWeight","bold","HorizontalAlignment","center");
    end
end
for plot_row = 1:4
    for plot_col = plot_row:4
        delete(AX(plot_row,plot_col));
    end
end
title("Correlation coefficients");
```

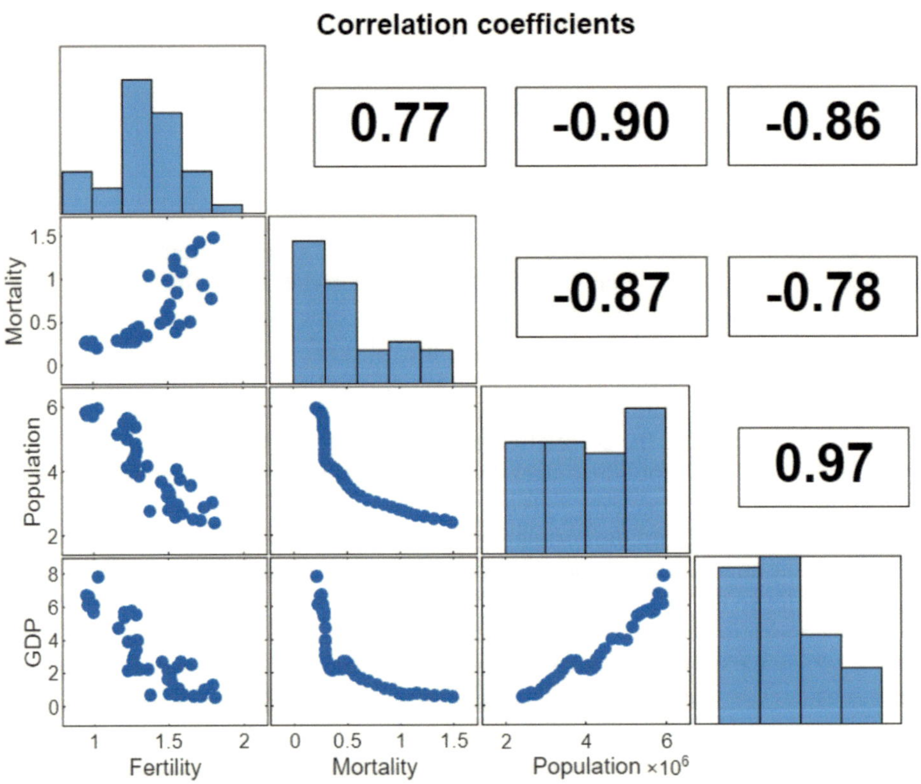

Fig. 15.28 Histograms and scatter plots with black-boxed numeric labels placed in an empty upper triangle (4-row, 4-column layout)

15.6 Questions

Question 1

Create Fig. 15.29 using the **plotmatrix** function with ⌜[BEL_Pop, BEL_Fertility, BEL_Mortality]⌟ as the first argument. In detail, set the markers for the plots to upward-pointing triangles and use solid green curves. Additionally, label the y-axis of the plot in the first row, first column with ⌜Population⌟ and the y-axis of the plot in the third row, first column with ⌜Mortality⌟. Note that no label should be added to the plot in the second row, first column. Finally, use the title function to set the title to ⌜plotmatrix⌟.

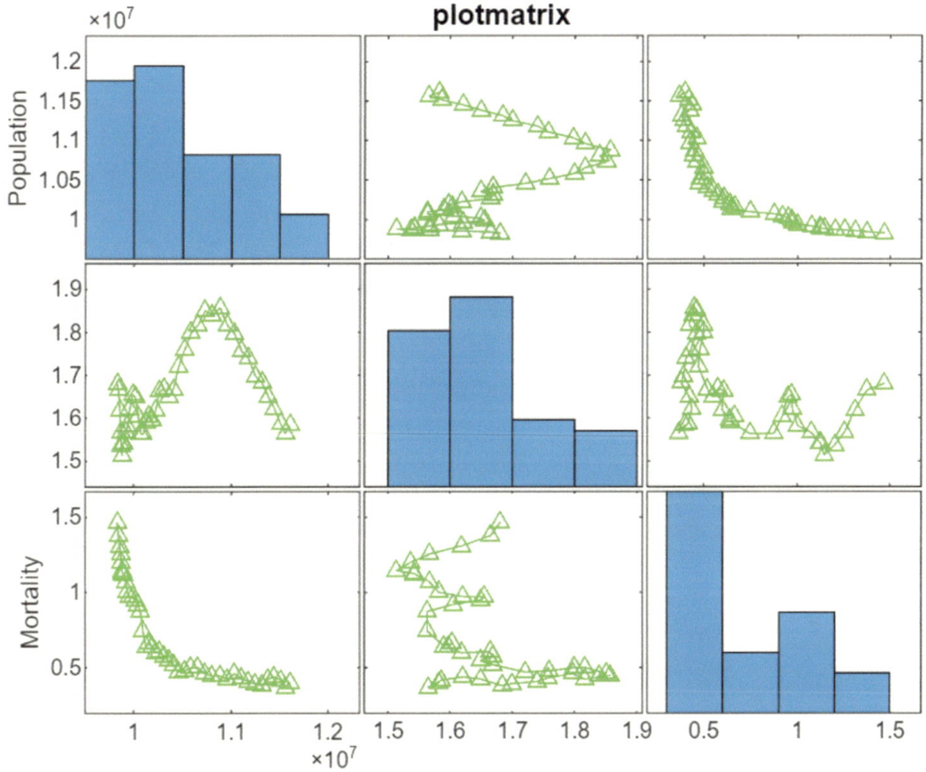

Fig. 15.29 Histograms and scatter plots (Question 1)

Answer 1

```
figure
[~,AX] = plotmatrix([BEL_Pop,BEL_Fertility,BEL_Mortality],"^-g");
ylabel(AX(1,1),"Population");
ylabel(AX(3,1),"Mortality");
title("plotmatrix");
```

Question 2

Create Fig. 15.30 using the **plotmatrix** function with 「[BEL_Pop, BEL_Fertility, BEL_Mortality]」 as the first argument. Remove the plots in the lower-left (the plots in the second row, first column; the third row, first column; and the third row, second column).

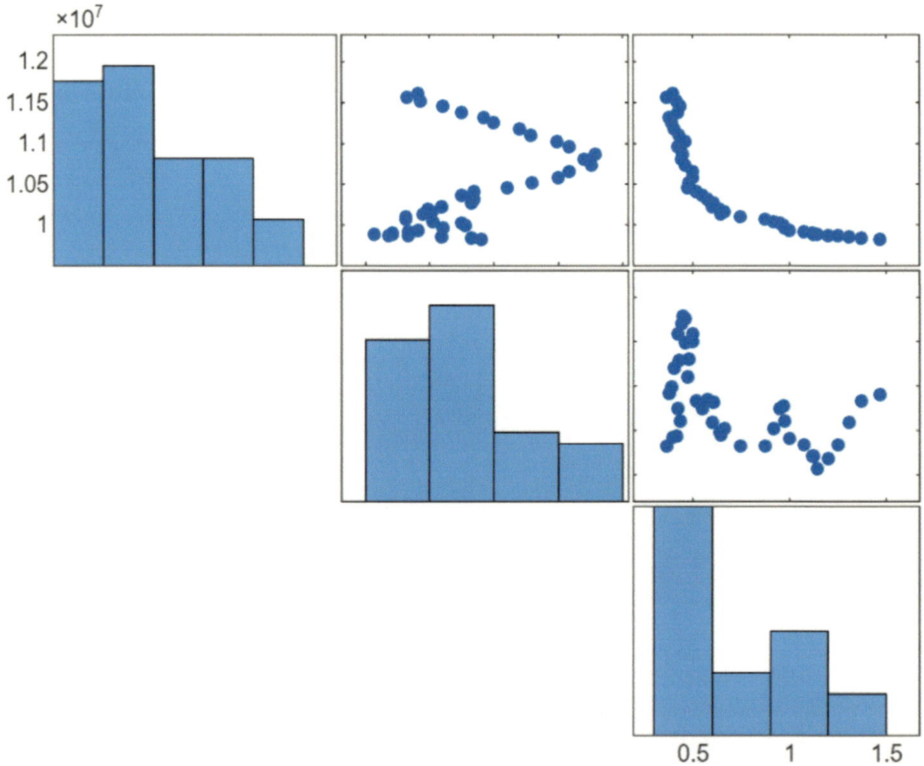

Fig. 15.30 Histograms and scatter plots (Question 2)

Answer 2

```
figure
[~,AX] = plotmatrix([BEL_Pop,BEL_Fertility,BEL_Mortality]);
for plot_row = 2:3
    for plot_col = 1:plot_row-1
        delete(AX(plot_row,plot_col));
    end
end
```

Question 3

Create Fig. 15.31 using the **plotmatrix** function with 「[BEL_Pop, BEL_Fertility, BEL_Mortality]」 as the first argument. Additionally, create a blue filled histogram with high transparency and a thick outline.

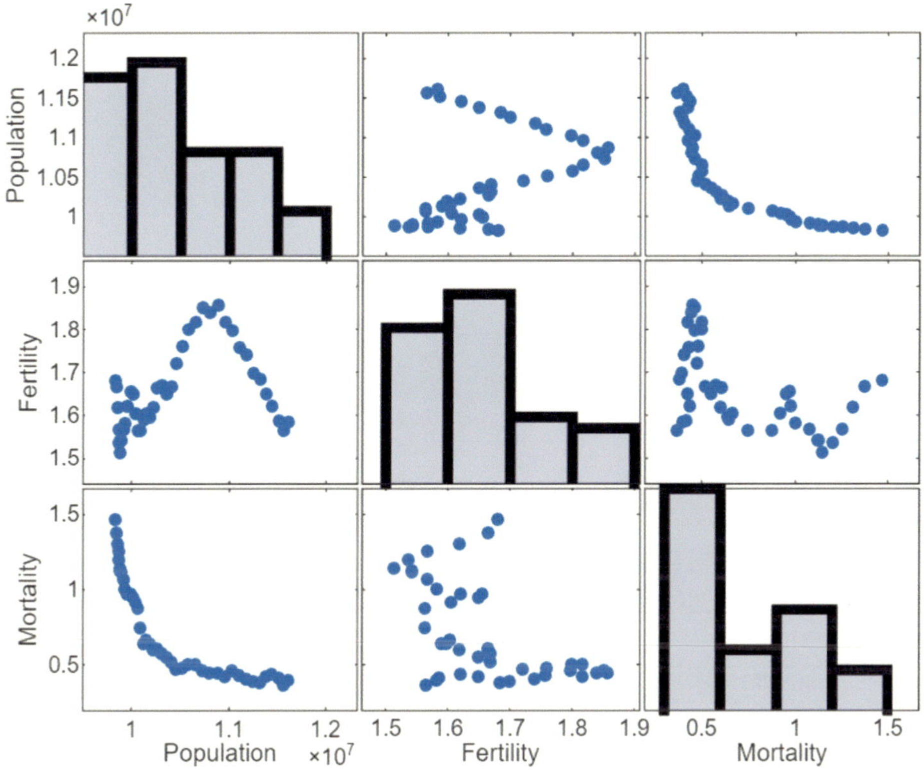

Fig. 15.31 Histograms and scatter plots (Question 3)

Answer 3

```
figure
labels = {"Population","Fertility","Mortality"};
[~,AX,~,H] = plotmatrix([BEL_Pop,BEL_Fertility,BEL_Mortality]);
for i = 1:3
    H(i).FaceColor = "blue";
    H(i).FaceAlpha = 0.2;
    H(i).LineWidth = 4;
    xlabel(AX(3,i),labels{i});
    ylabel(AX(i,1),labels{i});
end
```

Bubblechart Function

Contents

16.1 Details of the Data Used in this Chapter – 498

16.2 Excel Data of ▶ Section 16.1 – 499

16.3 Basic Syntax of the Bubblechart Function – 499

16.4 Setting and Controlling the Bubblechart Function – 503

16.5 Combining Optional Functions and the Bubblechart Function – 506

16.6 Questions – 528

Supplementary Information The online version contains supplementary material available at https://doi.org/10.1007/978-981-96-6213-5_16.

Y. Nakamoto, *A Comprehensive Guide to Creating Graphics with MATLAB Programming*, Classroom Companion: Economics, https://doi.org/10.1007/978-981-96-6213-5_16

Learning Objectives

This chapter creates bubble diagrams using the **bubblechart** function. By reading this chapter, you will learn how to use the **bubblechart** function. Additionally, you will learn how to convey three-variable data (i.e., three-dimensional information) in a 2D scatter plot using bubble sizes.

Overview

In this chapter, we create a circles (bubbles) diagram using the **bubblechart** function. The charts created by the **bubblechart** function are a type of scatter plot. Specifically, the size of the bubbles can represent quantities of data, allowing us to incorporate an additional dimension of data. Thus, compared to the **scatter** function (▶ Chap. 3), the **bubblechart** function is suitable for representing multiple variables simultaneously, allowing for a visual understanding of data relationships.

16.1 Details of the Data Used in this Chapter

In this chapter, we use data on land area, population, and population density for various countries located mainly in Asia. The variable 「country」 represents the names of the countries, 「land_area」 represents the area (in thousands of square kilometers), 「pop」 represents the population (in thousands), and 「pop_den」 represents the population density (people per km²).

```
country = {"United Arab Emirates","Iran","India","Uzbekistan",...
    "Qatar","Cambodia","Kuwait","Saudi Arabia","Thailand","Korea",...
    "China","Turkey","Japan","Nepal","Pakistan","Philippines",...
    "Vietnam","Malaysia","Myanmar","Mongolia"}';

 land_area = [71,1631,3287,449,12,181,18,2207,513,...
     100,9600,784,378,147,796,300,331,331,677,1564]';

pop = [9365,87923,1407564,34081,2688,16589,4250,35950,71601,51830,...
    1425893, 84775,123224,30035,231402,113880,97468,33574,53798,3348]';

pop_den = [132,54,428,76,231,92,239,16,140,...
    517,149,108,326,204,291,380,294,102,80,2]';

tbl_Chap16 = table(country,land_area,pop,pop_den);
```

16

Next, the data for 「land_area」, 「pop」, and 「pop_den」 are each divided into two parts. The variable 「landA」 contains the first half of the 「land_area」 data, while the variable 「landB」 contains the second half. Similarly, 「popA」 and 「popB」 contain the first and second halves of the 「pop」 data, respectively. Likewise, 「pop_denA」 and 「pop_denB」 contain the first and second halves of the 「pop_den」 data, respectively.

```
landA = land_area(1:10);

popA = pop(1:10);

pop_denA = pop_den(1:10);

landB = land_area(11:end);

popB = pop(11:end);

pop_denB = pop_den(11:end);
```

16.2 Excel Data of ► Section 16.1

We load data from an Excel file named 「alldata.xlsx」, specifically from the sheet 「Chap16_Asia_pop」, and store it in a table variable called 「tbl_Chap16」 using the **readtable** function. Next, we use the **renamevars** function to rename 「Land_area_1000」, 「Population_1000」, and 「Population_density」 to 「land_area」, 「pop」, and 「pop_den」, respectively. Finally, as in ► Sect. 16.1, we create the six variables: 「landA」, 「landB」, 「popA」, 「popB」, 「pop_denA」, and 「pop_denB」, respectively.

```
tbl_Chap16 = readtable("alldata.xlsx","Sheet","Chap16_Asia_pop");
```

```
Warning: Column headers from the file were modified to make them valid MATLAB identifiers before creating
variable names for the table. The original column headers are saved in the VariableDescriptions property.
Set 'VariableNamingRule' to 'preserve' to use the original column headers as table variable names.
```

```
tbl_Chap16 = renamevars(tbl_Chap16,...
    ["Land_area_1000","Population_1000","Population_density"],...
    ["land_area","pop","pop_den"]);
country = tbl_Chap16.Country;
land_area = tbl_Chap16.land_area;
pop = tbl_Chap16.pop;
pop_den = tbl_Chap16.pop_den;
landA = land_area(1:10);
popA = pop(1:10);
pop_denA = pop_den(1:10);
landB = land_area(11:end);
popB = pop(11:end);
pop_denB = pop_den(11:end);
```

16.3 Basic Syntax of the Bubblechart Function

Basic Syntax:

The basic syntax 「bubblechart(x,y,sz)」 , enclosed in the black box in Figure 16.1, creates a bubble chart in MATLAB. The variable 「x」 in the first argument of the **bubblechart** function is used as the x-coordinates of the data points, while 「y」 in the second argument is used as the y-coordinates. The variable 「sz」 in the third argument determines the size of the bubbles corresponding to each data point at the (x,y) coordinates. Specifically, larger values in 「sz」 result in larger bubbles in the chart.

Syntax

```
bubblechart(x,y,sz)
bubblechart(x,y,sz,c)

bubblechart(tbl,xvar,yvar,sizevar)
bubblechart(tbl,xvar,yvar,sizevar,cvar)

bubblechart(ax, __ )
bubblechart( __ ,Name,Value)
bc = bubblechart( __ )
```

Fig. 16.1 The syntax of the bubblechart function

Here, the numbers 2, 4, and 5 are used in the first, second, and third arguments, respectively, as displayed in 「bubblechart(2,4,5)」. This creates a single bubble with the x-coordinate of 2, the y-coordinate of 4, and a bubble size of 5. See Fig. 16.2.

```
bubblechart(2,4,5);
```

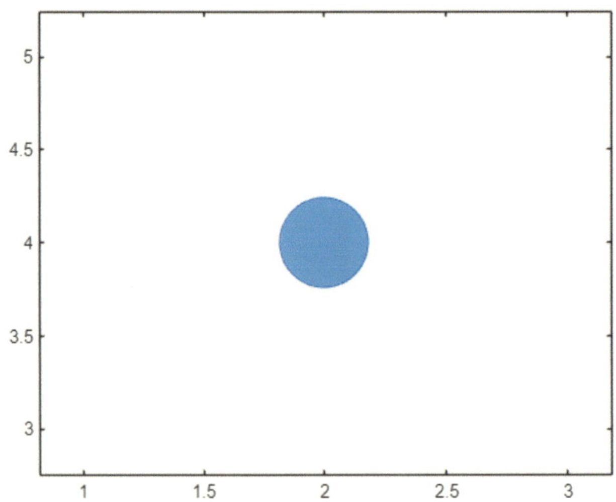

Fig. 16.2 A single bubble created by the bubblechart function

Here, based on 「bubblechart(x,y,sz)」, the variables 「x」 and 「sz」 are defined as 1-by-4 vectors, while the variable 「y」 is defined as a 2-by-4 matrix. The two rows in 「y」 create two groups. Specifically, the first group has 「x=[1 2 3 4]」, 「y=[4 5 6 7]」, and 「sz=[1 2 3 4]」, while the second group has 「x=[1 2 3 4]」, 「y=[10 8 6 4]」, and 「sz=[1 2 3 4]」. These two groups are automatically color-coded. See Fig. 16.3.

```
bubblechart([1 2 3 4],[4 5 6 7; 10 8 6 4],[1 2 3 4]);
```

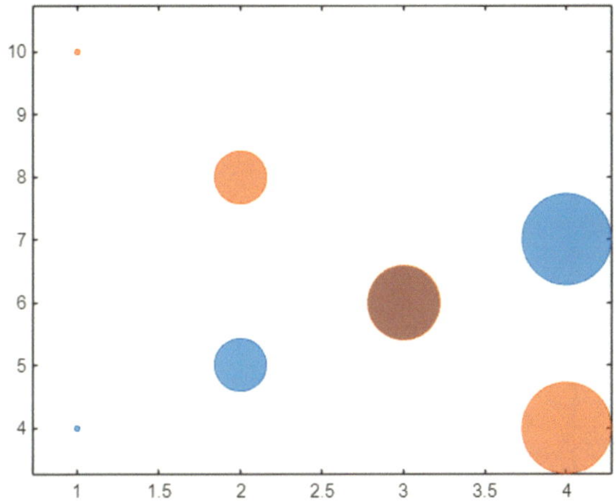

Fig. 16.3 Two grouped bubbles

Here, each of the variables 「x」, 「y」, and 「sz」 are defined as 4 rows and 2 columns of data. Two columns create two groups. While the previous program in Fig. 16.3 created groups based on two rows, the following program creates groups based on columns, not rows. In particular, the first group has 「x=[1;2;3;4]」, 「y=[4;5;6;7]」, and 「sz=[1;2;3;4]」, and the second group has 「x=[20;18;16;14]」, 「y=[10;15;20;25]」, and 「sz=[4;3;2;1]」. Compare the previous program that created Fig. 16.3 with the following program. The resulting figure is presented in Fig. 16.4.

```
x = [1 20; 2 18; 3 16; 4 14];
y = [4 10; 5 15; 6 20; 7 25];
sz = [1 4; 2 3; 3 2; 4 1];
bubblechart(x,y,sz);
```

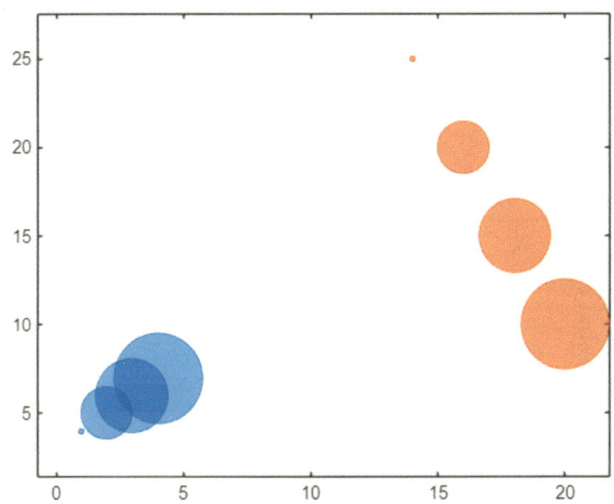

Fig. 16.4 Two grouped bubbles (cf. Fig. 16.3)

Here, using the variables ⌜land_area⌟, ⌜pop_den⌟, and ⌜pop⌟, which are 20-by-1 vectors, we create a bubble chart. In particular, the horizontal axis represents ⌜land_area⌟, the vertical axis represents ⌜pop_den⌟, and the size of the bubbles repres-ents ⌜pop⌟. The resulting figure is shown in ◘ Fig. 16.5.

```
bubblechart(land_area,pop_den,pop);
```

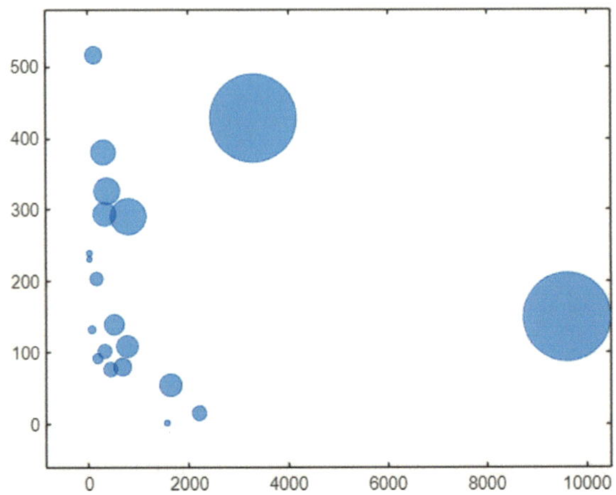

◘ **Fig. 16.5** Bubbles from three variables

Changing the order of variables may help you understand the relationships among them. Here, we rearrange the arguments, unlike the previous program that created ◘ Fig. 16.5. The first argument is now ⌜pop_den⌟, and the second argument is ⌜pop⌟. Compare ◘ Fig. 16.5 with ◘ Fig. 16.6.

```
bubblechart(pop_den,pop,land_area);
```

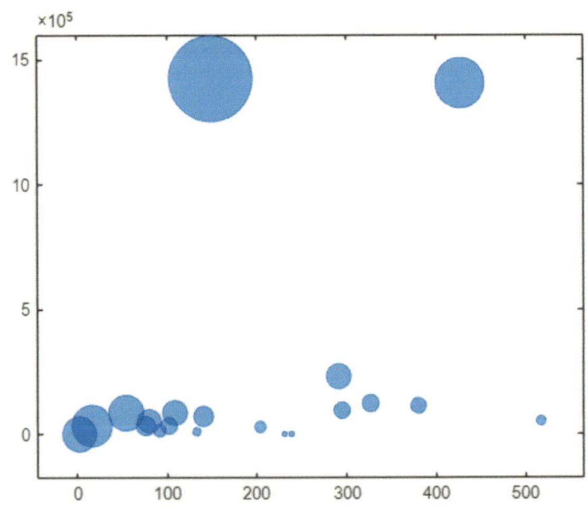

◘ **Fig. 16.6** Bubbles from three variables (cf. ◘ Fig. 16.5)

Attention

Error Program: If you specify only one or two variable names, as in the following three programs, the same error occurs. This is because the **bubblechart** function requires three variables. See Fig. 16.7 for the concrete error message.

```
%%% Error program
%bubblechart(land_area,pop_den);
```

```
%%% Error program
% bubblechart(tbl_Chap16);
```

```
%%% Error Program
% bubblechart(tbl_Chap16,"Land_area_1000","Population_density");
```

```
Error using bubblechart
Not enough input arguments.
```

Fig. 16.7 Error output of the bubblechart function: Not enough input variables

16.4 Setting and Controlling the Bubblechart Function

Here, we use the syntax ⌜bubblechart(x,y,sz,c)⌟, where ⌜x⌟, ⌜y⌟, and ⌜sz⌟ are the same variables as those defined in the previous program that created Fig. 16.4. Additionally, we specify the colors of the bubbles using the variable ⌜c⌟. Specifically, for the first group, ⌜c=[1 0 0]⌟ is used, and for the second group, ⌜c=[0 1 1]⌟ is used. See Fig. 16.8.

```
x = [1 20; 2 18; 3 16; 4 14];
y = [4 10; 5 15; 6 20; 7 25];
sz = [1 4; 2 3; 3 2; 4 1];
c = [1 0 0; 0 1 1];
bubblechart(x,y,sz,c);
```

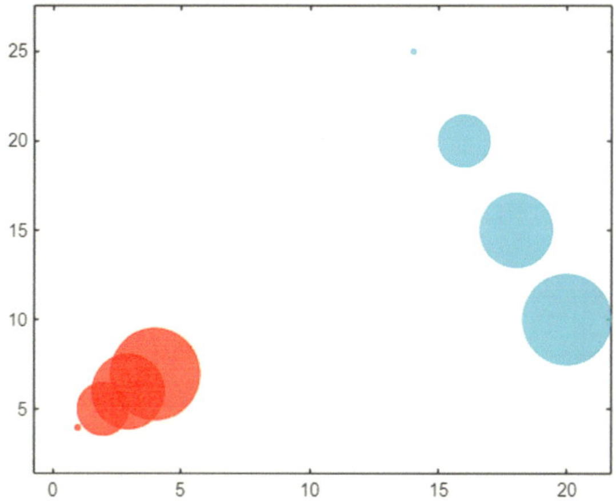

Fig. 16.8 Two grouped bubbles in red and cyan

Here, we use the syntax 「bubblechart(x,y,sz,c)」. Unlike the previous program that created Fig. 16.8, we add the third group in black. See Fig. 16.9.

```
x = [1 20 30; 2 18 35; 3 16 40; 4 14 45];
y = [4 10 30; 5 15 30; 6 20 30; 7 25 30];
sz = [1 4 0.5; 2 3 0.5; 3 2 0.5; 4 1 0.5];
c = [1 0 0; 0 1 1; 0 0 0];
bubblechart(x,y,sz,c);
```

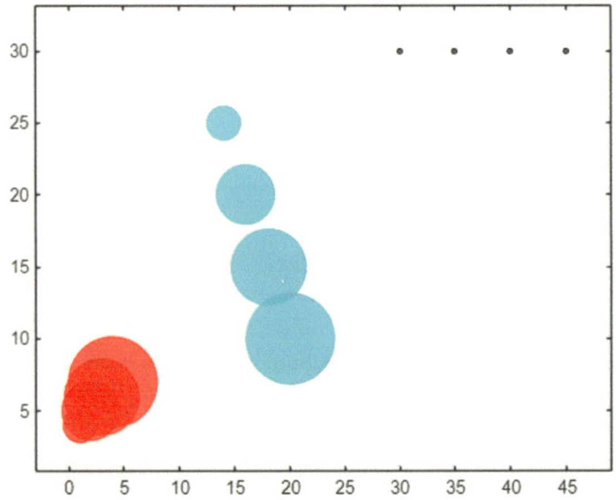

Fig. 16.9 Three grouped bubbles in red, cyan, and black

Here, we create four bubbles and specify the color of each bubble. See Fig. 16.10.

```
x = [1 2 3 4];
y = [4 5 6 7];
sz = [1 2 3 4];
c = [0 1 0; 1 0 0; 0.5 0.5 0.5; 0.6 0 1];
bubblechart(x,y,sz,c);
```

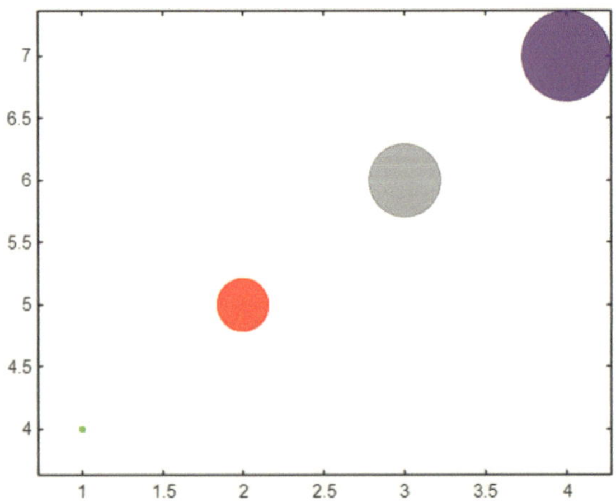

Fig. 16.10 Single grouped bubbles with various colors

16

By using the **bubblechart** function, we store its properties in the variable 「bch」. Then, we set 「MarkerFaceColor」 to 「green」, 「MarkerEdgeColor」 to 「black」, and 「LineWidth」 to 5, creating green bubbles with thick black outlines. See ◘ Fig. 16.11.

```
bch = bubblechart(pop_den,pop,land_area);
    bch.MarkerFaceColor = "green";
    bch.MarkerEdgeColor = "black";
    bch.LineWidth = 5;
```

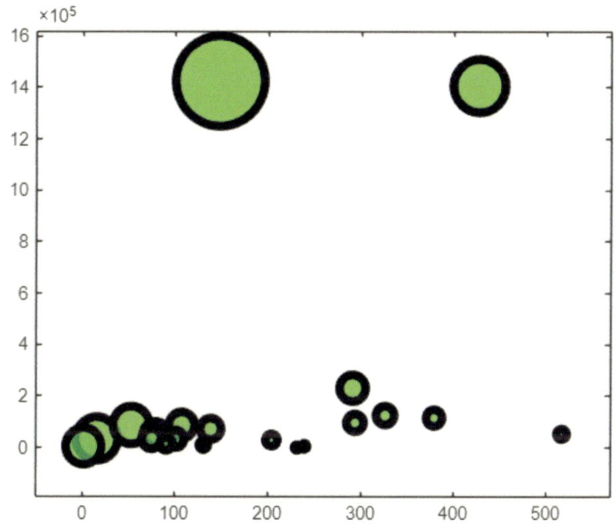

◘ **Fig. 16.11** Green bubbles with thick black edges

By using the **bubblechart** function, we store its properties in the variable 「bch」. Then, we set 「MarkerFaceColor」 to 「blue」, 「MarkerEdgeColor」 to 「black」, 「LineWidth」 to 1, and 「MarkerFaceAlpha」 to 0.1, creating blue bubbles with black outlines and high transparency. See ◘ Fig. 16.12 for the created figure.

```
bch = bubblechart(pop_den,pop,land_area);
    bch.MarkerFaceColor = "blue";
    bch.MarkerEdgeColor = "black";
    bch.LineWidth = 1;
    bch.MarkerFaceAlpha = 0.1;
```

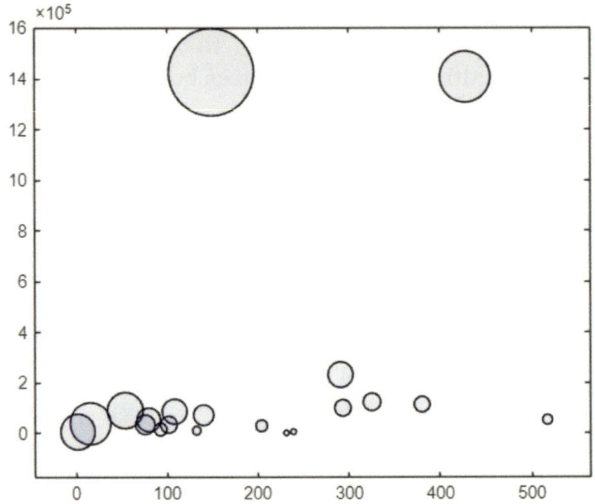

Fig. 16.12 Light purple bubbles with thin black edges

16.5 Combining Optional Functions and the Bubblechart Function

Here, we use the **title** function and store its properties in the variable ｢t｣. Then, we set ｢FontSize｣ to 20 and ｢FontWeight｣ to ｢bold｣, creating a large, bold title. See Fig. 16.13.

```
bubblechart(land_area,pop_den,pop);
t = title("Bubble of population size");
    t.FontSize = 20;
    t.FontWeight = "bold";
```

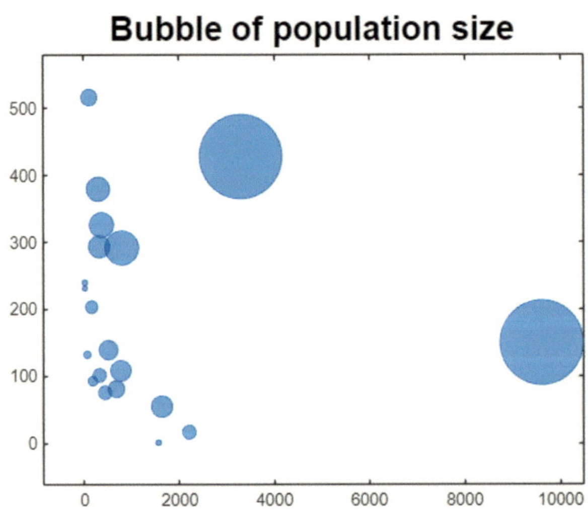

Fig. 16.13 Bubbles with a large bold title

Here, we obtain the data range for the smallest and largest bubbles. First, the created figure is shown in Fig. 16.14. After that, as seen in Fig. 16.15, we use the **bubblesize** function to confirm that the size of the smallest bubble is 3 and the size of the largest bubble is 50.

```
bubblechart(land_area,pop_den,pop);
```

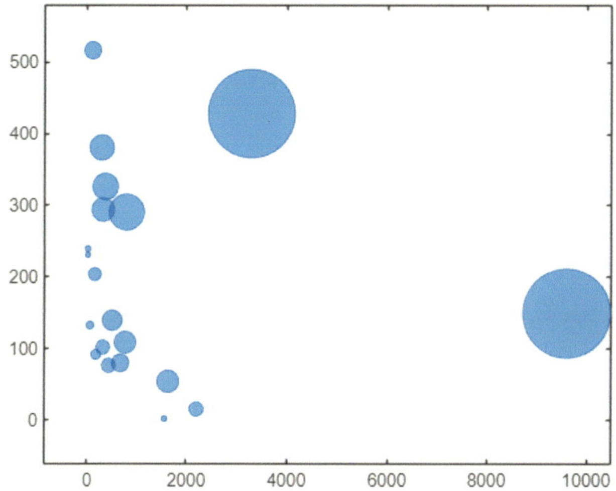

Fig. 16.14 Use of the bubblechart function before applying the bubblesize function

```
bubblesize
```

```
ans = 1×2
     3    50
```

Fig. 16.15 Results of the bubblesize function

We change the minimum and maximum bubble sizes. In particular, by using 「bubblesize([3 100])」, we set the minimum size to 3 and the maximum size to 100. Therefore, compared to the previous program that created Fig. 16.14, where the minimum bubble size is the same as 3, but the maximum size is 50 (see Fig. 16.15), the size of the largest bubbles is larger, resulting in overall larger bubble sizes. Compare Fig. 16.14 with Fig. 16.16.

```
bubblechart(land_area,pop_den,pop);
bubblesize([3 100]);
```

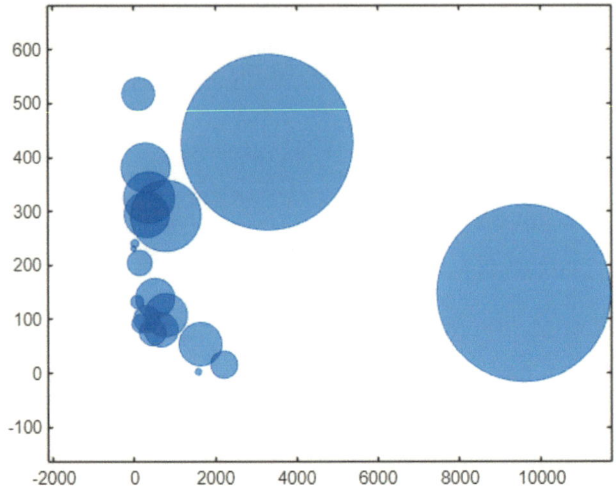

Fig. 16.16 Bubbles with large bubble sizes

Using the **bubblesize** function, we set the minimum bubble size to 1 and the maximum bubble size to 3. See **Fig. 16.17**.

```
bubblechart(land_area,pop_den,pop);
bubblesize([1 3]);
```

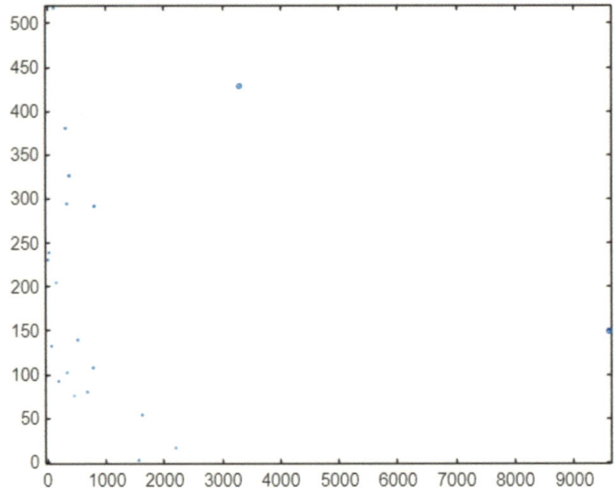

Fig. 16.17 Bubbles with small bubble sizes

16

Here, we first create a bubble chart using the **bubblechart** function. The resulting figure is shown in **Fig. 16.18**. Next, we use the **bubblelim** function to confirm the minimum and maximum values of the variable that determines the size of the bubbles. In this case, as 「pop」 is used in the third argument of the **bubblechart** function, the variable used for the bubble size is 「pop」. The minimum value of 「pop」, which determines the size of the smallest bubble, is 2688 (country: Qatar), and the maximum value is 1425893 (country: China). See ◘ Fig. 16.19.

```
bubblechart(land_area,pop_den,pop);
```

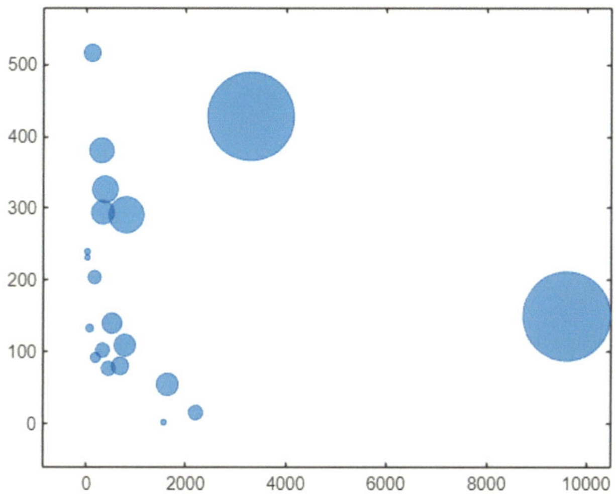

◘ **Fig. 16.18** Use of the bubblechart function before applying the bubblelim function

```
bubblelim
```

```
ans = 1×2
        2688     1425893
```

◘ **Fig. 16.19** Results of the bubblelim function

Here, we use the **bubblelim** function to change the value that determines the largest bubble size. Specifically, the original bubble range has a minimum of 2688 and a maximum of 1425893. We adjust the maximum size to 50000, while keeping the minimum size unchanged. Therefore, for all data points where the value of 「pop」 exceeds 50000, the maximum bubble size is applied. See ◘ Fig. 16.20.

```
bubblechart(land_area,pop_den,pop);
bubblelim([2688 50000]);
```

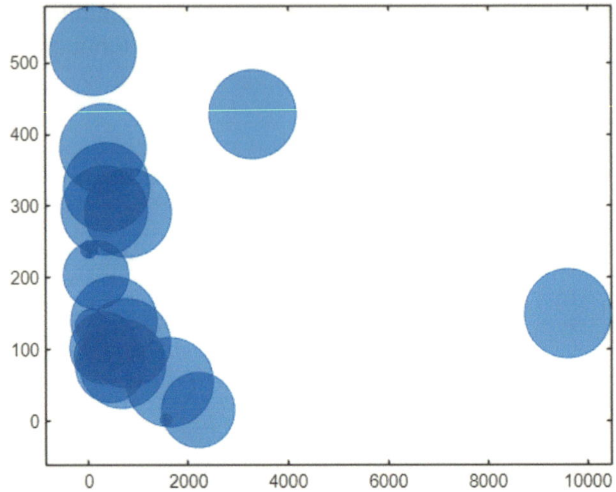

■ **Fig. 16.20** Bubbles with adjusted upper bubble size limit

Here, we use the **bubblelim** function to set the minimum value to 1400000 and the maximum value to 1500000. As a result, any value below 1400000 will have the smallest bubble size. Furthermore, because the maximum bubble size is set to 1500000, the bubble size gradually increases for values between 1400000 and 1500000. Compare ■ Fig. 16.20 with ■ Fig. 16.21.

```
bubblechart(land_area,pop_den,pop);
bubblelim([1400000 1500000]);
```

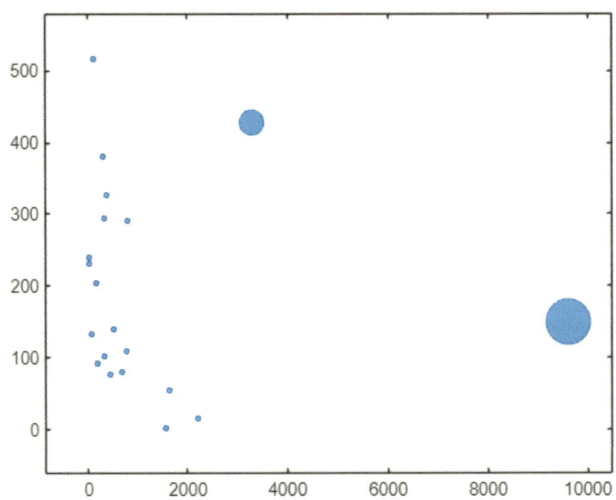

■ **Fig. 16.21** Bubbles with adjusted lower and upper bubble size limits

Here, we use the same **bubblelim** range as in the previous program that created ◘ Fig. 16.21. Next, we add annotations to two bubbles that exceed 1400000 using the **annotation** function. Specifically, we use the **annotation** function twice. In the first **annotation** function, the location of the text arrow is specified by 「x1」 and 「y1」, and in the second **annotation** function, the location is specified by 「x2」 and 「y2」. Additionally, in the second **annotation** function, the color of the arrow and text is changed to red by setting 「Color」 to 「Red」. For more details on the **annotation** function, see ► Sect. 18.1. The resulting figure is shown in ◘ Fig. 16.22.

```
bubblechart(land_area,pop_den,pop);
bubblelim([1400000 1500000]);
x1 = [0.3 0.4];
y1 = [0.85 0.75];
annotation("TextArrow",x1,y1,"String","India");
x2 = [0.7 0.8];
y2 = [0.4 0.38];
ta = annotation("TextArrow",x2,y2,"String","China");
    ta.Color = "Red";
```

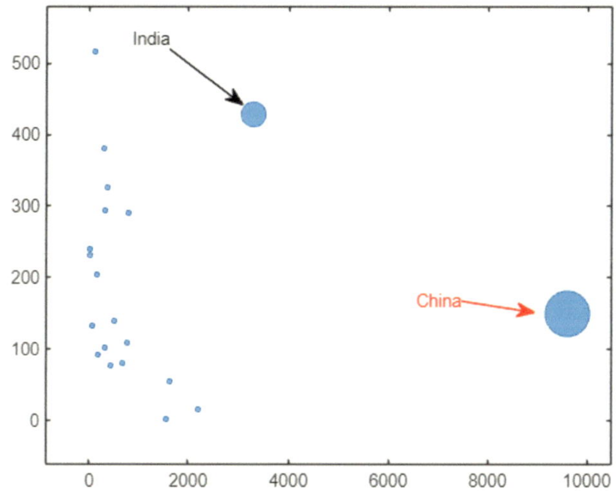

◘ **Fig. 16.22** Bubbles with black and red text arrows

Here, we use the **annotation** function to add a text box in the lower-right corner of the figure. The location and size of the text box are specified by the variable 「dim」. Specifically, 「dim=[x y w h]」 is defined as a 4-element vector, where the elements 「x」 and 「y」 determine the position of the box, and 「w」 and 「h」 determine its size. Additionally, setting 「FixBotToText」 to 「on」 makes the box fit tightly around the text. For more details on the **annotation** function, see ► Sect. 18.1. See ◘ Fig. 16.23.

```
bubblechart(land_area,pop_den,pop);
bubblelim([1400000 1500000]);
dim = [0.65 0.05 0.15 0.15];
str = "Data source: AABB";
annotation("TextBox",dim,"String",str,"FitBoxToText","on");
```

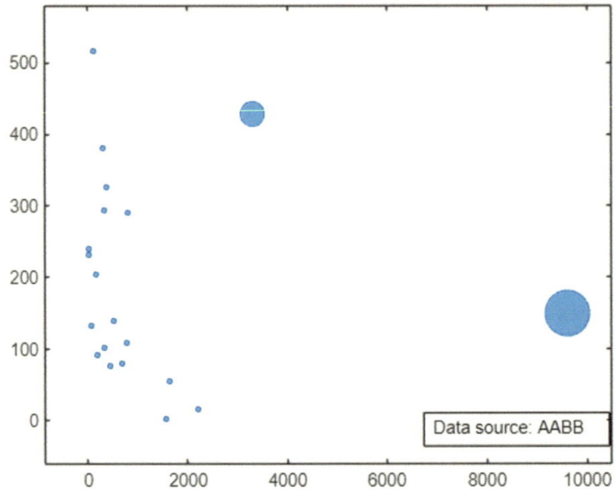

Fig. 16.23 Bubbles with a size-fitted textbox in the lower right

Here, we use the **annotation** function to create two types of annotations. The first specifies a text box and places 「Data source: AABB」 in the upper right corner. The second is used to draw a rectangle around the rightmost bubble, where 「dim2」 specifies both the location and size of the rectangle. Specifically, we set 「FaceColor」 to 「red」 and 「FaceAlpha」 to 0.1, creating a red rectangle with high transparency. For more details on the **annotation** function, see ► Sect. 18.1. See Fig. 16.24.

```
bubblechart(land_area,pop_den,pop);
bubblelim([1400000 1500000]);
dim1 = [0.65 0.75 0.15 0.15];
str1 = "Data source: AABB";
annotation("TextBox",dim1,"String",str1,"FitBoxToText","on");
dim2 = [0.75 0.3 0.15 0.15];
annotation("Rectangle",dim2,"FaceColor","red","FaceAlpha",0.1);
```

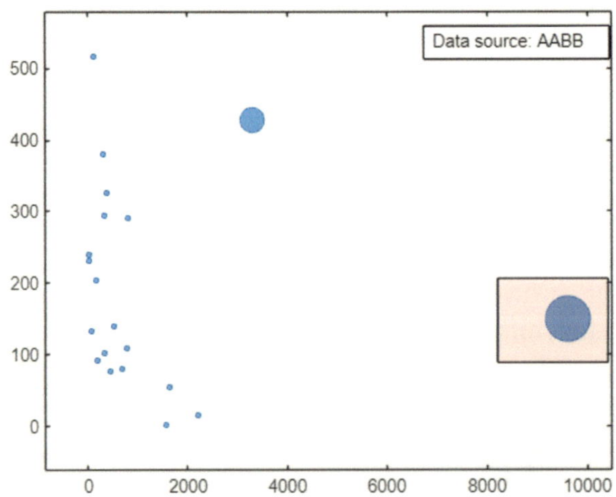

Fig. 16.24 Bubbles with a size-fitted textbox in the upper right and a semi-transparent red rectangle

16

Here, we store the properties of the **bubblechart** function in the variable 「bh」 and set 「MarkerFaceColor」 to 「r (=red)」, filling the bubbles with red. In addition, we use the **xregion** function to create a filled region on the x-axis in the range from 0 to 2000. See ◘ Fig. 16.25.

```
bh = bubblechart(land_area,pop_den,pop);
    bh.MarkerFaceColor = "r";
xregion(0,2000);
```

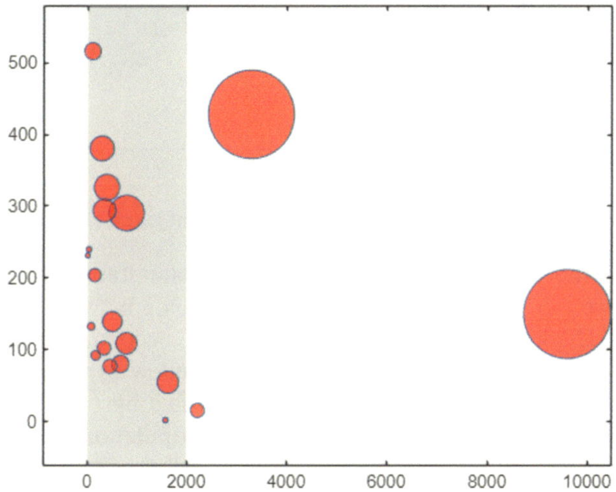

◘ **Fig. 16.25** Bubbles with a vertical gray shaded area

Here, we use 「bubblesize([1 100])」 to change the minimum bubble size to 1 and the maximum bubble size to 100. Additionally, note that in the **bubblechart** function, unless axis ticks are explicitly specified, they are generated automatically.

Next, we specify 「xr=xregion(-Inf,0)」 to create a filled region for the area below 0 on the x-axis. Furthermore, we set 「FaceColor」 to 「b (=blue)」 and 「EdgeColor」 to 「k (=black)」, filling the region with blue and changing the surrounding lines to black.

We also set 「DisplayName」 to 「Outside」, labeling this region as 「Outside」. Consequently, when we use the **legend** function to display a legend, 「Outside」 appears as the label for the filled region. See ◘ Fig. 16.26.

```
bubblechart(land_area,pop_den,pop);
bubblesize([1 100]);
xr = xregion(-Inf,0);
    xr.FaceColor = "b";
    xr.EdgeColor = "k";
    xr.DisplayName = "Outside";
legend;
```

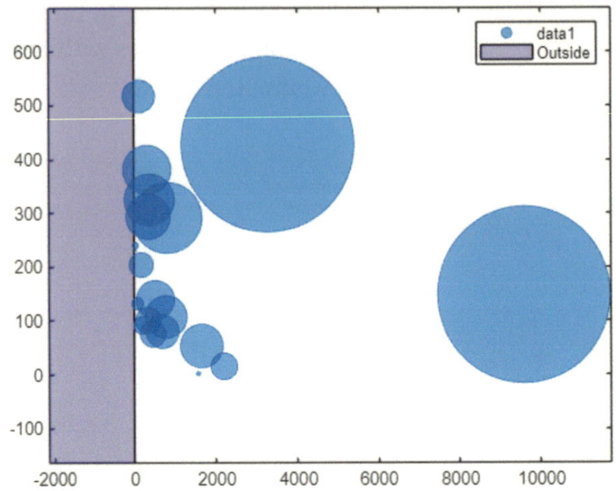

Fig. 16.26 Bubbles with adjusted sizes, a legend, and a vertical purple shaded area

Here, using the **xregion** and **yregion** functions simultaneously, we fill specific regions. First, we use 「xregion(0,2000)」 to fill the region between 0 and 2000 on the x-axis and set 「DisplayName」 to 「Small land area」. Second, we use 「yregion(400, Inf)」 to fill the area above 400 on the y-axis with yellow by setting 「FaceColor」 to 「Yellow」. Additionally, 「DisplayName」 is specified by 「High population density」. In the legend, 「Small land area」 and 「High population density」 appear. See Fig. 16.27.

```
bh = bubblechart(land_area,pop_den,pop);
    bh.MarkerFaceColor = "r";
xr = xregion(0,2000);
    xr.DisplayName = "Small land area";
yr = yregion(400,Inf);
    yr.DisplayName = "High population density";
    yr.FaceColor = "Yellow";
legend([xr yr]);
```

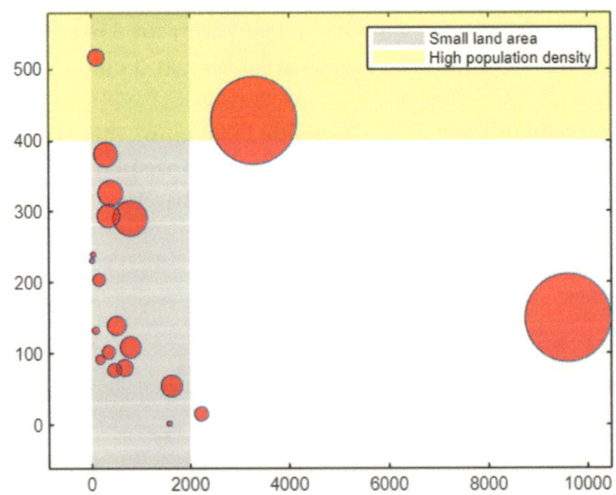

Fig. 16.27 Bubbles with horizontal yellow and vertical gray shaded areas, and a legend

16

Here, using the **xline** and **yline** functions, we add vertical and horizontal lines to the figure. Specifically, we create a labeled dash-dotted line using the **xline** function. By specifying 0 as the first argument, we draw a dash-dotted line at x=0. Additionally, we set 「Color」 to 「red」 and 「LineWidth」 to 5, making the dash-dotted line red and thicker. Similarly, we use the **yline** function to create a labeled dash-dotted line. For more details on the **xline** and **yline** functions, see ▶ Sect. 18.3. The resulting figure is presented in ◘ Fig. 16.28.

```
bubblechart(land_area,pop_den,pop);
bubblelim([1400000 1500000]);
xl = xline(0,"-.",{"Land area"});
    xl.Color = "red";
    xl.LineWidth = 5;
yl = yline(0,"--",{"Population Density"});
    yl.LabelHorizontalAlignment = "center";
```

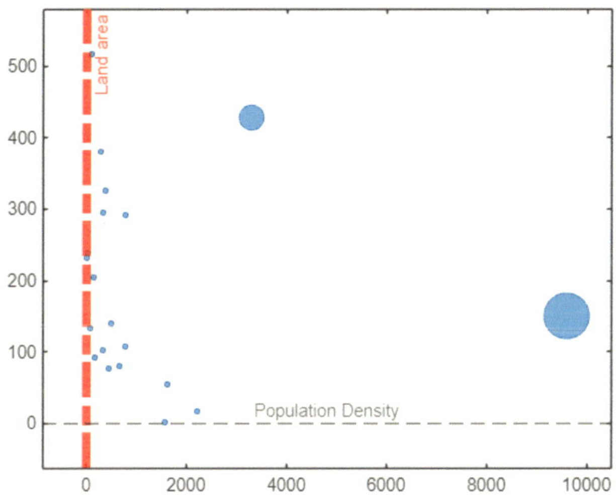

◘ **Fig. 16.28** Bubbles with a vertical red dash-dotted line and a horizontal black dashed line

Here, we use the **xline** and **yline** functions to create labeled dash-dotted and dashed lines. Specifically, by setting the first argument of the **xline** function to 8000 and the second argument to 「-.」, a dash-dotted line is drawn vertically at x=8000. Additionally, by setting 「LabelVerticalAlignment」 to 「middle」, the text label (「Large land area」) is centered. Note that the label spans two lines. Similarly, the **yline** function draws a dashed line horizontally at y=100. For the dashed line at y=100, we set 「Color」 to 「blue」 and 「LineWidth」 to 5, creating a thick blue dashed line. For more details on the **xline** and **yline** functions, see ▶ Sect. 18.3. The resulting figure is presented in ◘ Fig. 16.29.

```
bh = bubblechart(land_area,pop_den,pop);
    bh.MarkerFaceColor = "green";
    bh.MarkerFaceAlpha = 0.1;
xl1 = xline(8000,"-.",{"Large","land area"});
    xl1.LabelVerticalAlignment = "middle";
yl1 = yline(100,"--",{"Low","population density"});
    yl1.LabelVerticalAlignment = "bottom";
    yl1.LabelHorizontalAlignment = "center";
    yl1.Color = "blue";
    yl1.LineWidth = 5;
```

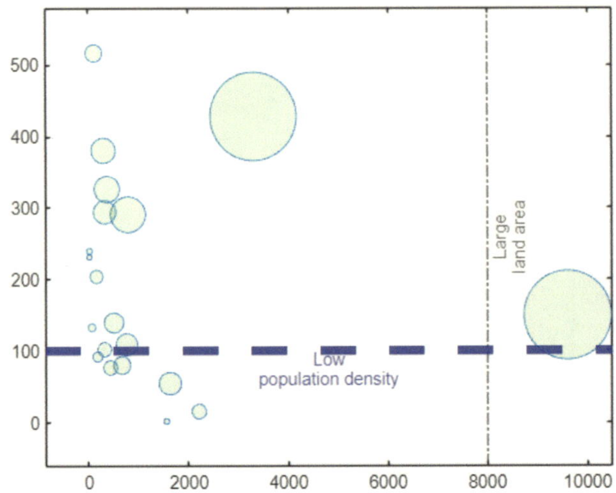

Fig. 16.29 Bubbles with a horizontal thick blue dashed line and a vertical black dash-dotted line

Here, we use the **xline**, **yline**, **xregion**, and **yregion** functions simultaneously to fill the area below 0. Additionally, we create a single legend using the **legend** function. Specifically, we use the **xline** function to create a dash-dotted line at x=0, and the **yline** function to create a dashed line at y=0. For more details of the **xline** and **yline** functions, see ▶ Sect. 18.3. Next, we use the **xregion** and **yregion** functions to fill the areas below 0 on the x-axis and y-axis, respectively. By storing the properties of the **xregion** function in the variable 「xr」, we create a legend labeled 「outside」. The resulting figure is shown in Fig. 16.30.

16

```
bubblechart(land_area,pop_den,pop);
bubblelim([100000 1500000]);
xline(0,"-.",{"Land area"});
yline(0,"--",{"Population Density"});
xr = xregion(-Inf,0);
yregion(-Inf,0);
legend(xr,"outside");
```

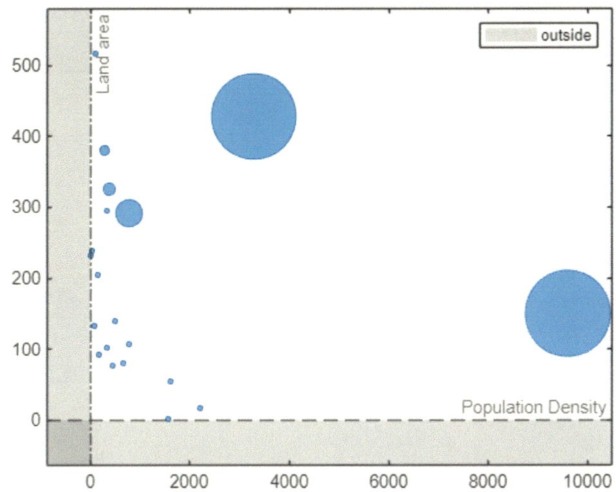

Fig. 16.30 Bubbles with horizontal and vertical gray shaded areas below zero, a legend, and two lines

Here, we use the **xregion** function to create four filled regions, and store the properties of the **xregion** function in ⌜xr⌟. In particular, ⌜xr(1)⌟ corresponds to the re gion from x=-Inf to x=0, ⌜xr(2)⌟ corresponds to the region from x=0 to x=2000, ⌜xr(3)⌟ corresponds to the region from x=2000 to x=4000, and ⌜xr(4)⌟ corresponds to the region from x=4000 to x=Inf. Additionally, by using the **legend** function and setting ⌜NumColumns⌟ to 4, we create a horizontal legend with 1 row and 4 columns. See Fig. 16.31.

```
bh = bubblechart(land_area,pop_den,pop);
    bh.MarkerEdgeColor = "black";
    bh.LineWidth = 3;
grid minor
xr=xregion([-Inf,0,2000,4000],[0,2000,4000,Inf]);
    xr(1).DisplayName = "outside";
    xr(2).FaceColor = "red";
    xr(2).DisplayName = "Small";
    xr(3).FaceColor = "blue";
    xr(3).DisplayName = "Middle";
    xr(4).FaceColor = "yellow";
    xr(4).DisplayName = "Large";
title("Land area, pop density, and pop size in Asian countries");
lgd = legend(xr);
    lgd.TextColor = "blue";
    lgd.NumColumns = 4;
```

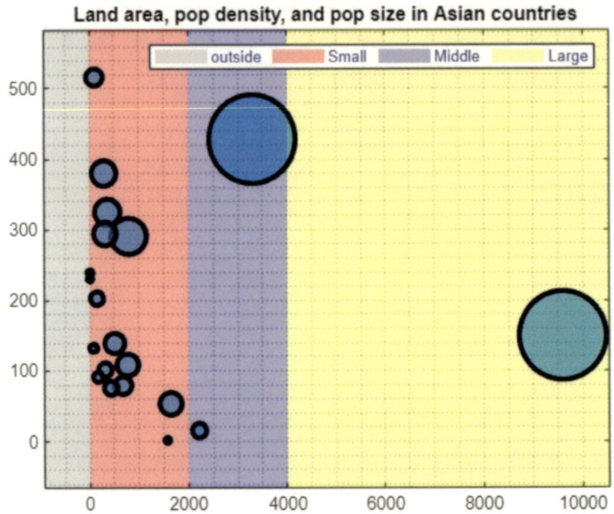

Fig. 16.31 Bubbles with several vertical shaded areas and a horizontal legend

Because the bubble sizes created by the **bubblechart** function do not have labels that explain the scale, it is useful to use the **bubblelegend** function, which creates a bubble legend that represents the sizes of the bubbles. The bubble legend is a special function that can be used with bubble charts. See Fig. 16.32.

```
bubblechart(land_area,pop_den,pop);
bubblelegend("Population in Asia");
```

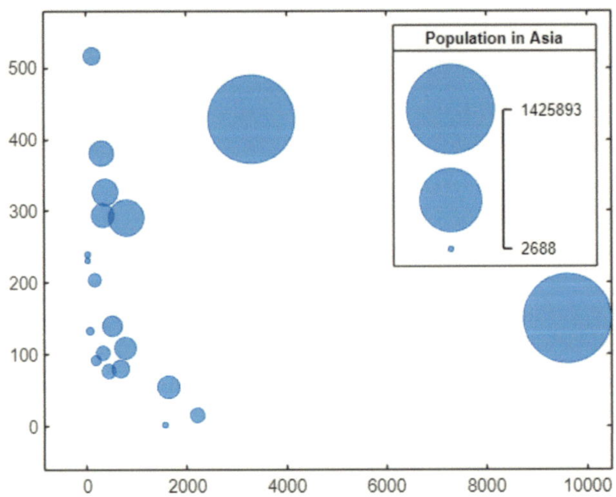

Fig. 16.32 Bubbles with a bubble legend

Here, we use the **bubblelegend** function to create a bubble legend. Specifically, we set 「NumBubbles」 to 2, changing the number of bubbles in the bubble legend to 2. Compare the number of bubbles in the bubble legend in ◘ Fig. 16.33 with that in ◘ Fig. 16.32. The default value for 「NumBubbles」 is 3. Note that in MATLAB R2024b, it can be changed to either 2 or 3. Additionally, by setting 「LineWidth」 to 3 and 「MarkerEdgeColor」 to 「black」 in the **bubblechart** function, we modify the appearance of the bubbles. Note that this change is reflected in the bubble legend. See ◘ Fig. 16.33.

```
bh = bubblechart(land_area,pop_den,pop);
    bh.LineWidth = 3;
    bh.MarkerEdgeColor = "black";
bubblelegend("Population in Asia","NumBubbles",2);
```

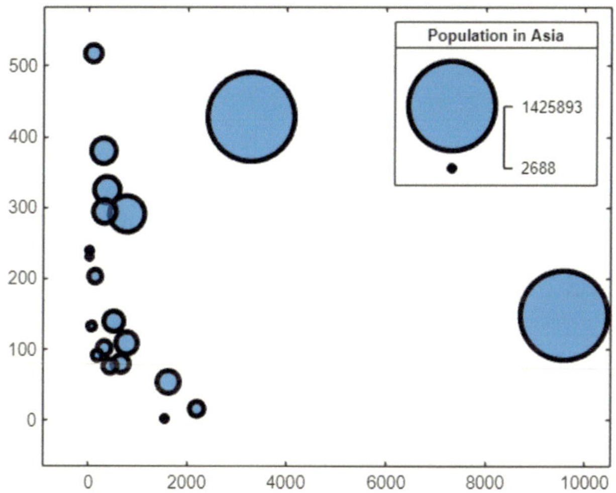

◘ **Fig. 16.33** Bubbles with a two-size bubble legend

Here, by using the **bubblelegend** function, we display a bubble legend and specify its location as 「eastoutside」, creating a legend outside the bubble chart. We also use the **annotation** function to create two text arrows. Specifically, in the second use of the **annotation** function, the arrow points to the location of the legend created outside the bubble chart. For more details on the **annotation** function, see ▶ Sect. 18.1. See ◘ Fig. 16.34 for the created figure.

```
bubblechart(land_area,pop_den,pop);
bubblelegend("Population in Asia","Location","eastoutside");
x1 = [0.3 0.3];
y1 = [0.88 0.82];
annotation("textarrow",x1,y1,"String","India");
x2 = [0.8 0.8];
y2 = [0.88 0.75];
annotation("textarrow",x2,y2,"String","Legend");
```

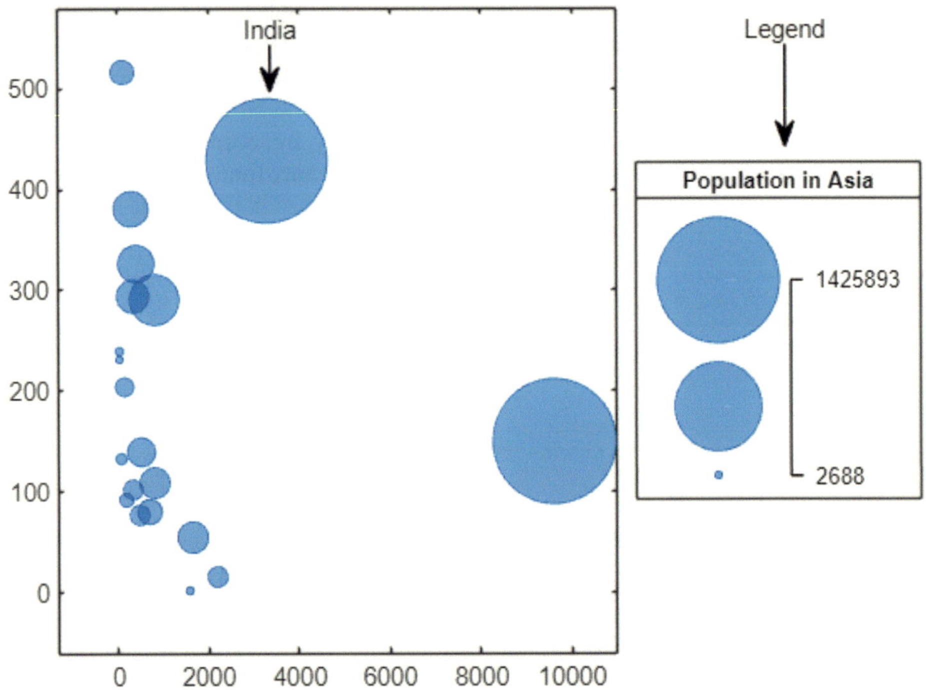

Fig. 16.34 Bubbles with a bubble legend (outside-right position) and two text arrows

Here, using the **bubblelim** function, we modify the bubble sizes. Specifically, the **bubblelim** function sets the minimum bubble size for values less than or equal to 1400000. There are two differences between Fig. 16.34 and Fig. 16.35.

First, the overall size of the bubbles in the bubble chart is smaller than in Fig. 16.34. For example, looking at the point (India) indicated by the text arrow created by the **annotation** function, the bubble in the previous figure is larger. This is caused by the use of the **bubblelim** function.

Second, in the bubble legend created by the **bubblelegend** function, the smallest size here is labeled 「≤1400000」, while in Fig. 16.34, it was labeled 「2688」. Compare Fig. 16.34 with Fig. 16.35. For more details of the **annotation** function, see ► Sect. 18.1.


```
bubblechart(land_area,pop_den,pop);
bubblelegend("Population in Asia","Location","eastoutside");
bubblelim([1400000 1500000]);
x1 = [0.29 0.29];
y1 = [0.85 0.76];
annotation("textarrow",x1,y1,"String","India");
x2 = [0.8 0.8];
y2 = [0.88 0.75];
annotation("textarrow",x2,y2,"String","Legend");
```

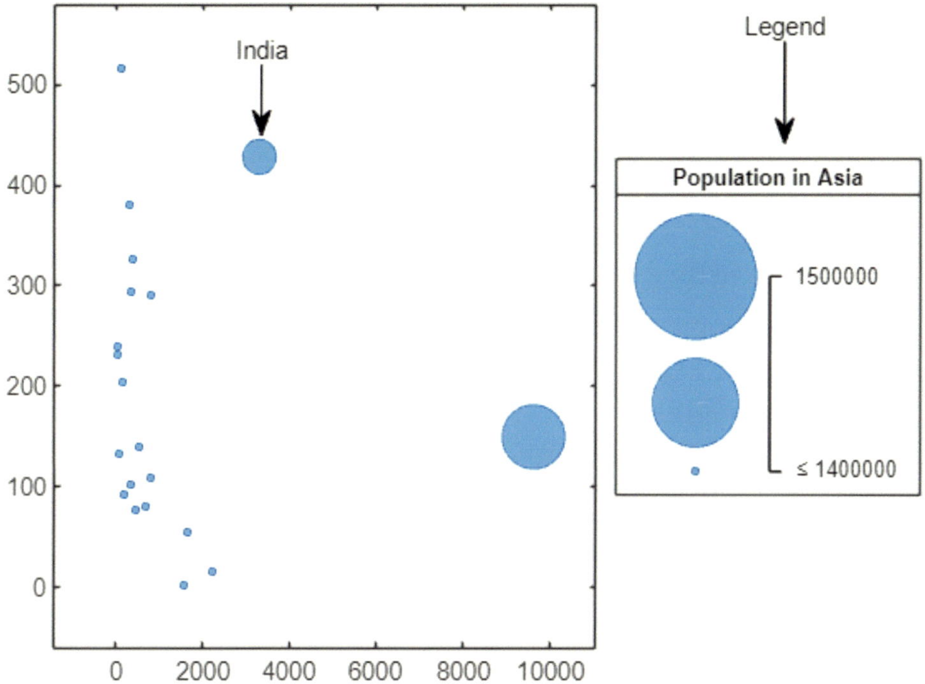

Fig. 16.35 Bubbles with small bubble sizes, a bubble legend (outside-right position), and two text arrows (cf. Fig. 16.34)

Here, using the **title**, **xlabel**, and **ylabel** functions, we create the title, x-axis label, and y-axis label. These labels are positioned in correspondence with the figure generated by the **bubblechart** function. Specifically, the title 「Bubblechart」 and the x-axis label 「Land area」 are centered on the bubble chart figure, rather than on the figure space. See Fig. 16.36.

```
bh = bubblechart(land_area,pop_den,pop);
    bh.MarkerFaceColor = "red";
bubblelegend("Population in Asia","Location","eastoutside");
bubblelim([1400000 1500000]);
t = title("Bubblechart");
    t.Color = "blue";
    t.FontSize = 16;
xlabel("Land area");
ylabel("Populoation density");
```

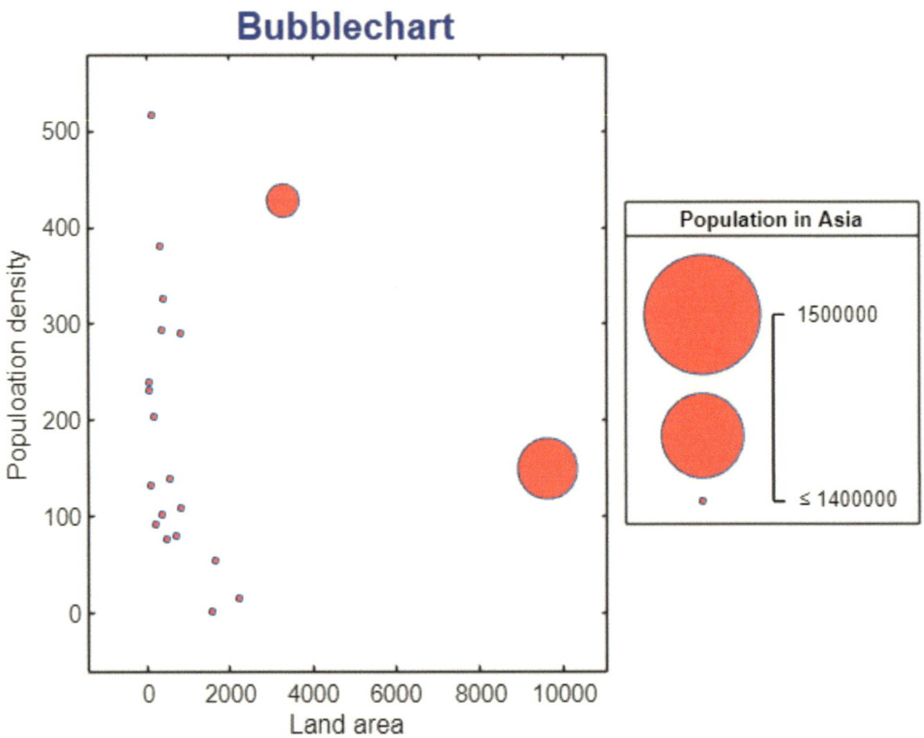

Fig. 16.36 Red bubbles with a bubble legend (outside-right position), a title, and axis labels (cf. Fig. 16.35)

Here, we store the properties of the **bubblechart** function in the variable 「bh」. Then, we set 「MarkerFaceColor」 to 「r (=red)」, 「MarkerFaceAlpha」 to 0.1, and 「LineWidth」 to 3, creating red bubbles with thick outlines and high transparency. Additionally, after retrieving the properties of the axes with 「ax=gca」, we set 「TitleHorizontalAlignment」 to 「left」, positioning the text created by the **title** function on the left side of the bubble chart. See Fig. 16.37.

```
bh = bubblechart(land_area,pop_den,pop);
    bh.MarkerFaceColor = "red";
    bh.MarkerFaceAlpha = 0.1;
    bh.LineWidth = 3;
bubblelegend("Population in Asia","Location","eastoutside");
bubblelim([1400000 1500000]);
t = title("Bubblechart");
    t.Color = "blue";
    t.FontSize = 16;
ax = gca;
    ax.TitleHorizontalAlignment = "left";
xl = xlabel("Land area");
    xl.Color = "red";
ylabel("Population density");
```

16

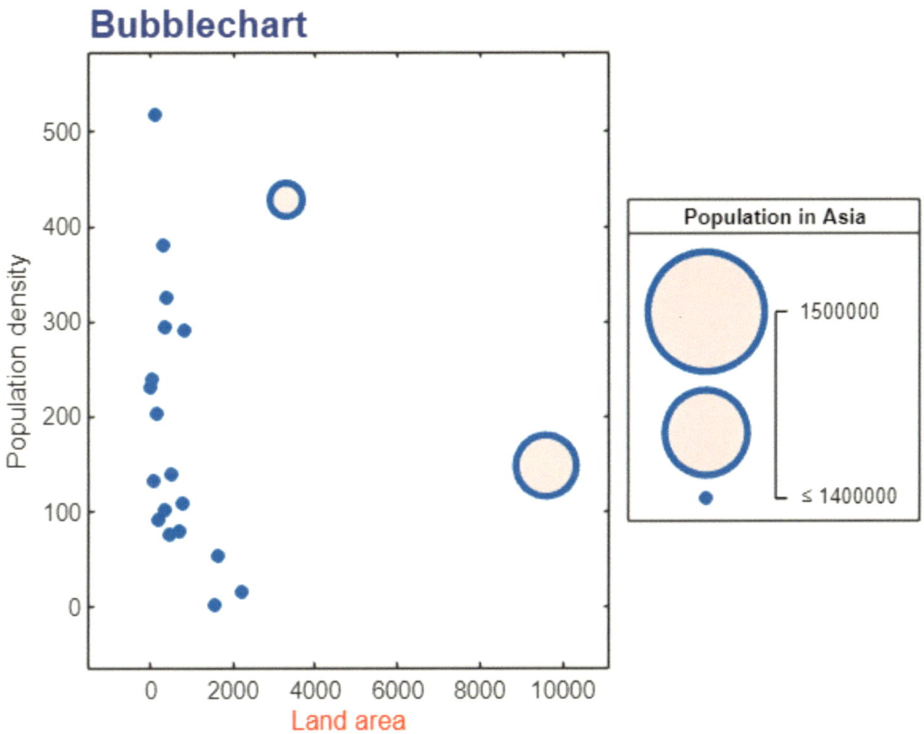

Fig. 16.37 Pink bubbles with thick blue edges, a bubble legend (outside-right position), a title, and axis labels

Here, we use the **title** and **subtitle** functions to create the title labels. Additionally, by specifying 「xticks([0 5000 10000])」 with the **xticks** function, the x-axis has three ticks at 0, 5000, and 10000. The resulting figure is presented in Fig. 16.38.

```
bh = bubblechart(land_area,pop_den,pop);
    bh.MarkerEdgeColor = "black";
    bh.LineWidth = 2;
bubblelegend("Population in Asia","Location","eastoutside");
bubblesize([1 100]);
t = title("bubblechart");
    t.FontSize = 20;
st = subtitle("Asian countries");
    st.Color = "magenta";
xticks([0 5000 10000]);
```

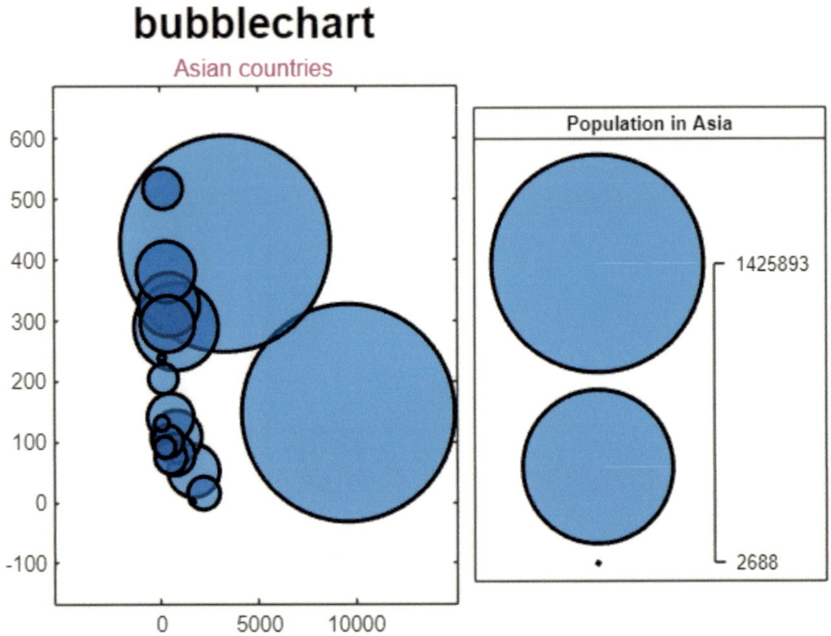

■ **Fig. 16.38** Bubble chart with large bubble sizes, a title, a subtitle, a bubble legend (outside-right position), and modified x-axis ticks

Here, using the **bubblelegend** function, we set 「Style」 to 「telescopic」, creating a bubble legend different from ■ Figs. 16.32 through 16.38. The resulting figure is shown in ■ Fig. 16.39.

```
bubblechart(land_area,pop_den,pop,"MarkerFaceColor","g");
bubblelegend("Population in Asia","Style","telescopic");
bubblesize([1 100]);
```

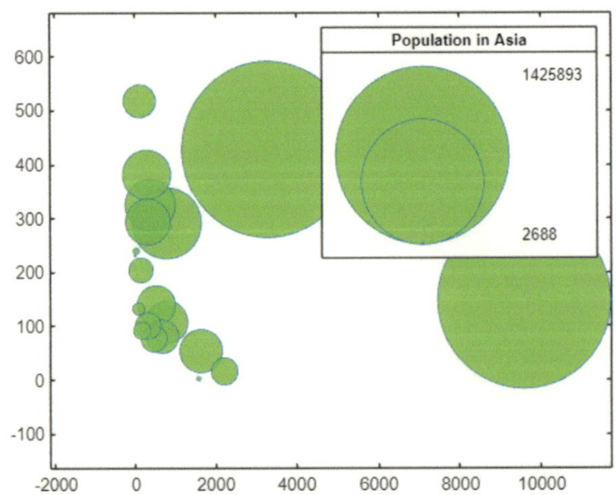

■ **Fig. 16.39** Bubbles with a bubble legend in telescopic style

Here, for the bubble legend, we set 「Style」 to 「horizontal」, creating a bubble legend where the bubbles are arranged horizontally. Compare ◘ Fig. 16.40 with the bubble legends in ◘ Figs. 16.32 through 16.39. Additionally, we use the **xlim** and **ylim** functions to change the ranges of the x-axis and y-axis. Specifically, by setting 「xlim([0 Inf])」 and 「ylim([0 Inf])」, the range below 0 is not displayed in the figure. As a result, some of the bubbles are cut off. See ◘ Fig. 16.40.

```
bubblechart(land_area,pop_den,pop);
bubblelegend("Population in Asia","Style","horizontal");
bubblesize([1 100]);
xlim([0 Inf]);
ylim([0 Inf]);
```

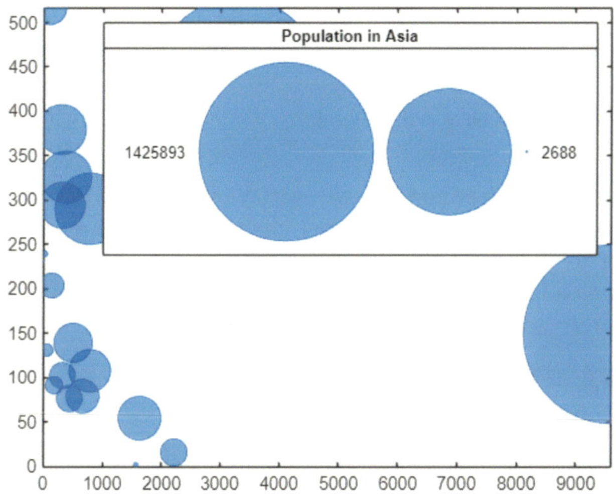

◘ **Fig. 16.40** Bubbles with a horizontal bubble legend

Using the **hold on** command, we overlay two types of bubble charts. In particular, using the three variables 「landA」, 「pop_denA」, and 「popA」 in the **bubblechart** function, we create the first bubble chart. By setting 「MarkerEdgeColor」 to 「black」 and 「LineWidth」 to 3, we change the bubble outlines to thick black lines.

Next, using the three variables 「landB」, 「pop_denB」, and 「popB」, we create the second bubble chart. When setting 「MarkerFaceAlpha」 to 0.1 and 「Marker FaceColor」 to 「cyan」, we create highly transparent bubbles filled with cyan. Finally, we use the **legend** function to display a legend. See ◘ Fig. 16.41 for the resulting figure.

```
bh1 = bubblechart(landA,pop_denA,popA);
    bh1.MarkerEdgeColor = "black";
    bh1.LineWidth = 3;
hold on
bh2 = bubblechart(landB,pop_denB,popB);
    bh2.MarkerFaceAlpha = 0.1;
    bh2.MarkerFaceColor = "cyan";
hold off
legend("Group A","Group B");
```

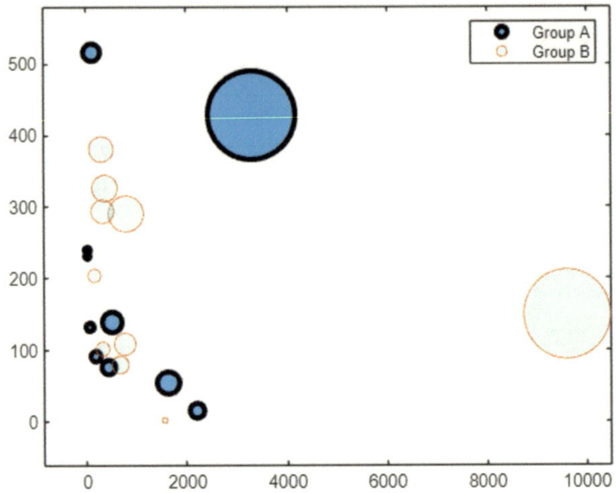

Fig. 16.41 Two grouped bubbles with a legend

Here, using the **hold on** command, we overlay two types of bubble charts. In addition, we use the **xregion** and **yregion** functions to specify the areas below 0 as filled regions, similar to Fig. 16.30. We set 「Location」 to 「southeast」 in the **legend** function, creating a legend in the lower-right part of the figure. See Fig. 16.42 for the created figure.

```
bubblechart(landA,pop_denA,popA);
hold on
bubblechart(landB,pop_denB,popB);
hold off
xregion(-Inf,0);
yregion(-Inf,0);
legend("Group A","Group B","Location","southeast");
```

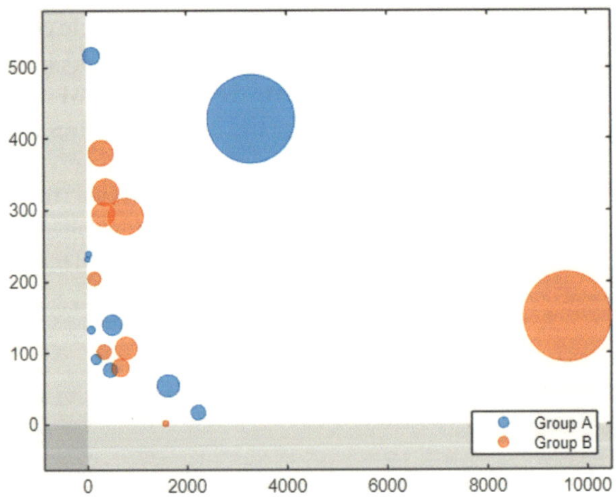

Fig. 16.42 Two grouped bubbles with a legend (bottom-right position), and gray shaded areas under zero on both axes

16

In addition to the legend created with the **legend** function, we also create a bubble legend with the **bubblelegend** function. By setting 「Location」 to 「eastoutside」 in both the **legend** and **bubblelegend** functions, the legend generated by the **legend** function and the bubble legend generated by the **bubblelegend** function are aligned side-by-side next to the bubble chart. See ◘ Fig. 16.43.

```
bubblechart(landA,pop_denA,popA);
hold on
bubblechart(landB,pop_denB,popB);
hold off
legend("Group A","Group B","Location","eastoutside");
bubblelegend("Population in Asia","Location","eastoutside");
grid on
```

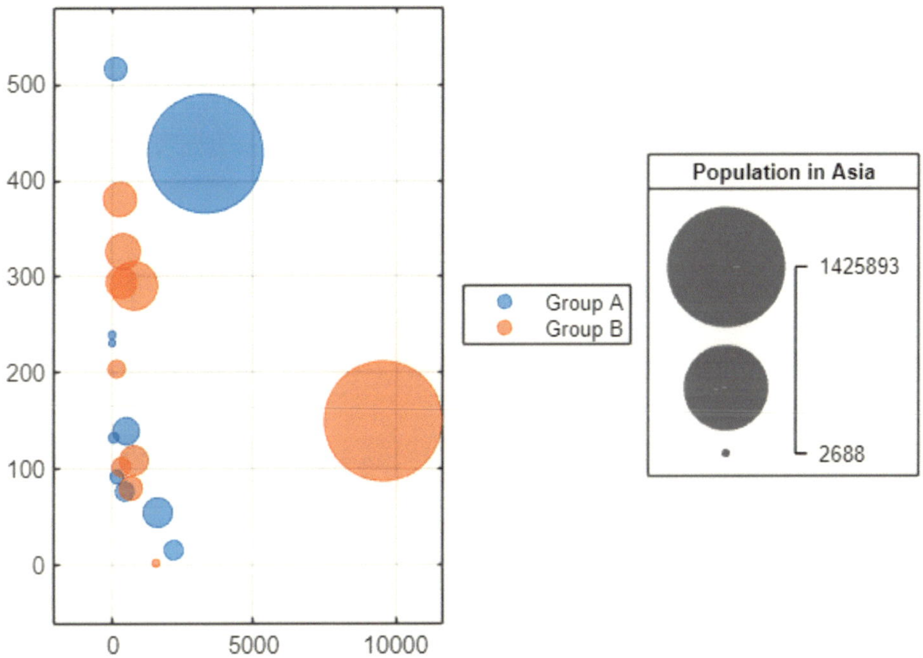

◘ **Fig. 16.43** Two grouped bubbles with horizontally aligned group and bubble legends (outside-right position)

Here, by specifying 「tiledlayout(1,1)」, we create a 1-by-1 tiled layout. The reason for using 「tiledlayout(1,1)」 is to change the positions of the legend and the bubble legend. Specifically, by setting 「Layout.Tile」 to 「east」 in the **legend** and **bubblelegend** functions, we position them vertically on the right side of the chart, which differs from the horizontally arranged position in ◘ Fig. 16.43. Compare ◘ Fig. 16.43 with ◘ Fig. 16.44.

```
tiledlayout(1,1);
nexttile;
bubblechart(landA,pop_denA,popA);
hold on
bubblechart(landB,pop_denB,popB);
hold off
lh = legend("Group A","Group B");
blh = bubblelegend("Population in Asia");
    lh.Layout.Tile = "east";
    blh.Layout.Tile = "east";
grid minor
```

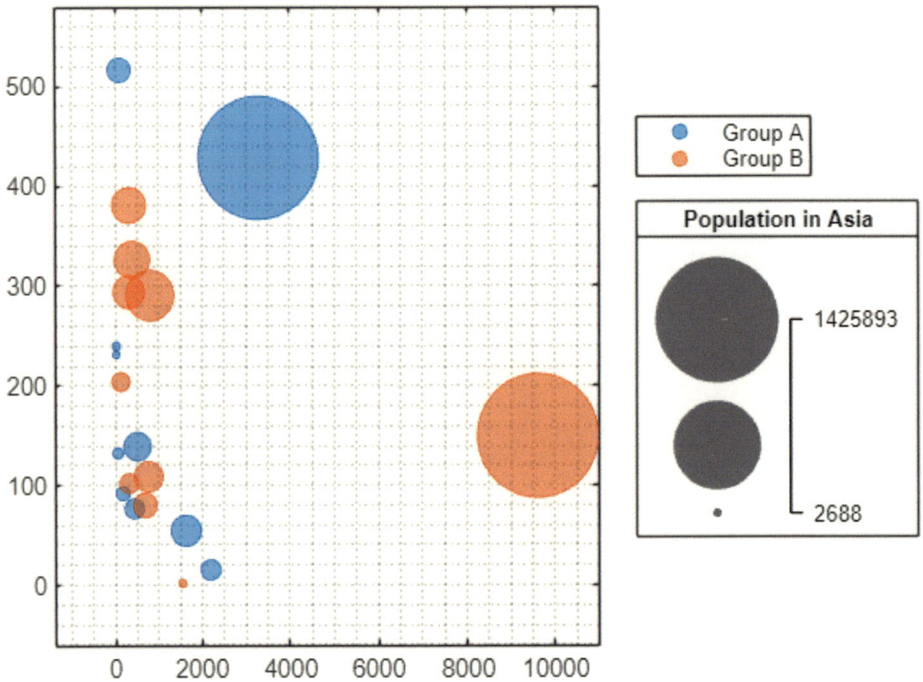

Fig. 16.44 Two grouped bubbles with vertically aligned group and bubble legends (outside-right position)

16.6 Questions

16

Question 1

Create Fig. 16.45 using the **bubblechart** function with the variables 「land_area」, 「pop_den」, and 「pop」. Concretely, store the data from the first to fifth elements of the variables 「land_area」, 「pop_den」, and 「pop」 in the variables 「land_area1」, 「pop_den1」, and 「pop1」, respectively. Additionally, store the data from the sixteenth to twentieth elements of the variables 「land_area」, 「pop_den」, and 「pop」 in the variables 「land_area2」, 「pop_den2」, and 「pop2」.

To create the first bubble chart, use (X,Y,Z)=(「land_area1」, 「pop_den1」, 「pop1」). Similarly, to create the second bubble chart, use (X,Y,Z)=(「land_area2」, 「pop_den2」, 「pop2」).

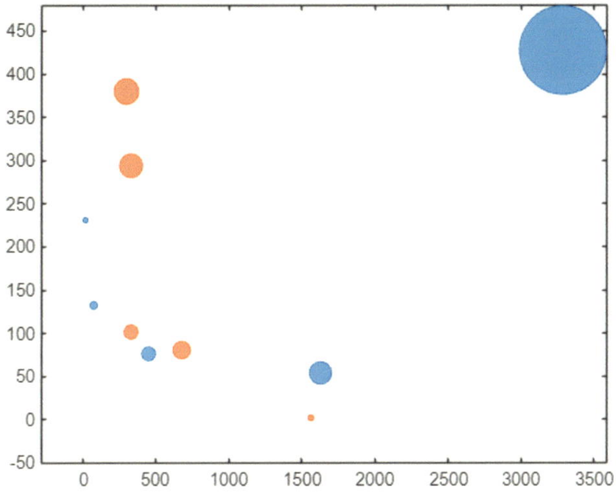

Fig. 16.45 Bubbles (Question 1)

Answer 1

```
figure
land_area1 = land_area(1:5);
pop_den1 = pop_den(1:5);
pop1 = pop(1:5);
land_area2 = land_area(16:20);
pop_den2 = pop_den(16:20);
pop2 = pop(16:20);
bubblechart([land_area1 land_area2],[pop_den1 pop_den2],[pop1 pop2]);
```

Question 2

Create Fig. 16.46 with red bubbles by using 「land_area」, 「pop_den」, and 「pop」 as the arguments for the **bubblechart** function. Additionally, use the **bubblelegend** function to place a bubble legend outside to the left of the figure, and set the title of the bubble legend to 「Population in Asia」. Use the **title** function to create a bold title, 「Bubblechart」, with a larger font size. Finally, use the **bubblelim** function to set the minimum bubble size to 0 and the maximum bubble size to 100000.

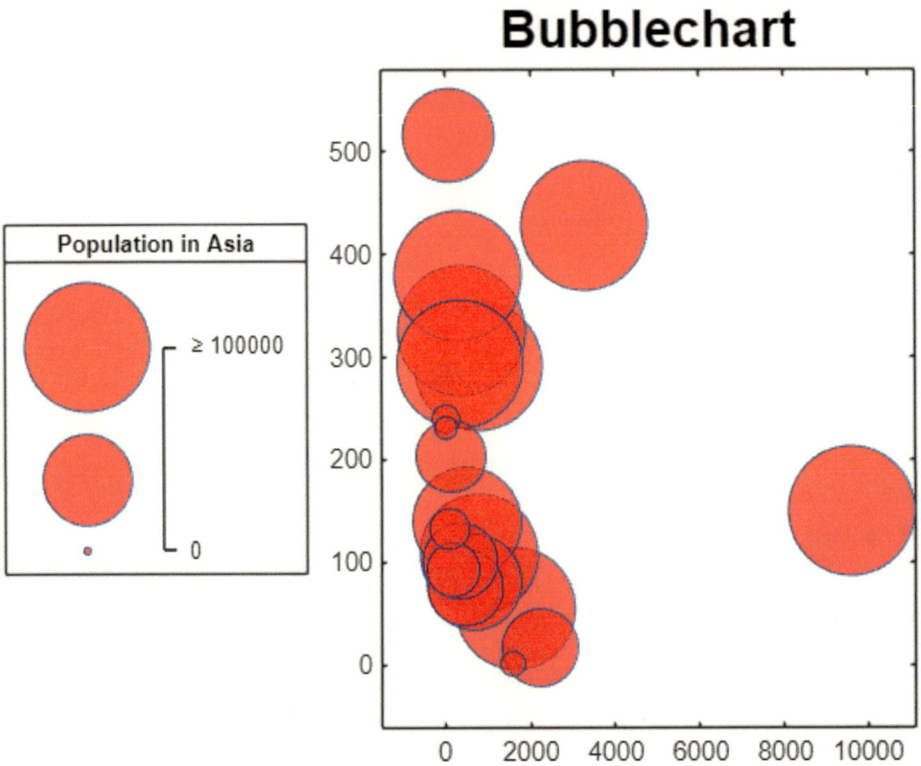

Fig. 16.46 Bubbles (Question 2)

Answer 2

```
figure
Ex2 = bubblechart(land_area,pop_den,pop);
    Ex2.MarkerFaceColor = "red";
bubblelegend("Population in Asia","Location","westoutside");
t = title("Bubblechart");
    t.FontSize = 20;
    t.FontWeight = "bold";
 bubblelim([0 100000]);
```

16

Question 3

Create Fig. 16.47 by overlaying two bubble charts using the **hold on** command. For the first bubble chart, use ⌜landA⌟ as the first argument, ⌜pop_denA⌟ as the second argument, and ⌜popA⌟ as the third argument in the **bubblechart** function. For the second bubble chart, use ⌜landB⌟ as the first argument, ⌜pop_denB⌟ as the second argument, and ⌜popB⌟ as the third argument in the **bubblechart** function. Additionally, use the **annotation** function to add a text box with the text ⌜Data source: AABB⌟, and use the **legend** function to create a legend.

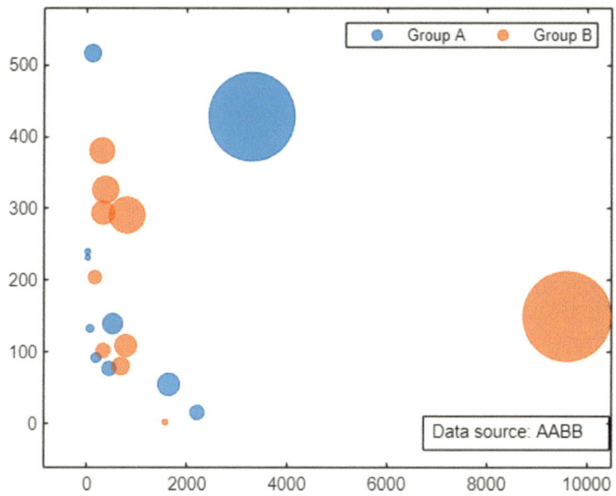

Fig. 16.47 Bubbles (Question 3)

Answer 3

```
figure
bubblechart(landA,pop_denA,popA);
hold on
bubblechart(landB,pop_denB,popB);
hold off
lgd = legend("Group A","Group B");
    lgd.NumColumns = 2;
dim = [0.65 0.05 0.15 0.15];
str = "Data source: AABB";
annotation("TextBox",dim,"String",str,"FitBoxToText","on");
```

Bubblechart3 Function

Contents

17.1 Details of the Data Used in this Chapter – 534

17.2 Excel Data of ► Section 17.1 – 535

17.3 Basic Syntax of the Bubblechart3 Function – 536

17.4 Setting and Controlling the Bubblechart3 Function – 539

17.5 Combining Optional Functions and the Bubblechart3 Function – 542

17.6 Questions – 553

Supplementary Information The online version contains supplementary material available at https://doi.org/10.1007/978-981-96-6213-5_17.

Y. Nakamoto, *A Comprehensive Guide to Creating Graphics with MATLAB Programming*, Classroom Companion: Economics, https://doi.org/10.1007/978-981-96-6213-5_17

Learning Objectives

This chapter creates 3D bubble diagrams using the **bubblechart3** function. By reading this chapter, you will learn how to use the **bubblechart3** function and how to represent four-variable (4D) data in 3D figures by encoding the fourth variable in the size of the bubbles.

Overview

The **bubblechart3** function creates 3D bubble charts. The **bubblechart3** function requires not only 3D data for the x-axis, y-axis, and z-axis, but also a fourth piece of information about the bubble size. The advantage of this function is that it allows the visual representation of four-dimensional data at once, making it effective for understanding relationships between different variables. However, because it deals with 3D visualization, it can sometimes be difficult to interpret visually, which is a drawback. It may be necessary to combine it with the optional functions introduced in ▶ Sect. 17.5.

17.1 Details of the Data Used in this Chapter

In this chapter, we use data of land area, population, and population density for countries located mainly in Asia. 「country」 represents the names of the countries, 「land_area」 represents the area, 「pop」 represents the population, and 「pop_den」 represents the population density. These data are used in ▶ Chap. 16. Additionally, 「gdp」 shows the data of GDP per capita in 2021. These data are presented as column vectors.

```
country = {"United Arab Emirates","Iran","India","Uzbekistan","Qatar",...
    "Cambodia","Kuwait","Saudi Arabia","Thailand","Korea","China",...
    "Turkey","Japan","Nepal","Pakistan","Philippines","Vietnam",...
    "Malaysia","Myanmar","Mongolia"}';

land_area = [71,1631,3287,449,12,181,18,2207,513,100,9600,784,378,...
    147,796,300,331,331,677,1564]';

pop = [9365,87923,1407564,34081,2688,16589,4250,35950,71601,51830,...
    1425893,84775,123224,30035,231402,113880,97468,33574,53798,3348]';

pop_den = [132,54,428,76,231,92,239,16,140,...
    517,149,108,326,204,291,380,294,102,80,2]';

gdp = [43439,3410,2238,2014,65381,1680,28834,28215,7227,35142,...
    12572,9664,39933,1241,1566,3576,3753,11476,1217,4483]';

tbl_Chap17 = table(country,land_area,pop,pop_den,gdp);
```


Next, the vectors 「land_area」, 「pop」, 「pop_den」, and 「gdp」 are divided into two parts: the first 10 countries and the remaining countries. This is mainly to use the overlay of the bubble charts. Specifically, the variables 「landA」, 「popA」, 「pop_denA」, and 「gdpA」 contain the first half of the 「land_area」 data, 「pop」 data, 「pop_den」 data, and 「gdp」 data. Next, 「landB」, 「popB」, 「pop_denB」, and 「gdpB」 represent the second half of their respective data.

```
landA = land_area(1:10);

popA = pop(1:10);

pop_denA = pop_den(1:10);

landB = land_area(11:end);

popB = pop(11:end);

pop_denB = pop_den(11:end);

gdpA = gdp(1:10);

gdpB = gdp(11:end);
```

17.2 Excel Data of ▶ Section 17.1

Here, we load the data from an Excel file. Specifically, we read the data from the sheet 「Chap17_Asia_gdp」 in the Excel file 「alldata.xlsx」 and store it in the table variable 「tbl_Chap17」. Next, using the **renamevars** function, we rename the variables in the table variable 「tbl_Chap17」. In detail, we change 「Land_area_1000」, 「Population_1000」, 「Population_density」, and 「GDP_per_capita」 to 「land_area」, 「pop」, 「pop_den」, and 「gdp」, respectively. Then, we store the respective columns of the table variable into the variables 「country」, 「pop」, 「pop_den」, and 「gdp」. Then, we split each variable into two parts, as we did in ▶ Sect. 17.1.

```
tbl_Chap17 = readtable("alldata.xlsx","Sheet","Chap17_Asia_gdp");
```

```
Warning: Column headers from the file were modified to make them valid MATLAB identifiers before creating
variable names for the table. The original column headers are saved in the VariableDescriptions property.
Set 'VariableNamingRule' to 'preserve' to use the original column headers as table variable names.
```

```
tbl_Chap17 = renamevars(tbl_Chap17,["Country","Land_area_1000",...
    "Population_1000","Population_density","GDP_per_capita"],...
    ["country","land_area","pop","pop_den","gdp"]);
country = tbl_Chap17.country;
land_area = tbl_Chap17.land_area;
pop = tbl_Chap17.pop;
pop_den = tbl_Chap17.pop_den;
gdp = tbl_Chap17.gdp;
landA = land_area(1:10);
popA = pop(1:10);
pop_denA = pop_den(1:10);
landB = land_area(11:end);
popB = pop(11:end);
pop_denB = pop_den(11:end);
gdpA = gdp(1:10);
gdpB = gdp(11:end);
```

17.3 Basic Syntax of the Bubblechart3 Function

Basic Syntax:

Here, we explain the basic syntax of the **bubblechart3** function. In 「bubblechart3(x,y,z,sz)」, which is enclosed in the black rectangle in Figure 17.1, the variables 「x」, 「y」, and 「z」 represent the data corresponding to the x-axis, y-axis, and z-axis, respectively. Additionally, the fourth argument 「sz」 represents the data corresponding to the size of the bubbles.

Syntax

```
bubblechart3(x,y,z,sz)
bubblechart3(x,y,z,sz,c)

bubblechart3(tbl,xvar,yvar,zvar,sizevar)
bubblechart3(tbl,xvar,yvar,zvar,sizevar,cvar)

bubblechart3(ax, __ )
bubblechart3( __ ,Name,Value)
bc = bubblechart3( __ )
```

Fig. 17.1 The syntax of the bubblechart3 function

Here, based on the basic syntax of 「bubblechart3(x,y,z,sz)」, we enter a single number for each argument. The following program creates a bubble at the position (x,y,z)=(2,3,4), with a size of 100. See Fig. 17.2.

```
bubblechart3(2,3,4,100);
```

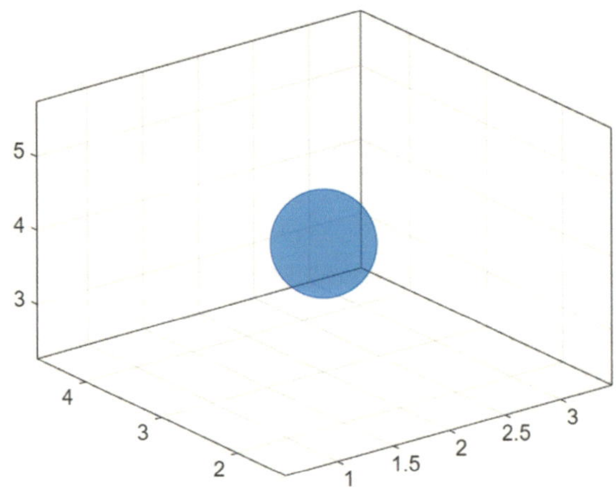

Fig. 17.2 A single 3D bubble created by the bubblechart3 function

Here, the variables 「x」 and 「sz」 are 1-by-4 vectors, while the variables 「y」 and 「z」 are 2-by-4 matrix data. Since both the variables 「y」 and 「z」 have two rows, two groups of bubble charts are created. The first group, represented by blue bubbles, is defined by 「x=[1 2 3 4]」, 「y=[1 2 3 4]」, 「z=[0 5 10 15]」, and 「sz=[1 2 3 4]」. The second group, represented by red bubbles, is defined by 「x=[1 2 3 4]」, 「y=[7 8 9 10]」, 「z=[14 15 16 17]」, and 「sz=[1 2 3 4]」. See Fig. 17.3.

```
x = [1 2 3 4];
y = [1 2 3 4; 7 8 9 10];
z = [0 5 10 15; 14 15 16 17];
sz = [1 2 3 4];
bubblechart3(x,y,z,sz);
```

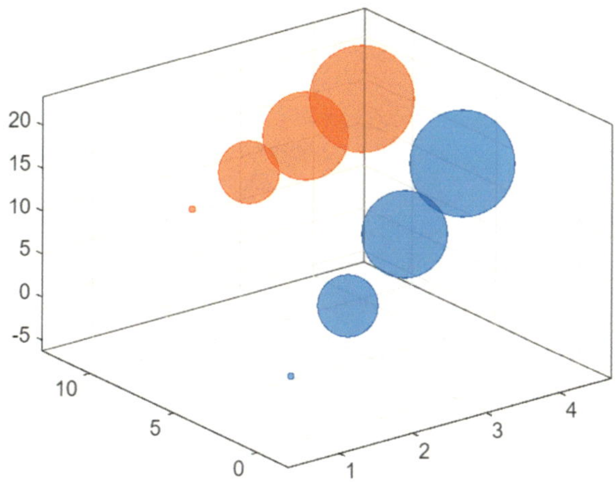

Fig. 17.3 Two grouped 3D bubbles

Here, variables are used in each argument. Specifically, the first argument is 「pop」, the second is 「pop_den」, the third is 「land_area」, and the fourth is 「gdp」. Therefore, the sizes of the bubbles are based on the values of 「gdp」. See Fig. 17.4.

```
bubblechart3(pop,pop_den,land_area,gdp);
```

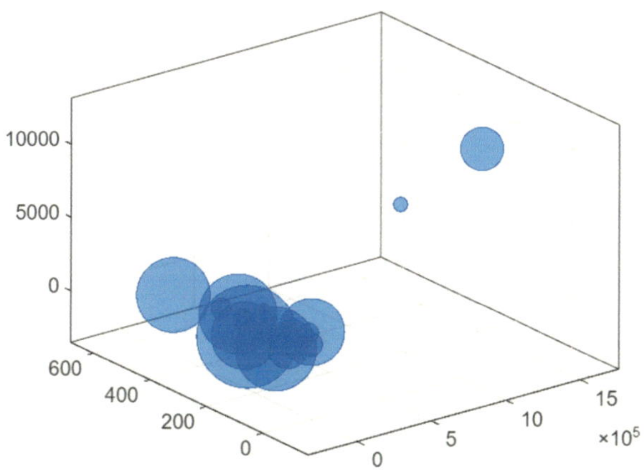

Fig. 17.4 3D bubbles from four variables

Here, we use specific elements of each variable. In the third argument, we use 「land_area(1:5)」, which means that we use the first through fifth elements of the variable 「land_area」. Additionally, 「[pop(1:5) pop(6:10)]」 and 「[pop_den(1:5) pop_den(6:10)]」 in the first and second arguments create 2-by-5 matrices. These matrices create a bubble chart with two groups. See Fig. 17.5.

```
bubblechart3([pop(1:5) pop(6:10)],[pop_den(1:5) pop_den(6:10)],...
    land_area(1:5),gdp(1:5));
```

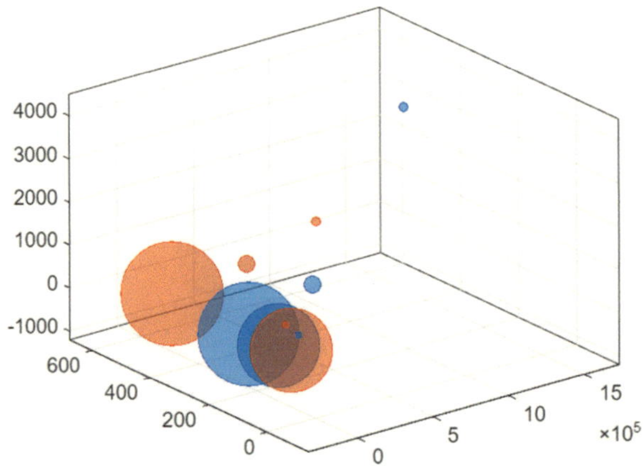

Fig. 17.5 Two grouped 3D bubbles from variable subsets

As in the previous program that created Fig. 17.5, we use matrix data. In particular, each argument 「[popA,popB]」, 「[pop_denA,pop_denB]」, 「[landA,landB]」, and 「[gdpA,gdpB]」 are 10-by-2 matrices, creating a bubble chart with two groups. See Fig. 17.6.

```
bubblechart3([popA,popB],[pop_denA,pop_denB],[landA,landB],[gdpA,gdpB]);
```

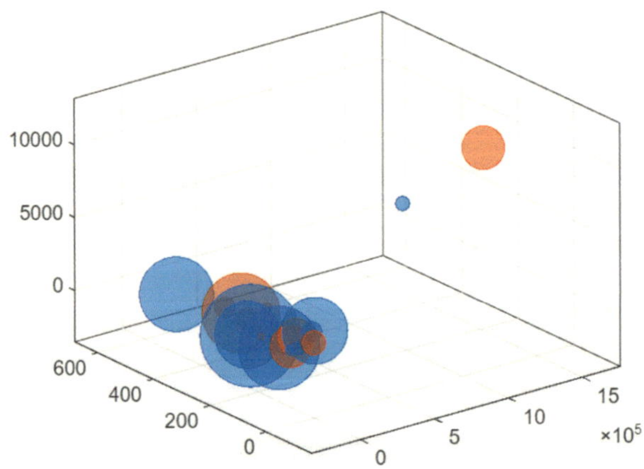

Fig. 17.6 Two grouped 3D bubbles from matrix data

17.4 Setting and Controlling the Bubblechart3 Function

By specifying 「red」 as the fifth argument, we change the color of the bubbles to red. See Fig. 17.7.

```
bubblechart3(pop,pop_den,land_area,gdp,"red");
```

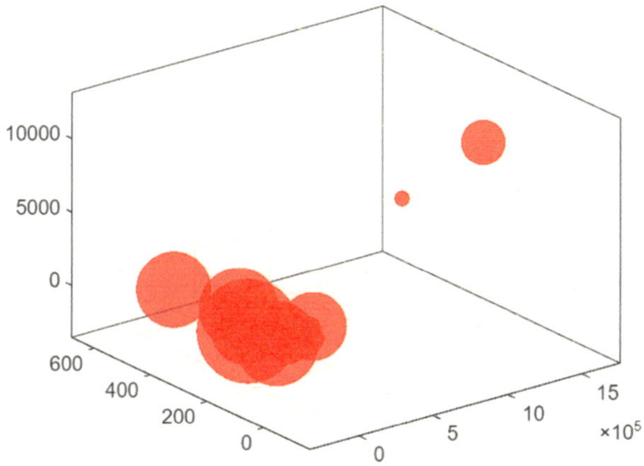

Fig. 17.7 Red 3D bubbles

Here, we set the fifth argument of the **bubblechart3** function to 「green」 and store its properties in the variable 「bbc3」. After that, we set 「LineWidth」 to 2 and 「MarkerEdgeColor」 to 「black」, creating green bubbles with a thick black outline. See Fig. 17.8.

```
bbc3 = bubblechart3(pop,pop_den,land_area,gdp,"green");
    bbc3.LineWidth = 2;
    bbc3.MarkerEdgeColor = "black";
```

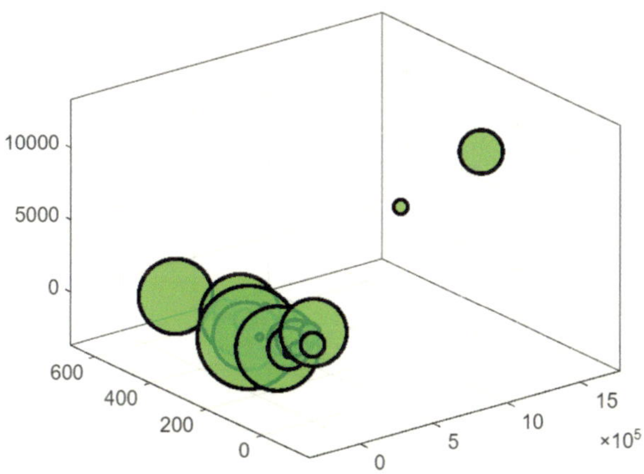

Fig. 17.8 Green 3D bubbles with thick black edges

The following program is the same as ◘ Fig. 17.8, except for one point. The difference is that we set 「bbc3.MarkerEdgeAlpha」 to 0.2, creating bubbles with high transparency. Compare ◘ Fig. 17.8 with ◘ Fig. 17.9.

```
bbc3 = bubblechart3(pop,pop_den,land_area,gdp,"green");
    bbc3.LineWidth = 2;
    bbc3.MarkerEdgeColor = "black";
    bbc3.MarkerEdgeAlpha = 0.2;
```

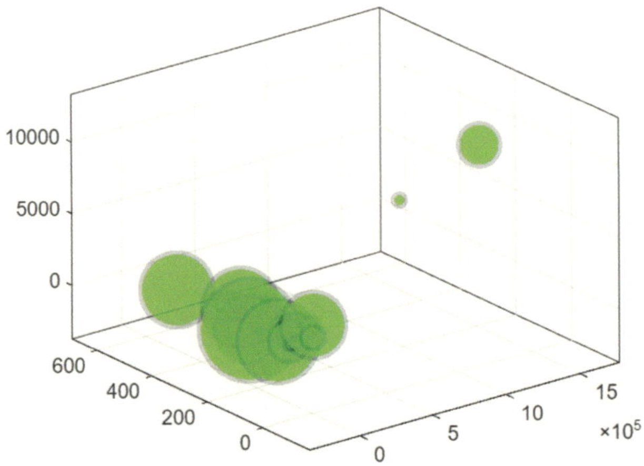

◘ **Fig. 17.9** Green 3D bubbles with semi-transparent black edges

In the following program, we use the **bubblechart3** function to create a bubble chart with two groups. The properties of the chart are stored in the variable 「bbc3」. Then, by specifying 「bbc3(1)」, we modify the properties of only the first group. In detail, we set 「MarkerFaceColor」 to 「green」, 「MarkerEdgeColor」 to 「black」, 「MarkerFaceAlpha」 to 0.2, and 「LineWidth」 to 2, creating semi-transparent green bubbles with a thick black outline. The resulting figure is presented in ◘ Fig. 17.10.

```
bbc3 = bubblechart3([popA,popB],[pop_denA,pop_denB],[landA,landB],[gdpA,gdpB]);
    bbc3(1).MarkerFaceColor = "green";
    bbc3(1).MarkerEdgeColor = "black";
    bbc3(1).MarkerFaceAlpha = 0.2;
    bbc3(1).LineWidth = 2;
```

17

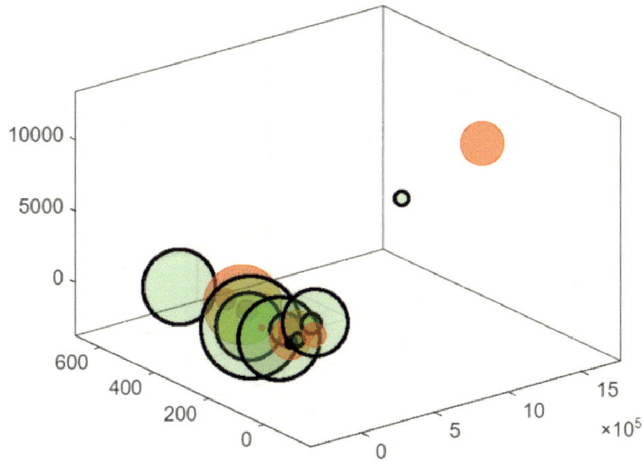

■ **Fig. 17.10** Two grouped 3D bubbles with one black-edged group in green

Here, for the first group (「bbc3(1)」), we set 「MarkerEdgeColor」 to 「blue」 and 「LineWidth」 to 1, creating bubbles with a slightly thick blue outline. Next, for the second group (「bbc3(2)」), we set 「MarkerFaceColor」 to 「none」 and 「LineWidth」 to 4, creating transparent bubbles with a thick outline. See ■ Fig. 17.11.

```
bbc3 = bubblechart3([popA,popB],[pop_denA,pop_denB],[landA,landB],[gdpA,gdpB]);
    bbc3(1).MarkerEdgeColor = "blue";
    bbc3(1).LineWidth = 1;
    bbc3(2).MarkerFaceColor = "none";
    bbc3(2).LineWidth = 4;
```

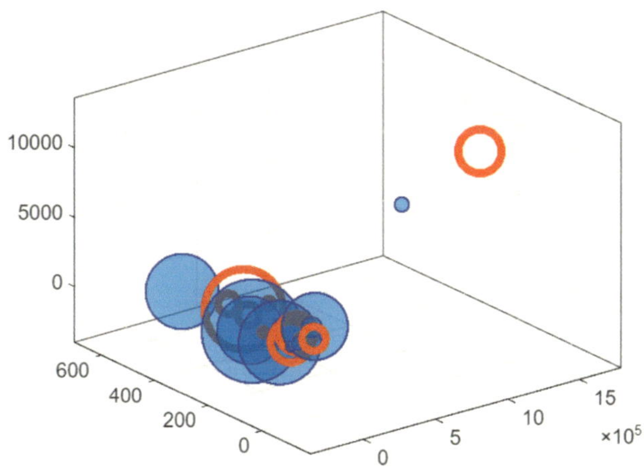

■ **Fig. 17.11** Two grouped 3D bubbles with one blue-edged group and one transparent red-edged group

17.5 Combining Optional Functions and the Bubblechart3 Function

Here, we use the **xlabel**, **ylabel**, and **zlabel** functions simultaneously. Additionally, we use the **box on** command to display the axis frame. See ◘ Fig. 17.12.

```
bubblechart3(land_area,pop_den,pop,gdp);
xlabel("land area");
ylabel("population density");
zlabel("population");
box on
```

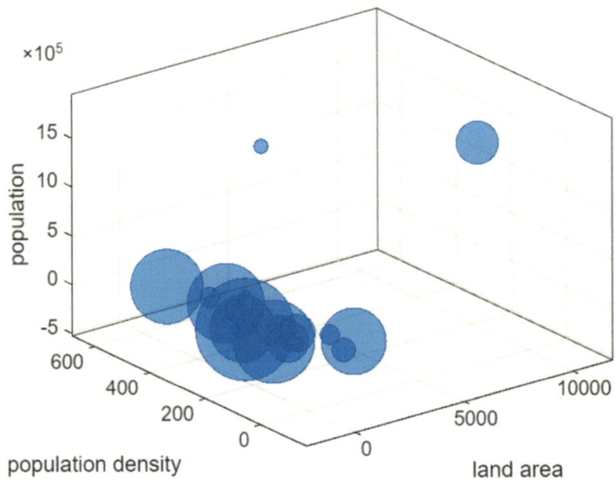

◘ **Fig. 17.12** 3D bubbles with X-, Y-, and Z-axis labels

Here, using the **gca** command, we store the axis properties in the variable 「ax」. Then, we set 「XColor」 to 「red」, 「YColor」 to 「blue」, and 「ZColor」 to 「green」, changing the x-axis color to red, the y-axis color to blue, and the z-axis color to green. Additionally, by setting 「GridColor」 to 「red」, we create red grid lines. Finally, we set 「LineWidth」 to 2, making the axis frame lines thicker. See ◘ Fig. 17.13.

17

```
bubblechart3(land_area,pop_den,pop,gdp);
ax = gca;
    ax.XColor = "red";
    ax.YColor = "blue";
    ax.ZColor = "green";
    ax.GridColor = "red";
    ax.LineWidth = 2;
```

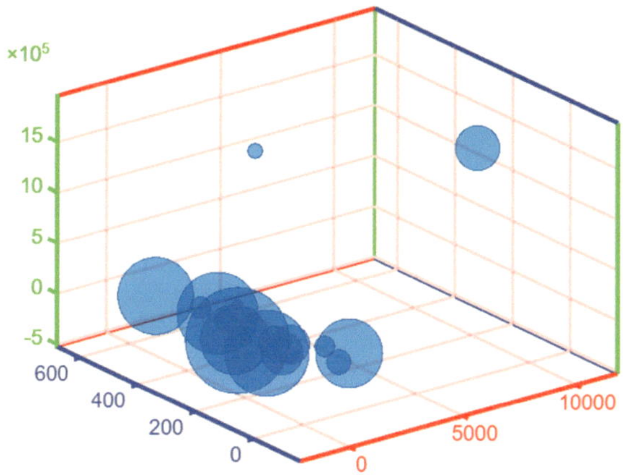

Fig. 17.13 3D bubbles with red X-, blue Y-, green Z-axes, and red gridlines

Here, we use the **gca** command to store the axis properties in the variable ⌜ax⌟. Then, by setting ⌜GridColor⌟ to ⌜red⌟, ⌜GridLineStyle⌟ to ⌜--⌟, and ⌜Line Width⌟ to 2, we create thick red dashed grid lines. See Fig. 17.14.

```
figure
bubblechart3(land_area,pop_den,pop,gdp);
ax = gca;
    ax.GridColor = "red";
    ax.GridLineStyle = "--";
    ax.LineWidth = 2;
```

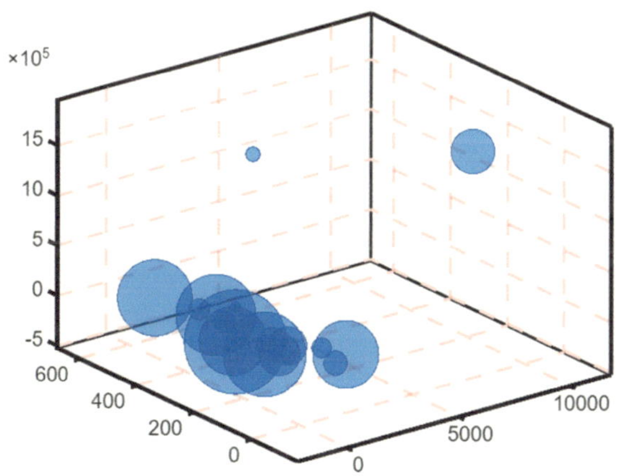

Fig. 17.14 3D bubbles with red dashed gridlines and thick black axes

In the following program, we first use the **gca** command to store the axis properties in the variable ⌜ax⌟. Then, by setting ⌜GridColor⌟ to ⌜none⌟, we remove the grid lines. Next, we adjust the tick marks on the x-axis and y-axis using the **xticks** and **yticks** functions. Compare the tick marks in Fig. 17.15 with those in Fig. 17.12.

```
bubblechart3(land_area,pop_den,pop,gdp);
ax = gca;
    ax.GridColor = "none";
xticks(0:1000:10000);
yticks(0:100:600);
```

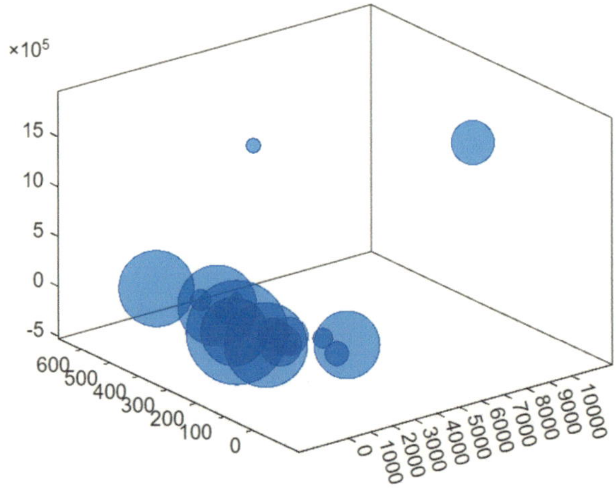

Fig. 17.15 3D bubbles with no gridlines and adjusted X- and Y-axis ticks

The following program is the same as the one that created **Fig. 17.15**, except for one point: the use of the **view** function. Here, by using 「view([30 120])」, we adjust the viewing angle. Compare **Fig. 17.15** with **Fig. 17.16**.

```
bubblechart3(land_area,pop_den,pop,gdp);
ax = gca;
    ax.GridColor = "none";
xticks(0:1000:10000);
yticks(0:100:600);
view([30 120]);
```

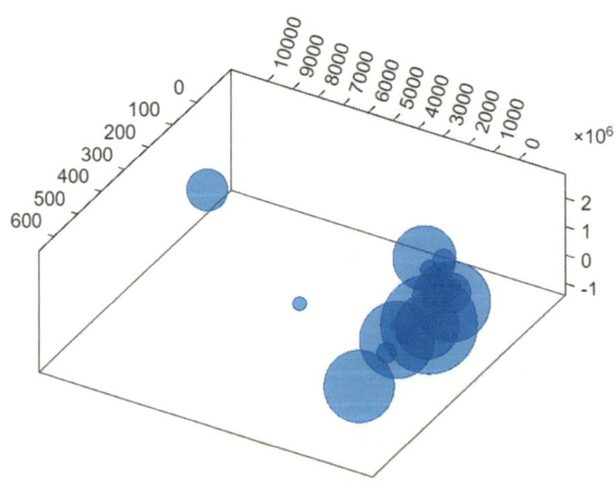

Fig. 17.16 3D bubbles from a different angle, with no gridlines and adjusted X- and Y-axis ticks

17

Here, we first use the **view** function to adjust the viewing angle. Then, by using the **bubblelegend** function, we create a legend for the bubbles with the title label 「GDP per capita」. Note that the size of the bubbles is created by the values of the variable 「gdp」. See ◘ Fig. 17.17.

```
bubblechart3(land_area,pop_den,pop,gdp);
view([150 50]);
bubblelegend("GDP per capita");
```

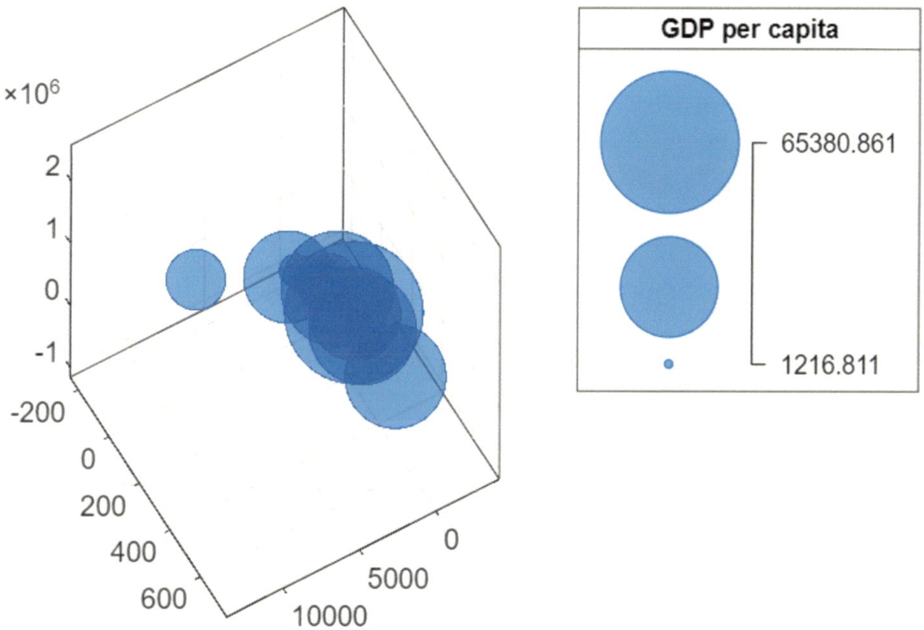

◘ **Fig. 17.17** 3D bubbles from a different angle, with a bubble legend

Here, by creating a 1-by-20 vector for the variable 「c」, we use it as the fifth argument of the **bubblechart3** function. The variable 「c」 determines the color of the bubbles. Specifically, each of the 20 bubbles is assigned a different color based on the values in the variable 「c」. Since the colors are not uniform, the bubbles are shown in dark gray in the legend created by the **bubblelegend** function. See ◘ Fig. 17.18.

```
figure
c = 1:20;
bubblechart3(land_area,pop_den,pop,gdp,c);
bubblelegend("GDP per capita");
```

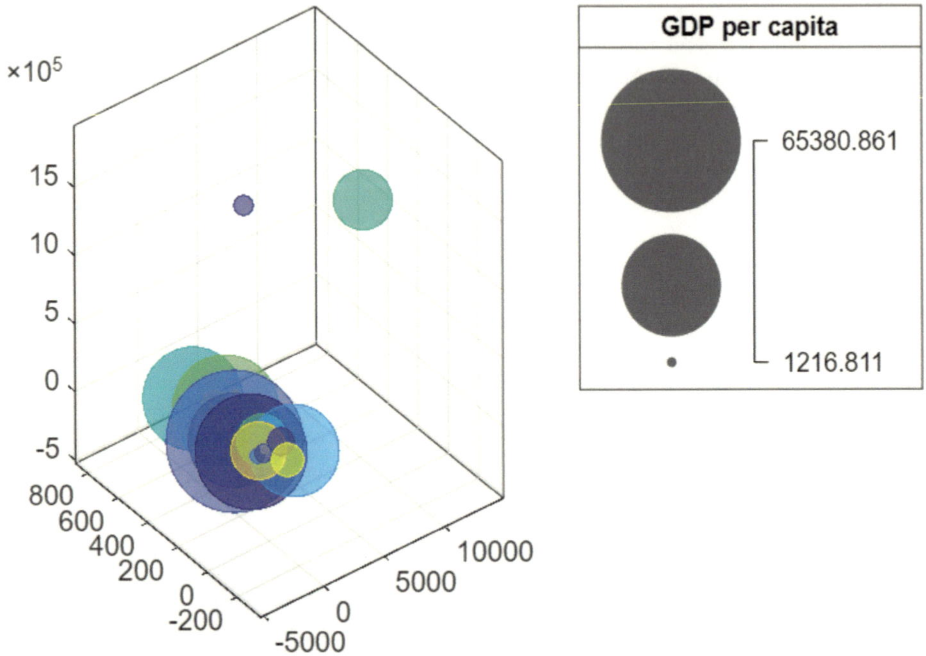

Fig. 17.18 3D bubbles in various colors, with a bubble legend

Here, by setting the fifth argument of the **bubblechart3** function to 「red」, we change the color of the bubbles to red and store the properties of the 3D bubbles in the variable 「bh3」. Then, we set 「MarkerFaceAlpha」 to 0.1 and 「LineWidth」 to 2, creating semi-transparent bubbles with a thick outline. Additionally, by using the **xlim, ylim**, and **zlim** functions to set the limits above 0, we create a chart that does not display any ranges below 0. The resulting figure is shown in Fig. 17.19.

```
figure
bh3 = bubblechart3(land_area,pop_den,pop,gdp,"red");
    bh3.MarkerFaceAlpha = 0.1;
    bh3.LineWidth = 2;
bubblelegend("GDP per capita");
xlim([0 Inf]);
ylim([0 Inf]);
zlim([0 Inf]);
```

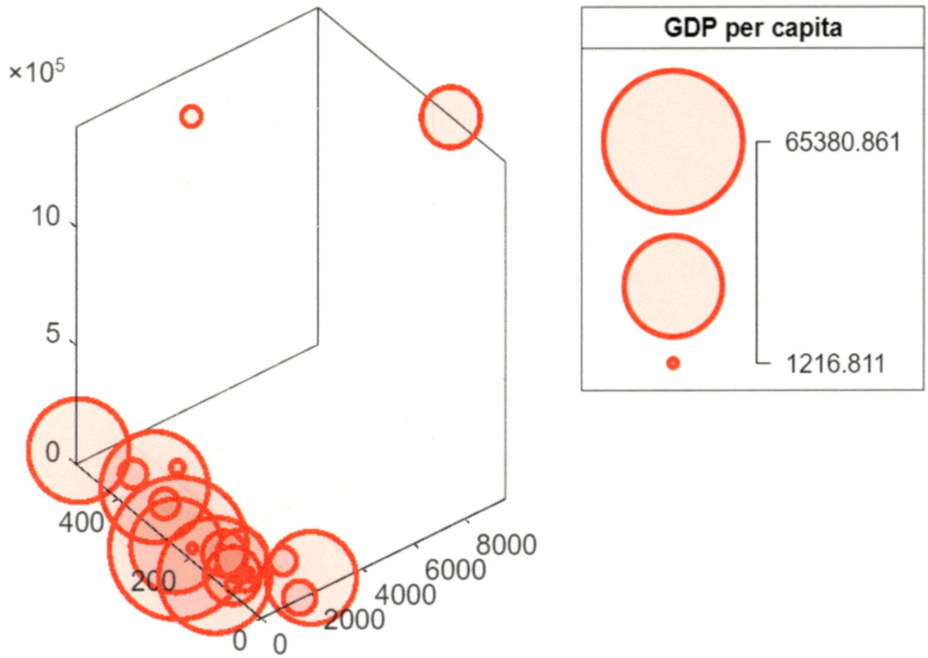

Fig. 17.19 3D bubbles with some bubbles outside of the frame and a bubble legend

Here, we use the **hold on** command to overlay two bubble charts created by the **bubblechart3** function. Additionally, we create a legend using the **legend** function and store its properties in the variable 「lgd」. Then, by using 「title(lgd,"Asian countries")」, we add a title to the legend. Moreover, by using 「title("The bubblechart3 function")」, we create the title for the entire figure. See Fig. 17.20.

```
bubblechart3(landA,pop_denA,popA,gdpA);
hold on
bubblechart3(landB,pop_denB,popB,gdpB);
hold off
grid minor
lgd = legend("Group A","Group B");
title(lgd,"Asian countries");
t = title("The bubblechart3 function");
    t.Color = "red";
    t.FontSize = 20;
```

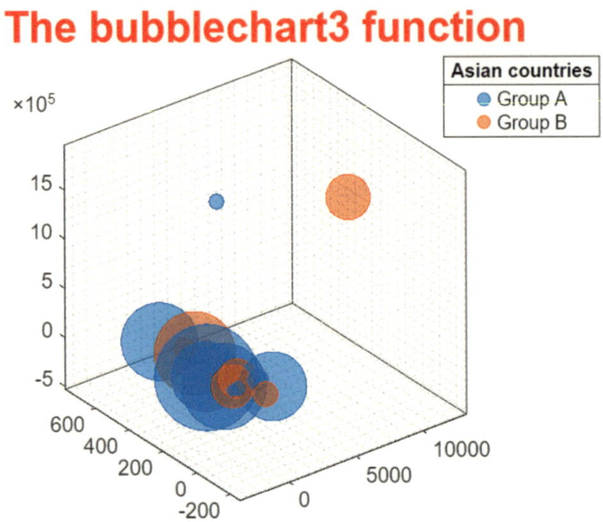

Fig. 17.20 3D bubbles with fine gridlines, a large red title, and a legend

In the following program, we create a 3D bubble chart and draw descending lines from specific bubbles. Specifically, the variables 「x_fixed_3」 and 「y_fixed_3」 store the third data of the variables 「land_area」 and 「pop_den」, respectively. Similarly, the variables 「x_fixed_11」 and 「y_fixed_11」 store the eleventh data of the variables 「land_area」 and 「pop_den」, respectively.

Then, using the **hold on** command and the **line** function, we overlay the lines connecting the specified coordinates. In the first **line** function, we draw a line connecting the two points 「(X,Y,Z)=(x_fixed_3,y_fixed_3,0)」 and 「(X,Y,Z)=(x_fixed_3, y_fixed_3, z)」. Next, in the second **line** function, we draw a line connecting the two points 「(X,Y,Z)=(x_fixed_11,y_fixed_11,0)」 and 「(X,Y,Z)=(x_fixed_11,y_fixed_11,z)」. See Fig. 17.21.

```
figure
bubblechart3(land_area,pop_den,pop,gdp);
x_fixed_3 = land_area(3);
y_fixed_3 = pop_den(3);
x_fixed_11 = land_area(11);
y_fixed_11 = pop_den(11);
z = max(pop);
zlim([0 z]);
z_range = [0,z];
hold on
line([x_fixed_3 x_fixed_3],[y_fixed_3 y_fixed_3],z_range,...
    "Color","blue","LineWidth",2,"LineStyle","--");
line([x_fixed_11 x_fixed_11],[y_fixed_11 y_fixed_11],z_range,...
    "Color","red","LineWidth",2,"LineStyle",":");
hold off
```

17

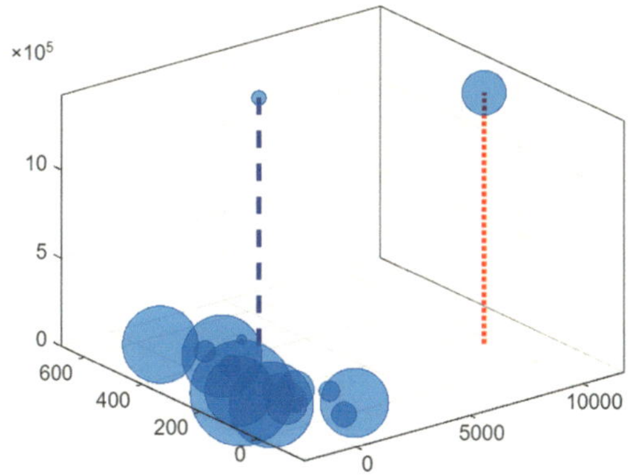

Fig. 17.21 3D bubbles with a blue dashed line and a red dotted line

Here, in addition to a 3D bubble chart, we use the **patch** function to overlay a surface. Specifically, we create a surface based on the data points of the third and eleventh bubbles. As seen in the program 「patch(x_patch,y_patch_3,z_patch, "blue","FaceAlpha",0.3,"EdgeColor","black","LineWidth",2)」, we specify the x-coor dinate, y-coordinate, and z-coordinate of the vertices of the patch using the variables 「x_patch」, 「y_patch_3」, and 「z_patch」. For example, we specify the x-co ordinates of the four corners of the surface by 「x_patch=[0,x,x,0]」. Similarly, we specify the y-coordinates and z-coordinates of the four corners, and then create the surface using the **patch** function.

Finally, by using the **text** function, we display text. Specifically, as confirmed in 「text(0,y_fixed_3,z,sprintf("Y=%d",y_fixed_3),"background","yellow","Fon tSize",14)」, we specify the x-coordinate, y-coordinate, and z-coordinate such as (x,y,z)=(0,y_fixed_3,z). Additionally, by setting 「background」 to 「yellow」, we set the background color of the text to yellow. For more details on the **text** function, see ► Sect. 18.2. See Fig. 17.22 for the resulting figure.

```
figure
bubblechart3(land_area,pop_den,pop,gdp);
hold on
x = max(land_area);
xlim([0 x]);
y = max(pop_den);
ylim([0 y]);
z = max(pop);
zlim([0 z]);
y_fixed_3 = pop_den(3);
y_fixed_11 = pop_den(11);
x_patch = [0,x,x,0];
z_patch = [0,0,z,z];
y_patch_3 = [y_fixed_3,y_fixed_3,y_fixed_3,y_fixed_3];
y_patch_11 = [y_fixed_11,y_fixed_11,y_fixed_11,y_fixed_11];
patch(x_patch,y_patch_3,z_patch,"red","FaceAlpha",0.3,...
    "EdgeColor","black","LineWidth",2);
patch(x_patch,y_patch_11,z_patch,"blue","FaceAlpha",0.3,...
    "EdgeColor","black","LineWidth",2);
text(0,y_fixed_3,z,sprintf("Y= %d",y_fixed_3),...
    "background","yellow","FontSize",14);
text(0,y_fixed_11,z,sprintf("Y= %d",y_fixed_11),...
    "background","yellow","FontSize",14);
hold off
```

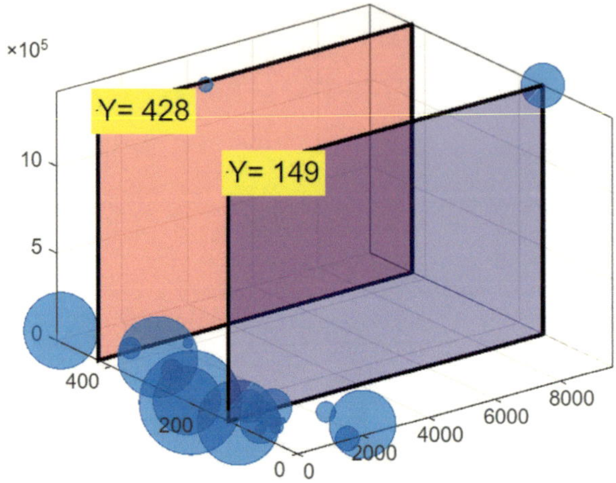

Fig. 17.22 3D bubbles with vertical surfaces in red and purple

Here, by specifying 「rectangle」 in the first argument of the **annotation** function, we enclose the third bubble with a rectangle. Note that both the position and size of the rectangle are specified by the variable 「dim」. Next, we use the **text** function to display the GDP value above the third bubble. For more details on the **annotation** and **text** functions, see ▶ Sects. 18.1 and 18.2. The resulting figure is shown in Fig. 17.23.

```
bubblechart3(land_area,pop_den,pop,gdp);
title("Title: Bubblechart3");
dim = [0.33 0.63 0.1 0.1];
annotation("rectangle",dim,"FaceColor","yellow","FaceAlpha",0.1);
text(land_area(3),pop_den(3)+300,pop(3),sprintf("GDP: %d",gdp(3)),...
    "FontSize",12,"Color", "black","background","yellow");
```

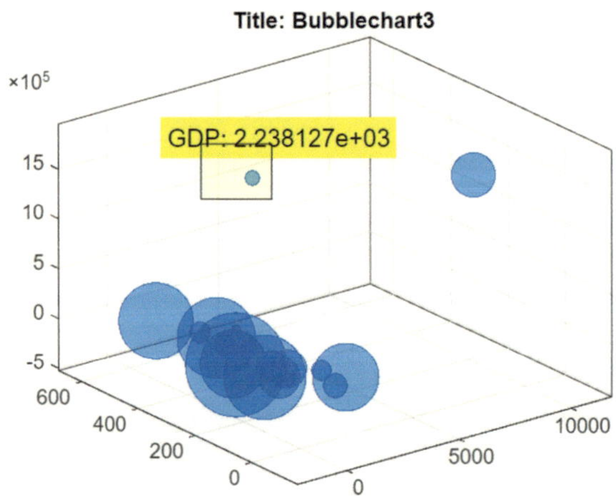

Fig. 17.23 3D bubbles with a yellow text label and a rectangle

17

Using the vectors ⌜land_area(1:5)⌟, ⌜pop_den(1:5)⌟, ⌜pop(1:5)⌟, and ⌜gdp (1:5)⌟ in each argument of the **bubblechart3** function, here we create five bubbles. Then, within a **for** loop, we retrieve the data values of ⌜land_area⌟, ⌜pop_den⌟, and ⌜pop⌟ and store them in the variables ⌜x_val⌟, ⌜y_val⌟, and ⌜z_val⌟. For instance, when ⌜i=1⌟, it can be seen that ⌜x_val=land_area(1)⌟, ⌜y_val=pop_ den(1)⌟, and ⌜z_val=pop(1)⌟. After that, these variables, ⌜x_val⌟, ⌜y_val⌟, and ⌜z_val⌟, are used as the x-coordinate, y-coordinate, and z-coordinate data in the **text** function.

Finally, in the **text** function, to move the text from the center of the bubbles, we set the first argument to ⌜x_val+5⌟, the second argument to ⌜y_val+5⌟, and the third argument to ⌜z_val+5⌟. For more details on the **text** function, see ▶ Sect. 18.2. ◘ Fig. 17.24 shows the resulting figure.

```
figure
bubblechart3(land_area(1:5),pop_den(1:5),pop(1:5),gdp(1:5));
for i = 1:1:5
    x_val = land_area(i);
    y_val = pop_den(i);
    z_val = pop(i);
    gdp_val = gdp(i);
    text(x_val+5,y_val+5,z_val+5,sprintf("GDP: %d",gdp_val),...
        "FontSize",12,"Color","black");
end
view([50 80]);
```

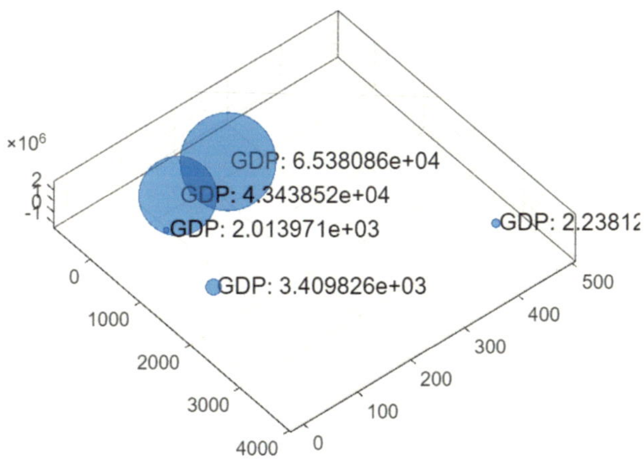

◘ **Fig. 17.24** 3D bubbles from a different angle, with text labels added by a for loop

Here, by using ⌜tiledlayout(2,1)⌟, we create a 2-by-1 tiled layout. Then, in the first row and first column, we create a 3D bubble chart using ⌜bubblechart3(pop_ denA,landA,popA,gdpA)⌟. Next, in the second row and first column, we create another 3D bubble chart using ⌜bubblechart3(pop_denB,landB,popB,gdpB)⌟. See ◘ Fig. 17.25.

```
figure
tiledlayout(2,1);
nexttile;
bubblechart3(pop_denA,landA,popA,gdpA);
title("gdp in Group A");
xlabel("Land area");
ylabel("Population density");
zlabel("Population");
nexttile;
bubblechart3(pop_denB,landB,popB,gdpB);
title("gdp in Group B");
xlabel("Land area");
ylabel("Population density");
zlabel("Population");
```

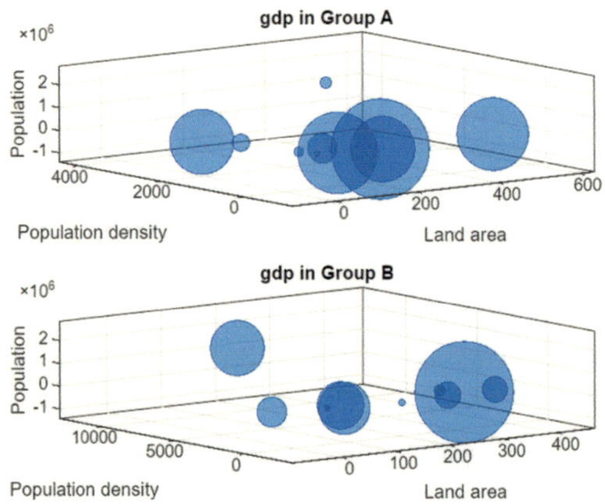

Fig. 17.25 3D bubbles (2-row layout)

In Fig. 17.25, the 2-by-1 tiled layout was created. Here, by using 「tiledlayout(1,2)」, we create a 1-by-2 tiled layout. Compare Fig. 17.25 with Fig. 17.26.

```
figure
tiledlayout(1,2);
nexttile;
bubblechart3(pop_denA,landA,popA,gdpA);
title("gdp in Group A");
xl1 = xlabel("Land area");
    xl1.FontSize = 6;
yl1 = ylabel("Population density");
    yl1.FontSize = 6;
zl1 = zlabel("Population");
    zl1.FontSize = 6;
nexttile;
bh3 = bubblechart3(pop_denB,landB,popB,gdpB);
    bh3.MarkerFaceColor = "red";
    bh3.MarkerFaceAlpha = 0.1;
    bh3.LineWidth = 1;
title("gdp in Group B");
xl2 = xlabel("Land area");
    xl2.FontSize = 6;
yl2 = ylabel("Population density");
    yl2.FontSize = 6;
zl2 = zlabel("Population");
    zl2.FontSize = 6;
```

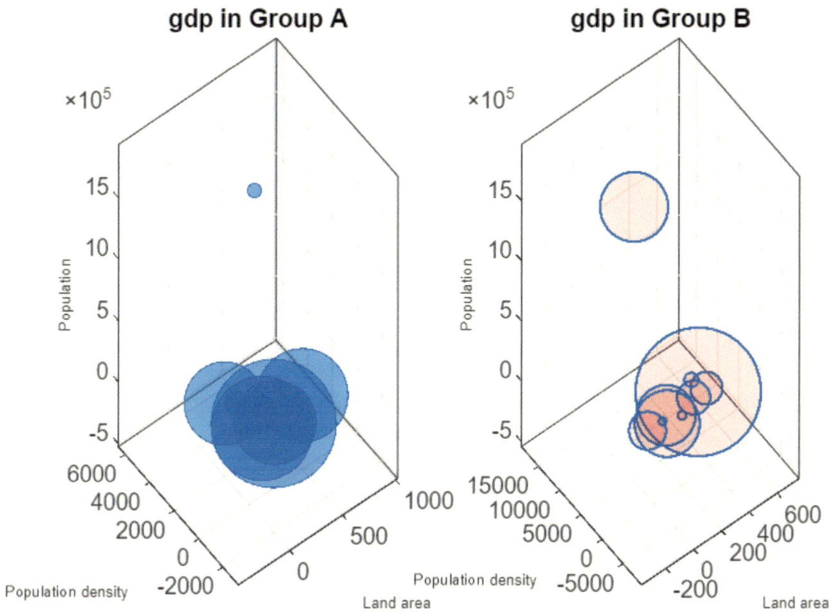

Fig. 17.26 3D bubbles (2-column layout)

17.6 Questions

Question 1

Create Fig. 17.27 using the **bubblechart3** function with 「[popA,popB]」 as the first argument, 「[pop_denA,pop_denB]」 as the second argument, 「[landA,landB]」 as the third argument, and 「[gdpA,gdpB]」 as the fourth argument. Then, store the propert ies of the **bubblechart3** function in the variable 「Ex1」. For the first group, set 「MarkerEdgeColor」 to 「black」, 「MarkerFaceAlpha」 to 0.5, and 「LineWidth」 to 2. For the second group, set 「MarkerFaceColor」 to 「none」, 「MarkerEdgeCo lor」 to 「red」, and 「LineWidth」 to 2.

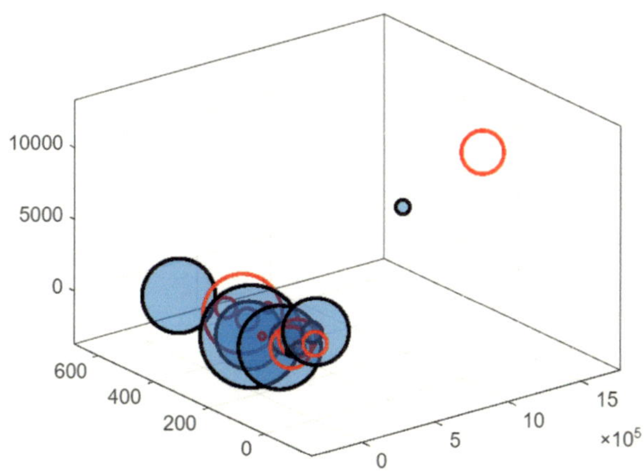

Fig. 17.27 3D bubbles (Question 1)

Answer 1

```
figure
Ex1 = bubblechart3([popA,popB],[pop_denA,pop_denB],[landA,landB],[gdpA,gdpB]);
    Ex1(1).MarkerEdgeColor = "black";
    Ex1(1).MarkerFaceAlpha = 0.5;
    Ex1(1).LineWidth = 2;
    Ex1(2).MarkerFaceColor = "none";
    Ex1(2).MarkerEdgeColor = "red";
    Ex1(2).LineWidth = 2;
```

Question 2

Create Fig. 17.28 using the **tiledlayout** and **bubblechart3** functions. Concretely, in the first row, first column space, create a bubble chart with the first argument as 「pop_denA」, the second argument as 「landA」, the third argument as 「popA」, and the fourth argument as 「gdpA」, and store the properties in a variable with an arbitrary name. After that, make the bubbles red with a thick blue outline. In the second row, first column space, create a bubble chart with the first argument as 「pop_denB」, the second argument as 「landB」, the third argument as 「popB」, and the fourth argument as 「gdpB」. Make the bubbles yellow with a thick black outline.

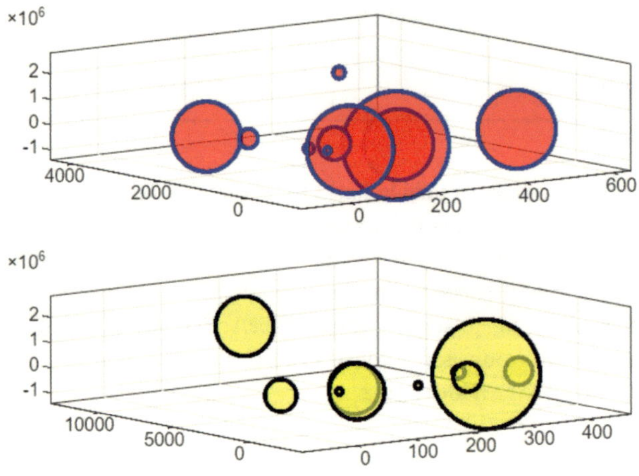

Fig. 17.28 3D bubbles (Question 2)

17

Answer 2

```
figure
tiledlayout(2,1);
nexttile;
Ex2no1 = bubblechart3(pop_denA,landA,popA,gdpA,"red");
    Ex2no1.MarkerEdgeColor = "blue";
    Ex2no1.LineWidth = 2;
nexttile;
Ex2no2 = bubblechart3(pop_denB,landB,popB,gdpB,"yellow");
    Ex2no2.MarkerEdgeColor = "black";
    Ex2no2.LineWidth = 2;
```

Question 3

Based on the program in Fig. 17.24, create Fig. 17.29. There are three changes from the program used for Fig. 17.24. First, set the first argument of the **bubblechart3** function to 「land_area」, the second argument to 「pop_den」, the third argument to 「gdp」, and the fourth argument to 「pop」. Second, use the sixteenth to twelfth values of each variable (「land_area」, 「pop_den」, 「gdp」, and 「pop」). Lastly, change the values displayed next to the bubbles to 「pop」.

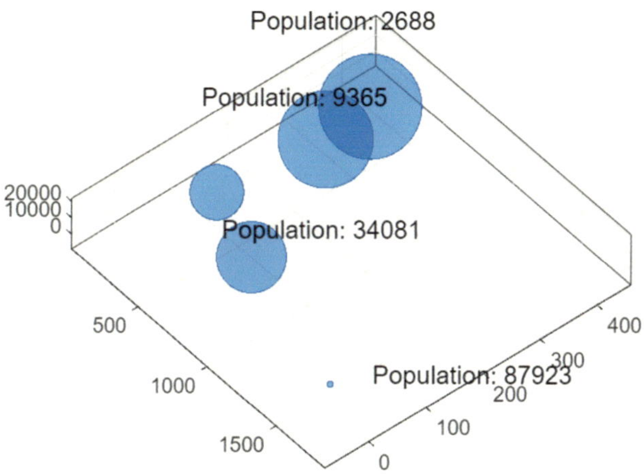

Fig. 17.29 3D bubbles (Question 3)

Answer 3

```
figure
bubblechart3(land_area(16:20),pop_den(16:20),gdp(16:20),pop(16:20));
for i = 1:1:5
    x_val = land_area(i);
    y_val = pop_den(i);
    z_val = gdp(i);
    pop_val = pop(i);
    text(x_val+5,y_val+5,z_val+5,sprintf("Population: %d",pop_val),...
        "FontSize",12,"Color","black");
end
view([50 80]);
```

Useful Optional Functions

Contents

18.1 Annotation Function – 558

18.2 Text Function – 566

18.3 Xline and Yline Functions – 572

18.4 Questions – 578

Y. Nakamoto, *A Comprehensive Guide to Creating Graphics with MATLAB Programming*, Classroom Companion: Economics, https://doi.org/10.1007/978-981-96-6213-5_18

Learning Objectives

This chapter explains four functions (**annotation**, **text**, **xline**, and **yline**). By reading this chapter, you will learn how to use these functions and gain a deeper understanding of how they are used throughout ► Chaps. 3–17.

Overview

This chapter explains four functions: the **annotation**, **text**, **xline**, and **yline** functions. Each of these functions can be used in combination with many of the functions from ► Chaps. 3–17; however, note that some functions cannot be used with these optional functions at the same time. For example, in MATLAB version 2024b, the **text** function cannot be used with the **piechart** and **donutchart** functions (► Chap. 14).

► Section 18.1 introduces the **annotation** function, which allows you to add annotations to a figure by inserting arrows, lines, or rectangles. This function is useful for highlighting specific data points or areas within the figure. ► Section 18.2 introduces the **text** function. This function enables you to embed text characters within a figure, which is beneficial for labeling data. By positioning text at specific locations, you can add valuable information directly to the figure. ► Section 18.3 explains the **xline** and **yline** functions, which are used to draw vertical or horizontal reference lines at specified positions. While the syntax for the **xline** and **yline** functions is essentially the same, it is important to note how to use 「LabelVerticalAlignment」 and 「LabelHorizontalAlig nment」.

18.1 Annotation Function

Basic Syntax:

We explain the basic syntaxes of the **annotation** function, enclosed in the black boxes in Figure 18.1. The syntax 「annotation(lineType,x,y)」 represents that 「lineType」 specifies the style of the line, while 「x」 and 「y」 represent the coordinates of the start and end points of the line. Next, the syntax 「annotation(shapeType,dim)」 uses 「shapeType」 to specify the type of shape, and 「dim」 to specify the position and size of the shape.

Syntax

```
annotation(lineType,x,y)
annotation(lineType)

annotation(shapeType,dim)
annotation(shapeType)

annotation(___,Name,Value)
annotation(container,___)

an = annotation(___)
```

Fig. 18.1 The syntax of the annotation function

In the following program, we use four different annotations: 「TextArrow」, 「Arrow」, 「doubleArrow」, and 「line」, all of which follow the same syntax, 「annotation(lineType,x,y)」. In particular, when the annotation uses 「TextArrow」, we can add a label (「TextArrow 1」) at the end of the arrow, as confirmed in ◘ Fig. 18.2. Additionally, each annotation creates an arrow or line parallel to the x-axis. This is because the third argument, 「y」 in 「annotation(lineType,x,y)」, consists of the same value for both the start and end points. For example, in the first **annotation**, the third argument is 「[0.3 0.3]」, which positions the annotation at y=0.3. Similarly, the second **annotation** uses 「[0.9 0.9]」, the third **annotation** uses 「[0.2 0.2]」, and the fourth **annotation** uses 「[0.7 0.7]」. See ◘ Fig. 18.2.

```
scatter(1,1,"filled");
annotation("TextArrow",[0.4 0.8],[0.3 0.3],"String","TextArrow 1");
annotation("Arrow",[0.8 0.3],[0.9 0.9]);
annotation("doubleArrow",[0.3 0.5],[0.2 0.2]);
annotation("line",[0.8 0.5],[0.7 0.7]);
```

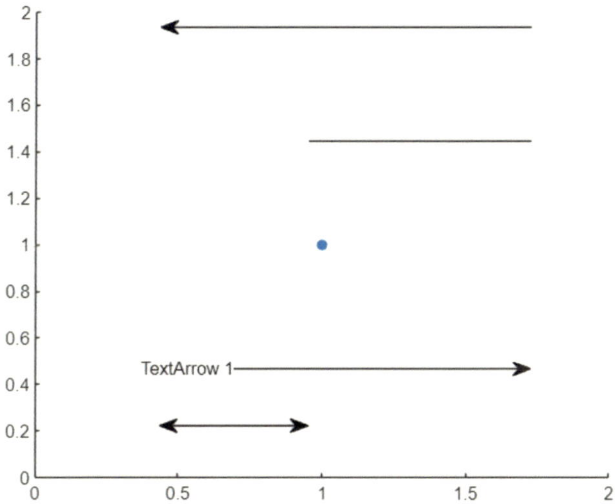

◘ **Fig. 18.2** Single-point scatter with multiple arrows and a line

Here, we apply the same annotations as in ◘ Fig. 18.2 to a 3D figure. The syntax remains the same as 「annotation(lineType,x,y)」, even for a 3D figure. This means that when using annotations, the position specification is considered in 2D. See ◘ Fig. 18.3. For a detailed description of annotations, see ◘ Fig. 18.2.

```
scatter3(1,1,1,"filled");
annotation("TextArrow",[0.4 0.8],[0.3 0.3],"String","TextArrow 1");
annotation("Arrow",[0.8 0.3],[0.9 0.9]);
annotation("doubleArrow",[0.3 0.5],[0.2 0.2]);
annotation("line",[0.8 0.5],[0.7 0.7]);
```

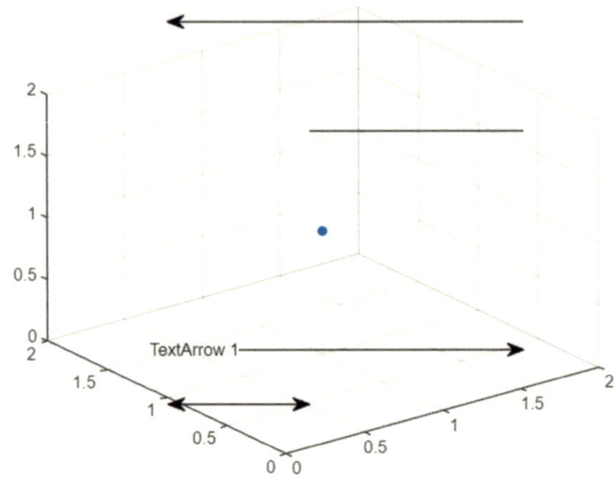

Fig. 18.3 Single-point 3D scatter with multiple arrows and a line

Here, we apply the same annotations used in Fig. 18.2 to a donut chart. See Fig. 18.4.

```
donutchart([1 2 3 4]);
annotation("TextArrow",[0.4 0.8],[0.3 0.3],"String","TextArrow 1");
annotation("Arrow",[0.8 0.3],[0.9 0.9]);
annotation("doubleArrow",[0.3 0.5],[0.2 0.2]);
annotation("line",[0.8 0.5],[0.7 0.7]);
```

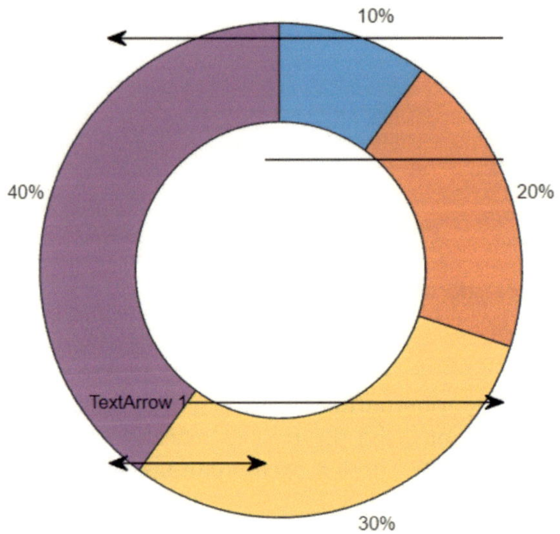

Fig. 18.4 Donut chart with multiple arrows and a line

Here, we create an arrow with text using 「TextArrow」 for a 3D figure. The syntax is the same as 「annotation(lineType,x,y)」 in Fig. 18.2. We store the properties of the **annotation** function in the variable 「an」. Then, we set 「FontSize」 to 20,

「Color」 to 「blue」, 「LineStyle」 to 「--」, and 「LineWidth」 to 3, creating a thick blue dashed arrow with text (「Point」). See Fig. 18.5.

```
scatter3(1,1,1,"filled");
an = annotation("TextArrow",[0.4 0.5],[0.2 0.48],"String","Point");
    an.FontSize = 20;
    an.Color = "blue";
    an.LineStyle = "--";
    an.LineWidth = 3;
```

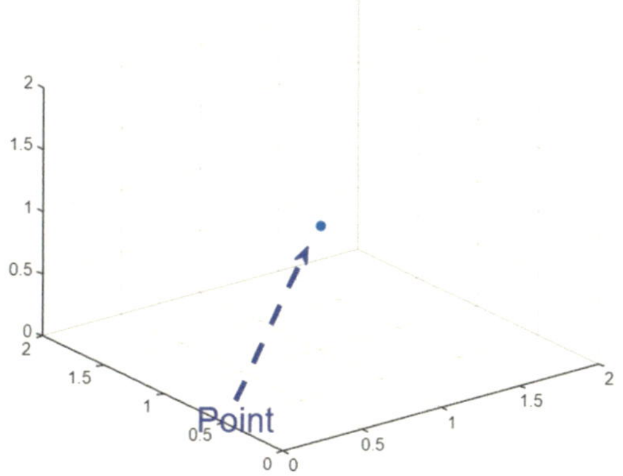

Fig. 18.5 Single-point 3D scatter with a blue dashed text arrow

Here, we create 「doubleArrow」 inside the center hole of the donut chart. In particular, we set 「LineStyle」 to 「:」, 「Color」 to 「red」, and 「LineWidth」 to 3, creating a thick red dotted double arrow. The resulting figure is shown in Fig. 18.6.

```
donutchart([1 2 3 4]);
da = annotation("doubleArrow",[0.4 0.6],[0.4 0.7]);
    da.LineStyle = ":";
    da.Color = "red";
    da.LineWidth = 3;
```

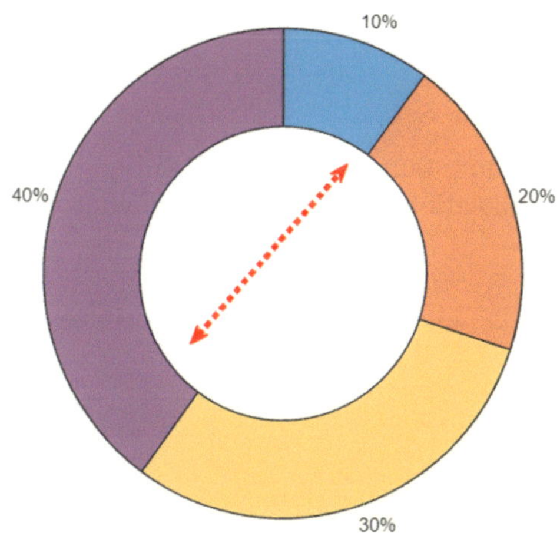

Fig. 18.6 Donut chart with a red dotted double-headed arrow in the central hole

Here, we use the syntax 「annotation(shapeType,dim)」 six times to illustrate the concept of 「dim」. For the first three **annotation** functions, the position for 「dim」 is set to 「0.4 0.6」, meaning that the starting points for all three rectangles are the same. Then, the size of each rectangle is gradually increased. Specifically, the first rectangle is set to 「0.1 0.1」, the second to 「0.2 0.2」, and the third to 「0.3 0.3」, making each rectangle larger than the previous one. Next, we use 「ellipse」 to create ellipses, positioning them at 「0.15 0.2」. Again, we increase the size incrementally. The resulting figure is shown in ◘ Fig. 18.7.

```
scatter(1,1,"filled");
annotation("rectangle",[0.4  0.6 0.1 0.1]);
annotation("rectangle",[0.4  0.6 0.2 0.2]);
annotation("rectangle",[0.4  0.6 0.3 0.3]);
annotation("ellipse",[0.15  0.2 0.1 0.1]);
annotation("ellipse",[0.15  0.2 0.2 0.2]);
annotation("ellipse",[0.15  0.2 0.3 0.3]);
```

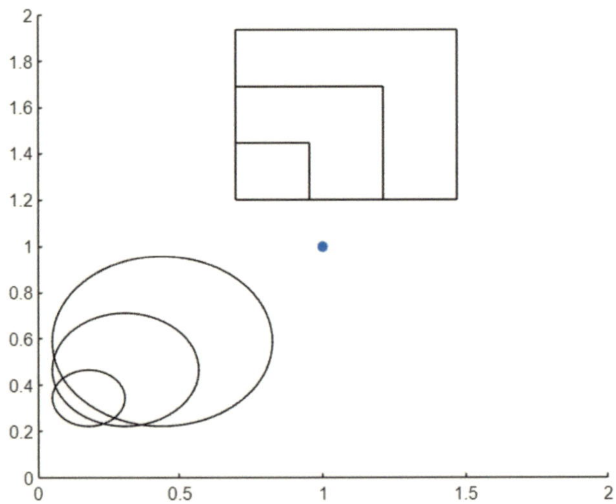

◘ **Fig. 18.7** Single-point scatter with multiple rectangles and ellipses

Here, we apply the annotations created in ◘ Fig. 18.7 to a 3D figure. See ◘ Fig. 18.8. See the explanation in ◘ Fig. 18.7 for details.

```
scatter3(1,1,1,"filled");
annotation("rectangle",[0.4  0.6 0.1 0.1]);
annotation("rectangle",[0.4  0.6 0.2 0.2]);
annotation("rectangle",[0.4  0.6 0.3 0.3]);
annotation("ellipse",[0.15  0.2 0.1 0.1]);
annotation("ellipse",[0.15  0.2 0.2 0.2]);
annotation("ellipse",[0.15  0.2 0.3 0.3]);
```

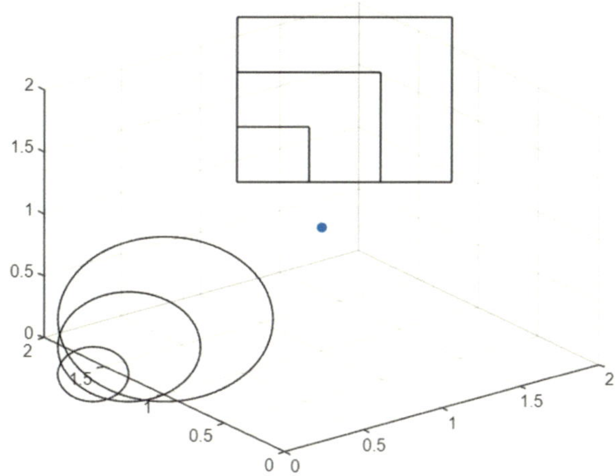

Fig. 18.8 Single-point 3D scatter with multiple rectangles and ellipses

Here, we specify ⌜rectangle⌟ with the **annotation** function and store its properties in the variable ⌜rec⌟. Then, we set ⌜Color⌟ to ⌜red⌟, ⌜LineWidth⌟ to 1.5, ⌜LineStyle⌟ to ⌜-.⌟, ⌜FaceColor⌟ to ⌜blue⌟, and ⌜FaceAlpha⌟ to 0.4. This creates a rectangle filled with semi-transparent blue and an outline in red, styled by a dash-dotted line. See Fig. 18.9.

```
scatter(1,1,"filled");
rec = annotation("rectangle",[0.4 0.6 0.5 0.1]);
    rec.Color = "red";
    rec.LineWidth = 1.5;
    rec.LineStyle = "-.";
    rec.FaceColor = "blue";
    rec.FaceAlpha = 0.4;
```

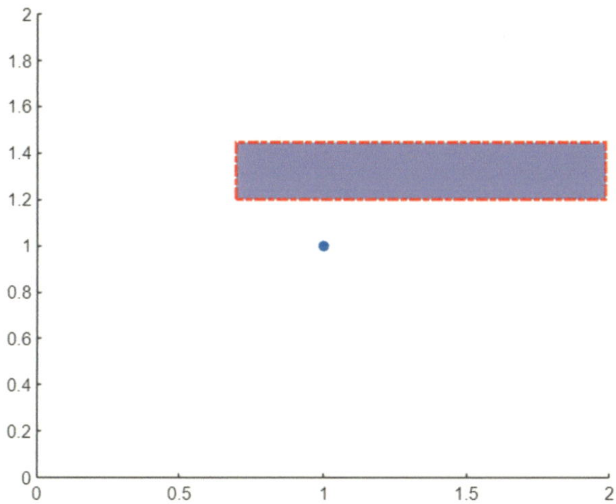

Fig. 18.9 Single-point scatter with a blue-filled rectangle and red dotted edge

Here, we specify 「textbox」 with the **annotation** function and store its properties in the variable 「rec」. Then, we set 「Color」 to 「red」, 「LineWidth」 to 1.5, 「LineStyle」 to 「-.」, 「FaceAlpha」 to 0.4, and 「FontSize」 to 25, creating a box with a dash-dotted outline and placing red text (「textbox」) inside the figure. See ◘ Fig. 18.10.

```
scatter(1,1,"filled");
rec = annotation("textbox",[0.4 0.6 0.5 0.1],"String","textbox");
    rec.Color = "red";
    rec.LineWidth = 1.5;
    rec.LineStyle = "-.";
    rec.FontSize = 25;
```

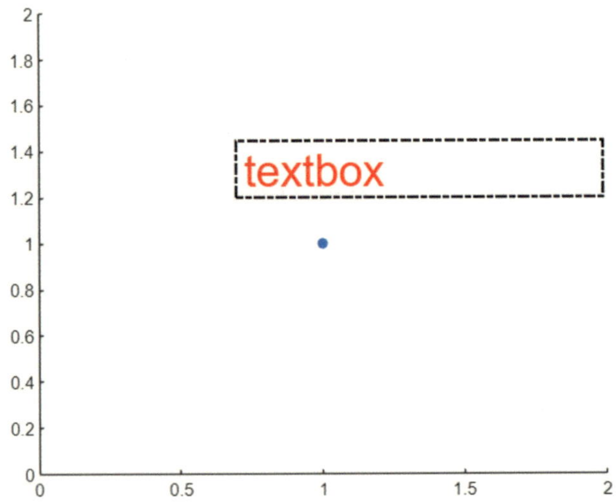

◘ **Fig. 18.10** Single-point scatter with a black dash-dot textbox containing red text

Here, the program is the same as the one used to create ◘ Fig. 18.10, except that 「FitBoxToText」 is set to 「on」. As a result, the box size is automatically adjusted to fit the size of the text string. See ◘ Fig. 18.11.

```
scatter(1,1,"filled");
txt = annotation("textbox",[0.4 0.6 0.5 0.1],...
    "String","textbox","FitBoxToText","on");
    txt.Color = "red";
    txt.LineWidth = 1.5;
    txt.LineStyle = "-.";
    txt.FontSize = 25;
```

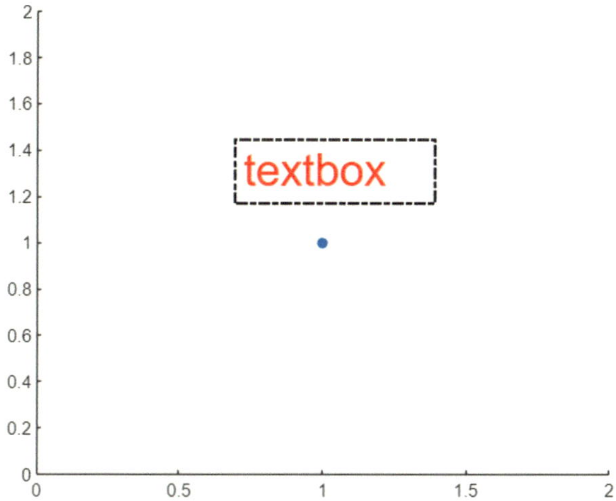

Fig. 18.11 Single-point scatter with a fitted textbox (cf. Fig. 18.10)

Here, we use the **annotation** function twice. First, we run the previous program that created Fig. 18.11. Second, we create a rectangle filled with semi-transparent yellow that surrounds a text box created with 「textbox」. See Fig. 18.12.

```
scatter(1,1,"filled");
txt = annotation("textbox",[0.4 0.6 0.5 0.1],...
    "String","textbox","FitBoxToText","on");
    txt.Color = "red";
    txt.LineWidth = 1.5;
    txt.LineStyle = "-.";
    txt.FaceAlpha = 0.4;
    txt.FontSize = 25;
rec = annotation("rectangle",[0.4 0.5 0.3 0.3]);
    rec.FaceColor = "yellow";
    rec.FaceAlpha = 0.2;
```

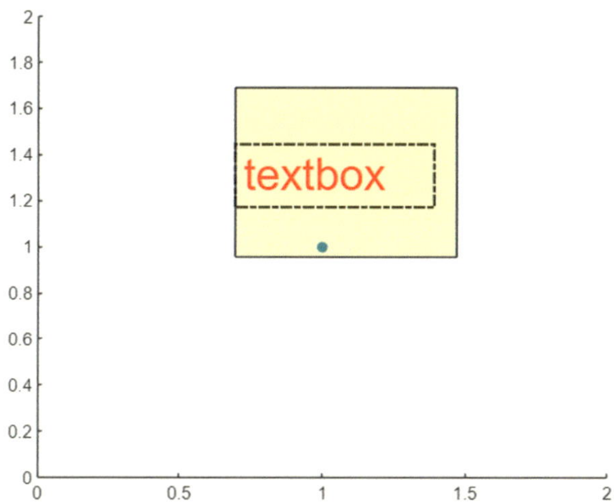

Fig. 18.12 Single-point scatter with red text in a dash-dot textbox over a yellow rectangle

18.2 Text Function

Basic Syntax:

Section 18.2 explains the **text** function, which allows you to place text labels in 2D or 3D space by specifying coordinates and the desired text content. See the two syntaxes enclosed in the black box in Figure 18.13. The first syntax, 「text(x,y,txt)」, places a text label at the 2D coordinates specified by 「x」 and 「y」. 「txt」 represents the text string to be displayed at this position (x,y). The second syntax, 「text(x,y,z,txt)」, allows you to place a text label in a 3D space. Here, the x-coordinate 「x」, y-coordinate 「y」, and z-coordinate 「z」 specify the location, while 「txt」 represents the text string.

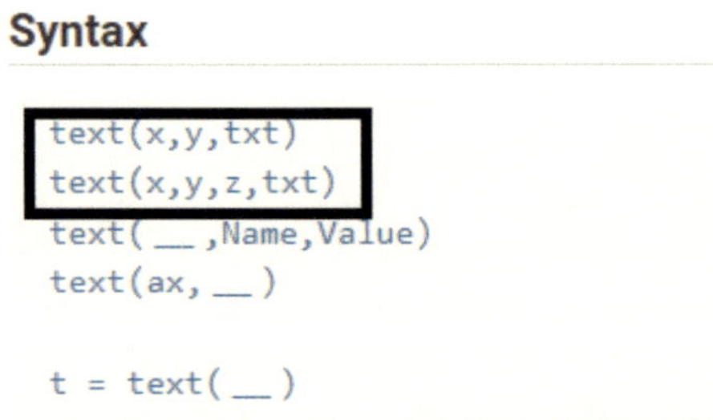

Fig. 18.13 The syntax of the text function

Here, we use the **text** function twice. In a 2D figure, we use 「text(x,y,txt)」. The text label 「corner」 is placed at (x,y)=(1,1). Next, the text label 「center」 is placed at (x,y)=(0.5,0.5). See Fig. 18.14.

```
figure
text(1,1,"corner");
text(0.5,0.5,"center");
```

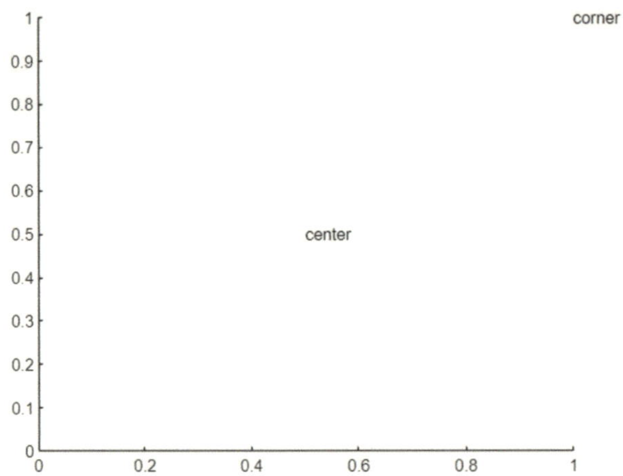

Fig. 18.14 Two text labels placed at center and top-right corner

Here, we place a text label at (x,y)=(0.5,0.5). Specifically, the text content is set to 「{"center 1", "center 2", "center 3"}」, displaying the text in three lines. See Fig. 18.15.

```
figure
text(0.5,0.5,{"center 1","center 2","center 3"});
```

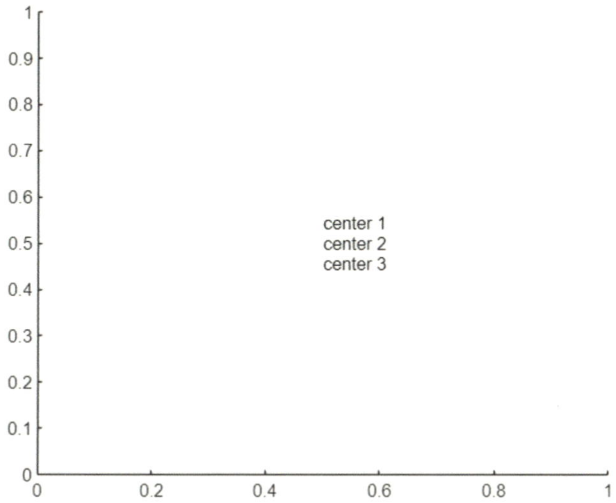

Fig. 18.15 Three text labels aligned vertically at center

Here, we introduce two types of spaces in the text characters. See the third argument of the **text** function. The first is 「"center 2"」, which adds spaces before the text 「center 2」. The second is 「""」, which creates a blank third line. See Fig. 18.16 for the created figure.

```
figure
txt = text(0.5,0.5,{"center 1","        center 2","","center 3"});
    txt.Color = "red";
    txt.FontSize = 20;
```

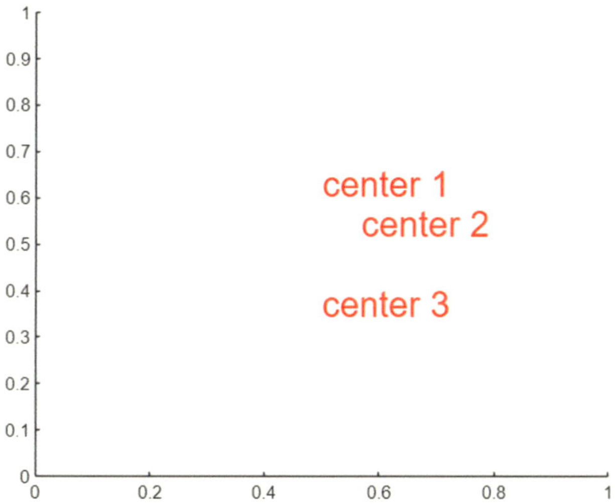

Fig. 18.16 Multiple red text lines at center

Here, we use 1-by-3 vectors for the first and second arguments in the **text** function. As a result, the text label 「points」 appears at (x,y)=(0.1,0.1), (x,y)=(0.3,0.3), and (x,y)=(0.5,0.5). See ◘ Fig. 18.17.

```
figure
text([0.1 0.3 0.5],[0.1 0.3 0.5],"points");
```

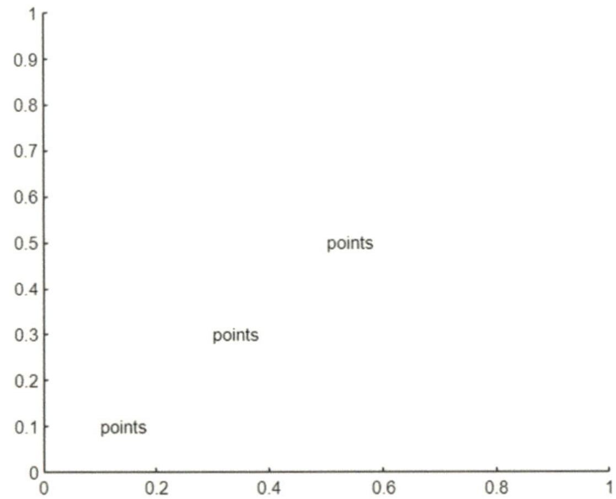

◘ **Fig. 18.17** Text labels placed on multiple diagonal coordinates

Here, we add two text labels to a 3D figure. Although the proper syntax for 3D text placement is 「text(x,y,z,txt)」, we use the 2D syntax 「text(x,y,txt)」 here. As a result, the text labels are placed at a height of z=0. See ◘ Fig. 18.18.

```
scatter3(1,1,1,"filled");
text(1,1,"corner");
text(0.5,0.5,"center");
```

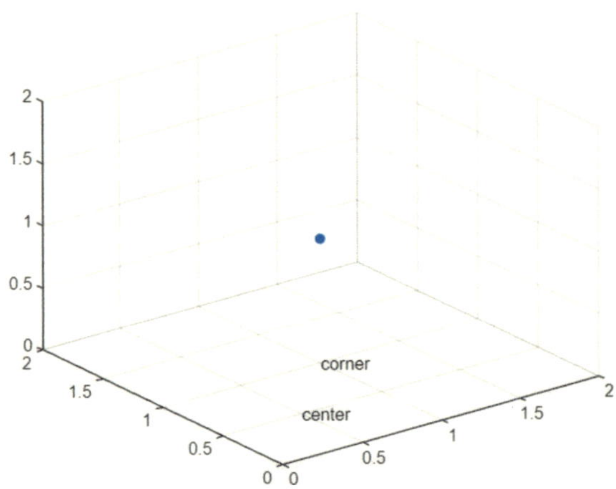

◘ **Fig. 18.18** Single-point 3D scatter with two text labels

Here, we use the syntax ⌜text(x,y,z,txt)⌟. The first text label, ⌜corner⌟, is placed at (x,y,z)=(2,2,2). The second text label, ⌜center⌟, is positioned at (x,y,z)=(0.5,0.5,0.5). See ◘ Fig. 18.19.

```
scatter3(1,1,1,"filled");
text(2,2,2,"corner");
text(0.5,0.5,0.5,"center");
```

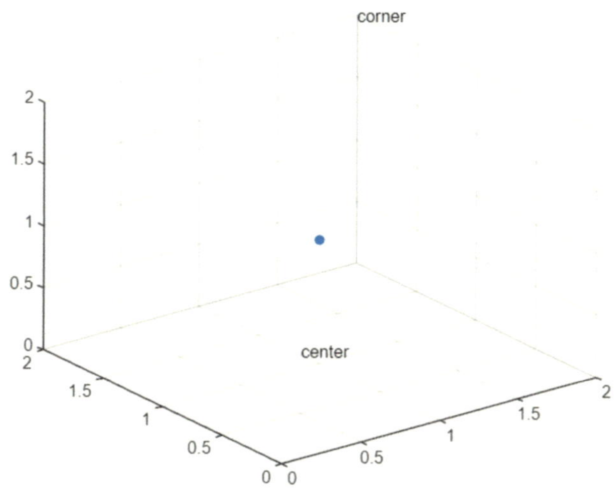

◘ **Fig. 18.19** Single-point 3D scatter with two text labels (cf. ◘ Fig. 18.18)

Three text labels are created. First, the text label ⌜corner 1⌟ is placed at (x,y,z)=(2,2,2), and its properties are stored in the variable ⌜txt1⌟. We then set ⌜FontSize⌟ to 25 and ⌜Color⌟ to ⌜red⌟, creating a large, red text label for ⌜corner 1⌟.

Next, by using ⌜text([0 0],[0 2],[0 2],"corner 2")⌟, we place the text label ⌜corner 2⌟ at two locations: (x,y,z)=(0,0,0) and (x,y,z)=(0,2,2). The properties of this **text** function are stored in the variable ⌜txt2⌟, which is a 2-by-1 text array. Using ⌜txt2(1)⌟ and ⌜txt2(2)⌟, we can adjust the properties of each ⌜corner 2⌟ label individually. See ◘ Fig. 18.20.

```
scatter3(1,1,1,"filled");
txt1 = text(2,2,2,"corner 1");
    txt1.FontSize = 25;
    txt1.Color = "red";
txt2 = text([0 0],[0 2],[0 2],"corner 2");
    txt2(1).FontSize = 25;
    txt2(2).FontSize = 25;
    txt2(1).FontWeight = "bold";
    txt2(2).FontWeight = "bold";
    txt2(1).Color = "blue";
```

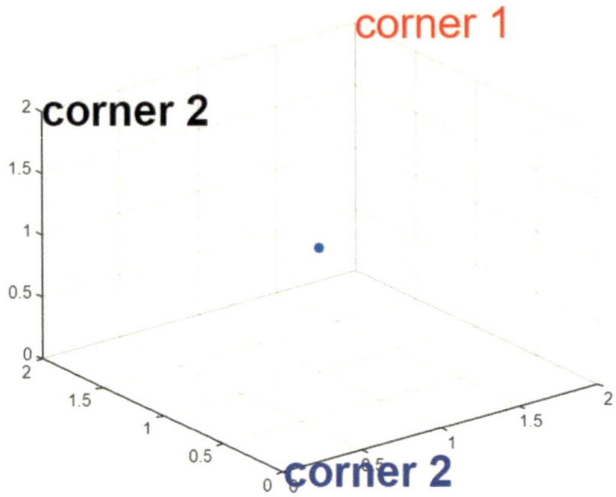

Fig. 18.20 Single-point 3D scatter with red, blue, and black text labels at three corners

Here, the **text** function is used with 「Interpreter」 set to 「latex」. As a result, a complex formula in purple is displayed at (x,y)=(0.5,0.1): 「$y=\int_{1}^{20}(x^{2}+\sqrt{x})dx$」, written in LaTeX format. Next, a text label 「Latex」 is placed at (x,y)=(0.5,0.9). Specifically, we set 「Color」 to 「red」, 「FontSize」 to 25, and 「Rotation」 to 45, creating a red, rotated text label with a large font size. In addition, in the first **text** function, we set 「HorizontalAlignment」 to 「right」, while in the second **text** function, we set 「HorizontalAlignment」 to 「center」. Although both text labels are positioned at x=0.5, their alignment is adjusted by the 「HorizontalAlignment」 setting. See Fig. 18.21.

```
figure
txt1 = text(0.5,0.1,"$y = ¥int_{1}^{20}(x^{2}+¥sqrt{x})dx$",...
    "Interpreter", "latex");
   txt1.Color = [0.4940 0.1840 0.5560];
   txt1.HorizontalAlignment = "right";
   txt1.FontSize = 25;
txt2 = text(0.5,0.9,"Latex");
   txt2.Color = "red";
   txt2.HorizontalAlignment = "center";
   txt2.FontSize = 25;
   txt2.Rotation = 45;
```

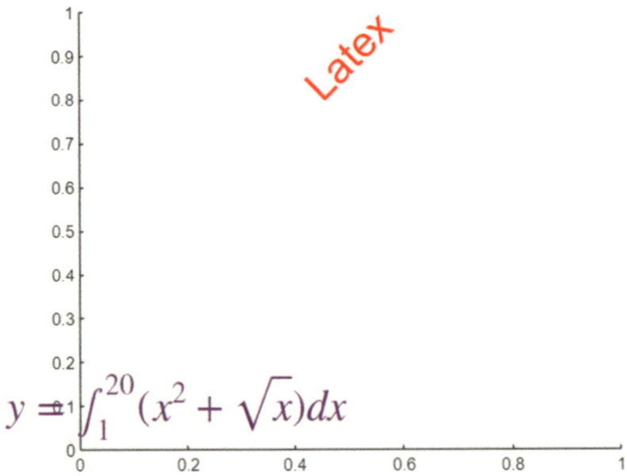

Fig. 18.21 LaTeX formula and rotated text

Here, we use a **for** loop to place five text labels. When 「i=1」 in the first iteration, the program 「text(i,i,i,num2str(i))」 becomes 「text(1,1,1,num2str(1))」. Then, the number 「1」 is placed at (x,y,z)=(1,1,1). When 「i=2」, the number 「2」 is placed at (x,y,z)=(2,2,2). This pattern continues from 「i=3」 to 「i=5」, placing each number at the corresponding coordinate. See Fig. 18.22.

```
figure
for i=1:5
    scatter3(i,i,i,"filled");
    hold on
    text(i,i,i,num2str(i));
end
hold off
```

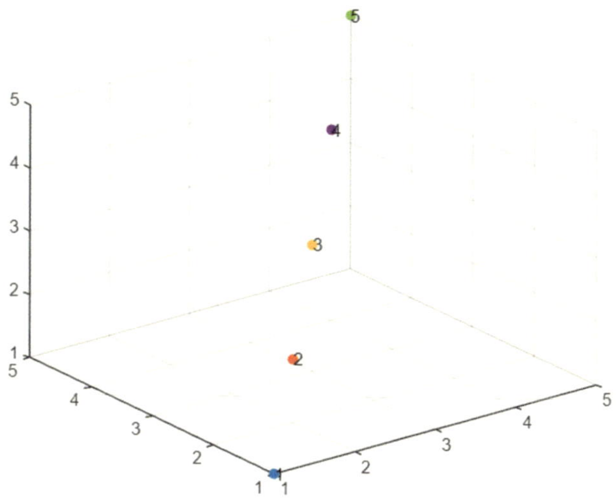

Fig. 18.22 3D scatter with numeric labels using a for loop

Here, we use a program similar to the one that created Fig. 18.22. In particular, when 「i=1」 in the first iteration, the **text** function is given by 「txt=text(1,1+0.5,1,sprintf("size=%d",1))」, which creates the label 「size=1」. Additionally, 「txt.FontSize=10」 is set. This pattern continues from 「i=2」 to 「i=5」, with each iteration increasing the font size of the text. See Fig. 18.23.

```
figure
for i=1:5
    scatter3(i,i,i,"filled");
    hold on
    txt=text(i,i+0.5,i,sprintf("size = %d",i));
        txt.FontSize = i*10;
end
hold off
```

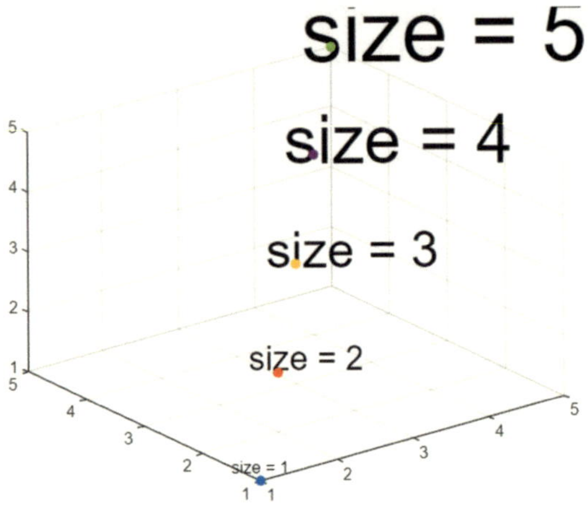

■ **Fig. 18.23** Scatter points with labels of increasing font size using a for loop

18.3 Xline and Yline Functions

Basic Syntax:

Section 18.3 introduces three main syntaxes for the **xline** function, enclosed in the black box in Figure 18.24. The syntax「xline(x)」creates a vertical line at the position specified by「x」on the x-axis.
Next,「xline(x,LineSpec)」creates a vertical line at「x」with a specific line style, color, or marker defined by「LineSpec」.
Moreover,「xline(x,LineSpec,labels)」creates a vertical line at「x」with the specified line style and displays a label next to the line. Similarly, Figure 18.25 shows the basic syntaxes for the **yline** function, which are fundamentally the same as those of the **xline** function.

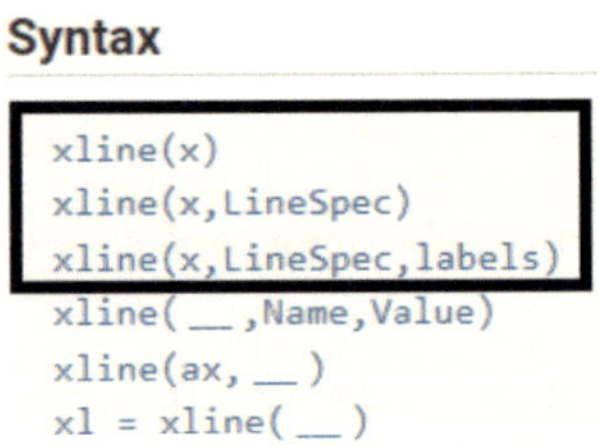
Syntax

```
xline(x)
xline(x,LineSpec)
xline(x,LineSpec,labels)
xline( __ ,Name,Value)
xline(ax, __ )
xl = xline( __ )
```

■ **Fig. 18.24** The syntax of the xline function

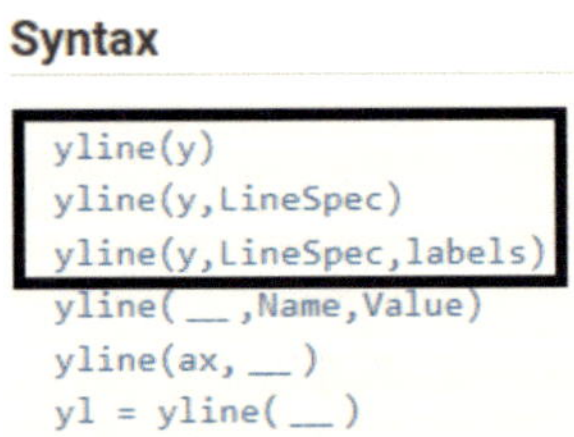
Syntax

```
yline(y)
yline(y,LineSpec)
yline(y,LineSpec,labels)
yline( __ ,Name,Value)
yline(ax, __ )
yl = yline( __ )
```


■ **Fig. 18.25** The syntax of the yline function

Here, four lines are drawn. Specifically,「xline(1)」creates a vertical line at x=1, and「yline(1)」creates a horizontal line at y=1. See ■ Fig. 18.26.

```
figure
xline(1);
xline(2);
xline(3);
yline(1);
xlim([0 5]);
```

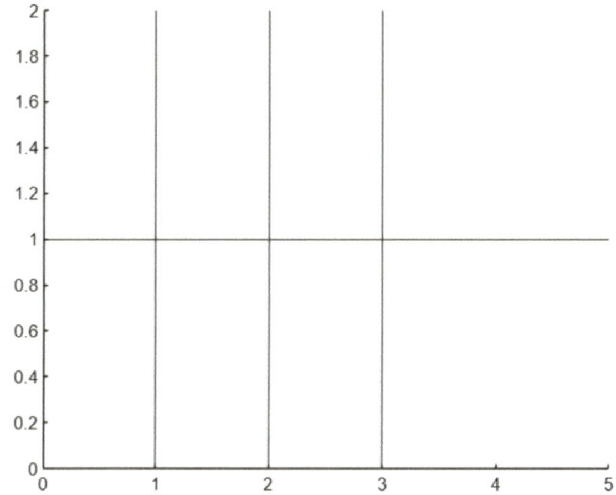

Fig. 18.26 Three vertical lines and one horizontal line

Here, four lines are drawn. Specifically, 「xline([1 2 3],"b--")」 is used to draw three vertical dashed lines in blue at x=1, x=2, and x=3. Additionally, 「yline(1,"r--")」 draws a single horizontal dashed line in red at y=1. See Fig. 18.27.

```
figure
xline([1 2 3],"b--");
yline(1,"r--");
xlim([0 5]);
```

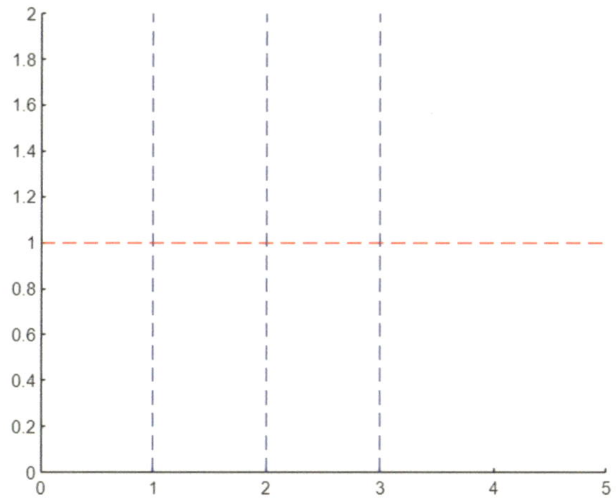

Fig. 18.27 Three vertical dashed lines in blue and one horizontal line in red

Here, four lines are drawn. Specifically, 「xline(1,"-","x=1")」 draws a vertical line at x=1 with the label 「x=1」. Next, 「xline([2 3],"r--",{"x=2", "x=3"})」 draws red dashed vertical lines at x=2 and x=3, with the labels 「x=2」 and 「x=3」, respectively. Finally, 「yline(1,":","y=1")」 creates a horizontal dotted line at y=1 with the label 「y=1」. See ◘ Fig. 18.28 for the created figure.

```
figure
xline(1,"-","x=1");
xline([2 3],"r--",{"x=2","x=3"});
yline(1,":","y=1");
xlim([0 5]);
```

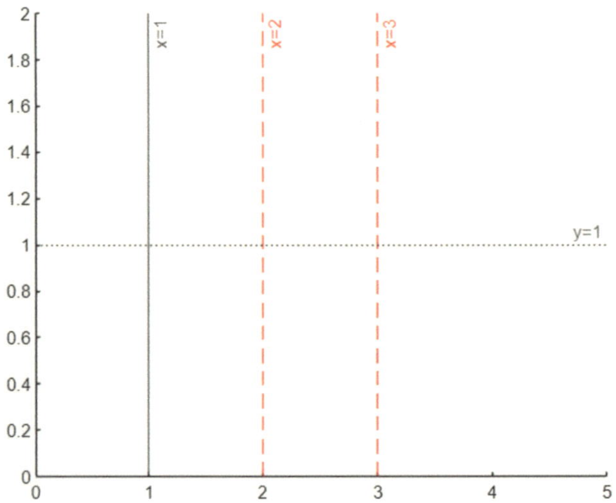

◘ **Fig. 18.28** Four lines with text labels

Here, we store the properties of the **xline** function in the variable 「xl」. Then, we set 「LabelVerticalAlignment」 to 「bottom」 and 「LabelHorizontalAlignment」 to 「left」, adjusting the position of the label 「x=1」. Compare ◘ Fig. 18.28 with ◘ Fig. 18.29.

```
figure
xl = xline(1,":","x=1");
    xl.LabelVerticalAlignment = "bottom";
    xl.LabelHorizontalAlignment = "left";
```

18

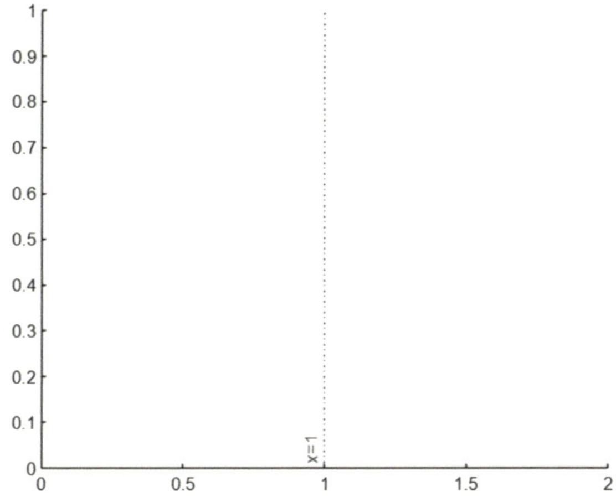

Fig. 18.29 One vertical line with a text label placed at bottom left (cf. Fig. 18.28)

Here, using the **yline** function, we set 「LabelVerticalAlignment」 to 「bottom」 and 「LabelHorizontalAlignment」 to 「left」, which adjusts the position of the label 「y=1」. Compare Fig. 18.29 with Fig. 18.30.

```
figure
yl = yline(1,"r:","y=1");
    yl.LabelVerticalAlignment = "bottom";
    yl.LabelHorizontalAlignment = "left";
```

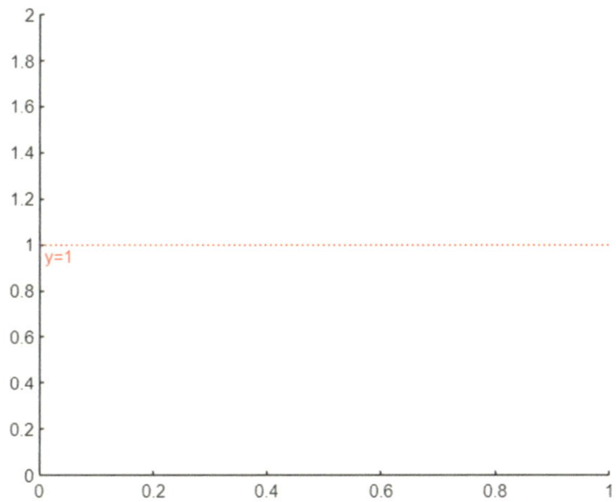

Fig. 18.30 One horizontal dotted line with a text label placed below its left end

Here, using the **xline** function, we draw three vertical lines with different styles: a solid line, a dashed line, and a dash-dotted line. Note that each line has a diff erent 「LabelVerticalAlignment」 setting: 「bottom」, 「middle」, and 「top」. See Fig. 18.31.

```
figure
xline(1,"-","x=1","LabelVerticalAlignment","bottom");
xline(2,"--","x=2","LabelVerticalAlignment","middle");
xline(3,"-.","x=3","LabelVerticalAlignment","top");
xlim([0 5]);
```

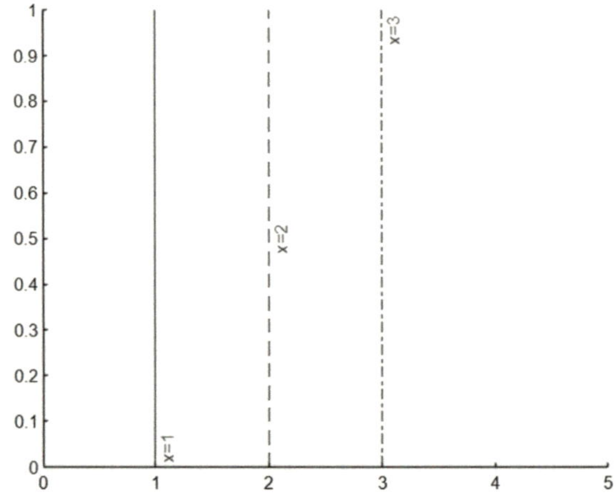

Fig. 18.31 Three vertical lines with text labels placed at bottom, middle, and top

In Fig. 18.31, 「LabelVerticalAlignment」 was set for each line, while in the following program, 「LabelHorizontalAlignment」 is set to 「left」, 「center」, and 「right」, respectively. Compare Fig. 18.31 with Fig. 18.32.

```
figure
xline(1,"b-","x=1","LabelHorizontalAlignment","left");
xline(2,"r--","x=2","LabelHorizontalAlignment","center");
xline(3,"-.","x=3","LabelHorizontalAlignment","right");
xlim([0 5]);
```

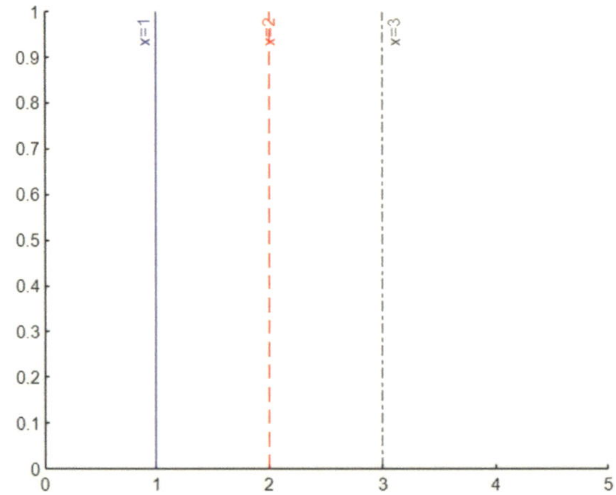

Fig. 18.32 Three vertical lines with text labels placed to the left, on, and to the right of the lines

18

!Attention

Error Program: Here, we attempt to plot multiple vertical lines with different styles at once. While the **xline** function in MATLAB allows multiple lines to be drawn at once, as shown in the program that created ◘ Fig. 18.27, note that it does not accept different line styles as an array. The error message is displayed in ◘ Fig. 18.33.

```
%%% Error program
% figure
% xline([1 2 3],{"-","--","-."});
```

```
Error using xline (line 35)
Invalid LineSpec.
```

◘ **Fig. 18.33** Error output of the xline function: Invalid specification of multiple line styles as an array

Here, each label is set to display in two lines. For instance, 「{"x=1","Line"}」 means that 「x=1」 is on the first line and 「Line」 is on the second line. Additionally, the font size, color, and position of each label are adjusted. See ◘ Fig. 18.34.

```
figure
xl = xline(1,"-",{"x=1","Line"});
    xl.LabelVerticalAlignment = "bottom";
    xl.LabelHorizontalAlignment = "left";
    xl.Color = "blue";
    xl.LineWidth = 2;
    xl.FontSize = 15;
xline(2,"--",{"x=2","Dashed"});
xline(3,"-.",{"x=3","Dashed-dotted"});
yl = yline(1,":",{"y=1","Dotted"});
    yl.LabelHorizontalAlignment = "left";
    yl.Color = "red";
    yl.LineWidth = 2;
    yl.FontSize = 15;
xlim([0 5]);
```

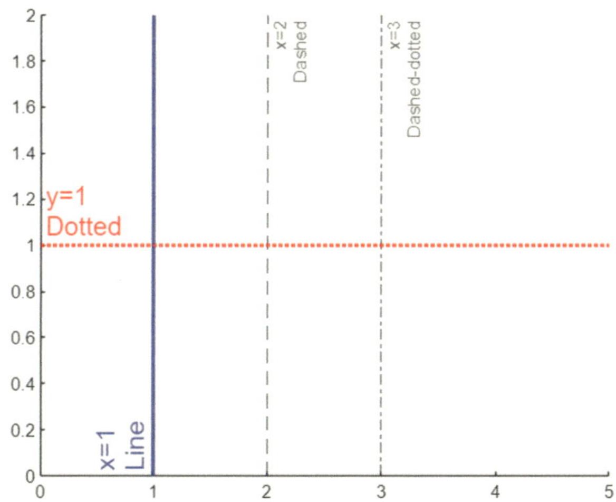

◘ **Fig. 18.34** Vertical and horizontal lines with two-line text labels

18.4 Questions

Question 1

Create Fig. 18.35. Specifically, use the **scatter** function to plot a point at 「(x,y)=(1,1)」. Then, use the **annotation** function to create four arrows pointing to that point.

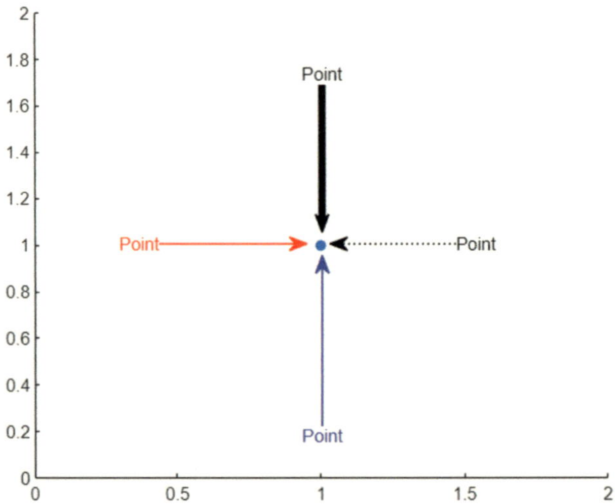

Fig. 18.35 Annotation function (Question 1)

Answer 1

```
figure
scatter(1,1,"filled");
an1 = annotation("TextArrow",[0.3 0.5],[0.52 0.52],"String","Point");
    an1.Color = "red";
an2 = annotation("TextArrow",[0.52 0.52],[0.2 0.5],"String","Point");
    an2.Color = "blue";
an3 = annotation("TextArrow",[0.7 0.53],[0.52 0.52],"String","Point");
    an3.LineStyle = ":";
an4 = annotation("TextArrow",[0.52 0.52],[0.8 0.54],"String","Point");
    an4.LineWidth = 4;
```

Question 2

Create Fig. 18.36 where you use **text** function only once. Refer to the program in Fig. 18.20 for guidance.

18

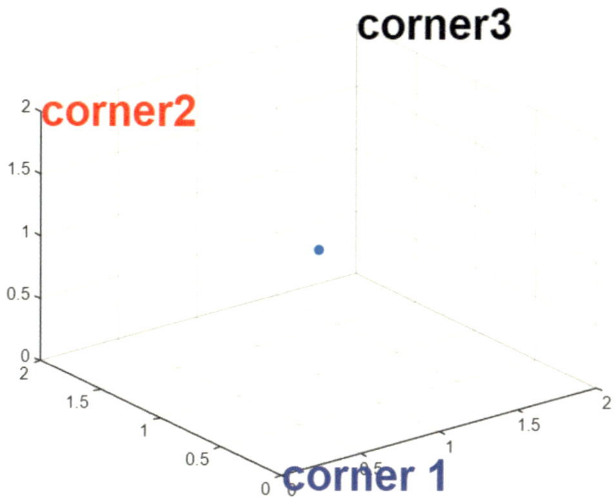

Fig. 18.36 Text function (Question 2)

Answer 2

```
figure
scatter3(1,1,1,"filled");
Ex_txt = text([0 0 2],[0 2 2],[0 2 2],{"corner 1","corner2","corner3"});
    Ex_txt(1).FontSize = 25;
    Ex_txt(1).FontWeight = "bold";
    Ex_txt(1).Color = "blue";
    Ex_txt(2).FontSize = 25;
    Ex_txt(2).FontWeight = "bold";
    Ex_txt(2).Color = "red";
    Ex_txt(3).FontSize = 25;
    Ex_txt(3).FontWeight = "bold";
```

Question 3

Create Fig. 18.37 using the **xline** and **yline** functions to display multi-line text. Use the **sprintf** function to generate multi-line text. Refer to the MATLAB documentation for details on the **sprintf** function.

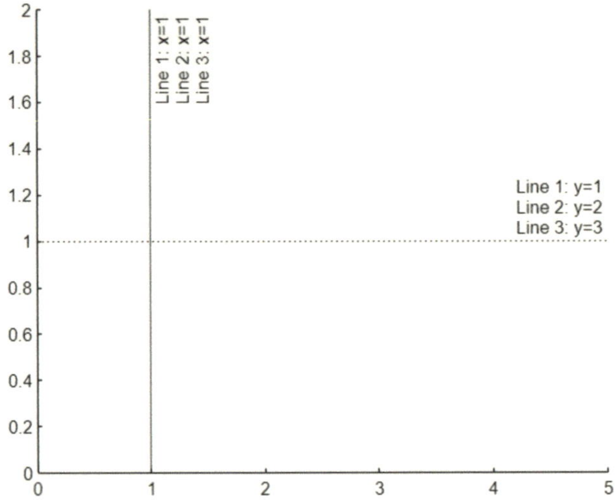

Fig. 18.37 Xline and yline functions (Question 3)

Answer 3

```
figure
xline(1,"-",{"Line 1: x=1","Line 2: x=1","Line 3: x=1"});
yline(1,":",sprintf(" Line 1: y=1\n Line 2: y=2\n Line 3: y=3"));
xlim([0 5]);
```

MIX
Papier aus verantwortungsvollen Quellen
Paper from responsible sources
FSC® C105338

If you have any concerns about our products,
you can contact us on
ProductSafety@springernature.com

In case Publisher is established outside the EU,
the EU authorized representative is:
Springer Nature Customer Service Center GmbH
Europaplatz 3, 69115 Heidelberg, Germany

Printed by Libri Plureos GmbH
in Hamburg, Germany